Oliver Botta · Jeffrey L. Bada · Javier Gomez-Elvira ·
Emmanuelle Javaux · Franck Selsis · Roger Summons
Editors

# Strategies of Life Detection

Previously published in *Space Science Reviews* Volume 135,
Issues 1–4, 2008

 Springer

Oliver Botta
International Space Science Institute (ISSI),
Bern, Switzerland

Jeffrey L. Bada
Scripps Institution of Oceanography,
University of California at San Diego,
La Jolla, CA, USA

Javier Gomez-Elvira
Robotics & Planetary Exploration Laboratory,
Centro de Astrobiología (INTA/CSIC),
Insituto Nacional de Técnica Aeroespacial,
Madrid, Spain

Emmanuelle Javaux
Department of Geology,
University of Liege,
Liege, Belgium

Franck Selsis
Observatoire de Bordeaux,
Floriac, France

Roger Summons
Department of Earth, Atmospheric and
Planetary Sciences,
Massachusetts Institute of Technology,
Cambridge, MA, USA

*Cover illustration:* The image shows a vertical profile of a large conical stromatolite with a small column developed on its flank, from the c. 3.4 Ga Strelley Pool Chert, Kelly Group, 'Trendall locality', northeast Pilbara, Western Australia. The small divisions on the scale are each 1 cm. Image courtesy of The Geological Survey of Western Australia, Department of Industry and Resources. © State of Western Australia 2008.

Library of Congress Control Number: 2008929533

ISBN-978-0-387-77515-9     e-ISBN-978-0-387-77516-6

Printed on acid-free paper.

© 2008 Springer Science+Business Media, BV

1

springer.com

# Contents

# Foreword

**J.L. Bada · J. Gomez-Elvira · E. Javaux · M. Rosing ·
F. Selsis · R. Summons · R.M. Bonnet · O. Botta**

Originally published in the journal Space Science Reviews, Volume 135, Nos 1–4.
DOI: 10.1007/s11214-008-9328-1 © Springer Science+Business Media B.V. 2008

Two of the overarching questions asked in the pursuit of scientific knowledge are: (1) Is there life outside the Earth? and (2) How did life originate on the Earth? Not coincidently, these questions are major milestones on the roadmap of the new interdisciplinary science field of Astrobiology. A significant part in the quest for answers to these questions requires the involvement of space exploration, either in the form of the deployment of planetary probes to various target objects in the Solar System or of the construction of large telescopes and spectrometers in various orbits around the Earth or the Sun. It does not come as a surprise that space agencies such as NASA or ESA have established programs in support of these missions as well as the development of instruments. In the case of planetary probes, which in the current epoch are mainly focusing on the exploration of Mars, but also include missions to asteroids and comets, the development of *in-situ* instruments to search for traces of past

J.L. Bada
Scripps Institution of Oceanography, UCSD, La Jolla, CA, USA

J. Gomez-Elvira
Centro de Astrobiología (INTA-CSIC), Madrid, Spain

E. Javaux
University of Liège, Liège, Belgium

M. Rosing
Geological Museum Copenhagen, Copenhagen, Denmark

F. Selsis
University of Lyon, Lyon, France

R. Summons
Massachusetts Institute of Technology (MIT), Cambridge, MA, USA

R.M. Bonnet · O. Botta (✉)
International Space Science Institute, Hallerstrasse 6, 3012 Bern, Switzerland
e-mail: oliver.botta@issibern.ch

or present life has been established. However, the answer to question of how to recognize traces of past or present life is not straightforward and requires multiple pieces of evidence. This is best illustrated by the difficulties to recognize the first traces of life on Earth, where a combination of morphologic, molecular and isotopic biosignatures is applied, even with the whole arsenal of state-of-the-art laboratory instrumentation available for sample analysis. These difficulties are significantly exaggerated when only limited resources in mass and power are available on a robotic lander or rover operating on another planet. In the case of searching for habitable planets around other stars, the problems are of course the large distances to the object and the immense brightness difference between the host star and the planets surrounding it. The major challenges here are to identify appropriate host stars, to acquire photons from the planets without the interference of the host star, and to identify spectroscopic biosignatures that would provide strong evidence for habitability or even the presence of life on the planet.

This volume is the result of an ISSI Workshop held in April 2006. An international group of 36 chemists, biologists, geoscientists, paleontologists, instrument developers and astronomers was invited to present their data, debate controversies and identify future challenges in an informal setting. The workshop was convened by Jeffrey L. Bada (Scripps Institution of Oceanography, UCSD), Javier Gomez-Elvira (Centro de Astrobiologìa), Emmanuelle Javaux (University of Liège), Minik Rosing (Geological Museum Copenhagen), Franck Selsis (then at the University of Lyon), Roger Summons (MIT), Roger M. Bonnet and Oliver Botta (both ISSI). It was our aim to foster interdisciplinary collaboration and exchange by combining information about results, interpretations and challenges in terrestrial paleo-biosignature research with the latest developments regarding *in-situ* and remote sensing instrument techniques. The volume is divided into five chapters. The Introduction and Chap. 2 provide an overview about life in general, theoretical concepts of life detection and some important aspects in Solar System exploration. Chapter 3 is subdivided into three sections on morphological, molecular and isotopic biosignatures, respectively, and discusses in great detail the challenges in the recognition of early life on Earth, in particular during the early Archean eon. The *in-situ* instrumentations are presented in Chap. 4, beginning with a look back to the results of the Viking Biology Experiments, followed by an overview of instruments currently under development and construction, as well as a preview of future life detection concepts. Chapter 5 is dedicated to remote sensing of extrasolar planets and discusses the scientific rationale behind the selection of target stars for future searches as well as current and future astronomical techniques to search and characterize terrestrial exoplanets.

With great pleasure we would like to thank all those who have contributed to this volume and to the workshops in general. First of all, we thank the authors for writing up their contributions. All papers were peer-reviewed by referees, and we would also like to thank all the reviewers for their critical reports. We also thank the directorate and staff of ISSI for selecting this topic for a workshop and for their support in making it happen, in particular Roger M. Bonnet, Brigitte Fasler, Andrea Fischer, Vittorio Manno, Saliba F. Saliba, Irmela Schweizer, and Silvia Wenger.

Group photograph (from left to right, nose tip counts): Thierry Fouchet, Jennifer Eigenbrode, Andrew Schuerger, Luc Arnold, Franck Selsis, Margaret Turnbull, Javier Gomez-Elvira, Valentine Wakelam, Paul Mahaffy, Gerhard Kminek, Emmanuelle Javaux, Wladyslaw Altermann, Minik Rosing, Jeffrey Bada, Victor Parro, Christophe Lovis, Peter Gogarten, Antonio Lazcano, Wolfgang Krumbein, Frances Westall, Stephane Udry, Malcolm Fridlund, Roger Bonnet, Francois Raulin, Andreas Quirrenbach, Lynn Rothschild, Beda Hofmann, David Catling, Jürgen Popp, Sara Seager, Hans Thiele, Shuhei Ono, Oliver Botta, Roger Summons, Alexander Pavlov, and Mark van Zuilen.
Missing: Chris McKay, Francois Forget, Laurence Barron, Nikos Prantzos, and Therese Encrenaz

# Introduction

# Towards a Definition of Life: The Impossible Quest?

Antonio Lazcano

Originally published in the journal Space Science Reviews, Volume 135, Nos 1–4.
DOI: 10.1007/s11214-007-9283-2 © Springer Science+Business Media B.V. 2007

**Abstract** "Life" is an empirical concept whose various definitions and phenomenological characterizations depend on historical frameworks. Although analysis of existing literature suggests that attempts to define life will remain, at best, a work in progress, the history of biology shows that some efforts have been more fruitful than others. There is a major distinction between natural selection—which is clearly a defining trait of biology—and the changes that result from purely physical chemical evolution, which can be observed in non-biological complex systems. Accordingly, it can be concluded that life cannot be understood without considering the presence of genetic material and Darwinian evolution. This shows the usefulness of the suggestion that life can be considered as a self-sustaining chemical system (i.e., one that turns environmental resources into its own building blocks) that is capable of undergoing natural selection.

**Keywords** Life's definition · Autopoiesis · Complexity · Natural selection

## 1 Introduction

Perhaps as never before in the history of science, "life" has been transformed into a value-ridden term that sits in the center of a tense debate, as shown by the (not always well informed) discussions on abortion, euthanasia, transgenic organisms, and synthetic biology, to name just a few. In spite of the spectacular developments in our understanding of the molecular processes that underlie biological phenomena, we still lack a generally agreed definition of life, and not for want of trying (see, e.g., Rizzoti 1996; Pályi et al. 2002). As Nietzsche once wrote, there are concepts that can be defined, whereas others only have a history. This is not surprising: as argued by Immanuel Kant, precise definitions are achievable in mathematics and philosophy, but empirical concepts such as "life" can only be made explicit (cf. Fry 2002) in ways that are strongly dependent on historical circumstances. Eighty years ago, for instance, when the role of nucleic acids was largely

A. Lazcano (✉)
Facultad de Ciencias, UNAM, Apdo. Postal 70-407, Cd. Universitaria, 04510 Mexico D.F., Mexico
e-mail: alar@correo.unam.mx

unknown, proposals on the emergence of life included a wide array of possibilities based on the random emergence of autocatalytic enzymes, on autotrophic "protoplasm," and on the step-wise evolution of heterotrophic microbes from gene-free coacervates (cf. Lazcano 1995).

It has been argued by many that attempts to define life may be a useless endeavor, bound to fail (Cleland and Chyba 2002). Indeed, attempts to address the definition of living systems have often led to nothing more than phenomenological characterizations of life, which are then reduced to a mere list of observed (or inferred) properties. These inventories are not only unsatisfactory from an epistemological viewpoint, but may also become easily outdated and may fail to provide criteria by which the issue of life (and its traces) can be defined (Oliver and Perry 2006). This can become a unsolved burden for biological sciences, as shown, for instance, by the intense debates on the ultimate nature of the microscopic structures in the Martian meteorite Allan Hills 84001, or those found in early Archean sediments and that not all accept as fossils.

## 2  Life as a Self-Sustaining System

Since the nineteenth century, metabolism has been recognized as a central trait of life, a conclusion that has led us to consider viruses and other subcellular biological entities as nonliving. The recognition that life's continuous production of itself is based on networks of anabolic/catabolic reactions and energy flow led Maturana and Varela (1981) to define life as an autopoietic system, i.e., as an entity defined by an internal process of self-maintenance and self-generation. A shown by Bernal's (1959) statement that "[life is] . . . the embodiment within a certain volume of self-maintaining chemical processes," the idea of autopoiesis is not without historical precedents. As discussed in the following, however, for Bernal and some of his contemporaries like Oparin, the ultimate nature of living systems could not be understood in the absence of an evolutionary perspective (Lazcano 2007).

Although autopoiesis refers and is limited to minimal life forms (Luisi et al. 1996), it is a concept largely dependent on the existence of metabolism, which is a trait common to all living beings. Cells and organisms made of cells are autopoietic and metabolize continuously, and in doing so continuously affect the chemical composition of their surroundings (Margulis and Sagan 1995). Multicellular organisms, on the other hand, consist of units that are living systems in themselves, and will remain so even if the entire system is destroyed (Szathmáry et al. 2005). This is illustrated, for instance, by the extraordinary success of organ transplants.

There are a number of physical and chemical analogues that have been considered autopoietic and that mimic some of the basic properties of life. One of the most enticing examples is that of the self-replicating micelles and liposomes described by Pier Luigi Luisi and his associates. For instance, synthetic vesicles formed by caprylic acid containing lithium hydroxide and stabilized by an octanoid acid derivative have been shown to catalyze the hydrolysis of ethyl caprylate. The resulting caprylic acid is incorporated into the micelle walls, leading to their growth and, eventually, to their fragmentation, during several "generations" (Bachmann et al. 2002).

However surprising, replicative micelles and liposomes do not exhibit genealogy or phylogeny. Albeit due to different processes, the same is true of prions, whose multiplication involves only the transmission of phenotypes due to self-perpetuating changes in protein conformations. As underlined by Orgel (1992) these systems replicate without transmission of information, i.e., they lack heredity. This is in sharp contrast to living beings. Organisms may be recognized as the ultimate example of autopoietic systems (Margulis and Sagan 1995). However, the properties that form the basis of the self-sustaining abilities of living beings are the outcome of historical processes, and it is somewhat difficult for biologists to accept a definition of life that lacks a Darwinian framework. Regardless of their complexity, all living beings have been shaped by a lengthy evolutionary history, and since life is neither the outcome of a miracle or of rare chance event, proper understanding of the minimal properties required for a system to be considered alive require the recognition of the evolutionary processes that led to it. The appearance of life was marked by the transition from purely chemical reactions to autonomous, self-replicating molecular entities capable of evolving by natural selection. How did this take place? At what point in time was the difference between a chemical system and the truly primordial, first organisms, established?

## 3 Life and the RNA World

The lack of an all-embracing, generally agreed definition of life sometimes gives the impression that what is meant by its origin is defined in somewhat imprecise terms, and that several entirely different questions are often confused. For instance, until a few years ago the origin of the genetic code and of protein synthesis was considered synonymous with the appearance of life itself. This is no longer a dominant point of view: four of the central reactions involved in protein biosynthesis are catalyzed by ribozymes, and their complementary nature suggest that they first appeared in an RNA world, i.e., that ribosome-catalyzed, nucleic acid-coded protein synthesis is the outcome of Darwinian selection of RNA-based biological systems, and not of mere physico-chemical interactions that took place in the prebiotic environment.

The discovery and development of the catalytic activity of RNA molecules, i.e., ribozymes, has given considerable support to the idea of the "RNA world," a hypothetical stage before the development of proteins and DNA genomes. During this stage, alternative life forms based on ribozymes existed. This does not imply that wriggling autocatalytic nucleic acid molecules were floating in the waters of the primitive oceans, ready to be used as primordial genes, or that the RNA world sprung completely assembled from simple precursors present in the prebiotic soup. In other words, the genetic-first approach to life's emergence does not necessarily imply that the first replicating genetic polymers arose spontaneously from an unorganized prebiotic organic broth due to an extremely improbable accident, or that the precellular evolution was a continuous, unbroken chain of progressive transformations steadily proceeding to the first living beings. Many prebiotic cul-de-sacs and false starts probably took place, with natural selection acting over populations of primordial systems based on genetic polymers simpler than RNA, in which company must have been kept by a large number of additional organic components such as amino acids, lipids and sugars of prebiotic origin, as well as a complex assemblies of clays, metallic ions, etc.

However, it is true that the arguments in favor of an RNA world have led many to argue that the starting point for the history of life on Earth was the de novo emergence of the RNA world from a nucleotide-rich prebiotic soup, or in the origin of cryptic and largely unknown pre-RNA worlds. Not all accept these possibilities: there is a group of scientists that favors the possibility that life is a self-maintaining emergent property of complex systems that may have started with the appearance of self-assembled autocatalytic metabolic networks initially lacking genetic polymers (Kauffman 1993).

These different viewpoints reflect a rather sharp division that emerged between those who favor (1) the idea that life is an emergent interactive system endowed with dynamic properties that exist in a state close to chaotic behavior, and (2) those who are reluctant to adhere to a definition of living systems lacking of a genetic component whose properties reflect the role that Darwinian natural selection and, in general, evolutionary processes, have played in shaping its the central characteristics. From a biologist's viewpoint, however, neither the nature of life nor its origin can be understood in the absence of an evolutionary approach.


## 4 Complexity and the Nature of Life

In a way, current attempts to explain the nature of life on the basis of complexity theory and self-assembly phenomena can be understood as part of the deeply rooted intellectual tradition that led physicists to search for all-encompassing laws that can be part of a grand theory, one that encompasses many, if not all, complex systems (Fox Keller 2002). Unfortunately, in some cases invocations of spontaneous generation appear to be lurking behind appeals to undefined "emergent properties" or "self-organizing principles" that are used as the basis for what many life scientists see as grand, sweeping generalizations with little, if any, relationship to actual biological phenomena (Fenchel 2002).

Self-assembly is not unique to biology, and may indeed be found in a wide variety of systems, including cellular automata, the complex flow patterns of many different fluids such as tornadoes, cyclic chemical phenomena (such as the Belousov–Zhabotinsky reaction, and the formose reaction, for instance), and in the autoorganization of lipidic molecules in bilayers, micelles, and liposomes. There are indeed some common features among these different self-organized systems, and it has been claimed by a number of theoreticians that they follow general principles that are in fact equivalent to universal laws of nature. Perhaps this is true. The problem is that such all-encompassing principles, if they exist at all, have so far remained undiscovered (Farmer 2005). This has not stopped a number of researchers from attempting to explain life as a continuously renewing, complex interactive system that emerged as self-organizing metabolic cycles that did not require genetic polymers. It is unfortunate that many proposals on an autotrophic origin of life and of living systems as complex systems on the verge of chaos have turned out to be creative guesswork or empty speculation.

However, complexity models have promised much but delivered little. Evidence for the spontaneous origin of catalytic system and of metabolic replication would indeed be exciting (Kauffman 1993) if it could be established. It is true that under given conditions the self-organization of lipidic molecules into liposomes, for instance, can lead to the spontaneous formation of microenvironments which may have had significant roles in the emergence of life. But they are not alive, even if they replicate.

Prebiotic organic compounds very likely underwent many complex transformations, but there is no evidence that metabolic cycles could spontaneously self-organize, much less replicate, mutate, and evolve. Theories that advocate the emergence of complex, self-organized biochemical cycles in the absence of genetic material are hindered not only by the lack of empirical evidence, but also by a number of unrealistic assumptions about the properties of minerals and other catalysts required to spontaneously organize such sets of chemical reactions (Orgel 2000). However complex, systems of chemical reactions such as the formose reaction are not adapted to ensure their own survival and reproduction; they just exist. Life cannot be reduced to one single molecule such as DNA or a population of replicating ribozymes, but current biology indicates that it could not have evolved in the absence of a genetic replicating mechanism ensuring the stability and diversification of its basic components.

## 5 The (Evolutionary) Emergence of Life

Following his 1946 conversations with Einstein on the underlying biochemical unity of the biosphere, John D. Bernal wrote that "... life involved another element, logically different from those occurring in physics at that time, by no means a mystical one, but an element of *history*. The phenomena of biology must be ... contingent on events. In consequence, the unity of life is part of the history of life and, consequently, is involved in its origin" (cf. Brown 2005). History, in biology, implies genealogy and, in the long term, phylogeny. This requires an intracellular genetic apparatus able to store, express and, upon reproduction, transmit to its progeny information capable of undergoing evolutionary change. The most likely candidates for this appear to be genetic polymers.

A good case can thus be made that Darwinian evolution is essential for understanding the nature of life itself. Accordingly, life could be defined as a self-sustaining chemical system (i.e., one that turns resources into its own building blocks) that is capable of undergoing Darwinian evolution (cf. Joyce 1994). Such tentative definition, which was the outcome of a discussion group convened by NASA in the early 1990s, has been rejected by a number of authors who argue on different grounds that a single definition is impossible (Luisi 1998; Cleland and Chyba 2002). Life cannot be defined on the basis of a single trait, but since natural selection is indeed a unique feature of living systems, the basic nature of living systems cannot be understood without it.

The suggestion that life can be understood as a self-sustaining chemical process capable of undergoing Darwinian evolution is consistent with the well-known fact that cyanobacteria, plants, and other autotrophs are not only self-sustaining, but also very much alive. But what about the first life forms? Clearly, if at its very beginning life was already a self-sustaining entity capable of turning external resources into its own building blocks, then it must have been endowed with primordial metabolic routes that allowed it to use as precursors environmental raw materials (such as $CO_2$ and $N_2$, for instance). This appears unlikely to many biologists. An alternative possibility is that the first living entities were systems capable of undergoing Darwinian evolution (i.e., endowed with genetic material capable of replication, change, and heredity) whose self-sustaining properties depended on the availability of organic molecules already present in the primitive environment. Although this can be read as an update of the hypothesis of the prebiotic soup and the heterotrophic origin of life, those involved in the study of emergence of living systems have to ponder not just on how replicative systems appeared, but also how they became encapsulated and how metabolic pathways evolved (Lazcano 2007).

## 6 Conclusions

Research into the origin and nature of life is doomed to remain, at best, a work in progress. It is difficult to find a definition of life accepted by all, but the history of biology has shown that some efforts are much more fruitful than others. As Gould (1995) once wrote, to understand the nature of life, we must recognize both the limits imposed by the laws of physics and chemistry, as well as history's contingency. It is easy to understand the appeal of autopoiesis and complexity theory when attempting to understand the basic nature of living systems. However, there is no evidence indicating how a system of large or small molecules can spontaneously arise and evolve into nongenetic catalytic networks. It is true that many properties associated with cells are observed in nonbiological systems, such as catalysis, template-directed polymerization reactions, and self-assemblage of lipidic molecules or tornadoes. Like fire, life can multiply and exchange matter and energy with its surroundings. It is true that living systems are endowed with properties of autopoeitic, self-organized replicative systems. However, there is a major distinction between purely physical–chemical evolution and natural selection, which is one of the hallmarks of biology. In spite of many published speculations, life cannot be understood in the absence of genetic material and Darwinian evolution.

## References

P.A. Bachmann, P.L. Luisi, J. Lang, Autocatalytic self-replicating micelles as models for prebiotic structures. Nature **357**, 57–59 (2002)

J.D. Bernal, The problem of stages in biopoiesis, in *The Origin of Life on the Earth*, ed. by Oparin et al. (Pergamon Press, London, 1959), p. 38

A. Brown, *J.D. Bernal: the Sage of Science* (Oxford University Press, Oxford, 2005)

C.E. Cleland, C.F. Chyba, Defining "life". Orig. Life Evol. Biosphere **35**, 333–343 (2002)

D.J. Farmer, Cool is not enough. Nature **436**, 627–628 (2005)

T. Fenchel, *Origin and Early Evolution of Life* (Oxford University Press, Oxford, 2002)

E. Fox Keller, *Making Sense of Life: Explaining Biological Development with Models, Metaphors, and Machines* (Harvard University Press, Cambridge, 2002)

I. Fry, *The Emergence of Life on Earth* (Rutgers University Press, New Brunswick, 2002)

S.J. Gould, "What is life?" as a problem in history, in *What is Life? The Next Fifty Years*, ed. by M.P. Murphy, L.A.J. O'Neill (Cambridge University Press, Cambridge, 1995), pp. 25–39

G.F. Joyce, Foreword, in *The Origin of Life: The Central Concepts*, ed. by D.W. Deamer, G. Fleischaker (Jones and Bartlett, Boston, 1994)

S.A. Kauffman, *The Origins of Order: Self Organization and Selection in Evolution* (Oxford University Press, New York, 1993)

A. Lazcano, Aleksandr I. Oparin, the man and his theory, in *Frontiers in Physicochemical Biology and Biochemical Evolution*, ed. by B.F. Poglazov, B.I. Kurganov, M.S. Kritsky et al. (Bach Institute of Biochemistry and ANKO, Moscow, 1995), pp. 49–56

A. Lazcano, Prebiotic evolution and the origin of life: is a system-level understanding feasible? in *Systems Biology. Volume I: Genomics*, ed. by I. Rigoutsos, G. Stephanopoulos (Oxford University Press, New York, 2007), pp. 57–78

P.L. Luisi, On various definitions of life. Orig. Life Evol. Biosphere **28**, 613–622 (1998)

P.L. Luisi, A. Lazcano, F.J. Varela, What is life? Defining life and the transition to life, in *Defining Life: The Central Problem in Theoretical Biology*, ed. by M. Rizzoti (University of Padova, Italy, 1996), pp. 149–165

L. Margulis, D. Sagan, *What is Life?* (Weidenfeld and Nicholson, London, 1995)

H.R. Maturana, F.J. Varela, *Autopoiesis and Cognition – The Realization of the Living* (Reidel, Boston, 1981)

J.D. Oliver, R.S. Perry, Definitely life but not definitively. Orig. Life Evol. Biosphere **36**, 515–521 (2006)

L.E. Orgel, Molecular replication. Nature **358**, 203–209 (1992)

L.E. Orgel, Self-organizing biochemical cycles. Proc. Natl. Acad. Sci. USA **97**, 12503–12507 (2000)

G. Pályi, C. Zucchi, L. Caglioti (eds.), *Fundamentals of Life* (Elsevier, Paris, 2002)

M. Rizzoti (ed.), *Defining Life: The Central Problem in Theoretical Biology* (University of Padova, Italy, 1996)

E. Szathmáry, M. Santos, C. Fernando, Evolutionary potential and requirements for minimal protocells. Top. Curr. Chem. **259**, 167–211 (2005)

# The Solar System

# Infrared Spectroscopy of Solar-System Planets

Thérèse Encrenaz

Originally published in the journal Space Science Reviews, Volume 135, Nos 1–4.
DOI: 10.1007/s11214-007-9230-2 © Springer Science+Business Media B.V. 2007

**Abstract** Most of our knowledge regarding planetary atmospheric composition and structure has been achieved by remote sensing spectroscopy. Planetary spectra strongly differ from one planet to another. $CO_2$ signatures dominate on Mars, and even more on Venus (where the thermal component is detectable down to 1 μm on the dark side). Spectroscopic monitoring of Venus, Earth and Mars allows us to map temperature fields, wind fields, clouds, aerosols, surface mineralogy (in the case of the Earth and Mars), and to study the planets' seasonal cycles. Spectra of giant planets are dominated by $H_2$, $CH_4$ and other hydrocarbons, $NH_3$, $PH_3$ and traces of other minor compounds like CO, $H_2O$ and $CO_2$. Measurements of the atmospheric composition of giant planets have been used to constrain their formation scenario.

**Keywords** Planetary atmospheres · Infrared spectroscopy

## 1 Introduction

Remote sensing spectroscopy is a powerful tool for investigating the atmospheres and surfaces of solar-system planets. Spectroscopic signatures of gaseous atmospheric components can be found over the whole range of the electromagnetic spectrum. In the UV, visible, and near-IR range—typically below 4 μm—the planetary spectrum corresponds to the reflected solar blackbody, peaking at 0.5 μm, over which planetary absorption features can be observed. These signatures allow us to determine the nature and the column density (the number of molecules integrated along the line of sight) of the different atmospheric constituents—the clouds, the aerosol particles and the characteristics of the surface, if there is any.

At longer wavelengths, the planetary spectrum corresponds to its thermal emission; its maximum depends upon the effective temperature of the planet, and is thus, to first order,

T. Encrenaz (✉)
LESIA, Observatoire de Paris, UPMC, Université Paris Diderot, CNRS, 92195 Meudon, France
e-mail: therese.encrenaz@obspm.fr

O. Botta et al. (eds.), *Strategies of Life Detection*. DOI: 10.1007/978-0-387-77516-6_3     11

**Fig. 1** The thermal structure of
planetary atmospheres

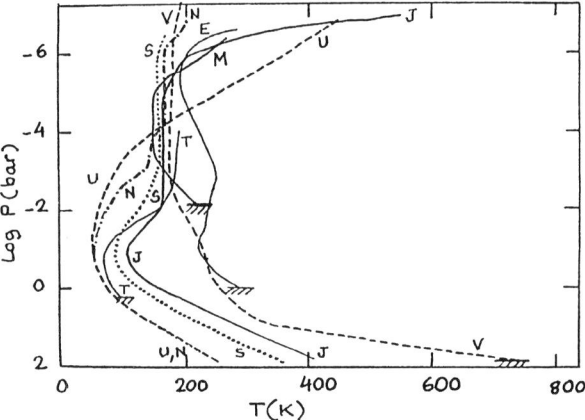

a function of its heliocentric distance. It should be noted that the real temperature also depends strongly on the planetary atmosphere and on internal heat source, and is not a clear function of distance. The Earth, with an effective temperature $Te$ of 288 K, shows a maximum blackbody emission around 10 μm, while Uranus and Neptune, for which $Te = 58$ K, have a peak of about 60 μm. Unlike its reflected solar spectrum, the thermal spectrum of a planet depends a great deal on its temperature structure. Its analysis provides information on the nature and the vertical distribution of the atmospheric constituents, but can also be used in some cases for a retrieval of its temperature structure. Depending upon the temperature lapse rate (Fig. 1), infrared lines can appear either in absorption (in the troposphere, where the gradient is negative) or in emission (in the stratospheres of the giant planets and Titan, where the gradient is positive). In some cases, gaseous atmospheric species can also be observed in fluorescence (usually resonant fluorescence by the solar flux). Finally, planetary spectra can also be diagnostic of solid or liquid signatures, either due to clouds and aerosols, suspended in the atmosphere, or to the surface itself.

Remote sensing spectroscopy, from the ground and/or from space (in particular with the Voyager 1 and 2 spacecrafts), has provided most of the information we have about the atmospheric composition and structure of the giant planets and Titan. Only recently have we gained access to in situ measurements, in the case of Jupiter with the Galileo probe, and in the case of Titan with the Huygens probe. Mars and Venus, in contrast, have been extensively explored with in situ spacecraft. In the case of Mars these spacecraft were Viking, Mars Pathfinder, and the rovers Spirit and Opportunity; in the case of Venus, Pioneer Venus and the Venera probes. Still, important information regarding their atmospheres and Mars' surface has been obtained by remote sensing spectroscopy, from the ground and from aboard orbiters (in particular the Venera spacecraft, Pioneer Venus Orbiter, Mariner 9, Viking, Mars Global Surveyor, Mars Odyssey, Mars Express, and, more recently, Mars Reconnaissance Orbiter and Venus Express).

This paper discusses the main questions dealing with the formation and evolution of planetary atmospheres, with special emphasis on those which have been addressed by remote sensing spectroscopy, either from the ground or from space. We will focus on infrared spectroscopy which has been the prime tool for investigating the physical and chemical properties of neutral atmospheres.

## 2 Terrestrial Planets and Giant Planets

This section briefly reviews the main properties of the solar-system planets. These planets fall naturally into two main categories (Table 1): the terrestrial planets (Mercury, Venus, Earth and Mars), and the giant planets (Jupiter, Saturn, Uranus and Neptune). The former "planet" Pluto, is now recognized as one of the biggest representatives of a new, recently discovered category, the trans-neptunian objects, which populate the Kuiper belt beyond the orbit of Neptune. The terrestrial planets, relatively close to the Sun ($Rh < 2$ AU), are characterized by a small size, a large density, and a small number of satellites. Their atmosphere, if present, is only a negligible fraction of their total mass. Beyond 5 AU, the giant planets are characterized by a large volume, a small density, and a large number of satellites; their atmosphere, by mass, is a significant fraction of their total mass. It is possible to understand these basic properties in the light of the formation model of the solar system, widely accepted today. This scenario, initiated by two pioneers, Kant and Laplace, is based on the observations of the planetary orbits, all almost coplanar and quasi-circular around the Sun, with planets orbiting counter-clockwise, as the Sun does.

These rotational properties strongly suggest the formation of planets within a disk, product of the gravitational collapse of a rotating nebula; this scenario is now supported by the observation of many protoplanetary disks around nearby stars. At the disk center, matter contracts to form the proto-Sun. Within the disk, solid particles accrete, following instabilities, and grow into aggregates through multiple collisions. The biggest objects grow further as they sweep the nearby material. After a few millions years (or a few tens of millions at

**Table 1** Orbital and physical properties of the solar-system planets

| Planet | $Rh$ (AU) | Mass ($M_E$) | Density (g/cm$^3$) | $Ts/Ps$ (K/bar) | Atmospheric composition |
|--------|-----------|--------------|--------------------|------------------|--------------------------|
| Mercury | 0.39 | 0.055 | 5.43 | 90–700/NA | N/A |
| Venus | 0.72 | 0.815 | 5.24 | 730/92 | $CO_2$, $N_2$+ traces $SO_2$, $H_2O$, Ar, CO, . . . |
| Earth | 1.00 | 1.000 | 5.52 | 288/1 | $N_2$, $O_2$, + $H_2O$ + traces Ar, $CO_2$, . . . |
| Mars | 1.52 | 0.107 | 3.93 | 150–300/ 0.006 | $CO_2$, $N_2$, Ar + traces $O_2$, CO, $H_2O$, . . . |
| Jupiter | 5.20 | 317.8 | 1.33 | 110 @ 0.1 | $H_2$, He, $CH_4$, + traces $NH_3$, $H_2O$, $PH_3$, . . . |
| Saturn | 9.54 | 95.2 | 0.69 | 90 @ 0.1 | $H_2$, He, $CH_4$, + traces $PH_3$, $H_2O$, . . . |
| Uranus | 19.19 | 14.4 | 1.32 | 55 @ 0.1 | $H_2$, He, $CH_4$, + traces $H_2O$, . . . |
| Neptune | 30.07 | 17.2 | 1.64 | 55 @ 0.1 | $H_2$, He, $CH_4$, + traces $H_2O$, . . . |

most), the planets are mostly formed, and the smallest particles of the disk are dissipated by the strong solar wind associated with the T-Tauri phase of the early Sun (so-called in reference to the star where this phenomenon was first observed).

How can we explain the difference between terrestrial and giant planets? As a first approximation, it is linked to the amount of solid material available at a given heliocentric distance to form a planet. The disk is made of primordial interstellar matter, that is, mostly hydrogen (and helium), and all other elements (altogether close to one percent in mass) with their cosmic abundances: first come O, C and N, and the heavier elements are less and less abundant. Near the Sun ($Rh < 2$ AU), where the temperature is several hundred K, only silicates and metals are in solid form; the solid mass available for planetary cores is thus limited. Thus, a few small, rocky planets, can be formed. In contrast, at larger heliocentric distances ($Rh > 4$ AU), the most abundant molecules (after $H_2$: $H_2O$, $CH_4$, $NH_3$, ...) are no more gaseous, but in condensed form. They are thus available to be incorporated into big nuclei, which can reach about 10 terrestrial masses. Then, theoretical models (Mizuno 1980; Pollack et al. 1996) predict the collapse of the surrounding gas, mostly $H_2$ and He. This collapse leads to the formation of big planets, of low density, surrounded by a system of satellites and rings formed within their own subnebulae. The limit between the two classes of planets is the "snow line", at the level of $H_2O$ condensation. It was probably at 4–5 AU at the time of planetary formation, when the disk was warmer than today; it is now at about 2 AU. Following the pioneering work of Mizuno (1980) and Pollack et al. (1996), many new models of giant planet formation have been proposed, some of them suggesting a different formation through disk instabilities which grow into protoplanets (Boss 2000, 2001, 2002), and most of the others still favouring the accretion-collapse model (Hersant et al. 2004; Owen and Encrenaz 2006) and including migration processes (Alibert et al. 2005).

The core-accretion formation scenario described earlier explains well why terrestrial planets are found in the vicinity of the Sun, while giant planets are found at greater heliocentric distances. It was thus a major surprise when, over the past decade, a large number of giant exoplanets were discovered in the immediate vicinity of their star. This strongly suggests that the formation scenario of the solar system is not ubiquitous in the Universe. Which mechanisms could explain how giant exoplanets are so close to their star? Most scientists now favor a formation at large astrocentric distance, in a cold environment, followed by a migration mechanism induced by interactions between the planet and the disk. This question is currently a subject of very active research.

The formation scenario of solar-system planets allows us to understand, to first order, the basic chemical composition of planetary atmospheres. In the protoplanetary disk, carbon and nitrogen can either be in the form of CO and $N_2$ (at high temperature and/or low pressure) or in the form of $CH_4$ and $NH_3$ (at low temperature and/or high pressure) (Fegley et al. 1991). $CH_4$ and $NH_3$ probably formed preferentially in the sub-nebulae of the giant planets, which explains why these components are observed in their atmospheres. In contrast, CO and $N_2$ were dominant in the environments of the terrestrial planets. In both cases, $H_2O$ must have been present. CO in turn reacted with $H_2O$ to form $CO_2$, which explains the main atmospheric composition of the terrestrial planets. There are still many unsolved questions, however. The origin of water in terrestrial planets is still unclear. A major question is the diverging evolution of the atmospheres of the terrestrial planets, with the disappearance of most of $H_2O$ on Mars and Venus, and the very tiny and dry present atmosphere of Mars.

An important parameter for the study of planet formation and evolution is the D/H ratio, which can be measured spectroscopically from the analysis of deuterated atmospheric species. In the case of terrestrial planets, D/H is retrieved from the HDO/$H_2O$ abundance ratio, and provides key information on the history of water in these planets (see Sect. 3.1).

In the case of the giant planets, D/H is inferred from the $HD/H_2$ and $CH_3D/CH_4$ abundance ratios, and provides information about the relative mass fraction of their icy core (see Sect. 4.3).

A special mention has to be made about Titan, Saturn's largest moon. Titan is the only satellite with a dense atmosphere; its surface pressure is 1.5 bar and its atmospheric composition is dominated by nitrogen. These similarities with the terrestrial atmosphere have raised considerable interest in this object, which has been explored in situ by the Cassini-Huygens space mission. The spectrum of Titan is discussed in Sect. 4.

## 3  The Terrestrial Planets

Mercury, the smallest terrestrial planet, is too small and too close to the Sun to retain a permanent atmosphere: indeed, the escape velocity is small because of its small gravity field, and the thermal velocity of molecules is large because of its high temperature. The three other terrestrial planets, Venus, the Earth and Mars, show a striking similarity in their primordial chemical composition (mostly $CO_2$ with a few percent of $N_2$), and probably in all cases a large fraction of $H_2O$. The current atmospheric composition of Venus and Mars is dominated by $CO_2$, with a few percent of $N_2$. There are important differences in the case of the Earth: water has been kept in the oceans where $CO_2$ was trapped, and oxygen has appeared as a result of the development of life. The physical properties of the atmospheres (Table 1) are very different, with a surface pressure ranging from about 90 bars (Venus) to 6 mbars (Mars), and a surface temperature ranging from 730 K (Venus) to about 210 K (Mars). In between, the Earth appears to have the best conditions ($Ps = 1$ bar, $Ts = 288$ K) to keep its water reservoir in the liquid form, a decisive condition for the development of life.

The high surface temperature of Venus is significantly above its effective temperature, that is, its expected equilibrium temperature in view of its heliocentric distance. Indeed, the surface and the lower atmosphere of Venus have been heated by a runaway greenhouse effect, initiated by the large amounts of gaseous $CO_2$ and $H_2O$ which were most likely present in the primordial atmosphere (see Sect. 3.1). In the case of Earth, water was trapped in the oceans and $CO_2$ was dissolved in the form of calcium carbonate ($CaCO_3$), so that the greenhouse effect remained moderate. With about 0.1 terrestrial mass, Mars probably had a primordial atmosphere which was less dense than the two others, although probably denser than today, as measured by D/H (see Sect. 3.1). A moderate greenhouse effect may have taken place in the first billion years, but strongly decreased afterward.

As a result, the three planets appear very different today. Venus is covered by a thick cloud deck of sulfuric acid $H_2SO_4$, at an altitude of 40–60 km, which prevents the visible observation of the surface. The Earth is the only planet to host water in its three phases, gaseous, solid and liquid; it is the only atmospheric condensible species. On Mars, $H_2O$ and $CO_2$ are present in the solid and gaseous form (with very tenuous traces of water vapor). Most of the ices are trapped in the polar caps, but clouds of $H_2O$ (and occasionally $CO_2$) are also present.

3.1  The Reflected Solar Spectrum

Reflected solar spectra provide information on atmospheric composition, but also on aerosol and surface properties. In the visible and the near-IR range, the reflected spectrum exhibits signatures of $CO_2$, $H_2O$ and CO, and traces of $CH_4$, $N_2O$ and $O_3$ in the case of Earth. In the

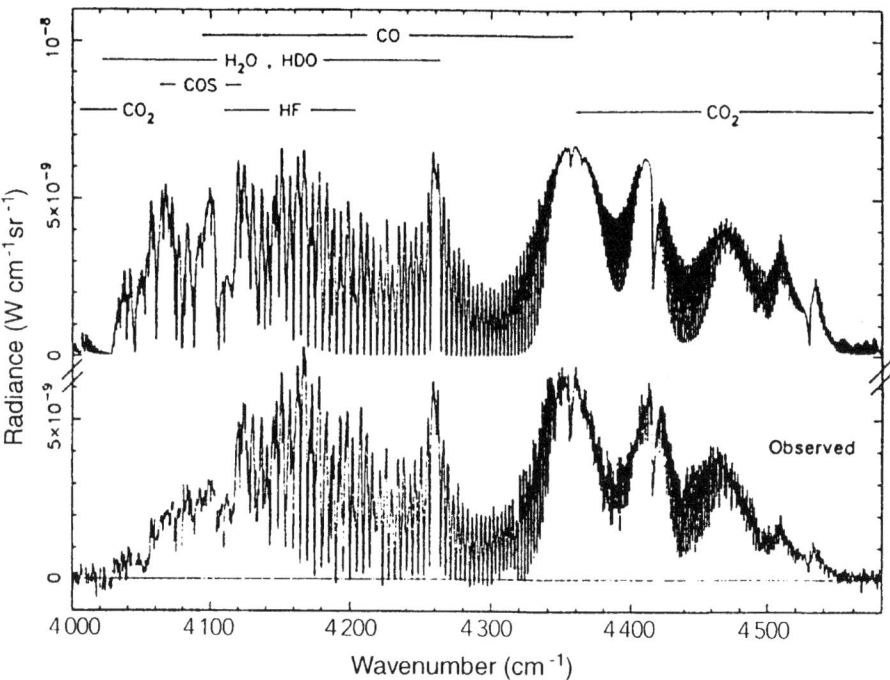

**Fig. 2** The 2.3-μm region in the dark side of Venus (Bézard et al. 1990). *Upper curve*: synthetic spectrum; *lower curve*: observed spectrum. Spectral signatures of $CO_2$, CO, $H_2O$, HDO, HF and OCS are detected

latter case, water vapor signatures are by far the strongest ones. Weaker absorptions of $CO_2$, $CH_4$, $O_3$, $N_2O$ and CO are also detected (Drossart et al. 1993), as well as a wide variety of signatures associated with surface mineralogy. With the increasing interest in the search for Earth-like exoplanets, terrestrial spectra have received a lot of attention from scientists (see Arnold, this issue).

The reflected solar spectrum of Venus, observed on the dayside, probes the middle atmosphere above the sulfuric cloud level, at a pressure of about 1 bar. In the case of Mars, the lines are very narrow, because the Lorentz broadening (proportional to the pressure) is very small. Thus, very high spectral resolution is required to search for minor species. For example, measurements of the D/H ratio on Mars (from $HDO/H_2O$) have been made by means of high-resolution ground-based spectroscopy (Owen et al. 1988; Krasnopolsky et al. 1997). The results indicated a deuterium enrichment by a factor 5–6, interpreted as the signature of a differential atmospheric escape over the planet's history (with HDO escaping less easily than $H_2O$), and thus a denser primitive atmosphere. Tentative detections of methane were reported in 2004 (Krasnopolsky et al. 2004; Mumma et al. 2004; Formisano et al. 2004), with two of them using high-resolution spectrographs on ground-based telescopes, and the third using PFS on Mars Express.

### 3.2 The Thermal Spectrum

Venus presents a very peculiar property: while planetary thermal spectra usually dominate beyond 3–4 μm, the thermal emission of Venus is detected at much smaller wavelengths, down to about 1 μm, because of its very high surface temperature. This emission, detectable

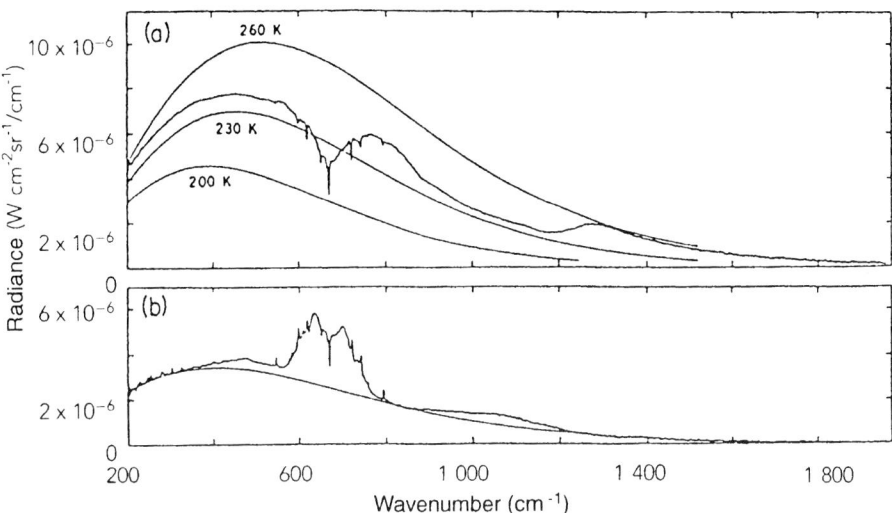

**Fig. 3** The thermal spectrum of Mars (Hanel et al. 1992). The main atmospheric feature is the $CO_2$ band at 15 μm. *Upper curves*: spectrum at mid-latitude, where the surface temperature is higher than the atmospheric one. *Lower curves*: Spectrum in a polar region, where the surface is colder than the lower atmosphere. The atmospheric temperature decreases with increasing altitude, which explains the absorption core inside the $CO_2$ band

only on the night side of the planet, provides an excellent tool for probing the chemical composition of the lower venusian atmosphere. This has been done, in particular, from high-resolution ground-based spectroscopy (Bézard et al. 1990) and led to the discovery and/or study of several minor constituents (CO, $H_2O$, HDO, $SO_2$, OCS, HCl, HF; Fig. 2). A very high value of D/H (see Sect. 2) was inferred from these data (120 times the terrestrial value), indicating the presence of an abundant primordial reservoir of water on Venus, and a very strong outgassing at the early stages of the planet's history. The near-IR thermal emission of Venus has also been observed by the NIMS imaging-spectrometer at the time of the Galileo flyby (Carlson et al. 1991), and is being presently monitored by the VIRTIS (infrared imaging spectrometer) and SPICAV (suite of UV and IR spectrometers) aboard the Venus Express orbiter.

Beyond 5 μm, the thermal spectra of Venus, the Earth and Mars are dominated by the strong $CO_2$ band at 15 μm. The Venus spectrum refers to the atmosphere above the sulfuric clouds. In addition, the terrestrial IR spectrum shows strong signatures of water vapor (very weakly visible on Mars and absent on Venus) and the strong signature of ozone $O_3$ at 9.7 μm. The far-IR spectrum of Mars shows an interesting property: the 15-μm band of $CO_2$ appears either in absorption or in emission, depending upon the temperature contrast between the atmosphere and the surface (Fig. 3). At mid-latitudes, the surface temperature is higher than the atmospheric temperature, and the band thus appears in absorption. The situation is inversed in the polar regions, where the surface (at 145 K) is colder than the atmosphere.

In the case of Mars, ground-based high-resolution spectroscopy in the IR and submillimeter range allowed us to detect two minor constituents: $O_3$ around 10 μm, and $H_2O_2$, in the submillimeter range (Clancy et al. 2004) and at 8 μm (Encrenaz et al. 2004a). At 8 μm, high-resolution imaging spectroscopy has led to a simultaneous mapping of $H_2O_2$ and $H_2O$ (Encrenaz et al. 2005).

## 4 The Giant Planets

A simple look at the total masses of the giant planets allows us to distinguish two categories. We have seen that the giant planets accreted most likely from an initial core of about 10 terrestrial masses. With masses of 318 and 95 terrestrial masses respectively, Jupiter and Saturn are mostly composed of primordial gas; they are called the gaseous giants. In contrast, Uranus and Neptune—with total masses of 14 and 17 terrestrial masses, respectively—are thought to have more than half of their mass made of their initial icy core; they are called the icy giants.

We have seen that the giant planets' outer envelopes are mostly composed of $H_2$ and He, with $CH_4$ and $NH_3$ as minor constituents. Other minor tropospheric constituents include (in the case of Jupiter and Saturn) $H_2O$, $PH_3$, $GeH_4$ and $AsH_3$. These species cannot be detected in the tropospheres of Uranus and Neptune because they condense around the tropopause level, due to its low temperature. Other minor stratospheric species include the products of methane photodissociation ($C_2H_2$, $C_2H_6$ and several other hydrocarbons), and a few oxygen species ($H_2O$, $CO_2$ and CO). The origin of the oxygen source is still under debate: it could be either local (from rings and/or satellites) or interplanetary (from comets and/or micrometeoroids); the recent collision of comet Shoemaker-Levy 9 with Jupiter in July 1994, illustrates that such events may feed the stratosphere with external species (Noll et al. 1996). CO is a special case, which may be of both internal and external origin: part of it could come from the deep interior and have been incorporated into the planetesimals, while the external component would come from the oxygen source. In the case of Neptune, both HCN and CO have been detected in the stratosphere, in unexpectedly high abundances; their origin is still under debate.

In the case of all giant planets and Titan, the thermal distribution is characterized by a troposphere, driven by convection, where the temperature gradient is close to adiabatic; the spectral lines formed in this region (those of the tropospheric species) are seen in absorption. In all cases, the minimum temperature, at the tropopause, occurs at a pressure level of about 100 mbars, with temperatures ranging from 110 K (Jupiter) to about 50 K (Uranus and Neptune; see Fig. 1). Above this minimum, the temperature increases again in the stratosphere, and the lines formed in this region (in particular the hydrocarbons and the oxygen species) are seen in emission.

What is the cloud structure of the giant planets? In the case of Jupiter and Saturn, an $NH_3$ cloud is found at about 150 K, and a lower cloud of $NH_4SH$ appears at about 200 K. An $H_2O$ ice cloud is expected at temperatures above 210 K. Note, however, that this main structure applies to the global cloud composition of the planets, and does not take into account local meteorological effects associated to convection (see the following). In the case of Uranus and Neptune, methane condensation takes place at about 80 K, and $H_2S$ could condense at about 120 K (unless some chemistry with $H_2O$ and $NH_3$ takes place at lower levels).

### 4.1 The Reflected Spectrum

Methane dominates the reflected spectrum of giant planets (Fig. 4). Signatures of $H_2$ and $NH_3$(on Jupiter and Saturn) were also detected several decades ago. These early ground-based measurements were used to infer the first abundance ratios (C/H, N/H) and it was soon realized that the C/H ratio in Uranus and Neptune was far above its solar abundance; this was the first argument in support of the nucleation model of the giant planets. Further evidence came later, as will be discussed in the following. In addition to $CH_4$, its deuterated species $CH_3D$ was also detected on Uranus and Neptune and led to the first estimates of D/H (from $CH_3D/CH_4$) in these two planets (see the following).

**Fig. 4** The near-infrared ground-based spectrum of the giant planets and Titan (Larson 1980). A stellar spectrum is shown to indicate the terrestrial atmospheric windows, and a $CH_4$ laboratory spectrum is shown for comparison. It can be seen that the planetary spectra are dominated by $CH_4$

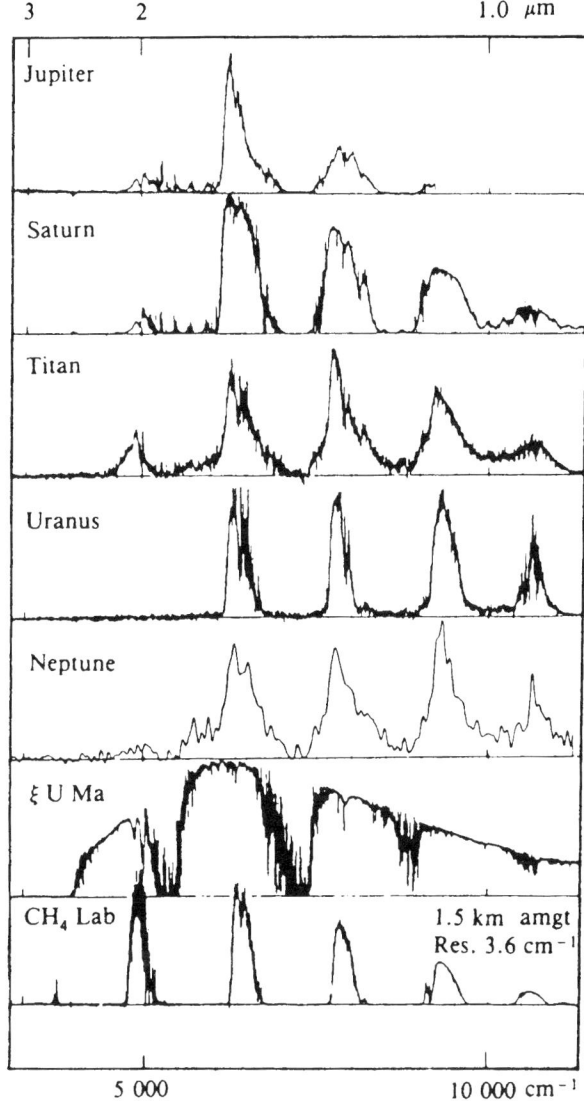

## 4.2 The Thermal Spectrum

The thermal spectrum of the giant planets (and also Saturn's major satellite Titan) has been extensively observed by the Voyager spacecraft, and later by the Infrared Space Observatory (ISO-SWS, Fig. 5). Between 7 and 12 μm, thermal spectra show a mixture of emission (stratospheric) and absorption (tropospheric) lines. All emission lines are due to hydrocarbons: $C_2H_2$, $C_2H_6$, and several less-abundant species. Many of them have been detected by ISO-SWS, and more recently by Spitzer in the case of Uranus and Neptune (Burgdorf et al. 2006). Ground-based observations have been performed in the 4–5 μm and 7–13 μm windows (Bézard et al. 2002; Encrenaz et al. 2004b).

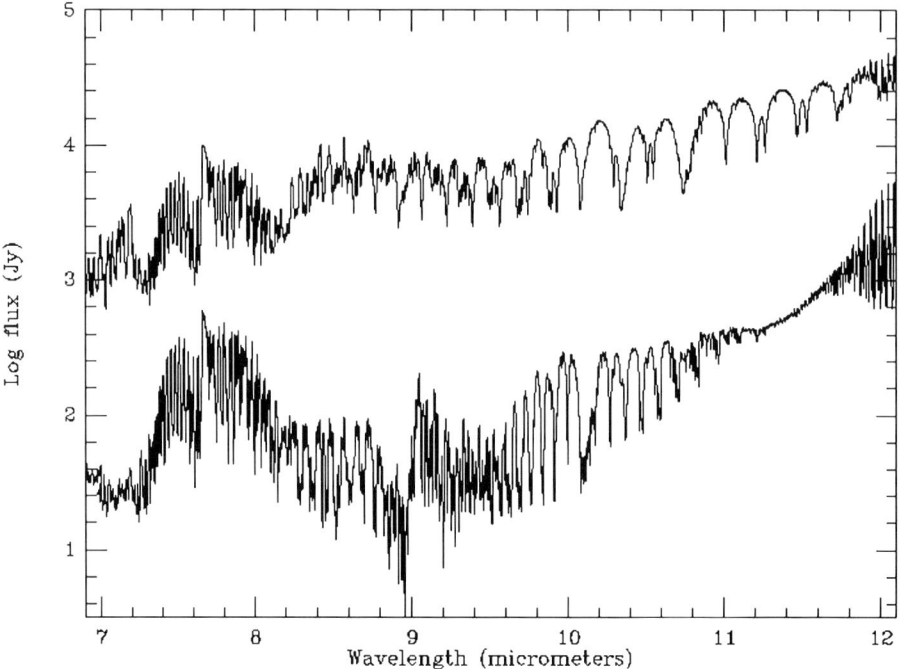

**Fig. 5** The ISO spectrum of Jupiter (*top*) and Saturn (*bottom*) between 7 and 12 µm. This region exhibits a mixture of absorption (tropospheric) features and emission (stratospheric) signatures. $CH_3D$ at 8 µm), $PH_3$ (at 9 µm) and $NH_3$ (around 10–11 µm) are in absorption in Jupiter; in Saturn, $PH_3$ is in absorption (at 9 µm and 10 µm). $CH_4$ (at 7.7 µm) and $C_2H_6$ (at 12 µm) are in emission on both planets. The figure is taken from Encrenaz (2003)

The 5-µm window is of special interest, as it allows us to probe the troposphere of Jupiter and Saturn below the clouds of ammonia ($NH_3$) and ammonia hydrosulfide ($NH_4SH$) (Uranus and Neptune are too cold for thermal emission to be detectable at 5 µm). Ground-based observations at 5 µm have shown that the IR radiation comes from localized regions, called "hot spots", where the cloud coverage is much thinner than elsewhere. The 5-µm window is the range where tropospheric $H_2O$ was detected, as well as germane ($GeH_4$) and arsine ($AsH_4$). An interesting result was the strong depletion of tropospheric water, as compared to its expected cosmic abundance (Drossart et al. 1982). The answer to this puzzling question was given by the Galileo probe which entered the jovian atmosphere in 1995. The probe entered one of the hot spots mentioned above, that is, a dry region of subsidence, almost free of clouds (Atreya et al. 1999). As a result, the oxygen content was found to be very low. The explanation of the oxygen depletion is thus linked to meteorological phenomena, driven by convection; in wet regions of upward motion, the water content was actually found to be higher. A similar depletion of oxygen in the hot spots was observed on Saturn with the Short Wavelength Spectrometer (SWS) of ISO (de Graauw et al. 1997).

### 4.3 Elemental and Isotopic Abundance Ratios

According to the nucleation model of the giant planet formation, planets accreted from an ice core made of heavy elements (with "heavy" meaning all elements heavier than helium). Following the gravitational collapse of the surrounding sub-nebula and homogenous mixing inside the protoplanet, one expects an enrichment of the heavy elements with regard to the solar abundances. Such enrichment has indeed been measured in the case of C/H, with increasing fractions from Jupiter (4) to Neptune (55). On the basis of this enrichment, it is possible to evaluate the mass of the initial core of the giant planets, assuming that all heavy elements are equally trapped (this assumption, however, may be questionable and will be discussed in the following). In all cases, it is found that the mass of the initial core for all four giant planets ranges from 8 to 13 terrestrial masses (Owen and Encrenaz 2006). In the case of Jupiter, in situ measurements from the Galileo probe mass spectrometer have led to the determination of a large number of abundance ratios (C, N, O, S and rare gases). With the exception of O (depleted because of local meteorological phenomena) and He and Ne (depleted by internal condensation processes), all heavy elements are enriched by a factor $4 \pm 2$ (Owen et al. 1999; Owen and Encrenaz 2003, 2006). The enrichment of N and Ar is surprising, because, according to laboratory measurements, these species are not expected to be trapped in ices (nor in clathrates) at temperatures above 40 K. Their presence in Jupiter seems to indicate that the planetesimals which formed Jupiter were accreted at very low temperature. There is no answer presently to this important question.

As discussed earlier, the D/H ratio in Mars and Venus, measured in water, was an important diagnostic of their atmospheric evolution. The D/H ratio can be measured in giant planets, either in methane or in hydrogen, their major constituent. The interest lies in the fact that deuterated species tend to be enriched in ices, as observed in the interstellar medium, and confirmed by laboratory measurements (Irvine and Knacke 1989). The enrichment is due to ion–molecule and molecule–molecule reactions which, at low temperature, favor the formation of D-bearing species, as compared to H-bearing species. As a result, the D/H ratio is an indicator of the formation temperature of the medium where it is measured. In the protosolar nebula, D/H is about $2 \times 10^{-5}$, as derived from solar wind measurements (Geiss and Gloeckler 1998). In the case of the giant planet Jupiter, one expects a small enrichment with respect to this value (since the icy core is only 3% the total mass). It should be moderate for Saturn (whose icy core is 10 percent of its mass) but significant for Uranus and Neptune, which are mostly made of their icy core. Spectroscopic measurements of HD and $H_2$ with ISO have confirmed these predictions: while Jupiter and Saturn have D/H ratios close to the protosolar value, Uranus and Neptune are enriched by a factor 2–3 (Feuchtgruber et al. 1999). Independent measurements of D/H in methane, obtained from the $CH_3D/CH_4$ ratio using Voyager and ground-based data, have also confirmed this result (Lellouch et al. 2001).

## 4.4 The External Oxygen Source

Another important discovery of ISO was the unexpected detection of water vapor emission lines in the thermal spectrum of all giant planets (Feuchtgruber et al. 1997). $CO_2$ was also detected by ISO on Jupiter, Saturn and Neptune, and also later on Uranus by Spitzer. Because of the low temperature at the tropopause, water has to come from an external source, the origin of which is still an open debate. The origin might be local (from rings and/or satellites), or interplanetary (in the form of comets or micro-meteorites), or a combination of both effects, depending of the planet (Lellouch et al. 2002). In the case of Jupiter, the collision of Shoemaker-Levy 9 with Jupiter in July 1994 fed the jovian stratosphere with minor species, formed by shock chemistry ($H_2O$, CO, CS, OCS, HCN, . . .), which, in some cases, survived for months or even for years.

A question linked to the oxygen source is the source of CO in the giant planets. In the case of Jupiter and Saturn, CO was first observed at 5 μm, with an abundance consistent with the thermochemical models; it was then believed that CO was of internal origin. However, large abundances of CO (by a factor a thousand) and HCN were detected in the stratosphere of Neptune (Rosenqvist et al. 1992; Marten et al. 1993). It was again suggested that, on Neptune, CO could be of internal origin, and is indicative of a formation process different from the three other giant planets. However, CO can also come from the external source of oxygen or be a reaction product of these oxygen species in the stratosphere. Recent ground-based observations of CO at high resolution seem to indicate that CO is both external and internal in Jupiter and Saturn (Bézard et al. 2002). Finally, CO was also detected in Uranus at 5 μm, but another emission mechanism is at work: the lines are formed by fluorescence, in the lower stratosphere, which apparently favors the external origin (Encrenaz et al. 2004b).

### 4.5 $H_3^+$ in the Giant Planets

$H_3^+$ is the only ion found so far in the giant planets. It was first detected on Jupiter, at 2 μm (Drossart et al. 1989), and later also on Saturn (Geballe et al. 1993) and Uranus (Trafton et al. 1993), at 2 and/or 4 μm. $H_3^+$ lines are formed very high in the upper stratospheres, at pressure levels on the order of a microbar. The formation mechanism could be either thermal emission or fluorescence. Thermal emission seems to be the most likely mechanism; at these levels, the temperature is high enough (600–1,000 K) for thermal emission to be detected.

## 5  Summary and Conclusions

This review illustrates the very large variety of spectra observed in the different solar-system planets, both in the reflected and the thermal regimes. In the latter case, this is partly due to different stratospheric structures, which strongly influence the observed spectrum. This variety of spectra is to be kept in mind when spectra of extrasolar planets become available.

Another conclusion to be drawn is the importance of remote sensing spectroscopy as diagnostic of chemical composition, thermal and cloud structure, elemental and isotopic composition, and even dynamical processes. Even in the case of terrestrial planets, extensively studied by in situ probes, remote sensing spectroscopic monitoring is essential for mapping temperature fields, wind fields, cloud and aerosol properties, and surface mineralogy.

Finally, recent detection of minor species, on both Mars ($H_2O_2$, tentatively $CH_4$) and the giant planets (CO) have illustrated the importance of ground-based high-resolution spectroscopy, in complement with space missions. Indeed, high spectroscopic resolution is essential for detecting and studying narrow lines, like the martian lines and the stratospheric lines of the giant planets. High resolution ($R > 10^6$) is achievable with heterodyne spectroscopy and has been mostly used in the millimeter range; its spectral range is now extended to the sub-millimeter range and the 10-μm region. In addition, imaging spectrometers with resolving powers of $10^4$–$10^5$ are available in the near-infrared and thermal ranges. Coupled with large telescopes, they allow us to map planetary disks at high spatial and spectral resolution, and will continue to offer a precious complement to in-orbit planetary remote sensing spectroscopy.

## References

Y. Alibert, C. Mordasini, W. Benz, C. Winisdoerffer, Astron. Astrophys. **434**, 343–353 (2005)
S.K. Atreya, M.H. Wong, T.C. Owen et al., Planet. Space Sci **47**, 1243–1262 (1999)
B. Bézard, C. de Bergh, D. Crisp, J.-P. Maillard, Nature **345**, 508–511 (1990)
B. Bézard, E. Lellouch, D.F. Strobel, J.-P. Maillard, P. Drossart, Icarus **159**, 95–111 (2002)

A.P. Boss, Astrophys. J. **536**, L101–L104 (2000)

A.P. Boss, Astrophys. J. **563**, L367–L371 (2001)

A.P. Boss, Astrophys. J. **567**, L149–L143 (2002)

M. Burgdorf, G.S. Orton, J. van Cleve, V. Meadows, J. Houck, Icarus **184**, 634–637 (2006)

R.W. Carlson et al., Science **253**, 1541–1548 (1991)

R.T. Clancy et al., Icarus **168**, 116–121 (2004)

T. de Graauw et al., Astron. Astrophys. **321**, L13–L16 (1997)

P. Drossart, T. Encrenaz, R. Hanel, R. Kunde, V. Hanel, M. Combes, Icarus **49**, 416–426 (1982)

P. Drossart et al., Nature **340**, 539–541 (1989)

P. Drossart et al., Planet. Space Sci. **41**, 551–561 (1993)

T. Encrenaz, Planet. Space Sci **51**, 89–103 (2003)

T. Encrenaz, P. Drossart, H. Feuchtgruber et al., Planet. Space Sci. **47**, 1225–1242 (1999)

T. Encrenaz et al., Icarus **170**, 424–429 (2004a)

T. Encrenaz, E. Lellouch, P. Drossart, H. Feuchtgruber, G. Ortron, S. Atreya, Astron. Astrophys. **413**, L5–L9 (2004b)

T. Encrenaz et al., Icarus **179**, 43–54 (2005)

B. Fegley, D. Gautier, T. Owen, R.G. Prinn, in *Uranus*, ed. by J.T. Bergstrahl et al. (University of Arizona Press, Tucson, 1991) pp. 147–203

H. Feuchtgruber, E. Lellouch, T. de Graauw, T. Bézard, B. Encrenaz, M. Griffin, Nature **389**, 159–162 (1997)

H. Feuchtgruber, E. Lellouch, B. Bézard, Th. Encrenaz, Th. de Graauw, G.R. Davis, Astron. Astrophys. **341**, L17–L21 (1999)

V. Formisano, S.K. Atreya, N. Encrenaz, T. Ignatiev, M. Giuranna, Science **306**, 1758–1761 (2004)

T.R. Geballe, M.-F. Jagod, T. Oka, Astrophys. J. **408**, L109–L112 (1993)

J. Geiss, G. Gloeckler, Space Sci. Rev. **85**, 241–252 (1998)

R.A. Hanel, B.J. Conrath, D.E. Jennings, R.E. Samuelson, *Exploration of the Solar System by Infrared Remote Sensing* (Cambridge University Press, Cambridge, 1992)

F. Hersant, D. Gautier, J.I. Lunine, Planet. Space Sci. **52**, 623–641 (2004)

W.M. Irvine, R.F. Knacke, in *Origin and Evolution of Planetary and Satellite Atmospheres*, ed. by S.K. Atreya, J.B. Pollack, M.S. Matthews (University of Arizona Press, 1989), pp. 3–34

V.A. Krasnopolsky, G.L. Bjoraker, M.J. Mumma, D.E. Jennings, J. Geophys. Res. **102**, 6525–6534 (1997)

V.A. Krasnopolsky, J.-P. Maillard, T. Owen, Icarus **172**, 537–547 (2004)

H.P. Larson, Ann. Rev. Astron. Astrophys. **18**, 43–75 (1980)

Lellouch et al., Astron. Astrophys. **370**, 610–622 (2001)

Lellouch et al., Icarus **159**, 112–131 (2002)

A. Marten, D. Gautier, T. Owen et al., Astrophys. J. **406**, 285–297 (1993)

H. Mizuno, Prog. Theor. Phys. **64**, 544–557 (1980)

M.J. Mumma et al., Bull. Am. Astron. Soc. **36**, 1127–1127 (2004)

K.S. Noll, H.A. Weaver, P.D. Feldman, *The Collision of Comet Shoemaker-Levy 9 and Jupiter* (Cambridge University Press, Cambridge, 1996)

T. Owen, T. Encrenaz, in *Solar System History from Isotopic Signature of Volatile Elements*, ed. by R. Kallenbach et al., Space Sci. Rev. **106**, 121–138 (2003)

T. Owen, T. Encrenaz, Plan. Space Sci. **54**, 1188–1196 (2006)

T. Owen, J.-P. Maillard, C. de Bergh, B.L. Lutz, Science **240**, 1767–1770 (1988)

T. Owen, P. Mahaffy, H.B. Niemann et al., Nature **402**, 269–270 (1999)

J.B. Pollack, O. Hubickyj, P. Bodenheimer, J. Lissauer, M. Podolak, Y. Greenzweig, Icarus **124**, 62–85 (1996)

J. Rosenqvist, E. Lellouch, P. Romani, G. Paubert, T. Encrenaz, Astrophys. J. **392**, L99–L102 (1992)

L.M. Trafton, T.R. Geballe, S. Miller, J. Tennyson, G.E. Ballester, Astrophys. J. **405**, 761–766 (1993)

# Extraterrestrial Organic Matter and the Detection of Life

Mark A. Sephton · Oliver Botta

Originally published in the journal Space Science Reviews, Volume 135, Nos 1–4.
DOI: 10.1007/s11214-007-9171-9 © Springer Science+Business Media, Inc. 2007

**Abstract** A fundamental goal of a number of forthcoming space missions is the detection and characterization of organic matter on planetary surfaces. Successful interpretation of data generated by in situ experiments will require discrimination between abiogenic and biogenic organic compounds. Carbon-rich meteorites provide scientists with examples of authentic extraterrestrial organic matter generated in the absence of life. Outcomes of meteorite studies include clues to protocols that will enable the unequivocal identification of organic matter derived from life. In this chapter we summarize the diagnostic abiogenic features of key compound classes involved in life detection and discuss their implications for analytical instruments destined to fly on future spacecraft missions.

**Keywords** Astrobiology · Solar system · Meteorites · Organic · Abiotic · Mars · Urey · SAM

## 1 Introduction

It is generally accepted that the origin of life on Earth was preceded by a period of chemical evolution utilising organic material that may have been generated in situ or inherited from the protoplanetary disk and its presolar starting materials. The relative importance of these sources has been debated for decades. However, a growing awareness that the Earth's early atmosphere contained too small a proportion of reducing gases for large-scale in situ production of organic matter (Kasting 1993) has encouraged many scientists to support theories of life's origins relying on organic matter supplied by extraterrestrial objects to the surface

M.A. Sephton (✉)
Impacts and Astromaterials Research Centre, Earth Science and Engineering, Imperial College London, South Kensington, SW7 2AZ, London, UK
e-mail: m.a.sephton@imperial.ac.uk

O. Botta
International Space Science Institute, Hallerstrasse 6, 3012 Bern, Switzerland
e-mail: oliver.botta@issibern.ch

of the early Earth (Oró 1961). Although recent reports do suggest that a Titan-like organic haze may have prevailed on the early Earth and aerosol production may have contributed organic material to the surface (e.g. Trainer et al. 2006).

The Earth-based record of pre-biotic chemical evolution has long-since been removed by geological processing. However, remains of the materials that would have been delivered to the early Earth are preserved in ancient asteroids, fragments of which are naturally-delivered to the Earth as meteorites. Carbonaceous chondrites are a particularly primitive class of meteorite that generally contain 2 to 5 wt.% carbon, most of which is present as organic matter. Much of our current understanding of meteoritic organic matter has come from investigations of the Murchison carbonaceous chondrite, approximately 100 kg of which fell in Australia in 1969. Indigenous organic matter in Murchison contains several classes of compounds that are important components in terrestrial organisms, further supporting the role of exogenous delivery for supplying life's starting materials.

Calculations of flux rates reveal that substantial amounts of organic matter could have been delivered to the early Earth by extraterrestrial infall. There are three main types of object that can deliver organic molecules intact to planetary surfaces: asteroids (or their meteoritic fragments), comets and interplanetary dust particles (IDPs) (Chyba and Sagan 1992). Using the very recent history of the Earth as an example, the major mass, by about two orders of magnitude, is estimated to come from the IDP flux. The total flux of 'giant' micrometeorites (particles in the size range of 100 $\mu$m to about 1 mm) before atmospheric entry was determined from direct impact crater counts on the metallic plates of the Long Duration Exposure Facility (LDEF) satellite, from which a global value of approximately $(4 \pm 2) \times 10^7$ kg yr$^{-1}$ of extraterrestrial matter was inferred for this source (Love and Brownlee 1993; Maurette et al. 2000). Assuming a carbon content of approximately 2.5 wt% (similar to carbonaceous chondrites), the annual accretion rate of carbon from micrometeorites and meteorites was estimated to be approximately $2.0 \times 10^5$ kg yr$^{-1}$, with meteorite-sized objects contributing a negligible $10^{-5}$ of that mass. Carbon delivery by comets is also relatively low and is thought to be three orders of magnitude less than for the IDPs both in modern and ancient times (Chyba and Sagan 1992).

## 2 Carbon in Carbonaceous Chondrites

The carbon in the carbonaceous chondrites is present in a number of forms. Organic matter is, quantitatively, the most important carbon bearing phase but can be subdivided into three phases defined by their physical and chemical responses to laboratory procedures (Sephton et al. 2003). The divisions rely on simple operational characteristics and although each organic fraction may respond to processing in a similar way, they may be composed of molecules from a number of extraterrestrial sources. *Free organic matter* is composed of various compound classes (such as amino acids, carboxylic acids, aromatic hydrocarbons, etc.) but all can be extracted by common organic solvents (Table 1). Macromolecular materials account for over 70% of the total organic matter in CI1 and CM2 meteorites and are structurally complex and relatively intractable. Some macromolecular materials can be broken down using heating techniques and these represent the second operational division, *labile organic matter*. In contrast, the third type of organic matter comprises macromolecular materials termed *refractory organic matter* that must be reacted with oxygen at high temperatures before degradation takes place.

The remaining carbon-bearing phases in carbonaceous meteorites are inorganic. Carbonates form a significant proportion of the carbon inventory in carbonaceous meteorites.

**Table 1** Types of mostly abiogenic organic matter in the Murchison (CM2) carbonaceous chondrite and their abundances

| Compounds | Abundances | | Reference |
|---|---|---|---|
| | % | $\mu$g g$^{-1}$ (ppm) | |
| Macromolecular material | 1.45 | | (Chang et al. 1978) |
| Carbon dioxide | | 106 | (Yuen et al. 1984) |
| Carbon monoxide | | 0.06 | (Yuen et al. 1984) |
| Methane | | 0.14 | (Yuen et al. 1984) |
| Hydrocarbons: | | | |
| aliphatic | | 12–35 | (Kvenvolden et al. 1970) |
| aromatic | | 15–28 | (Pering and Ponnamperuma 1971) |
| Acids: | | | |
| monocarboxylic | | 332 | (Lawless and Yuen 1979; Yuen et al. 1984) |
| dicarboxylic | | 25.7 | (Lawless et al. 1974) |
| $\alpha$-hydroxycarboxylic | | 14.6 | (Peltzer et al. 1984) |
| Amino acids | | 60 | (Cronin et al. 1988) |
| Diamino acids | | 0.04 | (Meierhenrich et al. 2004) |
| Alcohols | | 11 | (Jungclaus et al. 1976b) |
| Aldehydes | | 11 | (Jungclaus et al. 1976b) |
| Ketones | | 16 | (Jungclaus et al. 1976b) |
| Sugar-related compounds (polyols) | | ~24 | (Cooper et al. 2001) |
| Ammonia | | 19 | (Pizzarello et al. 1994) |
| Amines | | 8 | (Jungclaus et al. 1976a) |
| Urea | | 25 | (Hayatsu et al. 1975) |
| Basic N-heterocycles (pyridines, quinolines) | | 0.05–0.5 | (Stoks and Schwartz 1982) |
| Pyrimidines (uracil and thymine) | | 0.06 | (Stoks and Schwartz 1979) |
| Purines | | 1.2 | (Stoks and Schwartz 1981a) |
| Benzothiophenes | | 0.3 | (Shimoyama and Katsumata 2001) |
| Sulphonic acids | | 67 | (Cooper et al. 1997) |
| Phosphonic acids | | 1.5 | (Cooper et al. 1992) |

"Exotic" carbon (diamond, silicon carbide and graphite) is contained within the insoluble carbon but can be isolated by progressive oxidative techniques (e.g. dichromate and perchloric acids). However it is the free organic matter and macromolecular materials that can be solubilized by reactions with heat and water that is relevant for discussions of the Origin of Life.

## 3 Soluble Organic Compound Classes in Carbonaceous Chondrites

### 3.1 Amino Acids

The first compound class usually associated with life that was unambiguously identified and quantified in Murchison was the amino acids (Kvenvolden et al. 1970) and to date more than 80 different amino acids have been found in this meteorite. However, of these only eight $\alpha$-amino acids (glycine, alanine, aspartic acid, glutamic acid, valine, leucine, isoleucine, and proline) are the same as those used in terrestrial biology as constituents of proteins, e.g. are encoded in DNA and translated in the ribosome. A few others, including $\beta$-alanine, $\alpha$-aminoisobutyric acid (AIB) and sarcosine, have a very restricted biological occurrence on the Earth, e.g. as components of bacterial cell walls. The remainder are found naturally only in meteorites (i.e. they exist on Earth, but have to be synthesized in the laboratory).

Meteoritic amino acids can be mainly divided into two types, monoamino alkanoic acids (e.g. alanine) and monoamino alkandioic acids (e.g. aspartic acid), each of which can occur as N-alkyl derivatives or cyclic amino acids (Cronin and Chang 1993). One of the molecular characteristics of each type of compound is that they exhibit complete structural diversity, meaning that all isomeric forms of a certain amino acid are present in the meteorite, a fact that was experimentally confirmed for the $C_1$ to $C_7$ $\alpha$-amino acids (Cronin and Pizzarello 1986). Additional characteristics include a decrease in abundance of isomers in the order $\alpha > \gamma > \beta$, the predominance in abundance of branched carbon chain isomers over straight ones, and a smooth exponential decline in concentration with increasing carbon number within homologous series. Recently, diamino acids have also been detected in the Murchison meteorite (Meierhenrich et al. 2004).

In a typical amino acid analysis protocol (Fig. 1), the compounds are extracted with hot water, and a fraction of the extract is acid hydrolyzed and desalted in order to obtain critical information about the concentration ratio between the free and bound amino acids in the meteorite. Individual amino acids are then isolated through separation on a chromatographic column either in the gas phase (gas chromatography, GC) or liquid phase (high performance liquid chromatography, HPLC). Quantification occurs via detection by either a mass spectrometer or a fluorescence detector. In most cases, the amino acids have to be derivatized in order to be volatile enough to pass through the chromatographic system or to allow separation of the enantiomers (Botta and Bada 2002). This relatively straightforward protocol can be applied to biological samples, biogeophysical samples such as endolithic bacterial communities as well as abiotic material in meteorites to allow a direct comparison of the absolute and relative amino acid compositions. The assessment of the putative biogenicity of

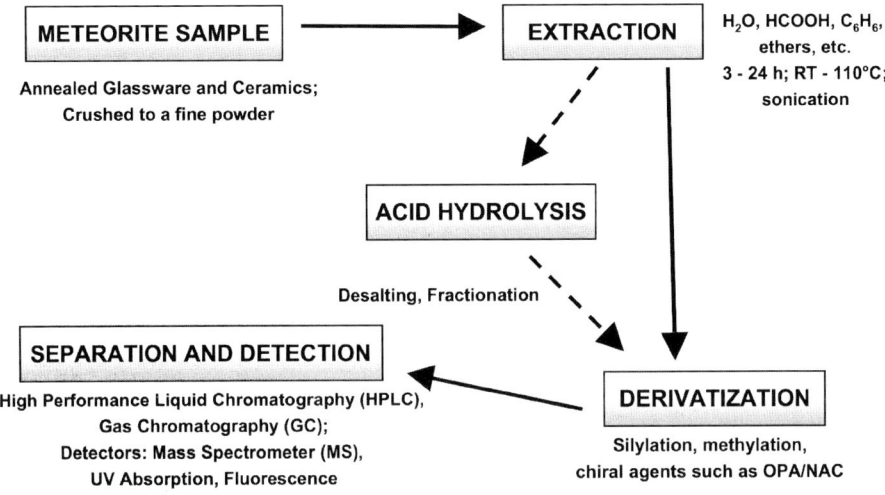

**Fig. 1** Flow diagram illustrating an extraction procedure for organic compounds in meteorites. *Solid lines* are mandatory steps; *dashed lines* indicated additional steps (for example acid hydrolysis to liberate bound amino acids). Abbreviations: OPA: o-phthaldialdehyde; NAC: N-acetyl-L-cysteine

meteoritic amino acids has been hampered by the absence of a quantitative metric to evaluate distribution variations arising from biotic and abiotic diagenesis as well as interdisciplinary differences in analytical methods (Vandenabeele-Trambouze et al. 2005). A combination of principal component analysis, hierarchical cluster analysis, and stochastic probabilistic artificial neural networks has been proposed as a powerful means to overcome this problem and to provide a quantitative means to assign the synthetic origin of these compounds (McDonald and Storrie-Lombardi 2006).

An important molecular signature of life is hosted by amino acids, namely the almost exclusive occurrence of only one (the left-handed) of the two possible enantiomers in all biosynthesized molecules. Although this has been considered to be a very powerful biosignature, because abiotic synthesis of chiral molecules always yields a 1 : 1, or *racemic*, mixture of the right- ($D$) and left- ($L$) handed enantiomers, there are two caveats that need to be evaluated very carefully. The first is the occurrence of enantiomeric excesses, or "*ee*", observed in the amino acids in the Murchison and Murray meteorites (Cronin and Pizzarello 1997; Engel and Nagy 1982; Pizzarello and Cronin 2000). In Murchison, L-excess was first recognised in five $\alpha$-H amino acids commonly found in proteins (alanine, glutamic acid, proline, aspartic acid and leucine) and *ee* ranged from 2.9% to 68% (Engel and Nagy 1982). Later studies recognised an *ee* in the two forms of 2-amino-2,3-dimethylpentanoic acid (*DL*-$\alpha$-methylisoleucine and *DL*-$\alpha$-methylalloisoleucine) as well as isovaline which displayed an *ee* of about 15% (Pizzarello et al. 2003). Both of these amino acids belong to the $\alpha$-methyl amino acid family and have only limited biological occurrence, making a terrestrial contamination source highly unlikely. On the other hand, $\alpha$-H amino acids such as alanine were found to be racemic in these analyses. The second caveat is the problem of racemization of chiral monomers such as amino acids because this process can take place over long periods of time (hundreds of millions of years to billions of years) even under dry and cold conditions, eventually hiding the potential biological signature.

In our quest to discriminate between biogenic and abiogenic sources of organic matter in extraterrestrial environments interpretation of amino acid data from spacecraft missions must take into account life's preference for a limited set of amino acids as well as the apparent abiogenic *ee* in $\alpha$-methyl amino acids. This means that instruments designed for the search for amino acids on the surface or subsurface of Mars need to be able not only to detect and characterize individual amino acids at very low abundances, but also have the critical capability to separate and accurately quantify the enantiomers of chiral amino acids (e.g. Skelley et al. 2004).

## 3.2 Nucleobases

The central role in biology for storage of genetic information is carried out by the nucleic acids, namely deoxyribonucleic acid (DNA) and its sister molecule ribonucleic acid (RNA). RNA is also responsible for the information transfer of the genetic code from DNA to the macromolecular cellular entity that is responsible for protein synthesis, the ribosome, where the code is translated into the primary sequence of amino acids that form the encoded protein. Nucleic acids are polymers comprised of monomeric units called nucleotides, which themselves are made out of three components: a nucleobase, a sugar (ribose or deoxyribose) and a phosphate residue. Nucleobases, for example adenine, guanine and uracil, are one- or two-ring aromatic molecules that have one or several carbon atoms in the ring replaced by nitrogen atoms. The molecules above, along the with structurally-related compounds xanthine and hypoxanthine, were discovered in the three carbonaceous chondrites Murchison (Fig. 2), Murray, Orgueil in abundances about ten times lower than the amino acids (Stoks

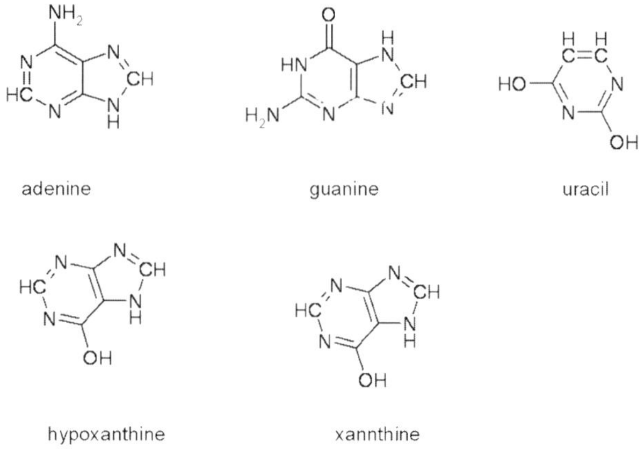

**Fig. 2** Murchison nucleobases and related compounds. Adenine, guanine and uracil are components of the genetic code in terrestrial life

and Schwartz 1979, 1981b). Because there are only such a small number of compounds present in the meteorites, the criteria of complete structural diversity can not be applied in this case. In addition, nucleobases are not chiral, which leaves only the measurement of stable isotope ratios for carbon or nitrogen as an option to make a distinction between terrestrial and extraterrestrial origins. Yet these measurements, at the level of individual molecules, pose significant difficulties even for meteoritic samples and sophisticated laboratory techniques in terrestrial laboratories, their application for planetary landers will be unavailable in the near future.

## 3.3 Sugar-Related Compounds

Sugars are also components of nucleic acids but this compound class has additional roles in biology. The storage of energy and the provision of structural support in organisms are roles undertaken by sugars of varying complexity. Sugars and related compounds are collectively called polyhydroxylated compounds or polyols, reflecting their structure where a number of hydroxyl groups are attached to a carbon skeleton. Sugars are readily produced abiogenically by the thermal polymerization of formaldehyde and this reaction, known as the formose reaction, results in a random distribution of a large number of sugars and other polyols. The fact that formaldehyde is abundant in the interstellar medium (Snyder et al. 1969) provides support that significant amounts of this sugar precursor compound were present in the planetesimals from which the planets and asteroids formed. Furthermore, several sugar-related compounds (sugar alcohols, sugar acids) and the simplest sugar molecule, dihydroxyacetone, were detected in Murchison (Cooper et al. 2001). For two compounds (glycerol and glyceric acid) abundances were determined on the same order of magnitude as for the amino acid glycine. Yet the simplicity and diversity of meteoritic polyols contrasts sharply with the high selectivity for their counterparts in terrestrial organisms, where deoxyribose and ribose are present in DNA and RNA respectively, glucose is the principle energy source and the highly ordered polysaccharide cellulose is the most abundant organic molecule in the biosphere. Consequently, the two sources are unlikely to be confused. However, due to their thermal lability polyols pose significant analytical challenges and the instruments that are currently under development will not have the capability to extract and detect sugar related molecules in situ.

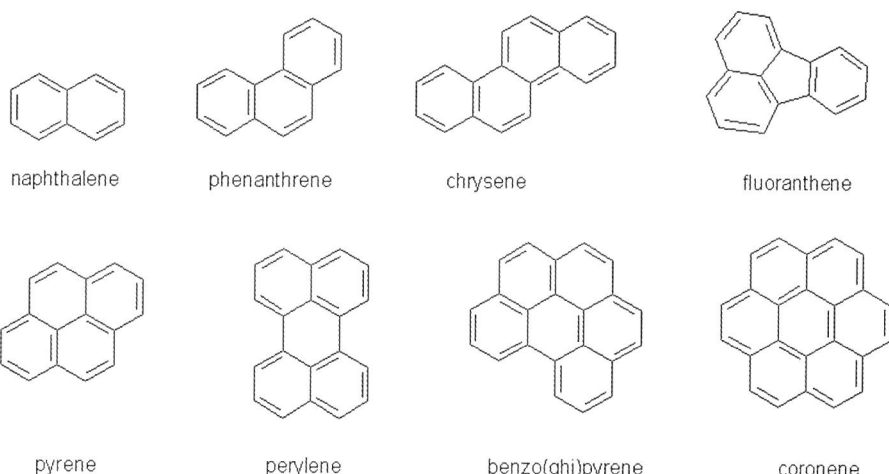

naphthalene                  phenanthrene                  chrysene                  fluoranthene

pyrene                  perylene                  benzo(ghi)pyrene                  coronene

**Fig. 3** Murchison PAH, the most abundant type of extraterrestrial organic compound in both meteorites and space (hydrogens not displayed)

## 3.4 Polycyclic Aromatic Hydrocarbons

The most dominant molecules in carbonaceous chondrites are the polycyclic aromatic hydrocarbons (PAH). PAH constitute over 60% of the organic matter in Murchison (Fig. 3), the vast majority of which is present as an intractable macromolecular component (e.g. Sephton et al. 2004). This molecular dominance should come as no surprise when it is recognised that PAH are the most abundant free organic molecules in space (d'Hendecourt and Ehrenfreund 1997). PAH are thought be mainly produced in space in the high-temperature, high-density ejecta of asymptotic giant branch (AGB) stars (Cherchneff et al. 1992) and are the molecular intermediaries in the soot formation process. On Earth, PAH can be produced from biological organic matter only through partial combustion or maturation following burial in the subsurface. Although PAH have no direct role in terrestrial biochemistry, they are excellent indicators of abiogenic matter and can support interpretations, that imply the absence of a biological contribution to organic mixtures. Hence, when found in close association with PAH, other less thermally stable compounds, such as those discussed above, should also be carefully examined for abiogenic signatures. PAH have, controversially, been used as indicators of biology in the Martian meteorite ALHA 84001 by McKay et al. (1996).

## 3.5 Macromolecular Materials

The major organic component of carbonaceous chondrites is a solvent-insoluble, high molecular weight macromolecular material that constitutes at least 70% of the total organic content in these meteorites. As the major organic component, the macromolecular material is key to theories associated with the origin of extraterrestrial organic matter and quantitative arguments for organic matter delivery to the early Earth.

A substantial proportion of the macromolecular material breaks down during heating to release a range of aromatic and polycyclic aromatic hydrocarbons (Sephton et al. 1998, 2004) including benzene, toluene, naphthalene, phenanthrene, carbazole, fluoranthene, pyrene, chrysene, perylene, benzoperylene and coronene units with varying degrees of alklyation (e.g. Fig. 4). Oxygen, sulphur and nitrogen containing compounds are also

Murchison (CM2) hydrous pyrolysate

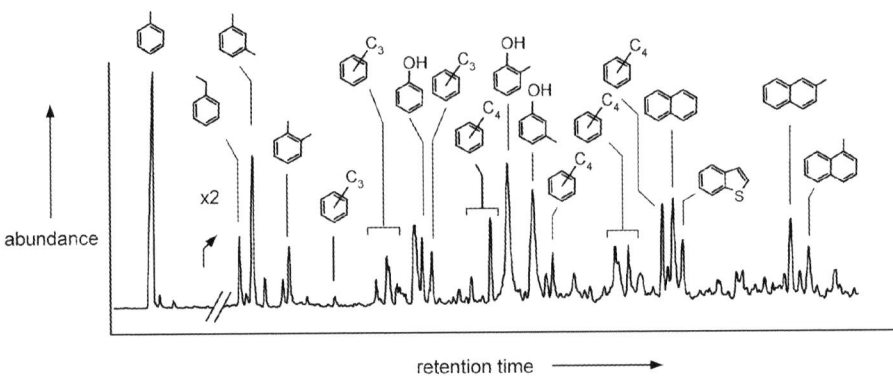

**Fig. 4** Some structural units released from the Murchison macromolecular material. In this case hydrous pyrolysis (pyrolysis in the presence of water) was used to generate the products (the pyrolysate). After Sephton et al. (1998)

released. This degradable portion of the macromolecular material has been termed labile organic matter. The structural units that make up macromolecular materials appear to be relatively conservative between the primitive low petrographic types (Sephton et al. 2000).

Although thermal degradation liberates substantial amounts of organic matter from the macromolecular material, over 50% of macromolecular carbon is resistant to degradation. Comparisons with experimental yields from terrestrial macromolecular materials indicate that this refractory residue probably consists of a network dominated by at least five- or six-ring PAH cross-linked together (Sephton et al. 2004). This non-degradable portion of the macromolecular material is referred to as refractory organic matter.

Owing to its molecular structure, the macromolecular material has a high survivability in the harsh environment of space. Its role in contributing to primitive living systems depends on types and amounts of fragments which it can shed during thermal and aqueous processing. A series of hydrous pyrolysis experiments have revealed that functionalised aromatic structures are readily liberated from the labile macromolecular material by the action of heat and water (Sephton et al. 1998, 2000). Intriguingly, recent analyses imply that molecular asymmetry is present in a macromolecular organic fraction that is removed by hydrothermal treatment (Kawasaki et al. 2006), most likely the labile organic material. Macromolecular components once liberated by aqueous processing may have had an inductive effect on chemical evolution on the early Earth.

## 4 Implications for Future In Situ Analytical Instruments

There are currently two instruments under development that, at least theoretically, have the capability to detect life on another planet. Scheduled to fly first, is Sample Analysis at Mars (SAM), the largest and heaviest science instrument on the NASA Mars Science Laboratory (MSL) mission. The other instrument, Urey, is a evolution of the Mars Organic Detector (MOD) that has been merged with the Mars Oxidant Instrument (MOI). Both instruments will be described in more detail by their principal investigators Paul Mahaffy (SAM) and Jeffrey L. Bada (Urey) and co-authors later in this volume, and the purpose of this section is to describe their capabilities with regard to the detection of biogenic organic compounds.

## 4.1 Sample Analysis at Mars (SAM)

First it should be noted that SAM, in analogy to the Viking gas chromatography-mass spectrometry (GC-MS) investigation, is not considered a life-detection-instrument, although it has some inherent capabilities that will allow it to find traces of life. The SAM instrument suite consists of three major components: A GC system with six columns, a quadrupole mass spectrometer (QMS) and a tunable laser spectrometer (TLS). Solid samples are either directly pyrolyzed under ambient atmosphere in one of two ovens to temperatures of up to 1100°C or first derivatized using a silylating reagent (N-methyl-N-(*tert.*-butyldimethylsilyl)-trifluoroacetamide (MTBSFA)) before stepwise heating up to 750°C. Evolved gases from both processes are separated in the GC columns and the components detected and analyzed using the QMS. Due to the harsh thermal conditions, the most likely class compounds to be found with direct pyrolysis-GC-MS is probably going to be aromatic and PAH, which, as discussed above, are probably abiotic in origin. It should be noted that the detection of such compounds on Mars are the primary goal of SAM and would be tremendous achievement in itself. With the application of derivatization with MTBSTFA, the chances of detection compound classes such as amino acids, nucleobases and carboxylic acids increase dramatically, and, provided abundances above the detection limit of the QMS, SAM would have the capability to provide information about absolute and relative abundances of these compound families, allowing the determination of their abiotic or biogenic origin through the application of the appropriate data analysis (McDonald and Storrie-Lombardi 2006). The derivatization of amino acids with *tert*-butyldimethylsilyl-groups does not introduce a second chiral center into the molecule, which makes chiral separation impossible unless a chiral column is used, which may be problematic on SAM due to stability issues with this type of column. The TLS has the capability to measure high-precision stable isotope ratios of $CO_2$ and $CH_4$ and may therefore detect isotopic differences between evolved gases of pyrolyzed soil samples and atmospheric gases. However, because there is no combustion chamber in the gas flow system between the GC column and the TLS (and the QMS), it will also not be possible with the SAM instrument suite to measure the isotopic composition of individual organic compounds.

## 4.2 Urey

The Urey instrument, selected to be part of the Pasteur payload of the European ExoMars mission, is also a suite of different components, including a sub-critical water extractor (SCWE), a sublimation and concentration chamber (based on the original MOD design), a micro-capillary electrophoresis system ($\mu$CE) with fluorescence detection and finally the two parts of the MOI. The SCWE is used to extract soluble organic compounds from solid samples in a less aggressive way than in SAM. Following the extraction, the target compounds, including amino acids, nucleobases and PAH, are sublimed inside the MOD and concentrated on disks for initial photometric detection. In case of a positive signal by either PAH or fluorescamine-derivatized amino acids, the sublimate is transferred into the $\mu$CE system for separation and quantification. Using this protocol, Urey has the same capabilities for determining amino acid distributions as SAM, but at a higher sensitivities. However, due to its extraction protocol, Urey can only target a limited set of organic compounds and does not have the capability to determine the exact mass of the detected organic compounds. The $\mu$CE on Urey allows the separation and quantification of amino acid enantiomers. Due to the high precision of the measurement, the instrument has the ability to distinguish prebiotic chemistry (for example accumulated amino acids that were delivered by carbonaceous meteorites to Mars), from both present and past life.

**Acknowledgements** The authors are grateful for the constructive comments of two anonymous reviewers.

# References

O. Botta, J.L. Bada, Surv. Geophys. **23**, 411–467 (2002)
S. Chang, R. Mack, K. Lennon, Lunar Planet. Sci. **9**, 157–159 (1978)
I. Cherchneff, J.R. Barker, A.G.G.M. Tielens, Astrophys. J. **401**, 269–287 (1992)
C. Chyba, C. Sagan, Nature **355**, 125–132 (1992)
G. Cooper, N. Kimmich, W. Belisle, J. Sarinana, K. Brabham, L. Garrel, Nature **414**, 879–883 (2001)
G.W. Cooper, W.M. Onwo, J.R. Cronin, Geochim. Cosmochim. Acta **56**, 4109–4115 (1992)
G.W. Cooper, M.H. Thiemens, T.L. Jackson, S. Chang, Science **277**, 1072–1074 (1997)
J.R. Cronin, S. Chang, in *Chemistry of Life's Origins*, ed. by J.M. Greenburg, V. Pirronello (Kluwer, Dordrecht, 1993), pp. 209–258
J.R. Cronin, S. Pizzarello, Geochim. Cosmochim. Acta **50**, 2419–2427 (1986)
J.R. Cronin, S. Pizzarello, Science **275**, 951–955 (1997)
J.R. Cronin, S. Pizzarello, D.P. Cruikshank, in *Meteorites and the Early Solar System*, ed. by J.F. Kerridge, M.S. Matthews (P. Univ. Ariz, Tucson, 1988)
L. d'Hendecourt, P. Ehrenfreund, *Life Sciences: Complex Organics in Space*. Advances in Space Research, vol. 19 (Pergamon Press Ltd, Oxford, 1997), pp. 1023–1032
M.H. Engel, B. Nagy, Nature **296**, 837–840 (1982)
R. Hayatsu, E. Anders, M.H. Studier, L.P. Moore, Geochim. Cosmochim. Acta **39**, 471–488 (1975)
G. Jungclaus, J.R. Cronin, C.B. Moore, G.U. Yuen, Nature **261**, 126–128 (1976a)
G.A. Jungclaus, G.U. Yuen, C.B. Moore, J.G. Lawless, Meteoritics **11**, 231–237 (1976b)
J.F. Kasting, Science **259**, 920–926 (1993)
T. Kawasaki, K. Hatasea, Y. Fujiia, K. Joa, K. Soaia, S. Pizzarello, Geochim. Cosmochim. Acta **70**, 5395–5402 (2006)
K. Kvenvolden, J. Lawless, K. Pering, E. Peterson, J. Flores, C. Ponnamperuma, I.R. Kaplan, C. Moore, Nature **228**, 928–926 (1970)
J.G. Lawless, G.U. Yuen, Nature **282**, 396–398 (1979)
J.G. Lawless, B. Zeitman, W.E. Pereira, R.E. Summons, A.M. Duffield, Nature **251**, 40–42 (1974)
S.G. Love, D.E. Brownlee, Science **262**, 550–553 (1993)
M. Maurette, J. Dupart, C. Engrand, M. Gounelle, G. Kurat, G. Matrajt, A. Toppani, Planet. Space Sci. **48**, 1117–1137 (2000)
G.D. McDonald, M.C. Storrie-Lombardi, Astrobiology **6**, 17–33 (2006)
D.S. McKay, E.K. Gibson, K.L. Thomas-Keprta, H. Vali, C.S. Romanek, S.J. Clemett, D.F. Chillier, C.R. Maechling, R.N. Zare, Science **273**, 924–930 (1996)
U.J. Meierhenrich, G.M. Munoz Caro, J.H. Bredehoft, E.K. Jessberger, W.H.P. Thiemann, Proc. Natl. Acad. Sci. USA **101**, 9182–9186 (2004)
J. Oró, Nature **190**, 389–390 (1961)
E.T. Peltzer, J.L. Bada, G. Schlesinger, S.L. Miller, Adv. Space Res. **4**, 69–74 (1984)
K.L. Pering, C. Ponnamperuma, Science **173**, 237–239 (1971)
S. Pizzarello, J.R. Cronin, Geochim. Cosmochim. Acta **64**, 329–338 (2000)
S. Pizzarello, X. Feng, S. Epstein, J.R. Cronin, Geochim. Cosmochim. Acta **58**, 5579–5587 (1994)
S. Pizzarello, M. Zolensky, K.A. Turk, Geochim. Cosmochim. Acta **67**, 1589–1595 (2003)
M.A. Sephton, C.T. Pillinger, I. Gilmour, Geochim. Cosmochim. Acta **62**, 1821–1828 (1998)
M.A. Sephton, C.T. Pillinger, I. Gilmour, Geochim. Cosmochim. Acta **64**, 321–328 (2000)
M.A. Sephton, A.B. Verchovsky, P.A. Bland, I. Gilmour, M.M. Grady, I.P. Wright, Geochim. Cosmochim. Acta **67**, 2093–2108 (2003)
M.A. Sephton, G.D. Love, J.S. Watson, A.B. Verchovsky, I.P. Wright, C.E. Snape, I. Gilmour, Geochim. Cosmochim. Acta **68**, 1385–1393 (2004)
A. Shimoyama, H. Katsumata, Chem. Lett. **3**, 202–203 (2001)
A.M. Skelley, J.R. Scherer, A.D. Aubrey, W.H. Grover, R.H.C. Ivester, P. Ehrenfreund, F.J. Grunthaner, J.L. Bada, R.A. Mathies, Proc. Natl. Acad. Sci. USA **102**, 1041–1046 (2004)
L.E. Snyder, D. Buhl, B. Zuckerman, P. Palmer, Phys. Rev. Lett. **22**(13), 679–681 (1969)
P.G. Stoks, A.W. Schwartz, Nature **282**(5740), 709–710 (1979)
P.G. Stoks, A.W. Schwartz, Geochim. Cosmochim. Acta **45**, 563–569 (1981a)
P.G. Stoks, A.W. Schwartz, in *Proc. 6th Internatl. Conf. on the Origin of Life*, ed. by Y. Wolman (Reidel, Dordrecht, 1981b), pp. 59–64

P.G. Stoks, A.W. Schwartz, Geochim. Cosmochim. Acta **46**, 309–315 (1982)
M.G. Trainer, A.A. Pavlov, H.L. DeWitt, J.L. Jimenez, C.P. McKay, O.B. Toon, M.A. Tolbert, Proc. Natl. Acad. Sci. USA **103**, 18035–18042 (2006)
O. Vandenabeele-Trambouze, M. Claeys-Bruno, M. Dobrijevic, C. Rodier, G. Borruat, A. Commeyras, L. Garrelly, Astrobiology **5**, 48–65 (2005)
G. Yuen, N. Blair, D.J. DesMarias, S. Chang, Nature **307**, 252–254 (1984)

# Astrobiology and Habitability of Titan

**Francois Raulin**

Originally published in the journal Space Science Reviews, Volume 135, Nos 1–4.
DOI: 10.1007/s11214-006-9133-7 © Springer Science+Business Media B.V. 2007

**Abstract** Largest satellite of Saturn and the only in the solar system having a dense atmosphere, Titan is one of the key planetary bodies for astrobiological studies, due to several aspects. (i) Its analogies with planet Earth, in spite of much lower temperatures, with, in particular, a methane cycle on Titan analogous to the water cycle on Earth. (ii) The presence of an active organic chemistry, involving several of the key compounds of prebiotic chemistry. The recent data obtained from the Huygens instruments show that the complex organic matter in Titan's low atmosphere is mainly concentrated in the aerosol particles. The formation of biologically interesting compounds may also occur in the deep water ocean, from the hydrolysis of complex organic material included in the chrondritic matter accreted during the formation of Titan. (iii) The possible emergence and persistence of Life on Titan. All ingredients which seem necessary for Life to appear and even develop – liquid water, organic matter and energy – are present on Titan. Consequently, it cannot be excluded that life may have emerged on or in Titan. In spite of the extreme conditions in this environment life may have been able to adapt and to persist. Many data are still expected from the Cassini-Huygens mission and future astrobiological exploration mission of Titan are now under consideration. Nevertheless, Titan already looks like another world, with an active organic chemistry, in the absence of permanent liquid water, on the surface: a natural laboratory for prebiotic-like chemistry.

**Keywords** Astrobiology · Cassini-Huygens · Prebiotic chemistry · Primitive Earth · Tholins · Titan

## 1 Introduction

Since the Voyager fly-by's of Titan in the early 1980's, our knowledge of this exotic place, the only satellite of the solar system having a dense atmosphere, has been largely improved. The

F. Raulin
Laboratoire Interuniversitaire des Systèmes Atmosphériques, LISA-UMR CNRS 7583, Universités Paris 7 et Paris 12, 61 Avenue du Général de Gaulle, F-94000 Créteil, France
e-mail: raulin@lisa.univ-paris12.fr

vertical atmospheric structure has been determined, and the primary chemical composition, trace compounds, and especially organic constituents described. Additional organics have also been identified later on by ground based observation and by the European Infrared Space Observatory satellite. Other ground based and Hubble observations have also allowed a first mapping of the surface, showing a heterogeneous milieu. However many questions still remained concerning Titan and its astrobiological aspects. What is the origin of its dense atmosphere? What is the source of methane? What is the chemical composition of the aerosols which are present in the atmosphere? What is the chemical composition of the surface? How complex is the organic matter on Titan? How close are the analogies between Titan and the primitive Earth? Is there life on Titan?

Bringing answers to these questions was among the main objectives of the Cassini-Huygens mission jointly designed and developed by NASA and ESA. Indeed, since the successful Saturn orbital insertion of Cassini on July 1st, 2004, and the descent of the Huygens probe in Titan's atmosphere on January 14th, 2005 (Lebreton *et al.*, 2005) many essential data have already been obtained which are of paramount importance for our understanding of Titan's astrobiology. These aspects of astrobiological importance of Titan are presented here on the basis of some of the new data provided by Cassini-Huygens.

## 2 Titan, an Earth-like Planetary Body

With a diameter of more than 5100 km, Titan is the largest moon of Saturn and the second largest moon of the solar system (Table 1). It is also the only one to have a dense atmosphere. Clearly evidenced by the presence of haze layers, Titan's atmosphere extends to approximately 1500 km (Fulchignoni *et al.*, 2005). The atmosphere of Titan is mainly composed of molecular nitrogen, $N_2$, like the Earth's atmosphere. The other main constituents are methane, ($CH_4$, 1.4% to 2.0% in the stratosphere (Flasar *et al.*, 2005; Niemann *et al.*, 2005)) and molecular hydrogen ($H_2$, approximate 0.1%). With a surface temperature of $\sim$94 K, and a surface pressure of 1.5 bar, Titan's atmosphere is 4.5 times denser than the Earth's.

Although much colder, Titan shows a vertical atmospheric structure qualitatively similar to that of the Earth (Table 1), with a troposphere ($\sim$94–$\sim$70 K), a tropopause (70.4 K) and a stratosphere ($\sim$70–175 K). Because of a larger scale height, in the case of Titan, the mesosphere extends to altitudes higher than 400 km (instead of only 100 km for the Earth), but the shape looks very much the same. This is related to the presence in both

**Table 1** Main characteristics of Titan (including the HASI-Huygens data)

| Surface radius | | | 2.575 km |
| Surface gravity | | | 1.35 m s$^{-2}$ (0.14 Earth's value) |
| Mean volumic mass | | | 1.88 kg dm$^{-3}$ (0.34 Earth's value) |
| Distance from Saturn | | | 20 Saturn radius ($\sim$1.2 $\times$ 10$^6$ km) |
| Orbit period around Saturn | | | $\sim$16 days |
| Orbit period around Sun | | | $\sim$30 years |
| | | | |
| Atmospheric data | | | |
| | Altitude (km) | Temperature (K) | Pressure (mbar) |
| Surface | 0 | 93.7 | 1470 |
| Tropopause | 42 | 70.4 | 135 |
| Stratopause | $\sim$250 | $\sim$187 | $\sim$1.5 $\times$ 10$^{-1}$ |
| Mesopause | $\sim$490 | $\sim$152 | $\sim$2 $\times$ 10$^{-3}$ |

atmospheres of greenhouse gases and anti-greenhouse elements. Methane, which exhibits strong absorption bands in the far infrared regions corresponding to the maximum of the infrared emission spectrum of Titan and is transparent in the near UV and visible spectral regions, is a very efficient greenhouse gas in Titan's atmosphere. Molecular hydrogen, which is also absorbing in the far infrared plays a similar role. In the pressure-temperature conditions of Titan's atmosphere, methane can condense but molecular hydrogen can not. Thus, on Titan, $CH_4$ and $H_2$ are equivalent to terrestrial condensable $H_2O$ and non-condensable $CO_2$, respectively. In addition the haze particles and clouds in Titan's atmosphere play an anti-greenhouse effect similar to that of the terrestrial atmospheric aerosols and clouds (McKay et al., 1991). Moreover, being photo-chemically produced, being UV-absorbing and warming the stratosphere, the haze also plays a role like ozone in Earth's atmosphere.

Moreover, methane on Titan seems to play the role of water on the Earth, with a complex cycle, which still has to be understood. One of the main questions concerns the sources of methane. The presence of hydrocarbon oceans on Titan's surface as main reservoir of methane (Lunine, 1993) is now ruled out (West et al., 2005). However the possibility that Titan's surface includes lakes of methane and ethane seems now confirmed with the very recent detection by the Cassini Radar of surface features near the North Pole which strongly suggest such liquid bodies (Figure 1). Moreover, the DISR instrument on Huygens has provided pictures of Titan's surface showing dentritic structures (Figure 2) which look like fluvial net, in a relatively young terrain, indicating recent liquid flow on the surface of Titan (Tomasko et al., 2005). In addition, the Huygens GC-MS data show that methane mole fraction increases in the low troposphere (up to 5%) and reaches the saturation level at approximately 8 km altitude, allowing the possible formation of clouds and rain (Niemann et al., 2005; Tokano et al., 2006). Furthermore, GC-MS analyses recorded a ~50% increase in the methane mole fraction at Titan's surface, suggesting the presence of condensed methane on the surface near the landed probe.

Other observations from the Cassini instruments show a very diversified surface including various features of different origins indicative of volcanic, tectonic, sedimentological and meteorological processes similar to those we find on Earth. The only noble gas detected in Titan's atmosphere (by Cassini-INMS and Huygens-GC-MS) is argon. Like on Earth, the most abundant isotope is $^{40}Ar$. Its stratospheric mole fraction is a few $10^{-5}$, as measured by GC-MS (Niemann et al., 2005). It should come from the radioactive decay of $^{40}K$. This indicates that Titan's atmosphere is of secondary origin, produced by the degassing of trapped gases. Since $N_2$ cannot be efficiently trapped in the icy planetesimals which accreted and

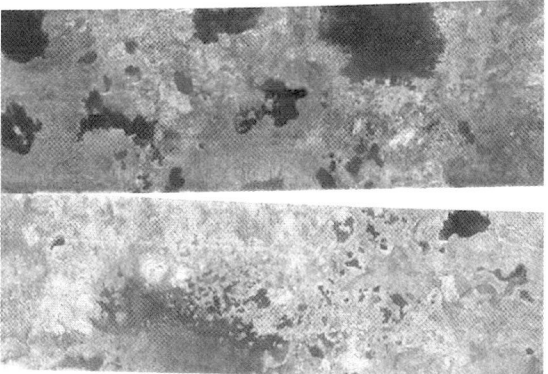

**Fig. 1** Radar image of Titan North Pole region showing several dozen dark features which could be lakes. The smallest are about 1 km, several are more than 30 km wide. The biggest lake is about 100 miles long. Courtesy NASA/JPL-Caltech

**Fig. 2** Channel networks, highlands and dark-bright interface seen by the DISR instrument on Huygens at 6.5 km altitude. Courtesy ESA/NASA/JPL-Caltech/ University of Arizona

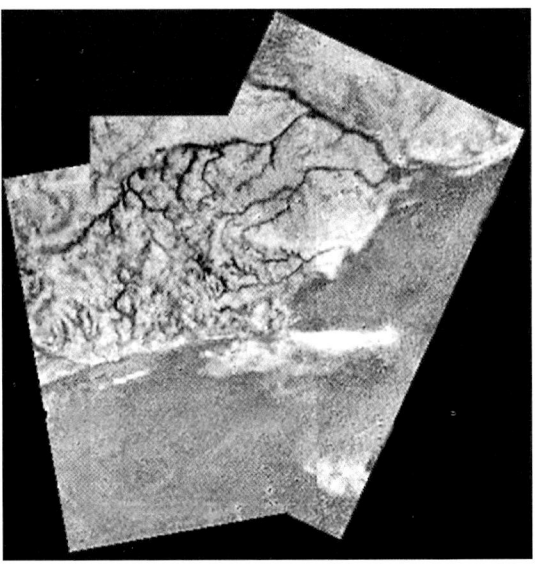

formed Titan, contrary to $NH_3$, this also indicates that its primordial atmosphere was initially made of $NH_3$. Ammonia was then transformed into $N_2$ by photolysis and/or impact driven chemical processes (Owen, 2000; Gautier and Owen, 2002). The $^{14}N/^{15}N$ ratio measured in the atmosphere by INMS and GC-MS (183 in the stratosphere) is 1.5 times less than primordial N, suggesting that the present mass of the atmosphere was probably lost several times during the history of the satellite (Niemann et al., 2005). Since such evolution may also imply methane transformation into organics, this may be also the indication of large deposits of organics on Titan's surface.

## 3 A Complex Organic Chemistry

There are also several analogies between the very active organic chemistry on Titan and the prebiotic chemistry which was active on the primitive Earth, in spite of the absence of permanent bodies of liquid water on Titan's surface. Several of the organic processes which are occurring today on Titan imply organic molecules, such as hydrogen cyanide (HCN) and cyanoacetylene ($HC_3N$), which are considered as key compounds in the prebiotic chemistry on the primitive Earth. In fact by its main composition ($N_2$ with a few % of $CH_4$), the atmosphere of Titan is one of the most favourable for atmospheric prebiotic syntheses. Such an atmosphere may be even closer to that of the primitive Earth than it was supposed until now. Indeed, very recent model of hydrogen escape in the primitive atmosphere of the Earth suggests that it may have been much richer in hydrogen and methane than previously thought (Tian et al., 2005), implying that Titan maybe even more similar to the primitive Earth than we believed.

In the atmosphere of Titan, $CH_4$ chemistry is coupled with $N_2$ chemistry producing the formation of many organics – in the gas and particulate phases. Those are hydrocarbons, nitriles and complex refractory organics. Several photochemical models of the atmosphere of Titan have been published for the last 20 years (Lebonnois et al., 2001; Wilson and Atreya, 2004; Hébrard et al., 2005; and references therein). The chemical pathways start with the dissociation of $N_2$ and $CH_4$ through electron and photon impacts. The primary processes allow the formation of $C_2H_2$ and HCN in the high atmosphere. These two molecules then diffuse down to the lower atmospheric levels where they induce the production of many other hydrocarbons and nitriles. Additional $CH_4$ dissociation probably also occurs in the low stratosphere through photocatalytic processes induced by $C_2H_2$ and polyynes. Many organic species are also formed in the ionosphere.

Another approach to study Titan's organic chemistry is to mimic their processes through simulation experiments in the laboratory. These experiments, carried out in particular at LISA, produce all the gas phase organic species already detected in Titan's stratosphere. They produce many other organics which can be assumed to be also present in Titan's atmosphere. Thus, such simulation experiments can be used as very efficient guides for further searches (both by remote sensing & in situ observations) of new species in the gas as well as the aerosol phases. More than 150 different organic molecules have been detected in these experiments (Coll *et al.*, 1998, 1999) using an open reactor flown by a low pressure $N_2$-$CH_4$ gas mixture. The energy source is a cold plasma discharge producing mid-energy electrons (around 1–10 eV). The gas phase end products are analyzed by FTIR (Fourier Transform InfraRed spectroscopy) and GC-MS (Gas Chromatography and Mass Spectrometry) techniques. The transient species (radicals and ions) are determined by on line UV-visible spectroscopy. The identified organic products are mainly hydrocarbons and nitriles. Among the other organics formed in these experiments and not yet detected in Titan's atmosphere, one should note the presence of polyynes ($C_6H_2$, $C_8H_2$) and probably cyanopolyyne $HC_4$-CN. These compounds are also included in photochemical models of Titan's atmosphere, where they could play a key role in the chemical schemes allowing the transition from the gas phase products to the aerosols. These studies also show the formation of ammonia (Bernard *et al.*, 2003) at noticeable concentration. Although the mole fraction of ammonia, if present in Titan's stratosphere should be very low because of its low vapour pressure within Titan's conditions, its detection in laboratory simulations opens new avenues in the chemical schemes of Titan's atmosphere. Moreover, it should be noticed that ammonia has recently been detected in Titan's ionosphere by INMS, the mass spectrometer on Cassini (Vuitton *et al.*, 2006).

Simulation experiments also produce solid organics, as mentioned above, usually named tholins (Sagan and Khare, 1979). These "Titan tholins" are supposed to be laboratory analogues of Titan's aerosols. They have been extensively studied since the first work by Sagan & Khare more than 20 years ago (Khare and Sagan, 1984, 1986 and references therein). These laboratory analogues show very different properties depending on the experimental conditions (Cruikshank *et al.*, 2005). In particular, their average C/N ratio varies between less than 1 to more than 3, in the experiments the most relevant to Titan's conditions. Dedicated experimental protocols allowing a simulation closer to the real conditions have been developed at LISA using low pressure and low temperature conditions (Coll *et al.*, 1998, 1999) and recovering the laboratory tholins without oxygen contamination (from the air of the laboratory) in a glove box purged with pure $N_2$. Representative laboratory analogues of Titan's aerosols have thus been obtained and their complex refractive indices have been determined (Ramirez *et al.*, 2002), with – for the first time – error bars. These data can be seen as a new point of reference to modelers who compute the properties of Titan's aerosols. Systematic studies have been carried out on the influence of the pressure of the starting gas mixture on the elemental composition of the tholins. They show that two different chemical-physical regimes are involved in the processes, depending on the pressure, with a transition pressure around 1 mbar (Bernard *et al.*, 2002; Imanaka *et al.*, 2004).

The molecular composition of the Titan tholins is still poorly known. However it is well established that they are made of macromolecules of largely irregular structure. Gel filtration

**Table 2** Main composition of Titan's stratosphere (mixing ratios at mid-latitudes around 120 km, adapted from Bézard, 2006): trace components already detected and comparison with the products of laboratory simulation experiments (Maj = major product; ++: abundance smaller by one order of magnitude; +: abundance smaller by two orders of magnitude)

| Compounds | Mixing ratio | Ref. | Production in simulation experiments |
|---|---|---|---|
| Main constituents | | | |
| Nitrogen $N_2$ | 0.98 | Niemann et al., 2005 | |
| Methane $CH_4$ | 0.014 | ibid. | |
| Hydrogen $H_2$ | $\sim$0.001 | Courtin et al., 1995 | Maj. |
| Hydrocarbons | | | |
| Ethane $C_2H_6$ | $1 \times 10^{-5}$ | Coustenis et al., 2003, 2006; Vinatier et al., 2006 | Maj. |
| Acetylene $C_2H_2$ | $2 \times 10^{-6}$ | ibid. | Maj. |
| Ethylene $C_2H_4$ | $4 \times 10^{-7}$ | ibid. | ++ |
| Propane $C_3H_8$ | $5 \times 10^{-7}$ | ibid. | ++ |
| Propyne $C_3H_4$ | $8 \times 10^{-9}$ | ibid. | + |
| Diacetylene $C_4H_2$ | $1 \times 10^{-9}$ | ibid. | + |
| Benzene $C_6H_6$ | $4 \times 10^{-10}$ | ibid. | + |
| N-Organics | | | |
| Hydrogen cyanide, HCN | $1 \times 10^{-7}$ | Marten et al., 2002; Teanby et al., 2006 | Maj. |
| Cyanoacetylene $HC_3N$ | $1 \times 10^{-9*}$ | ibid. | ++ |
| Acetonitrile $CH_3CN$ | $2 \times 10^{-8*}$ | Marten et al., 2002 | ++ |
| Cyanogen $C_2N_2$ | $1 \times 10^{-9**}$ | Teanby et al., 2006 | + |
| Dicyanoacetylene $C_4N_2$ | Solid Phase | Samuelson et al., 1997 | + |
| O-Compounds/Noble gases | | | |
| Carbon monoxide CO | $4.5 \times 10^{-5}$ | De Kok et al., 2006 | |
| Carbon dioxide $CO_2$ | $1.6 \times 10^{-9}$ | ibid. | |
| Water $H_2O$ | $4 \times 10^{-10}$ | ibid., Coustenis et al., 1998 | |
| Argon $^{40}Ar$ | $\sim 3 \times 10^{-5}$ | Niemann et al., 2005 | |
| $^{36}Ar$ | $\sim 2 \times 10^{-7}$ | ibid. | |

*At 300 km altitude, disk-averaged; **At 60° N

chromatography of the water soluble fraction of Titan tholins shows an average molecular mass of about 500 to 1000 Dalton ((McDonald et al., 1994). Information on their chemical structure has been obtained from their IR and UV spectra and from analysis by pyrolysis-GC-MS techniques (Ehrenfreund et al., 1995; Coll et al., 1998; Imanaka et al., 2004; and references therein). The data show the presence of aliphatic & benzenic hydrocarbon groups, of CN, $NH_2$ and C=NH groups. Direct analysis by chemical derivatization techniques before and after hydrolysis allowed the identification of amino-acids or their precursors (Khare et al., 1986). Their optical properties have been determined (Khare et al., 1984; McKay, 1996; Ramirez et al., 2002; Tran et al., 2003; Imanaka et al., 2004), because of their importance for retrieving observational data related to Titan. Finally, it is obviously of astrobiological interest to mention that Stoker et al. (1990) demonstrated the nutritious properties of Titan tholins for micro-organisms.

Several organic compounds have already been detected in Titan's stratosphere (Table 2). The list includes hydrocarbons (both with saturated and unsaturated chains) and nitrogen-containing organic compounds, exclusively nitriles, as expected from laboratory simulation experiments. Most of these detections were performed by Voyager observations, with the

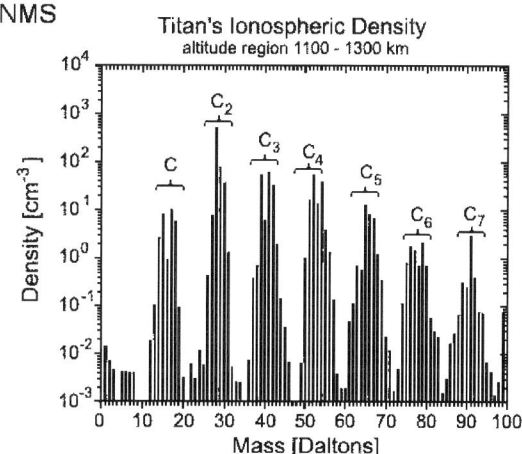

**Fig. 3** Mass spectrum of Titan's ionosphere near 1,200 km altitude. The spectrum shows signature of organic compounds including up to 7 carbon atoms. Credit: NASA/JPL/University of Michigan

exception of the $C_2$ hydrocarbons, which were observed previously, acetonitrile, which was detected by ground observation in the millimetre wavelength, and water and benzene, which were tentatively detected by ISO.

Since the Cassini arrival in the Saturn system, the presence of water and benzene has been unambiguously confirmed by the CIRS instrument. In addition, the direct analysis of the ionosphere by the INMS instrument during the low altitude Cassini fly-by's of Titan (Waite *et al.*, 2005; Vuitton *et al.*, 2006) shows the presence of many organic species at detectable levels (Figure 3), in spite of the very high altitude (1100–1300 km). The mass spectra collected by the GC-MS instrument during the descent of Huygens show that the medium and low stratosphere and the troposphere are poor in volatile organic species, with the exception of methane. Condensation of these species on the aerosol particles is a probable explanation for these atmospheric characteristics (Niemann *et al.*, 2005). These particles, for which no direct data on the chemical composition were available before, have been analyzed by the ACP instrument. Its results show that the aerosol particles are made of refractory organics which release HCN and $NH_3$ during pyrolysis at 600°C (Israel *et al.*, 2005). This strongly supports the tholin hypothesis: these first *in situ* measurement data strongly suggest that the aerosol particles are made of a refractory organic nucleus, covered with condensed volatile compounds (Figure 4). The nature of the pyrolysates indicates the potential presence of nitrile groups (—CN), amino groups (—$NH_2$, —NH- and —N<) and /or imino groups (—C — N ) in the nucleus. These particles sediment down to the surface where they likely form a deposit of complex refractory organics and frozen volatiles.

DISR collected the infrared reflectance spectra of the surface with the help of a lamp, illuminating the surface before the Huygens probe touched down. The data show the presence of water ice, which is also consistent with the observations of the SSP instrument (Zarnecki *et al.*, 2005), but no clear evidence – so far – of tholins. On the other hand, GC-MS analysis of the atmosphere near the surface shows the clear signature of many organics, including cyanogen, C3 – and C4 hydrocarbons and benzene, indicating that the surface is much richer in volatile organics than the low stratosphere and the troposphere (Niemann *et al.*, 2005). These observations are in agreement with the hypothesis that in the low atmosphere of Titan, most of the organic compounds are in the condensed phase.

Thus, altogether, these new data show the diversity of the locations where organic chemistry is taking place on Titan. The high atmosphere looks very active, with neutral and ion

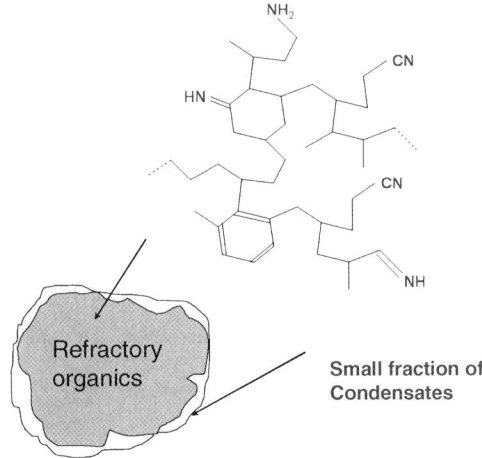

organic processes; the high stratosphere, where many organic compounds have already been detected before Cassini and since Cassini arrived in the Saturn system, also shows an active organic chemistry in the gas phase. In the lower atmosphere this chemistry seems mainly concentrated in the condensed phase. Titan's surface is probably covered with frozen volatile organics together with refractory, tholin-like, organic materials. Irradiating effects of cosmic rays reaching Titan's surface may induce additional organic syntheses of reactive compounds such as diazomethane and azides as well as the polymerization of HCN (Raulin *et al.*, 1995).

In spite of its low temperature, even the presence of liquid water is not excluded on Titan's surface. Cometary impacts on Titan, may melt surface water ice, offering possible episodes as long as $\sim$1000 years of liquid water (Obrien *et al.*, 2005). This provides conditions for short terrestrial-like prebiotic syntheses but at relatively low temperatures. Cryovolcanism, which has been clearly evidenced from the first images of Titan's surface provided by the VIMS, ISS and Radar instrument on Cassini (Sotin *et al.*, 2005), is another possibility for prebiotic chemistry in the presence of water. This has recently considered for dome Ganesa Macula on Titan, where liquid water-ammonia may have persisted for timescales up to $10^5$ years (Neish *et al.*, 2006). In addition, the possible presence of a water-ammonia ocean in the depths of Titan, as expected from models of its internal structure (Tobie *et al.*, 2005, and references therein), may also provide an efficient way to convert simple organics into complex molecules, and to reprocess chondritic organic matter into prebiotic compounds. These processes may have very efficiently occurred at the beginning of Titan's history (with even the possibility of the water-ammonia ocean exposed to the surface) allowing a CHNO prebiotic chemistry evolving to compounds of terrestrial biological interest.

Even if these liquid water scenarii are false, the possibility of a pseudo-biochemistry, occurring in the absence of a noticeable amount of O atoms, cannot be ruled out, with a N-chemistry, based on "ammono" analogues replacing the O-chemistry (Raulin and Owen, 2002). Such alternatives of terrestrial biochemistry, where, in particular the water solvent could be replaced by ammonia or other N-compounds, have also been recently re-examined by Benner (2002) and by Schulze-Makuch and Irwin (2004). Although it remains very speculative and has not been specifically studied experimentally, such a N-biochemistry involves chemical groups like the imino group, very reactive at room temperature, but more stable to the low temperatures of Titan.

## 4 Life on Titan?

Several ways can thus be considered in Titan's environment to drive chemistry to prebiotic chemistry and even to biotic systems. But if life emerged on this planetary body, are Titan's conditions compatible with the sustaining of life? The surface is too cold and not energetic enough to provide the right conditions. However, the (still hypothetical) subsurface oceans may be suitable for life. Fortes (2000) has shown that there are no insurmountable obstacles. With a possible temperature of this ocean as high as about 260 K and the possible occurrence of cryovolcanic hotspots allowing 300 K, the temperature conditions in Titan's subsurface oceans could allow the development of living systems. Even at depth of 200 km, the expected pressure of about 5 kbar is not incompatible with life, as shown by terrestrial examples. The expected pH of an aqueous medium made of 15% by weight of $NH_3$ is 11.5. Some bacteria can grow on Earth at pH 12. Even the limited energy resources do not exclude the sustenance of life.

Taking into account only the potential radiogenic heat flow ($\sim 5 \times 10^{11}$ W), Fortes (2000) estimated an energy flux available in the subsurface oceans of about $5 \times 10^8$ W. Such a flux corresponds to the production of about $4 \times 10^{11}$ mol of ATP per year and about $2 \times 10^{13}$ g of biomass per year. If we assume an average turn over for the living systems in the order of one year, the biomass density would be $1g/m^2$. This is very small compared to the lower limit of the value of the biomass for the Earth (about 1000 to 10000 $g/m^2$). Nevertheless, this indicates the possible presence of a limited but not negligible bioactivity on the satellite. The biota on Titan, if any, assuming that the living systems are similar to the ones we know on Earth, would thus be localised in the subsurface deep ocean. Several possible metabolic processes such as nitrate/nitrite reduction or nitrate/dinitrogen reduction, sulphate reduction and methanogenesis have been postulated (Simakov, 2001) as well as the catalytic hydrogenation of acetylene (Abbas and Schulze-Makuch, 2002; McKay and Smith, 2005).

As expected, no sign of macroscopic life has been detected by Huygens when approaching the surface or after it landed. This can be concluded in particular from the many pictures taken by DISR of the same location on Titan during more than one hour after landing. But this does not exclude the possibility of the presence of a microscopic life. The metabolic activity of the corresponding biota, even if it is localized far from the surface, in the deep internal structure of Titan, may produce chemical species which diffuse through the ice mantle covering the hypothetical internal ocean and feed the atmosphere. It has even been speculated that the methane present in the atmosphere today is the product of biological activity (Simakov, 2001; Schulze-Makuch and Irwin, 2004). If this was the case, the atmospheric methane would be notably enriched in light carbon. Indeed, on Earth, biological processes induce an isotopic fragmentation producing enrichment in $^{12}C$: $^{12}C/^{13}C$ increases from 89 (the reference value, in the Belemnite of the Pee Dee Formation) to about 91–94 depending on the biosynthesis processes. The $^{12}C/^{13}C$ ratio in atmospheric methane on Titan, as determined by the GC-MS instrument on Huygens is 82 (Niemann et al., 2005). Although we do not have a reference for $^{12}C/^{13}C$ on Titan, this low value suggests that the origin of methane is likely to be abiotic.

## 5 Conclusions

Although exotic life, like methanogenic life in liquid methane cannot be fully ruled out (McKay and Smith, 2005), a large presence of extent or extinct life on Titan seems unlikely. Nevertheless, with the new observational data provided by the Cassini-Huygens mission, the largest satellite of Saturn looks more than ever like a very interesting object for astrobiology. The several analogies of this exotic and cold planetary body with the Earth and the complex organic chemical processes which are going on now on Titan provide a fantastic means to better understand the prebiotic processes which are not reachable anymore on the Earth, at the scale and within the whole complexity of a planetary environment.

The origin and cycle of methane on Titan illustrate the whole complexity of the Titan's system. Methane may be stored in large amounts in the interior of the satellite, in the form of clathrates (methane hydrates) trapped during the formation of the satellite from the Saturnian subnebula where it was formed by Fisher-Tropsch processes (Sekine *et al.*, 2005). It may also be produced through high pressure processes, like serpentinization, allowing the formation of $H_2$ by reaction of $H_2O$ with ultramafic rocks, or by cometary impacts (Kress and McKay, 2004). Interestingly, those processes have rarely been considered in the case of the primitive Earth: an example of how Titan's study is indeed providing new insights into terrestrial chemical evolution.

In Titan's atmosphere, methane is photolysed by solar UV, producing mainly ethane and tholins-like organic matter. The net destruction rate of methane estimated in photochemical models is $1.3 \times 10^{10}$ (Toublanc *et al.*, 1995) and $6.5 \ 10^9$ molec. $cm^{-2} \ s^{-1}$ (Lebonnois *et al.*, 2001). The net production rate of the haze has been estimated to $10^{-14}$ g $cm^{-2} \ s^{-1}$ (McKay *et al.*, 1989), $3.2 \times 10^{-14}$ g $cm^{-2} \ s^{-1}$ (Wilson and Atreya, 2003) and $4 \times 10^{-14}$ g $cm^{-2} \ s^{-1}$ (Lebonnois *et al.*, 2002). If we assume that the aerosols are mainly made of C and H with a C/N ratio of 2 and use the maximum net production rate of the haze, then the flux of methane transformed into aerosols is $9 \times 10^8$ molec. $cm^{-2} \ s^{-1}$. This shows that only a small fraction of the photolysed methane is transformed into the haze: 7 to 14%. Since about 20% of methane is converted into ethane (Toublanc *et al.*, 1995), ethane is the dominant product. The resulting life time of methane in Titan's atmosphere is relatively short (about 10 to 30 myr). Thus methane stored in Titan's interior may be continuously replenishing the atmosphere, through degassing induced by cryovolcanism. In any case, the methane cycle should result in the accumulation of large amounts of complex organics on the surface and large amounts of ethane, which mixed with the dissolved atmospheric methane should form liquid bodies on the surface or in the near sub-surface of the satellite. Titan's surface looks very heterogeneous and it is likely that these two ingredients are not distributed uniformly on its surface. The dunes recently reported on the Cassini RADAR images of Titan (Lorenz *et al.*, 2006) maybe made of organic aerosols. The features observed in the north polar regions by the Cassini Radar may be lakes of liquid methane/ethane (which, however, are too small to completely solve the problem of methane/ethane reservoir).

The Cassini-Huygens mission is far from complete. It will continue its systematic exploration of the Saturnian system up to 2008, and probably 2012 if the extended mission is accepted. Numerous data of paramount importance for astrobiology are still expected from several of its instruments. The CIRS spectrometer should be able to detect new organic species in the atmosphere. ISS and VIMS should provide a detailed picture of Titan's surface revealing the complexity but also the physical and chemical nature of this surface and its diversity. Radar observation will also continue the systematic coverage of Titan's surface showing contrasted regions of smooth and rough areas, suggesting possible shorelines. The already available data of this new, exotic and astonishing world already show that a future mission to Titan is needed if we want to understand the prebiotic-like chemistry which is occurring, in particular on Titan's surface. Such a mission could include surface mobility, using ballooning, as proposed by Lorenz (2000) and surface sampling. It is essential for an astrobiological exploration of Titan that the scientific payload includes instruments capable to carry out a detailed chemical analysis of the atmospheric aerosols and the surface constituents. In particular the payload should analyse their organic fractions and determine their elemental and molecular composition and look for an eventual enantiomeric excess. Such data, which will not be obtained by Cassini, could provide crucial information for understanding the astrobiology of Titan.

**Acknowledgements** The author wishes to thank the Cassini-Huygens teams, particularly Jean-Pierre Lebreton, the Huygens project scientist and his colleague Olivier Witasse, and the PI's of the Huygens instruments, especially Guy Israel and Hasso Niemann, for making available several of the data used in this paper. Many thanks also to the anonymous referee for his suggestions to improve the manuscript and to Oliver Botta for the rereading. Final thanks go to the European Space Agency, ESA, and the French Space Agency, CNES, for financial support.

# References

O. Abbas, D. Schulze-Makuch, ESA SP **518**, 345 (2002)

S. Benner, *http://www7.nationalacademies.org/ssb/weirdlife.html* (2002)

J.-M. Bernard, P. Coll, F. Raulin, ESA SP **518**, 623 (2002)

J.-M. Bernard, P. Coll, A. Coustenis, F. Raulin, Planet. Space Sci. **51**, 1003 (2003)

B. Bézard, *European Planetary Science Congress 2006 Berlin,* Germany (2006)

P. Coll, D. Coscia, M.-C. Gazeau, F. Raulin, Origins Life Evol. Biosph. **28**, 195 (1998)

P. Coll, D. Coscia, N. Smith, M.-C. Gazeau, S.I. Ramirez, G. Cernogora, G. Israel, F. Raulin, Planet. Space Sci. **47**, 1331 (1999)

R. Courtin, D. Gautier, C. McKay, Icarus **114**, 144 (1995)

A. Coustenis, R.K. Achterberg, B.J. Conrath, D.E. Jennings, A. Marten, D. Gautier, C.A. Nixon, F.M. Flasar, N.A. Teanby, B. Bézard, R.E. Samuelson, R.C. Carlson, E. Lellouch, G.L. Bjoraker, P.N. Romani, F.W. Taylor, J.G.P. Irwin, T. Fouchet, A. Hubert, G.S. Orton, V.G. Kunde, S. Vinatier, J. Mondellini, M.M. Abbas, R. Courtin, Icarus, Submitted (2006)

A. Coustenis, A. Salama, E. Lellouch, Th. Encrenaz, G. Bjoraker, R.E. Samuelson, Th. De Graauw, H. Feuchtgruber, M.F. Kessler, Astron. Astrophys. **136**, L85 (1998)

A. Coustenis, A. Salama, B. Schulz, S. Ott, E. Lellouch, Th. Encrenaz, D. Gautier, H. Feuchtgruber, Icarus **161**, 383 (2003)

D.P. Cruikshank, H. Imanaka, C.M. Dalle Ore, Adv. Space Res. **36**, 178 (2005)

P. Ehrenfreund, J.P. Boon, J. Commandeur, C. Sagan, W.R. Thompson, B.N. Khare, Adv. Space Res. **15**(3), 335 (1995)

F.M. Flasar, R.K. Achterberg, B.J. Conrath, P.J. Gierasch, V.G. Kunde, C.A. Nixon, G.L. Bjoraker, D.E. Jennings, P.N. Romani, A.A. Simon-Miller, B. Bézard, P.J.G. Irwin, N.A. Teanby, J. Brasunas, J.C. Pearl, M.E. Segura, R.C. Carlson, A. Mamoutkine, P.J. Schinder, A. Barucci, R.Courtin, A. Coustenis, T. Fouchet, D. Gautier, E. Lellouch, A. Marten, R. Prangé, D.F. Strobel, S. Vinatier, S.B. Calcutt, P.L. Read, F.W. Taylor, N. Bowles, G.S. Orton, L.J. Spilker, T.C. Owen, R.E. Samuelson, J.A. Spencer, M.R. Showalter, C. Ferrari, M.M. Abbas, F. Raulin, S. Edgington, P. Ade, H.E. Wishnow, Science **308**, 975 (2005)

A.D. Fortes, Icarus **146**, 444 (2000)

M. Fulchignoni, *et al.,* Nature **438**, 785 (2005)

D. Gautier, T. Owen, Space Sci. Rev. **104**, 347 (2002)

E. Hébrard, Y. Bénilan, F. Raulin, Adv. Space Res. **36**, 268 (2005)

H. Imanaka, B.N. Khare, J.E. Elsila, E.L.O. Bakes, C.P. McKay, D.P. Cruikshank, S. Sugita, T. Matsui, R.N. Zare, Icarus **168**, 344 (2004)

G. Israël, C. Szopa, F. Raulin, M. Cabane, H.B. Niemann, S.K. Atreya, S.J. Bauer, J.-F. Brun, E. Chassefière, P. Coll, E. Condé, D. Coscia, A. Hauchecorne, P. Millian, M.J. Nguyen, T. Owen, W. Riedler, R.E. Samuelson, J.-M. Siguier, M. Steller, R. Sternberg, C. Vidal-Madjar, Nature **438**, 796 (2005)

B.N. Khare, C. Sagan, E.T. Arakawa, F. Suits, T.A. Callicott, W.M. Williams, Icarus **60**, 127 (1984)

B.N. Khare, C. Sagan, H. Ogino, B. Nagy, C. Er, K.H. Schram, E.T. Arakawa, Icarus **68**, 176 (1986)

R. de Kok, P.G.J. Irwin, N.A. Teanby, E. Lellouch, B. Bézard, S. Vinatier, C. Nixon, L. Fletcher, C. Howett, S.B. Calcutt, N.E. Bowles, F.M. Flasar, F.W. Taylor, Oxygen compounds in Titan's stratosphere as observed by Cassini CIRS. Icarus, submitted (2006)

M.E. Kress, C.P. McKay, Icarus **168**, 475 (2004)

S. Lebonnois, E.L.O. Bakes, C.P. McKay, Icarus **159**, 505 (2002)

S. Lebonnois, D. Toublanc, F. Hourdin, P. Rannou, Icarus **152**, 384 (2001)

J.P. Lebreton, O. Witasse, C. Sollazzo, T. Blancquaert, P. Couzin, A.-M. Schipper, J. Jones, D. Matson, L. Gurvits, D. Atkinson, B. Kazeminejad, M. Perez, Nature **438**, 758 (2005)

R.D. Lorenz, J. British Interplanet. Soc. **53**, 218 (2000)

R.D. Lorenz, S. Wall, J. Radebaugh, G. Boubin, E. Reffet, M. Janssen, E. Stofan, R. Lopes, R. Kirk, C. Elachi, J. Lunine, K. Mitchell, F. Paganelli, L. Soderblom, C. Wood, L. Wye, H. Zebker, Y. Anderson, S. Ostro, M. Allison, R. Boehmer, P. Callahan, P. Encrenaz, G.G. Ori, G. Francescetti, Y. Gim, G. Hamilton, S. Hensley, W. Johnson, K. Kelleher, D. Muhleman, G. Picardi, F. Posa, L. Roth, R. Seu, S. Shaffer, B. Stiles, S. Vetrella, E. Flamini, R. West, Science **312**, 724 (2006)

J.I. Lunine, Rev. Geophys. **31**(2), 133 (1993)

A. Marten, T. Hidayat, Y. Biraud, R. Moreno, Icarus **158**, 532 (2002)

G.D. McDonald, W.R. Thompson, M. Heinrich, B.N. Khare, C. Sagan, Icarus **108**, 137 (1994)

C.P. McKay, Planet. Space Sci. **44**, 741 (1996)

C.P. McKay, J.B. Pollack, C. Courtin, Icarus **80**, 23 (1989)

C.P. McKay, J.B. Pollack, C. Courtin, Science **253**, 1118 (1991)

C.P. McKay, H.D. Smith, Icarus **178**, 274 (2005)

C.D. Neish, R.D. Lorenz, D.P. O'Brien, Cassini RADAR Team, Int. J. Astrobiol. **5**(1), 57 (2006)

H.B. Niemann, S.K. Atreya, S.J. Bauer, G.R. Carignan, J.E. Demick, R.L. Frost, D. Gautier, J.A. Haberman, D.N. Harpold, D.M. Hunten, G. Israel, J.I. Lunine, W.T. Kasprzak, T.C. Owen, M. Paulkovich, F. Raulin, E. Raaen, S.H. Way, Nature **438**, 779 (2005)

D.P. O'Brien, R.D. Lorenz, J.I. Lunine, Icarus **173**, 243 (2005)

T. Owen, Planet. Space Sci. **48**, 747 (2000)

S.I. Ramirez, P. Coll, A. Da Silva, R. Navarro-Gonzalez, J. Lafait, F. Raulin, Icarus **156**, 515 (2002)

F. Raulin, P. Bruston, P. Paillous, R. Sternberg, Adv. Space Res. **15**(3), 321 (1995)

F. Raulin, T. Owen, Space Sci. Rev. **104**(1–2), 379 (2002)

C. Sagan, B.N. Khare, Nature **277**, 102 (1979)

R.E. Samuelson, L.A. Mayo, M.A. Knuckles, R.K. Khanna, Planet. Space Sci. **45**, 941 (1997)

D. Schulze-Makuch, L.N. Irwin, *Life in the Universe. Expectations and Constraints* (Springer, New York, 2004)

Y. Sekine, S. Sugita, T. Shido, T. Yamamoto, Y. Iwasawa, T. Kadono, T. Matsui, Icarus **178**, 154 (2005)

M.B. Simakov, ESA SP **496**, 211 (2001)

C. Sotin, R. Jaumann, B.J. Buratti, R.H. Brown, N.R. Clark, A.L. Soderblom, K.H. Baines, G. Bellucci, J.-P. Bibring, F. Capaccioni, P. Cerroni, M. Combes, A. Coradini, D.P. Cruikshank, P. Drossart, V. Formisano, Y. Langevin, D.L. Matson, T.B. McCord, R.M. Nelson, P.D. Nicholson, B. Sicardy, S. LeMouelic, S. Rodriguez, K. Stephan, C.K. Scholz, Nature **435**, 786 (2005)

C.R. Stoker, P.J. Boston, R.L. Mancinelli, W. Segal, B.N. Khare, C. Sagan, Icarus **85**, 241 (1990)

N.A. Teanby, P.G.J. Irwin, R. de Kok, C.A. Nixon, A. Coustenis, B. Bézard, S.B. Calcutt, N.E. Bowles, F.M. Flasar, L. Fletcher, C. Howett, F.W. Taylor, Icarus **181**, 243 (2006)

F. Tian, O.B. Toon, A.A. Pavlov, H. De Sterck, Science **308**, 1014 (2005)

G. Tobie, O. Grasset, J.I. Lunine, A. Mocquet, A. Sotin, Icarus **175**, 496 (2005)

T. Tokano, C.P. McKay, F. M. Neubauer1, S.K. Atreya, F. Ferri, M. Fulchignoni, H.B. Niemann, Nature **442**, 432 (2006)

M.G. Tomasko, B. Archinal, T. Becker, B. Bezard, M. Bushroe, M. Combes, D. Cook, A. Coustenis, C. de Bergh, L.E. Dafoe, L. Doose, S. Doute, A. Eibl, S. S. Engel, F. Gliem, B. Grieger, K. Holso, E. Howington-Kraus, E. Karkoschka, H.U. Keller, R. Kirk, R. Kramm, M. Küppers, P. Lanagan, E. Lellouch, M. Lemmon, J. Lunine, E. McFarlane, J. Moores, G.M. Prout, B. Rizk, M. Rosiek, P. Rueffer, S.E. Schröder, B. Schmitt, C. See, P. Smith, L. Soderblom, N. Thomas, R. West, Nature **438**, 765 (2005)

D. Toublanc, J.P. Parisot, J. Brillet, D. Gautier, F. Raulin, C.P. McKay, Icarus **113**, 2–26; ibidem, "errata", Icarus **117**, 218 (1995)

B.N. Tran J.P. Ferris, J.J. Chera, Icarus **162**, 114 (2003)

S. Vinatier, B. Bézard, T. Fouchet, N.A. Teanby, R. de Kok, P.G.J. Irwin, B.J. Conrath, C.A. Nixon, P.N. Romani, F.M. Flasar, A. Coustenis, Icarus, Submitted (2006)

V. Vuitton, V.R. Yelle, G.V. Anicich, Astrophys. J. **647**, L175 (2006)

H. Waite, H. Niemann, R.V. Yelle, T.W. Kasprzak, E.T. Cravens, J.G. Luhmann, R.L. McNutt, W.-H. Ip, D. Gell, V. De La Haye, I. Müller-Wordag, B. Magee, N. Borggren, S. Ledvina, G. Fletcher, E. Walter, R. Miller, S. Scherer, R. Thorpe, J. Xu, B. Block, K. Arnett, Science **308**, 982 (2005)

R.A. West, M.E. Brown, S.V. Salinas, A.H. Bouchez, H.G. Roe, Nature **436**, 670 (2005)

E.H. Wilson, S.K. Atreya, Planet. Space Sci. **51**, 1017 (2003)

E.H. Wilson, S.K. Atreya, J. Geophys. Res. - Planets **109**(E6), E06002 (2004)

J.C. Zarnecki, M.R. Leese, B. Hathi, A.J. Ball, A. Hagermann, M.C. Towner, R.D. Lorenz, J.A.M. McDonnell, S.F. Green, M.R. Patel, T.J. Ringrose, P.D. Rosenberg, K.R. Atkinson, M.D. Paton, M. Banaszkiewicz, B.C. Clark, F. Ferri, M. Fulchignoni, L.A.N. Ghafoor, G. Karg, H. Svedhem, J. Delderfield, M. Grande, D.J. Parker, G.P. Challenor, J.E. Geake, Nature **438**, 792 (2005)

# An Approach to Searching for Life on Mars, Europa, and Enceladus

Christopher P. McKay

Originally published in the journal Space Science Reviews, Volume 135, Nos 1–4.
DOI: 10.1007/s11214-007-9229-8 © Springer Science+Business Media B.V. 2007

**Abstract** Near-term missions may be able to access samples of organic material from Mars, Europa, and Enceladus. The challenge for astrobiology will be to determine if this material is the remains of dead microorganisms or merely abiotic organic material. The remains of life that shares a common origin with life on Earth will be straightforward to detect using sophisticated methods such as DNA amplification. These methods are extremely sensitive but specific to Earth-like life. Detecting the remains of alien life—that does not have a genetic or biochemical commonality with Earth life—will be much more difficult. There is a general property of life that can be used to determine if organic material is of biological origin. This general property is the repeated use of a few specific organic molecules for the construction of biopolymers. For example, Earth-like life uses 20 amino acids to construct proteins, 5 nucleotide bases to construct DNA and RNA, and a few sugars to construct polysaccharides. This selectivity will result in a statistically anomalous distribution of organic molecules distinct from organic material of non-biological origin. Such a distinctive pattern, different from the pattern of Earth-like life, will be persuasive evidence for a second genesis of life.

**Keywords** Life · Mars · Europa · Enceladus · Second genesis

## 1 Introduction

One of Astrobiology's key goals is to determine the diversity and distribution of life in the universe. In our Solar System, the most promising targets for a search for a second genesis of life are Mars, Europa, and Enceladus. It is important to appreciate that the question we are asking is not just "Was there life on Mars, Europa, or Enceladus?" Rather the question is "Was there a second genesis of life on Mars, Europa, or Enceladus?" Some have argued that we do not have a complete and compact definition of life and hence do not know how

C.P. McKay (✉)
Space Science Division, NASA Ames Research Center, Mail Stop 245-3, Moffett Field, CA 94035, USA
e-mail: cmckay@mail.arc.nasa.gov

to search for, or even recognize, alien life. Others have argued that life everywhere will be biochemically identical to life on Earth.

What is life? A concise definition still eludes us. Many definitions of life as a general phenomenon are a list of properties. Koshland (2002) listed seven features: (1) program (e.g. DNA), (2) improvisation (novel responses to the environment), (3) compartmentalization, (4) energy, (5) regeneration, (6) adaptability, and (7) seclusion (chemical control and selectivity). Davies (1999) has a similar list. Perhaps the most common definition of life is a physical system that undergoes Darwinian evolution, which, according to Chao (2000), is originally due to Muller (1966). Schrödinger (1945) defined life in the context of thermodynamics as "It feeds on negative entropy."

In the search for life on other worlds it is not clear that a definition of life is required or even helpful. More useful might be clarification of what constitutes evidence of life. There are several ways to search for life. First we can be searching for life as a collective general phenomenon. However, life might also be a single isolated organism. And that organism might be dead. A dead organism is a sign of life. Finally, signs of life may be fossils, artifacts, or other inorganic structures. In the search for life on other worlds, any of these would be of interest.

Definitions of life typically focus on the nature of the collective phenomenon. In general, such definitions are not useful in an operational search for life on other worlds. The one exception is Chao's (2000) proposal to modify the Viking Labeled Release (LR) experiment to allow for the detection of organisms that improve their capacity to use the provided nutrients. This would in principle provide a direct detection of Darwinian evolution and could unambiguously distinguish between biological metabolism and chemical reaction. Chao (2000) argued that Darwinian evolution is the fundamental property of life and other observables associated with life result from evolutionary selection. His method for searching for evolution would be practical if the right medium can be selected to promote the growth of alien microbes. Unfortunately, we now know that only a tiny fraction, $<1\%$, of microorganisms from an environmental sample grow in culture. This was not known at the time of the design of the Viking biology experiments, which were essentially culture experiments. The fact that most soils on Earth will grow up in a culture media is due to the vast diversity of soil microbes in these soils and not to the robustness of culturing as a way to detect organisms. We also now know that there are soils on Earth (e.g., Atacama Desert in Chile) where there are bacteria present in low numbers, but nothing grows in any known culture media (Navarro-Gonzalez et al. 2003).

Of course growth experiments of any kind do not detect dead organisms. Yet the remains of dead organisms are potentially important evidence of life on another planet, as are fossils. However, there is an important distinction between dead organisms and fossils. A fossil is evidence of past life but it does not reveal anything about the biochemical or genetic nature of that life. If we are searching for a second example of life, then we need to be able to compare the nature of that life to Earth life. For this an organism is needed, either dead or alive, but a fossil is not sufficient.

## 2  Mars, Europa, and Enceladus

A realistic assessment of what is possible on near-term missions suggests that the organic remains of past life are the most promising target for a search for signs of alien life on either Mars, Europa, or Enceladus. On Mars, the search for biological remains of past life is focused on the subsurface. The deep permafrost on Mars may hold remnants of past life

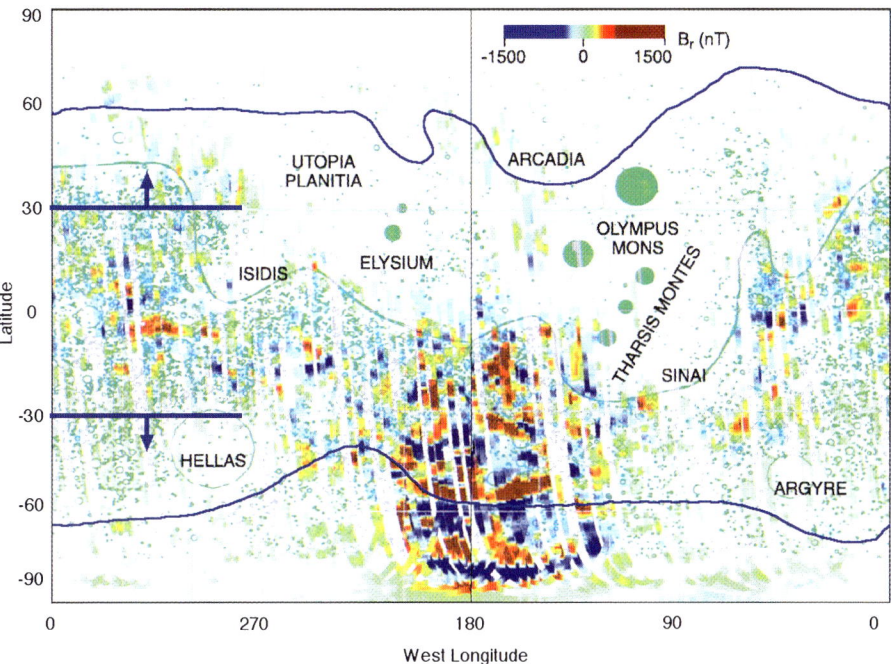

**Fig. 1** Maps showing crater distribution, ground ice, and crustal magnetism on Mars. *Each green dot* represents a crater with diameter greater than 15 km. The boundary between the smooth northern plains and the cratered southern highlands is shown with a *green line*. The crustal magnetism is shown as *red* for positive and *blue* for negative. Full scale is 1500 nT. The typical strength of Earth's magnetic field at the surface is 50,000 nT. The *solid blue lines* show the extent of near surface ground ice as determined by Odyssey mission. Ground ice is present near the surface polarward of these lines. Crater morphology indicates deep ground ice poleward of 30° (Squyres and Carr 1986), shown here by dark blue lines and arrows. The region between 60 and 80°S at 180°W is heavily cratered, preserves crustal magnetism, and has ground ice present. This is our suggested target site for drilling. This figure is adapted from Acuña et al. (1999), based on the crater distribution in Barlow (1997). The distribution of near-surface ground ice is from Feldman et al. (2002). Figure from Smith and McKay (2005)

(Smith and McKay 2005). Figure 1 shows a possible location in the southern hemisphere of Mars where we might find ancient frozen material. The high concentration of craters indicates that the surface is old and there is direct detection of ground ice. The presence of crustal magnetic fields indicates that the surface has been relatively undisturbed throughout Martian history. The organisms in any ancient ground ice on Mars are likely to be dead from accumulated radiation dose but their organic remains could be analyzed and compared to the biochemistry of Earth life.

On Europa the near-term target for a search for life is the surface. The linear features on the surface of Europa are generally thought to be cracks in the ice and may be locations where ocean water reached the surface, although it is not yet certain that this is the case. If there is life in the ocean of Europa, then organic remains of that life may be present at the surface cracks. Due to the high radiation dose received from Jupiter, it is unlikely that any organisms are alive at the surface but their organic remains could persist. Eventually the high radiation flux would destroy any biological signature in the organic remains as well.

Enceladus is one of the small icy moons of Saturn. Recently the Cassini spacecraft discovered $H_2O$ jetting from the south polar region of Enceladus (Porco et al. 2006). One possible source of this water jet is a subsurface aquifer powered by tidal or radiogenic heating. The water ice particles from Enceladus appear to be the source of the E ring of Saturn (Porco et al. 2006) implying that the activity has been ongoing for a considerable period of time. Trace gases in the icy plume include $N_2$ and $CO_2$ and $CH_4$ (Waite et al. 2006). These could be the products of a methanogenic life form in the subsurface aquifer. Thus, the collection of particles from the plume erupting from Enceladus could provide samples of any life in the subsurface aquifer.

## 3 Detection of Alien Life

Our most optimistic scenario is that we will find organic material in the ancient permafrost on Mars, in the surface ice of Europa, or in the plume of Enceladus. How can we determine if this organic material is of biological origin?

I have argued previously (McKay 2004) that one way to determine if a collection of organic material is of biological origin, is to look for a selective pattern of organic molecules similar to, but not necessarily identical with, the selective pattern of biochemistry in life on Earth. Pace (2001) argued that life everywhere will be life as we know it: "it seems likely that the basic building blocks of life anywhere will be similar to our own, in the generality if not in the detail." He contends that the biochemical system used by life on Earth is the optimal one and therefore evolutionary pressure will cause life everywhere to adopt this same biochemical system. It is instructive to consider this argument in the context of a conceptual organic phase space. If we imagine possible organic molecules as the dimensions of a phase space, then any possible arrangement of organic molecules is a point in that phase space. We can define biochemistries as those points in phase space that allow for life. The biochemistry of Earth life—life as we know it—represents one point in the organic phase space: we know that this one point represents a viable biochemistry. Pace's (2001) contention that biochemistry is universal is equivalent to stating that in the region of phase space of all possible biochemistries there is only one optimum biochemistry and thus any initial set of biochemical reactions comprising a system of living organisms will move toward that optimum as a result of selective pressure. This would imply that there is only one peak in the fitness landscape of biochemistry.

If Pace (2001) is correct, then the only variation between life forms that we can expect is that associated with chirality. As far as is known, the L and D forms of chiral organic molecules (such as amino acids and sugars) have no differences in their biochemical function. Life is possible that is exactly similar in all biochemical respects to life on Earth except that it has D instead of L amino acids in its proteins and L instead of D sugars in its polysaccharides. The question of the number of possible biochemistries consistent with life is an empirical one and can only be answered by observations of other life forms on other worlds, or by the construction of other life forms in the laboratory. The observation or construction of even one radically alien life form would suffice to show that biochemistry as we know it is not universal.

## 4 Life as Lego

The pattern of biochemistry of Earth life follows what I have called the "Lego" principle (McKay 2004). This is the straightforward observation that life uses a small set of molecules to construct the diverse structures that it needs. This is similar to the children's play blocks

**Fig. 2** Comparison of biogenic with nonbiogenic distributions of organic material. Nonbiological processes produce smooth distributions of organic material, illustrated here by the *curve*. Biology, in contrast selects and uses only a few distinct molecules, shown here as spikes (e.g., the 20 left-handed amino acids on Earth). Analysis of a sample of organic material from Mars, Europa, or Enceladus may indicate a biological origin if it shows such selectivity. Figure from McKay (2004)

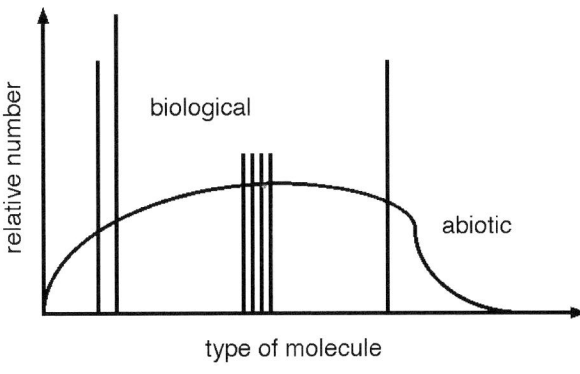

known as Legos, in which a few different units repeated over and over again are used to construct complex structures. The biological polymers that construct life on Earth are the proteins, the nucleic acids, and the polysaccharides. These are built from repeated units of the 20 L amino acids, the 5 nucleotide bases, and the D sugars. The use of only certain basic molecules allows life to be more efficient and selective.

Evolutionary selection on life anywhere is likely to result in the same selective use of a restricted set of organic molecules. As discussed earlier, it is premature to conclude that all life anywhere will use the same set of basic biomolecules. Thus I suggest that life will always use some basic set, but it may not be the same basic set used by life on Earth. This characteristic biogenic pattern of organic molecules would persist even after the organism is dead. Given our present state of understanding of biochemistry, we are not able to propose alternative and different biochemical systems that could be the basis for life, but that may reflect a failure of our understanding and imagination rather than a restriction on the possibilities for alien life.

A sample from the deep permafrost in the southern hemisphere of Mars, from a crack on the surface of Europa, or collected from the plume on Enceladus, could be analyzed for organic material with a fairly simple detection system. If organic material was detected, then it would be of interest to characterize any patterns in that organic material that would indicate a "Lego" principle pattern. Clearly one such pattern is the identical pattern of all Earth life; 20 L amino acids, the five nucleotide bases, A, T, C, G, and U, etc. However, more interesting would be a clear pattern different from the pattern known from Earth life. Figure 2 shows a schematic diagram of how a biological pattern would be different from a non-biological pattern. Abiotic sources result in smooth, not necessarily symmetric, distributions while biotic distributions are a series of spikes. Implementing this search in practical terms in near-term missions will require a sophisticated ability to separate and characterize organic molecules. Currently the instrument best suited for this task is a combined gas chromatograph and mass spectrometer with solvent extraction. However, new methods of fluorescence and Raman spectroscopy could provide similar information and may have a role in future mission applications.

Organisms will maintain their "anomalous" distribution of organics while they are alive. After death, however, physical factors will slowly degrade this distribution and over time will turn it into the statistically smooth distribution indistinguishable from abiotic sources. These physical factors are thermal decay and radiation damage. Examples of thermal decay are the racemization of amino acids and the spontaneous rearrangement of bonds. Radiation can break bonds as well. The level of radiation necessary to erase the biological information of dead microorganisms is uncertain but an upper limit would be the radiation dose that corresponds to every carbon bond being broken once. For the targets considered here—Mars, Europa, and Enceladus—this becomes a factor only on Europa due to the high levels of radiation from the Jovian magenetosphere.

## 5 Conclusions

Exploration of Mars, Europa, or Enceladus may give us our first sample of extraterrestrial organic material that may be of biological origin. Indeed it is more likely that we will find the organic remains of dead microorganisms on these worlds long before we find extant life—if we ever find extant life. Nonetheless, organic remains of organisms can be analyzed to determine if the now deceased life forms were related to Earth life. Life selects certain organic molecules and uses them to high degree while not using molecules of similar chemical structure. For example, life uses the left-handed (L) amino acids much more than the right-handed (D) amino acids. This selectivity will be evidence in the organic remains of dead organisms that are not yet degraded by time or radiation and can form the basis for a life search method applied to dead (as opposed to never alive) organic matter.

NASA's strategy in the search for life beyond the Earth has started with a strategy that is summarized by "follow the water." The next step is "find the organics" followed by "characterize the organics to determine possible biological origin."

## References

M.H. Acuña, J.E.P. Connerney, N.F. Ness, R.P. Lin, D. Mitchell, C.W. Carlson, J. McFadden, K.A. Anderson, H. Reme, C. Mazelle, D. Vignes, P. Wasilewski, P. Cloutier, Science **284**, 790–793 (1999)
N. Barlow, in *Encyclopedia of Planetary Sciences*, ed. by J.H. Shirley, R.W. Fairbridge (Chapman and Hall, London, 1997)
L. Chao, Bioscience **50**, 245–250 (2000)
P.C.W.D. Davies, *The Fifth Miracle: The Search for the Origin and Meaning of Life* (Simon and Schuster, NY, 1999)
W.C. Feldman, W.V. Boynton, R.L. Tokar, T.H. Prettyman, O. Gasnault, S.W. Squyres, R.C. Elphic, D.J. Lawrence, S.L. Lawson, S. Maurice, G.W. McKinney, K.R. Moore, R.C. Reedy, Science **297**, 75–78 (2002)
D.E. Koshland Jr., Science **295**, 2215–2216 (2002)
C.P. McKay, PLoS Biol. **2**, 1260–1263 (2004)
H.J. Muller, Am. Nat. **100**, 493–517 (1966)
R. Navarro-Gonzalez et al., Science **302**, 1018–1021 (2003)
N. Pace, Proc. Natl. Acad. Sci. USA **98**, 805–808 (2001)
C.C. Porco et al., Science **311**, 1393–1401 (2006)
E. Schrödinger, *What Is Life?* (Cambridge University Press, Cambridge, 1945)
H.D. Smith, C.P. McKay, Planet. Space Sci. **53**, 1302–1308 (2005)
S.W. Squyres, M.H. Carr, Science **231**, 249–252 (1986)
J.H. Waite Jr., et al., Science **311**, 1419–1422 (2006)

# Biosignatures: Morphological Biosignatures

## Accretion, Trapping and Binding of Sediment in Archean Stromatolites—Morphological Expression of the Antiquity of Life

### Wladyslaw Altermann

Originally published in the journal Space Science Reviews, Volume 135, Nos 1–4.
DOI: 10.1007/s11214-007-9292-1 © Springer Science+Business Media B.V. 2008

**Abstract** This paper reviews and discusses Archean stromatolite occurrences and their modes of growth in the context of sedimentary facies. Modes of sediment accretion and trapping and binding of sedimentary grains, together with the resulting morphology of stromatolites and microbial mats in the Archean are analysed, in order to show existing interaction between the growth patterns, morphology and facies association. Architectural elements of sediment arrangement in Archean stromatolites, together with the dependence of stromatolite distribution and morphology on sedimentary facies changes, clearly argue for a biological origin of stromatolitic lamination preserved in Archean cherts and carbonates. The observed sediment behaviour of laminae accretion and sediment precipitation, trapping and binding cannot be explained by abiogenic carbonate or silica precipitation from saturated solutions. The time-dependent, increasing complexity of stromatolitic structures in the Archean is an additional strong argument for biologic impact on stromatolite formation. Therefore, biogenic stromatolites and microbial mats were undoubtfully present at 3.5 Ga and occupied an increasingly wide range of sedimentary environments during the Archean.

**Keywords** Early life · Stromatolites · Sedimentary facies · Archean · Proterozoic · Precambrian

## 1 Introduction

With the questioning of the authenticity of some microfossils of Archean age (Altermann 2001; Brasier et al. 2002; Moorbath 2005), organo-sedimentary structures in Archean carbonate and chert rocks (stromatolites) were denigrated as nonbiogenic (Brasier et al. 2004). Despite the presence of typical (and otherwise highly improbable) modes of sediment accretion and trapping and binding of sedimentary grains by microbial mats, doubts were cast on the role of microbes in the production of the complex structures and an alternative genesis for the archetypal microbial laminations was proposed.

W. Altermann (✉)
Dept. Earth & Environmental Sciences, Geology and GeoBio-Center LMU, Luisenstr. 37, Munich 80333, Germany
e-mail: wlady.altermann@iaag.geo.uni-muenchen.de

Here, stromatolites in carbonate environments—the traditional setting of classical microbial mat studies since Hall (1883), Gürich (1906), Kalkowsky (1908), Mawson (1929), Black (1933), and many others in the last third of the past century are discussed, in order to depict their mode of sediment accretion, trapping and binding. Because carbonate deposits tend to lithify contemporaneously with deposition and as microbial mats actively contribute to these processes by calcification, the preservation potential of microbial mats in carbonate environments is several orders of magnitude higher than in siliciclastic environments (e.g., Noffke et al. 2006a, 2006b; Schieber et al. 2007). In carbonate environments the grains trapped and bound by microbial activity are mainly carbonate grains from surrounding carbonate build-ups. Waves, currents and winds may, however, deliver siliciclastic terrigenous detritus onto carbonate flats. Volcaniclastic influx is also common in many settings. All these sediments become bound by microbial mats, which are simultaneously passively or actively precipitating carbonate minerals, and provide evidence of the large variety of paleogeographic settings of ancient microbial mat formation.

In contrast to the Phanerozoic, most Precambrian, but especially Archean stromatolites, do not grow primarily by the mechanism of trapping and binding of fine sediment detritus, but instead, precipitate micritic carbonate along their laminae. Several reasons for this—at times contrasting—behaviour have been proposed. It seems, however, plausible that in the absence of higher organisms in the Precambrian, the variety of available carbonate detritus (and "bioclasts") was restricted to the microbial source (micrite) and to mechanical or bioerosion of neighbouring carbonate build-ups. Nevertheless, many Precambrian stromatolite examples show close affinity to carbonate clastic processes and are directly associated with "in situ clastic sediments" (*sensu* Miall 1984).

The intent of this contribution is to discuss examples of trapping and binding in Archean stromatolites and their significance to demonstrating biogenicity. Generally, there is no difference in the physical behaviour of siliciclastic, volcaniclastic and carbonate clastic detritus in microbialites. However, when microbial carbonates are heavily recrystallised, as in many Precambrian occurrences, the carbonate clastic influx is difficult to recognise and an allochthonous siliciclastic or volcaniclastic admixture (if present) facilitates the detection of detrital influence on stromatolite growth. In order to understand the relevance of trapping and binding of sediment in Archean stromatolites, it is further necessary to consider the topics of morphological classification and facies association of stromatolites. This is an equally important aim of this contribution.

## 1.1 Biogenicity of Ancient Stromatolites

Many explanations for "stromatolite" and related terms exist. According to most common definition (Awramik and Margulis, unpublished, in Walter 1976a, p. 1), stromatolites are lithified organo sedimentary structures, growing through accretion of laminae by the entrapment of sediment and by participation of carbonate, under active secretion or direct influence of micro-organisms. The above general definition is useful because it includes many others as, for example, the "microbialites" (Burne and Moore 1987) or "microbolites" (Dupraz and Strasser 1999). Consequently, *lithified* microbial mats—independent of their age, individual process of lithification, mineralogy or growth form and mat morphology—fall within the class of organo-sedimentary structures called stromatolites. Not all microbial mats are, however, automatically stromatolites, as not all of them have the potential for lithification (*sensu* "potential stromatolites", Krumbein 1983). The various aspects of stromatolite formation, classification and their role as "archives" of ancient life on Earth were extensively discussed by Riding (1999), Hofmann (2000) and Altermann (2002, 2004) and in many earlier papers.

As a result of the recent questioning of the authenticity of the Earth's oldest evidence of life (Brasier et al. 2004, 2006), irrefutable proof for the involvement of life in the formation of Earth's oldest, Archean stromatolites has been collected (Allwood et al. 2006; Kiyokawa et al. 2006; Noffke et al. 2006a, 2006b; Tice and Lowe 2006; Ueno et al. 2006a, 2006b; Lowe and Tice 2007; Schopf et al. 2007a; Sugitani et al. 2007). Additional evidence has been presented that the Earth's oldest (3.46 to 3.45 Ga) microfossils from the Apex Basalt chert (Schopf 1993), from the Barberton Mountain Land (Walsh 1992) and from the Dresser Formation (Ueno et al. 2001) are indeed cellularly preserved remains of Archean microbial life. They are closely associated with stromatolitic structures of the same ancient age and with various geochemical "biomarkers". Without a doubt, microbial life thrived on Earth well before 3.0 Ga (Schopf 2004, 2006; Altermann 2007; Allwood et al. 2007; Schopf et al. 2007b).

The discussion triggered by Brasier et al. (2002) on the authenticity of the Earth's oldest fossils ended in smoke, but introduced a high level of scepticism and a critical approach to any report of bodily preserved microbial remains from the Archean. It led to the introduction and development of new investigation techniques into Archean paleobiology, like Raman spectroscopy, analysis of molecular biomarkers, AFM techniques and new, stable isotope methods (e.g., N; Fe isotopes). The processes of biological isotope fractionation and of fossilisation of microbial organic matter are much better understood today. Van Zuilen et al. (2007) have shown for the various 3.4–3.2 Ga cherts of the Barberton greenstone belt (not using the original material of, e.g., Walsh 1992), that the organic matter in these cherts, although variably severely altered, in accordance to the different metamorphic grades, bears all the characteristics of metamorphosed biologic material. Nevertheless, it seems emblematic for the discussion of the authenticity of pre-3.0 Ga microbial remains, that the microfossils from the 3.45 Ga Apex chert (Schopf 1993) are still disputed by some researchers and regarded as not authentic. Simultaneously, M. Brasier and his group have seemingly forgotten their claims (Brasier et al. 2004) for nonexistence of life on Earth prior to c. 3.0 Ga and now declare to have found the oldest evidence of microbial endolithic life in microtubes in sandstone grains, in 3.4 Ga Strelley Pool Chert in Western Australia (Brasier et al. 2006).

The ability of some Precambrian stromatolites to trap and bind sedimentary grains influences their growth form and is thus important for their morphological classification. Logan et al. (1964), proposed a simple geometric description scheme for stromatolites. Hofmann (1973, 2000) proposed to classify stromatolites in a pyramidal diagram of triangular base, where each corner of the pyramid represents one of the four main stromatolite-forming processes. The triangular base of the pyramid is defined by the corners representing purely chemical precipitation, mechanical—clastic accretion and biological—nonskeletal accretion. The base includes all Archean stromatolites while younger stromatolites, displaying biological skeletal accumulation, form the peak of the pyramid. Although the morphology of early Archean stromatolites is, as a rule, less complicated and involves a lesser variety of internal structures than in the Neoarchean and Proterozoic, Allwood et al. (2006) demonstrated in a detailed study of the 3.5-Ga Strelley Pool Chert stromatolites from Pilbara, Western Australia, that stromatolitic reefs early in Earth's history nevertheless had a complicated, and rather complex morphology, shifting with depositional conditions. The presence of the adventitious branching structure and other microfeatures in these stromatolites was shown by Hofmann et al. (1999). As with many other Archean occurrences, the Strelley Pool Chert stromatolites are recrystallised and detail has been lost. Precambrian stromatolites are generally of little stratigraphic value (Altermann 2002), but even in relatively simple early Archean stromatolites, the increasing complexity of structures with time (compare Awramik

1992) is clear evidence of biological evolution of microbial consortia building mats and progressively conquering different habitats. Contrary to this development, chemical precipitates preserve an equal degree of complexity through geologic time.

In Archean (and Proterozoic) stromatolites, biogenically mediated precipitation was more significant than sediment trapping and binding processes (Walter 1976b; Schopf 1983; Schopf and Klein 1992; Riding and Awramik 2000). Most Precambrian, but especially Archean stromatolites precipitated micritic carbonate along their laminae (Grotzinger 1990; Kazmierczak and Altermann 2002; Altermann 2004). The modes of calcium carbonate precipitation might have been similar to those in the Phanerozoic (e.g., Pratt 2001, 2002; Altermann et al. 2006). The presence of sediment grains in Archean (and in many Proterozoic) stromatolites is not common, but examples that do show evidence of trapping and binding in the Archean are critical for facies interpretation and provide compelling evidence that microbial mats were an active component of the Archean biosphere. Restriction of Archean stromatolites to environments confined to shallow water (rarely below fair weather wave base), photic and typically "reef-growth-suitable" depositional conditions, is also an indicator of biogenicity. Rare occurrences of Archean microbial mats under allegedly deeper water conditions, below fair-weather wave base (Simonson and Carney 1999; Wright and Altermann 2000; Sumner 2002), all consist of much simpler, largely stratiform and flat-laminated morphology. These growth habits also argue for a biogenic influence on the growth morphology of stromatolitic structures. Obviously, in environments where phototactic growth was not advantageous, simplicity of growth structure prevails.

Where carbonate and terrigenous sediment grains are deposited, a purely chemical precipitation from the water column, suggested by some researchers as an alternative to biogenic stromatolite formation (Grotzinger and Rothman 1996; Sumner 2002; Brasier et al. 2004), is not possible. The growth morphology, especially the rough surface or delicate branching of columnar structures, contribute to the trapping of carbonate and siliciclastic and volcanic grains and ash (Lambert 1998). This is apparent from the concentration of such detritus and carbonate aggregates in depressions, in stromatolitic lamination and between stromatolitic columns. On the other hand, some of the Archean calcifying microorganisms (Kazmierczak and Altermann 2002; Kazmierczak et al. 2004; Altermann et al. 2006), most probably of close cyanobacterial affinity, must have had sticky, mucilaginous envelopes, just like some cyanobacteria today, that promoted trapping and binding of grains moved by currents, waves or winds on the surface of microbial mats. Such processes have been preserved in the fossil record and are visible in stromatolites that clearly overgrow and either partly obliterate or accentuate clastic sedimentary structures like ripple marks, crossbed foresets or desiccation cracks and teepees.

Stromatolites showing a vast range of shapes and sizes are the most striking property of Precambrian sedimentary carbonates. Stromatolites occur as small, patchy lithoherms or lithostromes[1] within widespread reefal build-ups of tens of metres in thickness and hundreds of kilometres of lateral extent (Eriksson and Truswell 1974; Beukes 1987; Grotzinger 1989; Altermann and Siegfried 1997). It is impossible to list all the large Precambrian stromatolite-bearing basins described in the literature in the last 60 to 70 years. To mention a few, the Mauritanian stromatolitic reefs extending for hundreds of kilometres of continuous exposure (Bertrand-Sarfati 1972), the stromatolites of the Western River

---

[1]In this contribution the terms "lithostrome" and "lithoherm" are used for lithified, ancient bedding parallel or lenticular—bulging upward stromatolite forms (stromatolite reefs) respectively. "Biostrome" and "bioherm" are used for such forms in a more general sense, also applying to recent, not fully lithified microbial strata and build-ups.

Formation of the Kilohigok Basin of Bathurst Inlet, Canada (Campbell and Cecile 1981), the Riphean stromatolites of Siberia (e.g., Serebryakov and Semikhatov 1974) or the stromatolitic reefs of the Great Slave basins (Hoffman 1974) are listed here. Abundant literature editions review the stromatolite biogeology, most extensive among them are Walter (1976b), Schopf (1983), Schopf and Klein (1992) or Riding and Awramik (2000). The 2.6 to 2.5 Ga old Carrawine Formation of the Pilbara Craton of Western Australia and the Campbellrand and Malmani carbonate platforms of South Africa represent the oldest, and most extensive, carbonate platforms of the world (Altermann and Nelson 1998; Nelson et al. 1999). Almost all of the carbonate deposition on these huge continent-based, shallow shelf platforms was confined to microbial activity and an overwhelmingly large portion of these platforms is occupied by stromatolites (Beukes 1987; Altermann et al. 2006). Older and less extensive but still large stromatolitic carbonate deposits encompass the 2.9 Ga Wit Mfolozi Formation of the Pongola Supergroup (Mason and von Brunn 1977), the lacustrine stromatolites of the Ventersdorp Supergroup (e.g., Buck 1980) or the Meentheena Carbonate Member of the Tumbiana Formation, Fortescue Group, 2.8 Ga (Awramik and Buchheim 2001), of possibly extensive lacustrine environment.

More than 800 morphological taxa of Precambrian stromatolites are known (Awramik 1992). Classification of stromatolites, however, has always been contentious. Hofmann (1969), Preiss (1972), Walter (1972), Walter et al. (1992) and many others distinguished diagnostic morphological criteria for stromatolite classification, such as:

- mode of occurrence (including: bioherm and biostrome forms),
- branching (including: bifurcation forms, like parallel or divergent),
- shape of columns and margin structure (including: bridging between columns, bumps, ribs or wall formation),
- shape of lamination (including: convex, rhombic or rectangular),
- ornamentation (including smooth, bumpy, lobate and others),
- lateral linkage between "cumulate stromatolites".

Bioherms of differing shapes are often occupied by similar microbial communities, in which different microbial species may be dominant (Schopf and Sovietov 1976; Awramik and Semikhatov 1979). Environmental impact on recent and ancient stromatolite microstructure and morphology was observed by Black (1933). Logan (1961), Horodyski (1977), Playford (1980), Altermann and Schopf (1995) and many other stromatolite researchers have suggested that current and wave energy and sediment influx strongly influence lamination and microstructure and thus the growth form of stromatolites (for recent overviews see Riding 1999; Hofmann 2000; and Altermann 2002, 2004; Allwood et al. 2007; Schopf et al. 2007a, 2007b). A strong environmental overprint, however, does not eradicate completely biogenetic inherent morphological features as observed, for example, by Grey (1994) in the Proterozoic Earaheedy Group of Western Australia. The geometry of the laminae and their vertical and lateral arrangement rule the inner structure and gross morphology of stromatolitic bioherms. Vice versa, purely inorganic sediment accumulations, chemical or clastic (carbonate sand grains and micrite), would behave in an extremely different manner, constructing rather regular sedimentary structures and smooth laminae of flat, botryoidal or crystal shapes, that are well documented from all aquatic environments. Thus, the immense variety of forms and the architecture of biosedimentary structures in stromatolites, when compared to purely physical or chemical sediment accumulation is itself an irrefutable argument for biogenicity (comp. Allwood et al. 2006).

| Basic form | Internal structure | Examples | Facies range & energy |
|---|---|---|---|
| Coniform | Isolated, solitary columns of circular or elongated base (a) / Mats of small conical columns, partly LL (b) / Pinnacle-like supports in microbial mats forming 3-D frames (c) | | **Types (a) and (b):** often no clastic input present and thus interpreted as quiet water conditions. If present, silt to sand sized siliciclastic sediments and carbonate sands evidence moderate to high energy conditions. The sediment usually coarsens with the increasing height of the columns, indicating increasing energy in lower intertidal to subtidal or wave agitated environment. **Type (c):** complete lack of sediment between laminae and in the large fenestral cavities is evidencing calm conditions during mat growth. Occasional slumps and tempestitic debris and mud intercalations cover lithostromes and hint towards storm events below fair weather wave-base. Deep subtidal shelf facies. Modern conical stromatolites in hot spring environments significantly differ from these Archean examples. **Type (d):** large conical columns, with elongated basal section indicate higher energy conditions in the deeper subtidal, long axis oriented subparallel to current flow direction. Carbonate sands are present between the columns. |
| Columnar (Can build large domes of circular and elongated base) | Bifurcating columns, (a) parallel; (b) divergent / Isolated, solitary columns, parallel, younging or widening upwards (c) / Sporadically laterally linked columns (LL) but mainly walled (d) | | **Types (a) and (b):** micrite, rare sparite, clay and marl, indicative of low hydrodynamic conditions, are the prevalent sediments associated with these stromatolites. Increasing density of columns is correlative to decreasing sediment grain size between the columns. Often, only cements are present and no detritus occurs between very densely spaced columns. Mainly subtidal environment. **Type (c):** sparitic and pelletal sediment between the columns indicates wave and current action, inclination indicates unidirectional prevalence of currents. Minimum depth of growth approximates the column height. Shallow subtidal to lower intertidal facies. **Type (d):** mainly micrite is present within laterally linked laminae. The sediment trapped between the columns becomes coarser with disappearing lateral linkage, indicating increasing energy. Upper subtidal to upper intertidal facies and alternating energetic conditions. |
| Laterally linked (LL) pseudocolumnar | Flat lamination (a) / Wavy lamination (b) / Curly lamination (c) | | Increasing degree of roughness of the laminae surface in laterally linked, pseudocolumnar stromatolites correlates with increasing grain size of sediment bound within the lamination and with increasing current and wave energy. Flat laminated stromatolite (a) will thus bind carbonate muds and silts and curly lamination (c) will bind mainly fine carbonate sand and silt. Thus, generally indicating gradually increasing hydrodynamic energy from **type (a) to type (c)**. Coarser detritus, in sand range, concentrates in depressions of the undulating lamination. Strong recrystallisation often obliterates sediment detritus. Wave agitated or shallow subtidal to upper intertidal facies. Domal lithoherms are rather of deeper facies than lithostromes and circular domes are rather less exposed to current energy than elongated domes, which are aligned subparallel to current flow direction. |
| Stratiform | Flat lamination (a) / Wavy lamination (b) / Curly lamination (c) | | As in laterally linked stromatolites, increasing degree of roughness of the laminae surface in flat, stratiform stromatolitic mats correlates with increasing grain size of sediment bound within the lamination. This indicates increase of energy with higher surface roughness. Flat stratiform stromatolites are often associated with desiccation structures (polygonal, stacked hemispheroids - inverted; SH-I) and are thus of upper intertidal to supratidal environment. In recent Shark Bay stromatolites, pustular or curly mats of **type (c)** trap coarse shell fragments. In Precambrian examples carbonate sands and aeolian terrigenous influx has been found in the lamination. Recrystallisation is pervasive and common in intertidal to supratidal facies. Deep subtidal flat laminites of **type (a)**, in the Precambrian, are barren of detrital sediment and display sometimes roll up structures. Coarse blocky cements are common in such roll up structures and compaction is usually relatively low. |

**Fig. 1** Stromatolite forms, associated sediments and facies range. The four basic stromatolite morphologies: coniform, columnar, laterally linked (pseudo-columnar, LL) and stratiform can be constructed of various types of internal lamination, as delineated in the examples. Typical for Precambrian stromatolites is rather the absence of sediment trapping and binding and pronounced biologically mediated precipitation of carbonate along the laminae. However, Precambrian stromatolites can show evidence of trapping and binding, but it can be obliterated by recrystallisation. The fifth basic stromatolite form, oncoidal structures (spheroidal stromatolites) is excluded here for the purpose of this contribution, as they volumetrically do not contribute significantly to rock formation and are not notably trapping and binding. Laterally linked pseudo-columnar stromatolites and columnar stromatolites can construct lithostromes and lithoherms. Lithostromes can be of huge lateral extent, up to tens of kilometres, and of tens of metres in thickness. Lithoherms (domes) can be ovoid, elongated, tens of metres long or circular, more than 10 m across and the domes can be up to 30 m high. Neighbouring large and small domes can form extensive lithoherms or lithostromes of kilometres of extend, i.e., very wide facies belts and stromatolitic reefs

## 1.2 Stromatolite Morphology and Facies Association

Although many authors vigorously defend the use of the International Code of Botanical Nomenclature in stromatolite classification, others are less strict on these rules (comp. Raaben and Sinha 1989; Raaben et al. 2001; Grey 1992). However, Altermann (2002 and discussion in the present publication) argue that because stromatolites are not fossils but fossilised organo-sedimentary structures and a large variety of combinations of sediment grains and precipitates, together with various prokaryotic (and eukaryotic) species are involved in their construction, the use of binominal biological nomenclature for stromatolite forms is not justified and confusing. Instead, a purely morphological, descriptive nomenclature (*sensu*, e.g., Logan et al. 1964), should be used.

Basically stromatolites have five major morphologies, as depicted in Fig. 1 (comp. Hofmann 1977):

- Coniform columns, solitary and colony-forming, where lamination can be laterally linked or the columns are walled. The conical laminae are vertically stacked inside each other and directed with the apices upward.
- Columnar stromatolites, where columns can be solitary or can form bioherms of various sizes and shapes and can be branching or nonbranching, slimming upwards or club shaped, with a narrow base and widening upwards or with straight, parallel walls, walled and nonwalled. They are formed by vertically stacked, convex upward, subspheroidal laminae.
- Laterally linked hemispheroids (LL-H) (Logan et al. 1964): Pseudo-columnar stromatolites with laterally linked, vertically stacked, alternating convex and concave subspheroidal lamination, which can be smooth, curly or crinkled.
- Stratiform, flat biostromes of bedding-parallel, flat-laminated stromatolites, with straight, smooth, wavy or wrinkled lamination.
- Spheroidal (oncoidal and oncolithic) stromatolites (not treated herein because of their scarcity to absence in the lower to middle Archean).

The first four of these morphological groups of stromatolites can largely overlap in their gross morphology and in their internal structures and pass laterally one into another. Normally, however, they vary within a set range of variation within facies change vertically and laterally. Often the vertical growth pattern consists of a base of flat-laminated forms, followed by domical growth and by branching forms (Walter 1972; Preiss 1972, 1973, 1974; Grey 1984), reflecting growth stages of colonialisation and diversification of a bioherm and shallowing or deepening conditions with increasing or decreasing energy and varying influx of detritus.

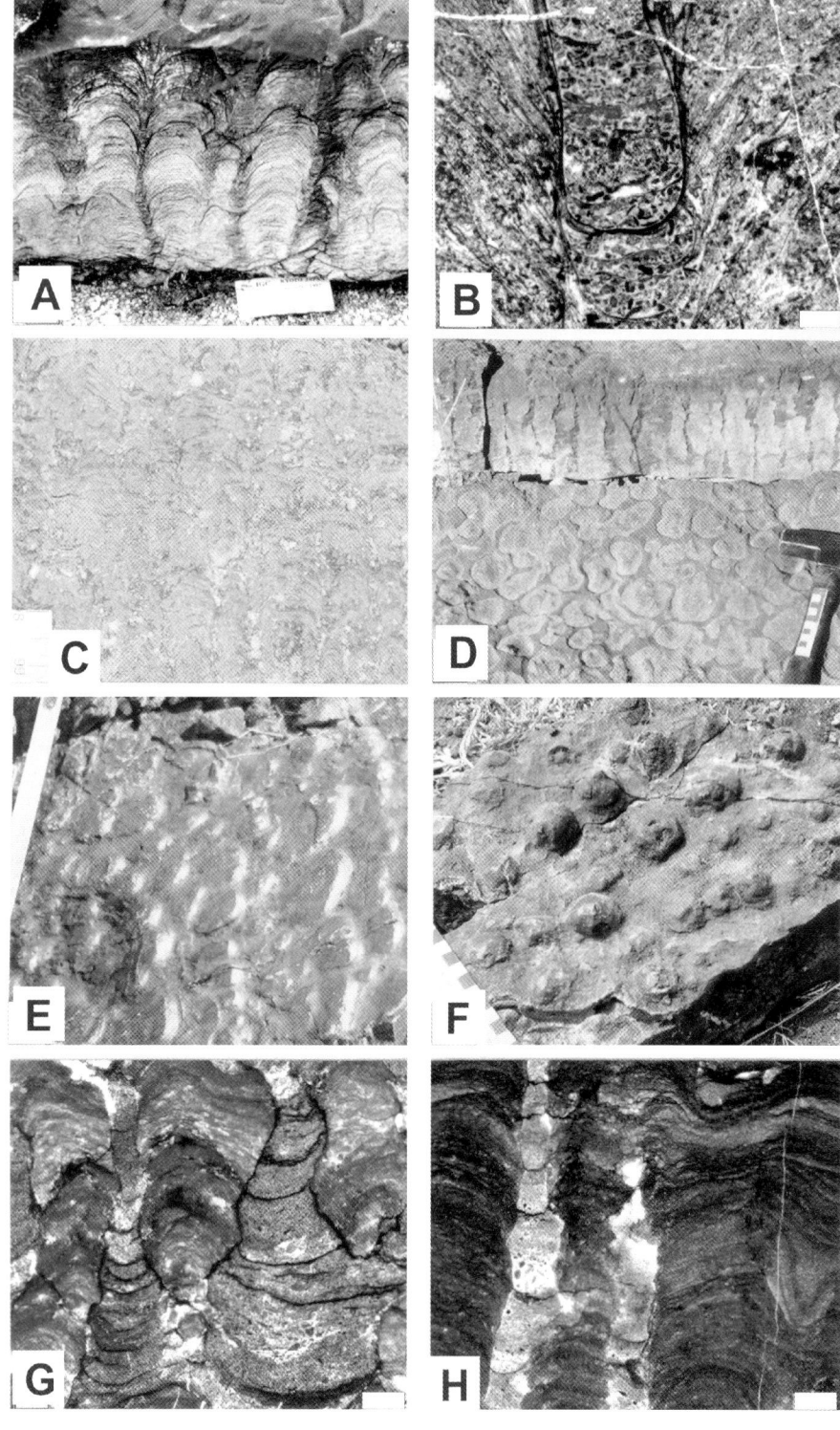

**Fig. 2** Examples of columnar stromatolites: **A** and **D** are intertidal, nonbranching but occasionally laterally linked or bridged columns, forming lithostromes. **A** is an outcrop photograph at Danielskuil, Griqualand West, South Africa, c. 2.52 Ga, scale bar = 10 cm; **D** is a similar lithostrome from Boetsap (Truswell and Eriksson 1973), more than 1,000 m further down the stratigraphic section. The small parallel branching columns in **C** belong to a large circular dome, over 1 m in diameter and about 40 cm of synoptic relief (outcrop photograph of a weathered surface at Reads Drift farm, south of Griquatown, Griqualand West, c. 2.55 Ga). In **E** an oblique surface view of the layer surface consisting of asymmetrical, elongated and inclined columns is shown (scale bar in cm). These are subtidal columns, that originated in current flow regime. The columns are about 6–8 cm in diameter and up to 20 cm high (Gamohaan Fm. Campbellrand Subgroup, Hopefield farm at Danielskuil, Hälbich et al. 1992). In **F**, small columns, 1–2 cm high and about 2–3 cm in diameter overgrow ripple mark crests in the upper intertidal facies; outcrop photograph of stromatolites in the Eccles Fm., Chuniespoort Group, Transvaal Supergroup, close to Ramotswa (Botswana), on the South African border side. In **B**, **G** and **H** micrographs of sediment binding and trapping columnar, partly laterally linked stromatolites are shown. Scale bars are 2.0 mm, transmitted light. Mainly micrite and organic matter delineate the indistinct laminae of walled columns. Very little stromatolitic debris is trapped between the columns and in **H**, graded bedding and fining upward to micrite is present in the sediment between the columns (white patches are diagenetic calcite cement). Dark, saucer-shaped laminae between the columns delineate periods of slow or nonsedimentation and probably of formation of microbial mat (bridging). In **B** the sediments (sapropel and fine stromatolitic debris) are trapped between the columns and within the laminae. All micrographs are of samples from intertidal and upper subtidal facies. **B** is a sample from the same outcrop as **D**; **G** and **H** are from Boetsap outcrops

The first four morphological groups above can form large bioherms of almost all possible internal combinations of columns, laterally linked (pseudo-columnar) structures and stratiform mats. However, coniform columns forming large domal lithoherms on the Archean stromatolitic platforms were not yet observed. Nevertheless, examples of coniform columns forming stromatolite domes can be found in the Mesoproterozoic Bangemall Group, Western Australia, where metre-high and up to 2 m wide cones cluster into enormous bioherms that probably have domed tops (Grey 1985), or in the Earaheedy Basin (Grey 1984, 1994).

In the following discussion examples of typical associations of stromatolite morphology and clastic sediments are described as characteristic of certain sedimentary facies realms and evidence of Archean life. In general, the four forms are preferably associated with specific hydrodynamic conditions, clastic carbonate and terrigenous influx, in which specific morphology results from the interplay of microbial binding and sediment accretion and the given hydrodynamic environment. The description is based mainly on my own, partly unpublished data from the Neoarchean stromatolitic carbonate platforms of South Africa and Western Australia, but also from older examples, going back to the 3.5 Ga stromatolites of the Pilbara craton, the 2.9 Ga Pongola microbial mats and 2.7 Ga Ventersdorp stromatolites.

*Columnar stromatolites*

i. Solitary, nonbranching columns with vertically equal width or upward widening club-shape and not forming domes but lithostromes (examples c in Fig. 1 and Fig. 2A,B) are usually associated with sparitic, pelletal, and stromatolitic debris wackestones to packstones, concentrated between and covering the columns. In situ reworked carbonate clastic sediment can also be bound along the laminae (Fig. 2B,G,H). The association of silt to fine sand-sized sediment with columns, spaced close to each other and of few cm to dm in width, provides evidence of wave- and current-agitated water. Current action may also be implied by uniform inclination of the columns and by elongated basal cross-sections (Fig. 2E). The column height indicates minimum water depth, but probably somewhat greater than that, when the columns do not exhibit internal disconformities (desiccation). Large lithostromes built of such columns (Fig. 2D) are traceable for kilometres in the Neoarchean Transvaal Supergroup, implying that shallow, wave-agitated facies belts were extremely wide within the Precambrian carbonate platforms. Lithoherms, like domal stromatolites behave differently and may be composed of fine, dark, submicron carbonate in what looks like a subtidal setting with little current activity (compare below and Grey 1985, Figs. 115, 116.)

ii. In diverging columnar stromatolites, branching can occur in a divergent-dendroid or parallel mode (examples a and b in Fig. 1). The branches are usually embedded in micrite or sparite with some admixture of clay and silt. The sediment is trapped between the columns, but micrite can also delineate the lamination (binding or precipitation). Shale may also cover the columns and lithoherms. Evidently, the water energy must have been weaker than in between nonbranching columns and typically these columns are smaller in size while the lamination is more delicate than in nonbranching types. Some small, delicate and closely branching columns lack any detrital sediment input and are embedded in pure blocky cements. Consequently, such stromatolites can be interpreted as a result of subtidal calm, lagoonal conditions. In some instances, however, the elongated domal shape of the lithoherms indicates recurring current action. The protected environment, in which fine sediment can be trapped between the columns or totally blocked out, has been presumably created by the dense columnar growth within the bioherm.

iii. Columnar stromatolites, with partly laterally linked laminae, transitional to laterally linked (LLH), pseudo-columnar stromatolites, and constructing lithostromes and domal lithoherms, occur with different sediment types or without any visible sediment trapping or binding. A decreasing degree of lateral linkage is conspicuously accompanied by increasing influx of sediment, up to carbonate arenites, in parts where the lateral linkage is only rudimentarily developed or absent. This behaviour of varying sediment trapping and binding activity with alternating mode of columnar growth (compare Fig. 2B,G and Fig. 3A,B; Fig. 8), can be interpreted as reflecting permanent changes in hydrodynamic conditions, probably within alternating lower intertidal to upper subtidal facies or in frequently wave agitated areas.

*Coniform stromatolites*

i. Small (centimetre sized) and large (decimetre to metre) solitary coniform columns (see examples a and b, coniform stromatolites, in Fig. 1) are usually associated with fine carbonate sands and muds. The lamination is crude and thick. Siliciclastic detritus is generally of silt to fine-sand-grain size. Smaller coniform columns, forming thin but laterally extensive lithostromes, are interpreted as forming in a calm, shallow subtidal to intertidal regime, whereas the larger (decimetre sized) coniform columns are found in deep subtidal conditions (Fig. 3E). Altermann and Herbig (1991) assigned small (centimetre range) conical stromatolites that form thin stratiform lithostromes, to intertidal facies (comp. Fig. 7B,C). The observed Archean conical stromatolites, independently of depositional depth, probably reflect an equilibrium of relatively fast sediment accumulation and growth velocity. Where sediment influx weakens, different growth forms develop (Fig. 8) and where the sediment influx exceeds the growth ability, stromatolites become buried. (This is strikingly different to modern conical stromatolites from hot spring environments, that have no visible sediment input; e.g., Walter et al. 1976).

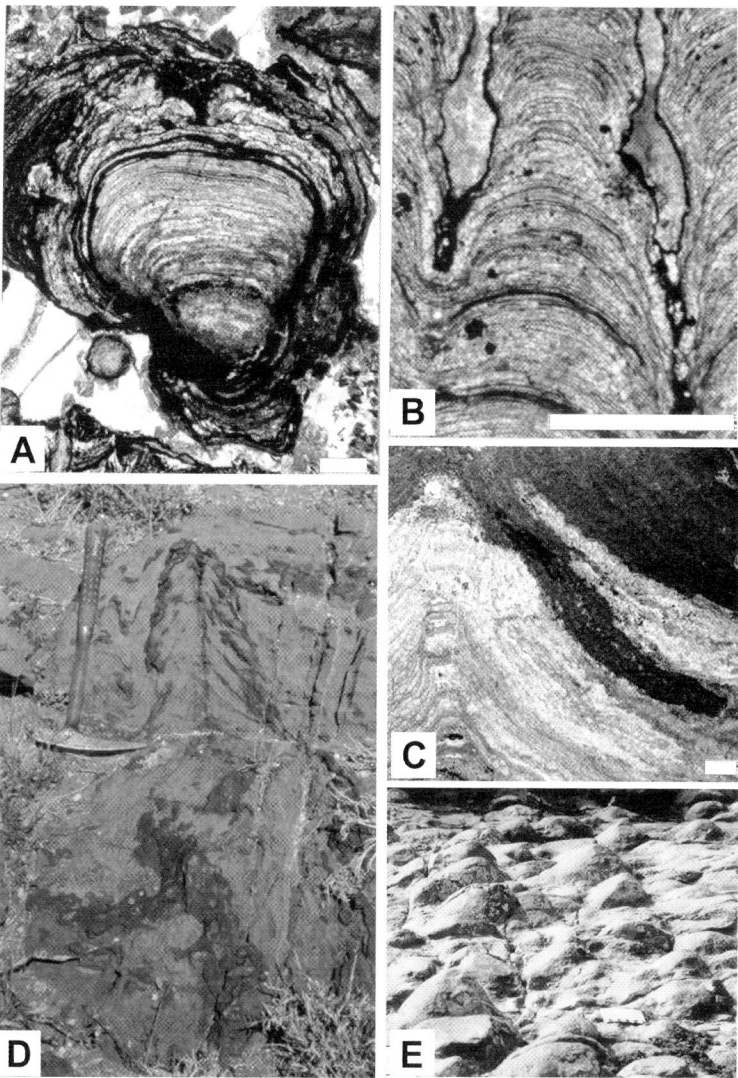

Fig. 3 **A**, **B** and **C** show nonsediment trapping and binding columnar, partly laterally linked stromatolites in thin section micrographs, in transmitted light. All scale bars = 2.0 mm. In **A**, a widening upward, small, walled column, embedded in recrystallised calcite and organic matter can be seen (Eastern Transvaal, Transvaal Supergroup, upper Malmani Dolomites). In **B** the lamination is extremely fine and of calcite and organic matter, but no sediment particles can be recognised (sample from the lower Reivilo Fm. Campbellrand Subgroup, North of the Orange River at Reads Drift farm). **C** is a micrograph of a conical column with partly indistinct lamination. Sedimentary particles are not present in the smooth, partly recrystallised laminae. Sample from the "Lime Acres Formation" at Lime Acres, Griqualand West. Further examples of coniform stromatolites are shown in **D** and **E**. In **D**, a large conical column, cut at an oblique view, with a pronounced axial zone and silicified walls and some distinct laminae is visible. These columns have a strongly elongated base oriented parallel to paleo-current direction (southern outcrops of the upper Campbellrand Subgroup, at Soetvlei farm, "Prieska facies"; Beukes 1987; Altermann and Herbig 1991). In **E**, a surface of a stromatolite lithostrome of conical columns at the upper Boetsap section, is visible. The tops of the columns with conical apices are weathering out and exhibit a slightly asymmetrical, inclined shape (current action from left to right). All examples are from the deep subtidal facies and between 2.52 to 2.55 Ga

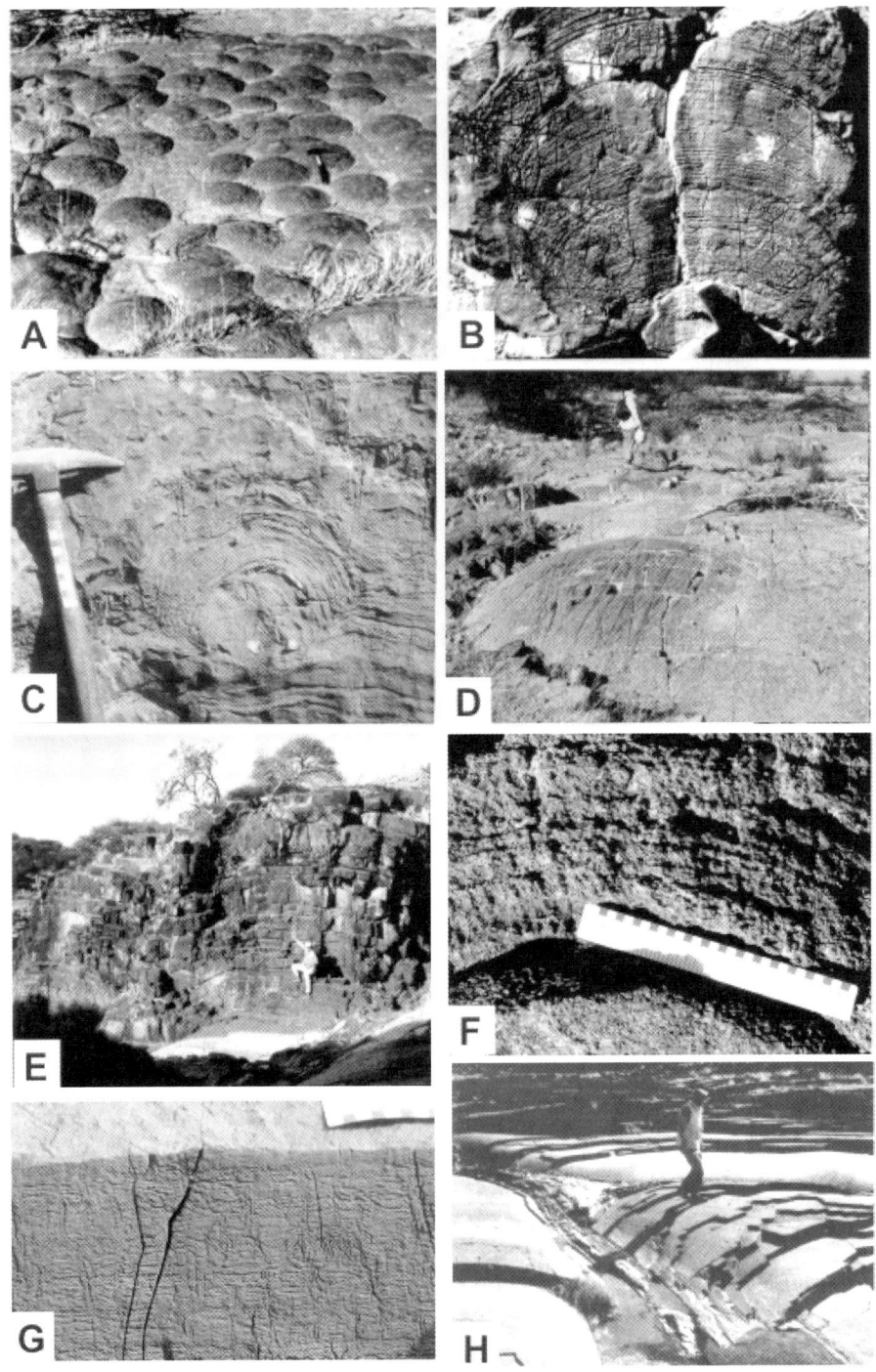

**Fig. 4** All examples are from the magnificent outcrops at the Boetsap waterfalls, Campbellrand Subgroup, Griqualand West, South Africa, 2.52 to 2.55 Ga (Truswell and Eriksson 1973; Eriksson and Truswell 1974). Domical stromatolites with a circular base: **A** and **B** are small domes, about 40 cm across and 40 cm high, of smooth to wrinkled lamination. In **C**, a smaller (20–30 cm) dome, with alternating wavy and smooth lamination is shown. In **D**, a large dome, about 3 m across is visible at the front of the photograph. **E**, **F**, **G** and **H** are examples of elongated domical stromatolites: **H** giant domes over 50 m long and of wavy irregular and crude lamination, as seen in **G**, are shown. The elongation of the domes is parallel to paleo-currents. **E** shows a cross-section view of such giant domes. In **F** the coarse, indistinct lamination of a smaller (about 1 m across and 3–4 m long), domical stromatolite can be seen. This is a rare example, where sediment binding in the lamination of a domical stromatolite can be demonstrated

ii. Large coniform columns from the Neoarchean "Nauga Formation", with elongated, ellipsoidal base, oriented parallel to each other (Fig. 1 coniform stromatolite, example d and Fig. 3D) are crudely laminated. They have sharp apices and walls and little sediment within the lamination (comp. Fig. 3C), but carbonate sands and muds are accumulated between the columns. The columns can attain 50 cm in height and 50 cm in basal length, and 30 cm in width, and have a prominent axial zone. From the facies development in the section, these conical columnar stromatolites are interpreted as deep subtidal, current influenced environment (Höferle et al. 2000). In the Strelley Pool Chert (c. 3.5 Ga, Kelly Group) large cones of rather circular base, are more than 1 m high and closely spaced, unlike the Nauga Formation columns. However, smaller, laterally linked coniform stromatolites occur as well (comp. Fig. 7C; Hofmann et al. 1999). Although carbonate recrystallisation obscures many details in these stromatolites, in the interbedded cherts sand-sized clasts of stromatolitic debris are abundant (Fig. 7G). The depositional environment ranges from high energy, wave- or tide-dominated shallow water sandstones and conglomerates passing upward to stromatolite reefs and lithostromes thriving under calm and probably deep conditions, that can be traced for tens of kilometres (Lowe 1983; Allwood et al. 2006; Van Kranendonk 2006).

*Domal stromatolitic bioherms with circular base and with laterally linked internal laminae, forming pseudo-columnar structures*

These stromatolites range in size from few decimetres to a few meters and 50 to 75 cm of synoptic relief (Fig. 4A–D).

i. Only domal bioherms with circular base, built up of small columns of about 1 cm in diameter, are clearly associated with fine carbonate sediment which is trapped between the columns. The sediment is always micritic, often with a strong clay admixture Thus, these stromatolites are interpreted as evidencing calm, subtidal conditions, perhaps below the fair weather wave base or in sheltered lagoons. The increasing size of the domes may reflect increasing depth.
ii. Domal bioherms with circular base and constructed of up-doming wavy laminae, may bind silt to fine sand size detritus in the lamination, probably evidencing wave agitated water without preferred wave direction (circular base). However, clear evidence of trapping and binding is rare and diagenetic recrystallisation is common in this facies. Rippled surfaces were not observed in these stromatolites (comp. below and Fig. 5C–E). Such lithoherms are often of lower synoptic relief than circular domes with columnar internal structure. Also in this case, however, the increasing size of the domes may reflect increasing depth of growth.

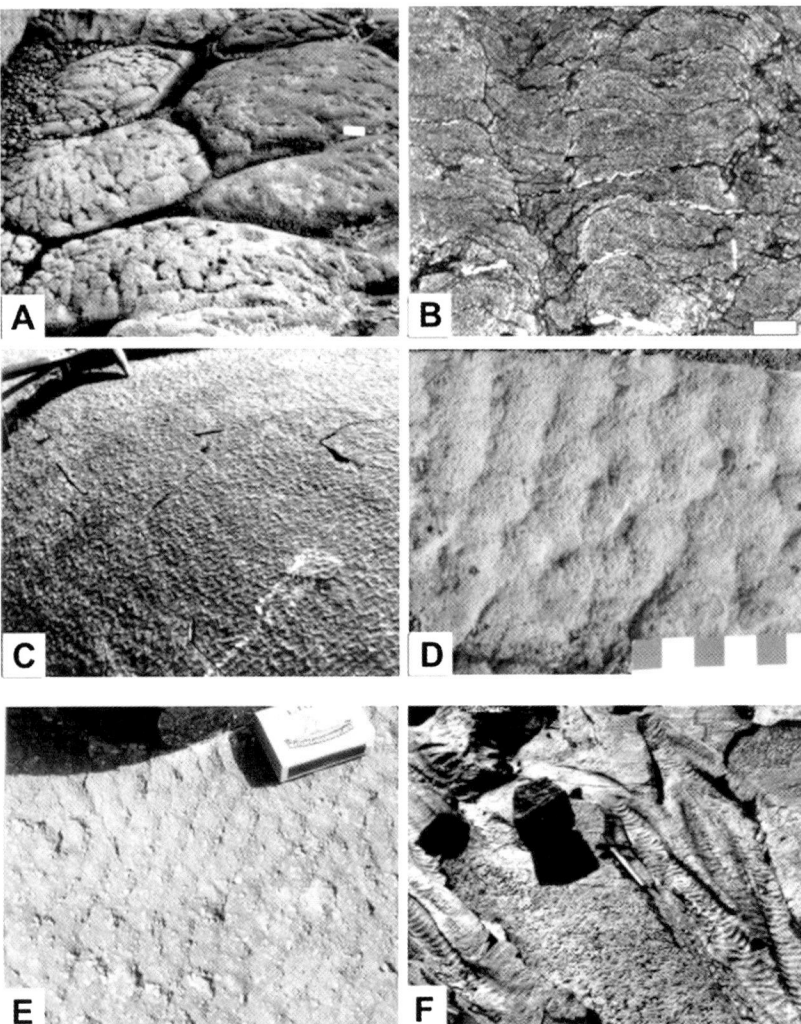

**Fig. 5** Examples of domical stromatolites: **A** are small domes from the lower Boetsap section, up to 1 m long, with an irregular, cortex-like or brain-like surface. The lithoherms have a sub-parallel arrangement and are embedded in shale. In **B**, thin-section micrograph from these bioherms, marl, organic matter and micrite constitute the laminae of the indistinct columns making up these stromatolites and the sediment fill between the columns. The lamination is slightly accentuated by compaction and dissolution of calcite (microstylolites). Scale bar is 2.0 mm. **C** to **F** are surfaces examples of domical stromatolites from Boetsap. In **C** the ripples are strongly accentuated by the bushy growth pattern of the stromatolitic mat, the regular and equidistant arrangement of the tufts however, strongly implies current activity rearranging the sediment bound by the mat. The surface in **D** exhibits more of a wave-ripple (symmetrical) morphology in contrast to **E**, where a current ripple morphology (asymmetrical) is implied by the sediment bound in the pustular mat. In **F**, erosional channels with current ripples, scoured across a domical stromatolite are visible. Examples **A** and **B** are of deep subtidal, low-energy environment and **C** to **F** are probably of shallow subtidal, high-energy environment

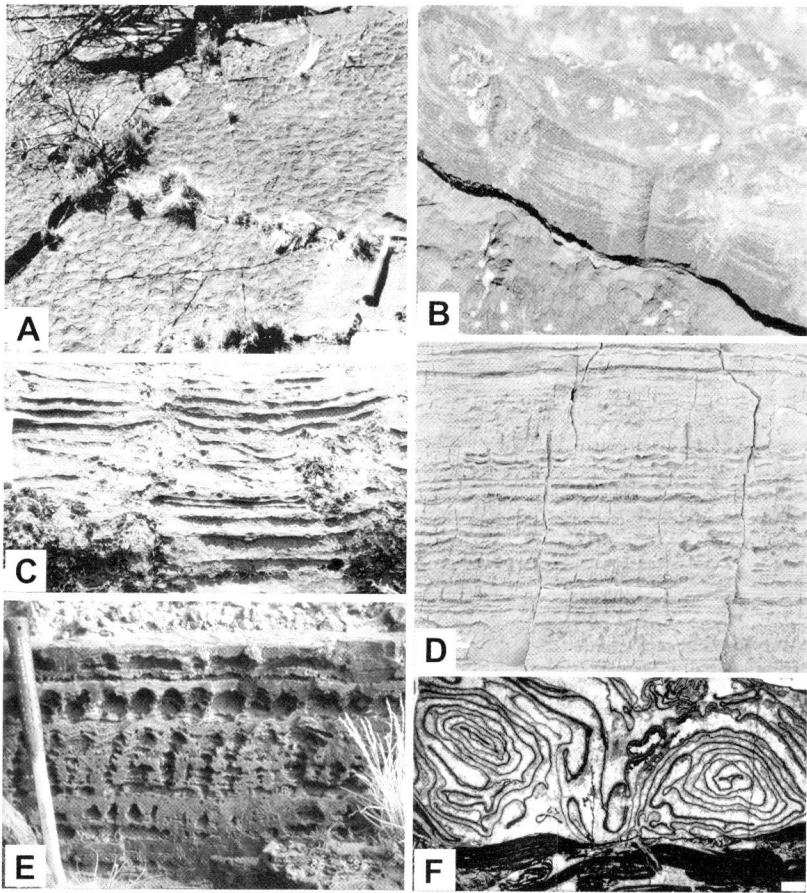

**Fig. 6** A, B, C and D are examples of fossilised Archean stratiform microbial mats in carbonate environments, in an intertidal to supratidal facies: A exhibits a surface of a stratiform stromatolite with polygonal desiccation structures. In B, the saucer-shape of the lamination is perceptible. A and B are from the northern banks of the Orange River at Prieska, "Prieska facies" Nauga Formation, 2.55 to 2.58 Ga (Beukes 1987; Altermann and Herbig 1991). C shows a cross-section of large, saucer-shaped structures forming polygons in plane view. The laminites are interbedded with carbonate sands and rare aeolian (?) lithic fragments (Kogelbeen Fm. Campbellrand Subgroup, Hopefield farm at Danielskuil, Hälbich et al. 1992). D is a fossilised wavy laminated stratiform mat of low intertidal facies. Geelbecksdam outcrops at Marydale, c. 2.59 Ga, "Prieska facies" (Beukes 1987; Altermann and Herbig 1991). E is an example of laterally linked pseudo-columnar stromatolites with regular linkage between the pseudo-columns and wavy laminites. The stromatolite is associated with coarse tuffaceous sand redeposited in wave agitated, lacustrine environment. Ventersdorp Supergroup, Oomdraaisvlei Farm, between Prieska and Britstown, South Africa, c. 2.71 Ga. F is an example of stratiform mat from subtidal facies: (thin-section micrograph, scale bar 2.0 mm). The thin, dark laminae are rich in organic matter and interlayered with light and thick laminae of fibrous and blocky calcite cements. No detritus is discernible in the lamination. The phase of biogenically dominated dark laminae may have taken as long as or longer to form than the intervening thicker sedimentary or precipitated phase. The structures are only weakly compacted, evidencing slow sedimentation and/or fast cementation (upper most "Prieska facies" at Prieska, Nauga Formation, Campbellrand Subgroup, 2.55 Ga)

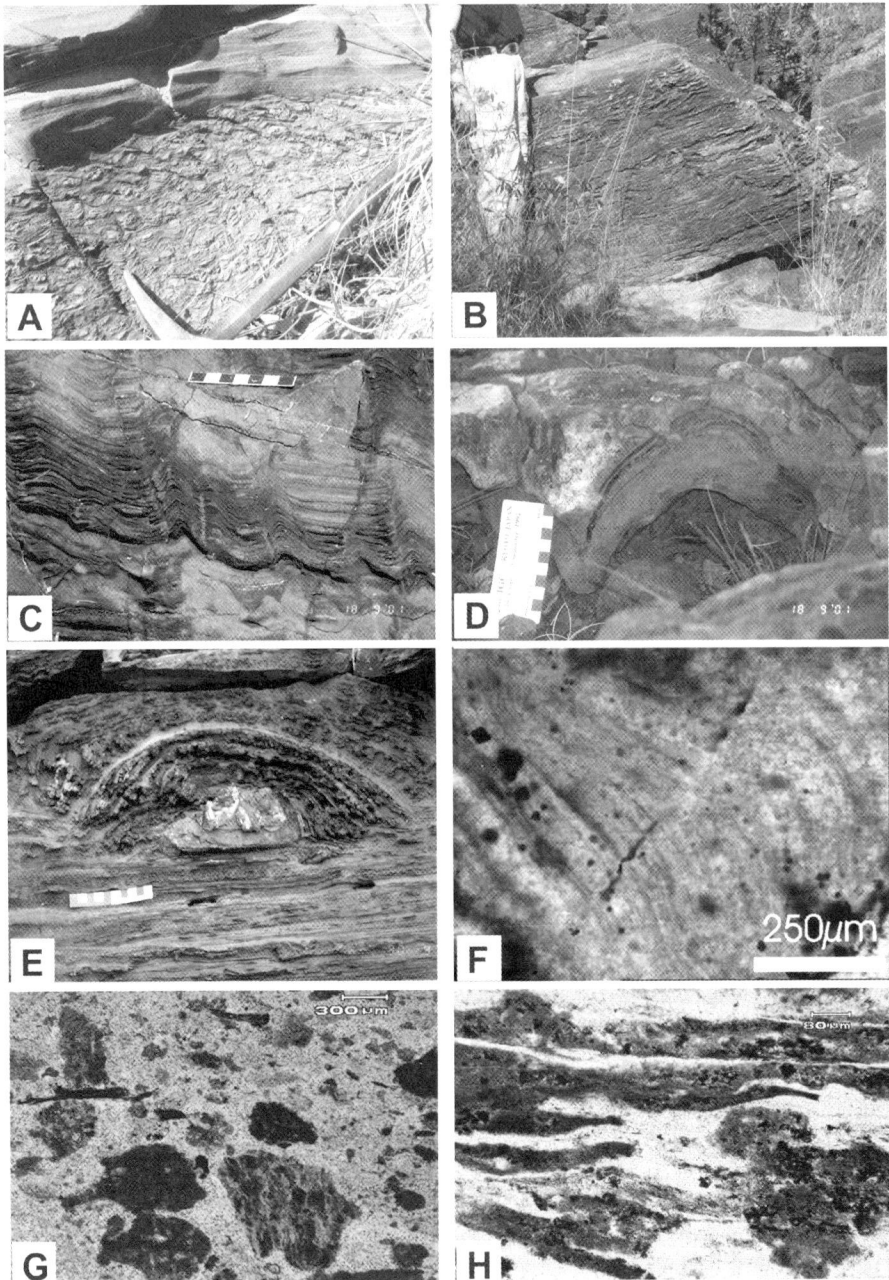

**Fig. 7** Examples of Archean stromatolites and microbial mats in various settings. **A** and **B** are from the Wit Mfolozi River section of the Pongola Group, c. 2.9 Ga, South Africa, at the road (R34) from Vryheid to Melmoth. In **A**, an intertidal intercalation of microbial mats and ripple marks, which are bound and stabilised by subsequent microbial mat overgrowth, is covered by a layer of carbonate sands with cross-bedding structures. In **B**, a chiefly, wavy stratiform stromatolite, but irregularly passing laterally into conical bulges, is visible. In between the lamina carbonate detritus is common in such stromatolites at this locality. **C**, **D**, **F** and **G** are Archean stromatolites (c. 3.49 to 3.45 Ga) form the Pilbara Craton, Western Australia. **D** is from the Dresser

Formation at North Pole locality. At this locality wrinkly stratiform stromatolites pass laterally into smooth domical forms. A small dome with silicified dark laminae is shown (for closer description see Van Kranendonk 2006, his Fig. 10). **C** is from the Strelley Pool Chert, Shaw River, so called Trendall locality (Hofmann et al. 1999) and exhibits coniform columns and bulges in a stratiform stromatolitic layer. The carbonate is heavily recrystallised and sediment trapping or binding cannot be detected; however, cherts intercalated with these stromatolites (comp. Van Kranendonk et al. 2001, Fig. 29, their "siliceous planar laminite"), contain sand- to silt-size stromatolite debris of angular erosional remnants of microbial mats, as visible in the thin-section micrograph, in **G**. **F** is a thin-section micrograph of the stromatolitic clast in the PPRG 2644 sample (original thin section of (Schopf 1993), deposited in the British Museum of National History and photographed by the present author at M. Brasier's laboratory, at Oxford University). The micrograph shows a part of a clast embedded in hydrothermal chert and displaying up-doming and, in the centre thickening, laminae with typical chert rich, silicified, recrystallised and thin, dark, carbon-rich, silicified and less recrystallised, alternating of microbial layers. Contrary to (Schopf 1993; Brasier et al. 2002), interpreted this clast as not stromatolitic. **E** shows a stromatolite in the Neoarchean Carrawine Dolomite (c. 2.6 Ga). The stromatolitic laminae overgrow a large clast resting on flat bedding. The entire structure forms a large protuberance on the otherwise flat surface. A normal sedimentation from suspension would be drastically different as thicker layers would have been deposited next to the clast and not on top of it, slowly levelling the topographic profile. Microbial laminae, however, tend to accentuate the topographic irregularity by adding equally thick or even thicker laminae on top of the bulging structure. **H** is a thin-section micrograph of a microbially laminated chert from the Barberton Greenstone Belt (sample WA03/26) from the Stolzberg Syncline (*fossiliferous chert of* Walsh 1992; uppermost chert in the Hoggenoeg Fm., c. 3.445–3.470 Ga). The chert displays strong hydrothermal recrystallisation in wavy patches and bands paralleling the bedding. The recrystallised, pale parts are coarser than the less recrystallised, dark (brown) chert patches. With advancing recrystallisation original texture is being replaced but in the less recrystallised parts evidence of bound kerogenous particles can be found

## *Elongated domal lithoherms, with columnar to laterally linked pseudo-columnar structure*

i. Elongated domes with smooth or wrinkled, relief-parallel, convex upward lamination can form the largest stromatolite lithoherms encountered. They can exceed 20 to 30 m in height and 50 m in length (Fig. 4E–H and Fig. 5). The synoptic relief can reach 2 m. Domical stromatolites of circular base with laterally linked lamination, subparallel to the synoptic relief and smooth or wrinkled, but wavy and convex-upward, are generally smaller. There is thus a salient correlation between the size of the dome and the type of the lamination in Neoarchean stromatolites. With increasing size of lithoherms the surface roughness of the lamination increases up to a relief of about 0.5 cm and wavelength of 2 cm in the largest domes. Rippled surfaces are common and the ripple crests are usually oriented perpendicular or oblique to the long axis of the domes (Fig. 5C,D,E). Such rippled surfaces are the only clear evidence of detrital sediment covering the laminae. Microscopically the sediment particles are commonly obliterated by recrystallisation. Obviously, the stromatolitic surface was periodically covered by thin veneers of carbonate sands and silts, which were subsequently stabilised by microbial growth. These domes must have originated under the influence of currents in subtidal realm and periodically have been exposed to waves. Channels with rippled beds have been carved in places across domal stromatolites (Fig. 5F). Such channels are up to 30 cm deep, up to 1 m wide and are oriented oblique to the long axis of the domes. They may have been formed by water run-off oblique to prevalent current direction, subsequently to moderate relative sea level drop.

ii. Domal stromatolites with columnar internal structure are often rich in organic matter and pelitic clay sediment. The lithoherms are up to 2–3 m long, 1–2 m wide and about 1 m high and the lamination is sometimes indistinct because of coalescing and agglutinating laminae around clots, giving the lithoherm a thrombolitic, clotted appearance (Clendenin 1989). The lithoherms are embedded in micrite, shale or marl (Fig. 5A,B).

The branching, if present, may be diverging dendroidal or parallel and branches are usually very fine, less than 1.0 cm across and densely growing, with clay or organic-rich sediment trapped between them. Erratic lateral linkage of laminae between the branches, i.e., bridging is common. The depositional environment indicated by such domal stromatolites is interpreted as of deep subtidal, below fair weather wave base. The water must have been rich in pelitic suspension deposited between the columns. The currents were probably weak because shale is present below and above these stromatolites and because their elongation is never as pronounced as in other domal lithoherms. The orientation of the long axes is also not strictly parallel to each other (Fig. 5A). The stromatolites exhibit a typical, irregular "cortex-like" or "brain-like" surface. The rich suspension sedimentation obviously hindered the stromatolitic growth and was trapped between the densely spaced columns.

*Stratiform stromatolites, of even, bed parallel internal laminae*

In these stromatolites, the thickness of the laminae varies between microns and millimetres and its recognition is very dependent on the preservational state and the degree of recrystallisation. Generally the wavy and wrinkled lamination is finer than in smooth laminae.

i. Wrinkled, wavy and smooth laminae in stratiform stromatolites (Fig. 6A–D) usually exhibit a thin veneer of carbonate sand and very rarely, a fine (silt) siliciclastic admixture. Fenestral and bird's-eye structures are common. Desiccation polygons occur in smooth laminites together with edgewise conglomerates and in situ flat pebble breccias. Occasionally, small symmetrical and asymmetrical ripple marks may cover the surfaces of stratiform stromatolites. The facies that is represented by these stromatolites is of intertidal to supratidal character (Altermann and Herbig 1991).

ii. Smooth, flat lamination, usually about 1 mm in thickness and consisting of micrite and shale, but often exhibiting spar-cements, is commonly associated with marls. There is conspicuous lack of sedimentary and diagenetic structures, such as bird's-eye and fenestrate structures. Roll-up structures, 2–3 cm in diameter and about 5–10 cm long occur sporadically (Fig. 6F). Roll-up structures in microbial mats were described from the 2.6 Ga Wittenoom Dolomite of the Pilbara Craton, Australia, by Simonson and Carney (1999). These authors assigned such structures from smooth stratiform microbial mats to deep water deposits, below the photic zone, because of intercalation with calciturbidites. Roll-up (slump) structures probably originated on an inclined unstable substrate, allowing for slumping and rolling up of unconsolidated microbial mats.

Donaldson (1976a, 1976b) attempted to correlate stromatolite morphology with water turbulence and subtidal to supratidal regimes, based on associated sediments and sedimentary structures, of the Middle to Upper Proterozoic. He has placed stratiform stromatolites, of even, bed parallel internal laminae in high to low water energy but chiefly supratidal environment. Various columnar stromatolites were assigned to chiefly intertidal to supratidal regime and low to rather high water turbulence. Diverse types of branching columns were interpreted as of low energy, subtidal to intertidal facies. Domical stromatolites were interpreted as of low turbulence and chiefly subtidal, but ranging into the supratidal, regime, similarly to laterally linked domes and columns, which were deduced as of low energy to moderately turbulent zones. Coniform stromatolites in these basins are interpreted as of exclusively low turbulence, subtidal setting because of lack of any sedimentary structures like ripple marks, cross-bedding, signs of desiccation or intraformational conglomerates. In the same volume, Hoffman (1976) placed diversely branching columnar stromatolites at a water depth between 0 and 10 m and coniform stromatolites between 10 and 100 m and below that. A general agreement as to the facies distribution of stromatolite forms can be deduced from this literature and the herein described Archean examples.

## 2 Conclusions

Most Precambrian stromatolite forms are typically not associated with clastic sediment grains and are not binding or trapping. Nevertheless, stromatolites in many cases can facilitate facies analysis and serve as key organo-sedimentary structures in facies identification. Even if stromatolites are not trapping and binding they are part of sedimentary environments and carbonate clastic and/or siliciclastic sediments can be found in their direct neighbourhood and be used for facies analysis. Stromatolites often exhibit vertical growth cycles which are expressions of migrating lateral facies changes. Thus, for example, a cycle starting with conical stromatolites in carbonate sand matrix passing upward to columnar and subsequently to laterally linked pseudo-columnar stromatolites, which are capped by stratiform mats and carbonate sands with bidirectional ripple marks on top, form a typical shallowing upward cycle (Fig. 8).

Various morphologies of Precambrian lithified microbial mats exhibit typical sedimentary facies associations. The above description of stromatolite morphology and related sediment is far from being complete and is not exclusive. Other associations of stromatolites and sediments exist, and with ongoing investigations may turn more representative for the Archean. For the Proterozoic, there is large number of stromatolite occurrences that may be different to the patterns outlined above for the Archean stromatolites. The problem is that many examples in the literature, particularly those of Russian stromatolites, are described in great detail morphologically, and even assigned a stratigraphically significant position. Their sedimentary facies context, however, is often not clear, nor the hierarchical size arrangement of the internal structure to gross morphology and to the trapped and bound sediment. Very often it is ambiguous, whether the described morpho-taxa, classified accordingly to the International Code of Botanical Nomenclature, are indeed individual "taxa" or a part of a larger domical lithoherm or of a lithostrome, or isolated build-ups. It can be thus disputed whether the actual identification of "taxa" is correct for these localities and the internal structure on centimetre scale and the metre scale morphology correspond to original taxonomic descriptions or have been misidentified and indeed can be found in similar or very different environments.

Although this contribution argues that the morphology of Archean stromatolites is a clear expression of the antiquity of life, the concept of stromatolite taxa classified accordingly to the International Code of Botanical Nomenclature is strongly questioned in favour of the (bio-)sedimentary facies concept. Biological influence on the combination of sedimentary facies with biosedimentary morphology is perceptible from stromatolitic associations and from their comparison to younger and modern examples. The conspicuous dependence of biosediment morphology and facies association cannot be explained by purely chemical precipitation processes. The architectural arrangement of varying stromatolite morphologies, their almost endless combination ranging from large, kilometre-scale to minute, sub-mm scale shapes is an expression of biology acting under differing physical (hydrodynamic) conditions. The association of sediment grains with a range of variously shaped laminations, that follows hydrodynamic laws, but also exhibits clear evidence for trapping and binding of grains, which under normal conditions, without the involvement of sticky bio-mats, would

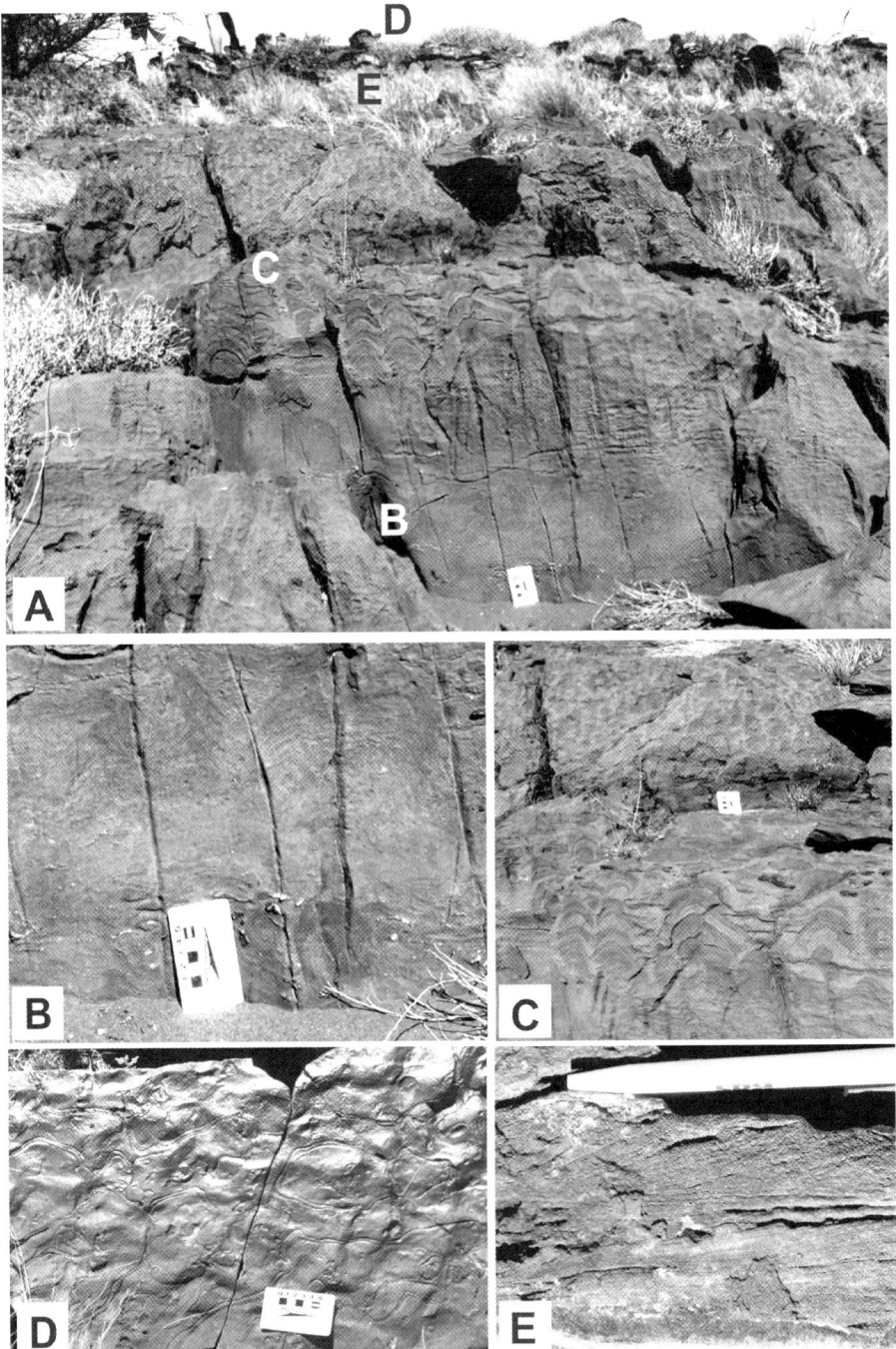

**Fig. 8** Stromatolitic cycle starting with conical stromatolites in carbonate sand matrix, passing upward to columnar and to laterally linked pseudo-columnar stromatolites, which decrease in size upward and are capped by stratiform mats and carbonate sands with bidirectional ripple marks on top of a typical shallowing upward cycle. Soetvlei Farm, southeast of Prieska, "Prieska facies", Nauga Formation, Campbellrand

Subgroup, 2.55 Ga (Altermann and Herbig 1991). **A** is an overview of the outcrop showing the entire cycle. **B**, **C**, **D** and **E** are locations of the enlargements and separate photographs **B**, **C**, **D** and **E** below. **B** shows a faint conical stromatolite next to small domical stromatolites, all embedded in carbonate-arenite matrix and trapping and binding such arenite. **C** displays the transition zone from, partly laterally linked, large columns to smaller columns of similar morphology. **D** exhibits a top view of the uppermost bed closing up this cycle with a stratiform microbial mat and superimposed bi-directional ripple marks, formed in extremely shallow water. In **E** a dolarenite bed with cross-bedding foresets, directly underlying the microbial mat shown in **D**, is visible

have been deposited only in the depressions between the laminae, are clear evidence for bio-genicity. Therefore, abiogenic mathematical or chemical models for the formation of such structures must be refuted based on the morphological complexity of the sediment accu-mulation, and the interplay of this complexity with sedimentary facies association. Even if such models or experiments are able to offer an apparently abiotic explanation for single shapes and accretion patterns in organo-sedimentary structures, and can produce such pat-terns under laboratory conditions, they can never simulate the affluence of the combination of structures as found in nature on geological time scales.

**Acknowledgements**    Research summarised in this contribution was supported through several grants by the DFG (German Research Foundation) to the author. Fieldwork in South Africa was actively supported by many students from the LMU and by the University of Pretoria through logistical support. Farmers and mining companies often provided generous shelter and access to the outcrops on their property. Brian Pratt and Hans Hofmann significantly improved an early version of this work and two very helpful anonymous reviewers contributed many important details and hints to the structure and organisation of this manuscript and to literature references.

# References

A.C. Allwood, M.R. Walter, B.S. Kamber, C.P. Marshall, I.W. Burch, Stromatolite reef from the Early Archean era of Australia. Nature **441**(8), 714–718 (2006)

A.C. Allwood, M.R. Walter, I.W. Burch, B.S. Kamber, 3.43 billion-year-old stromatolite reef from the Pilbara Craton of Western Australia: Ecosystem-scale insights to early life on Earth. Precambr. Res. **158**, 198–227 (2007)

W. Altermann, The oldest fossils of Africa – a brief reappraisal of reports from the Archean. J. Afr. Earth Sci. **33**, 427–436 (2001)

W. Altermann, The evolution of life and its impact on sedimentation, in *Precambrian Sedimentary Envi-ronments: A Modern Approach to Ancient Depositional Systems*, ed. by W. Altermann, P.L. Corcoran. Special Publication International Association of Sedimentologists, vol. 33 (IAS, Blackwell, 2002), pp. 15–32

W. Altermann, Precambrian stromatolites: Problems in definition, classification, morphology and stratigra-phy, in *The Precambrian Earth: Tempos and Events*, ed. by P.G. Eriksson, W. Altermann, D.R. Nelson, W. Mueller, O. Catuneanu. Developments in Precambrian Geology (Elsevier, Amsterdam, 2004), pp. 564–574

W. Altermann, The early Earth's record of enigmatic cyanobacteria and supposed extremophilic bacteria at 3.8 to 2.5 Ga, in *Algae and Cyanobacteria in Extreme Environment*, ed. by J. Seckbach. Cellular Origin, Life in Extreme Habitats and Astrobiology, vol. 11 (Springer, Heidelberg, 2007), Chap. 8

W. Altermann, J. Kazmierczak, A. Oren, D. Wright, Microbial calcification and its impact on the sedimentary rock record during 3.5 billion years of Earth history. Geobiology **4**, 147–166 (2006)

W. Altermann, H.G. Herbig, Tidal flats deposits of the Lower Proterozoic Campbell Group along the south-western margin of the Kaapvaal Craton, Northern Cape Province, South Africa. J. Afr. Earth Sci. **13**(3–4), 415–435 (1991)

W. Altermann, D.R. Nelson, Sedimentation rates, basin analysis, and regional correlations of three Neoarchean and Paleoproterozoic sub-basins of the Kaapvaal Craton as inferred from precise U-Pb zircon ages from volcaniclastic sediments. Sediment. Geol. **120**, 225–256 (1998)

W. Altermann, J.W. Schopf, Microfossils from the Neoarchean Campbell Group, Griqualand West Sequence of the Transvaal Supergroup, and their paleoenvironmental and evolutionary implications. Precambr. Res. **75**, 65–90 (1995)

W. Altermann, H.P. Siegfried, Sedimentology and facies development of an Archean shelf—carbonate platform transition in the Kaapvaal Craton, as deduced from a deep borehole at Kathu, South Africa. J. Afr. Earth Sci. **24**(3), 391–410 (1997)

S.M. Awramik, The history and significance of stromatolites, in *Early Organic Evolution: Implications for Mineral and Energy Resources*, ed. by M. Schidlowski, S. Golubic, M.M. Kimberley, D.M. McKirdy, P.A. Trudinger (Springer, Berlin, 1992), pp. 435–449

S.M. Awramik, H.P. Buchheim, Late Archaean lacustrine carbonates, stromatolites, and transgression, in Proceedings of the Fourth International Archaean Symposium Abstract (2001), pp. 222–223

S.M. Awramik, M.A. Semikhatov, The relationship between morphology, microstructure, and microbiota in three vertically intergrading stromatolites from the Gunflint Iron Formation. Can. J. Earth Sci. **16**, 2319–2330 (1979)

J. Bertrand-Sarfati, Les stromatolites du Précambrien supérieur du Sahara nord occidental; inventaire, morphologie et microstructures des laminations. Corrélations stratigraphiques. Centre de Recherches sur les Zones Arides, Paris. Public. CNRS, Geol. 14 (1972), 240pp

N.J. Beukes, Facies relations, depositional environments and diagenesis in a major Early Proterozoic stromatolitic carbonate platform to basinal sequence, Campbellrand Subgroup, Transvaal Supergroup. South. Afr. Sediment. Geol. **54**, 1–46 (1987)

M. Black, The algal sediments of Andros Island, Bahamas. R. Soc. Phil. Trans. B **122**, 169–192 (1933)

M. Brasier, O. Green, J. Lindsay, A. Steele, Earth's oldest (>3.5 Ga) fossils and the 'early Eden hypothesis': Questioning the evidence. Orig. Life Evol. Biosphere **34**, 257–269 (2004)

M. Brasier, N. McLoughlin, O. Green, D. Wacey, A fresch look at the fossil evidence for early Archean cellular life. Phil. Trans. R. Soc. **361**(B), 887–902 (2006)

M.D. Brasier, O.R. Green, A.P. Jephcoat, A.K. Kleppe, M.J. Van Kranendonk, J.F. Lindsay, A. Steele, N.V. Grassineau, Questioning the evidence for earth's oldest fossils. Nature **416**, 76–81 (2002)

S.G. Buck, Stromatolite and ooid depositswithin the fluvial and lacustrine sediments of the Precambrian Ventersdorp Supergroup of South Africa. Precambr. Res. **12**, 301–330 (1980)

R.V. Burne, L.S. Moore, Microbialites: Organosedimentary deposits of benthic microbial communities. Palaios **2**(3), 241–254 (1987)

F.H.A. Campbell, M.P. Cecile, Evolution of the Early Proterozoic Kilohigok Basin, Bathurst Inlet–Victoria Island, Northwest Territories. in *Proterozoic Basins of Canada*, ed. by F.H.A. Campbell. Geological Survey of Canada, paper 81-10, 1981, pp. 103–131

C.W. Clendenin, Tectonic influence on the evolution of the Early Proterozoic Transvaal sea. Unpubl. Ph.D. thesis, University Witwatersrand, 1989, 376pp

J.A. Donaldson, Aphebian stromatolites in Canada: Implications for stromatolite zonation, in *Stromatolites*, ed. by M.R. Walter. Developments in Sedimentology, vol. 20 (Elsevier, Amsterdam, 1976a), pp. 371–380

J.A. Donaldson, Paleoecology of Conophyton and associated stromatolites in the Precambrian Dismal Lakes and Rae Groups, Canada, in *Stromatolites*, ed. by M.R. Walter. Developments in Sedimentology, vol. 20 (Elsevier, Amsterdam, 1976b), pp. 523–534

C. Dupraz, A. Strasser, Microbialites and micro-encrusters in shallow coral bioherms (Middle to Late Oxfordian), Swiss Jura Mountains. Facies **40**, 101–130 (1999)

K.A. Eriksson, J.F. Truswell, Tidal flat associations from the lower Proterozoic carbonate sequence in South Africa. Sedimentology **21**, 293–309 (1974)

K. Grey, Biostratigraphic studies of stromatolites from the Proterozoic Earaheedy Group, Nabberu Basin, Western Australia. West. Aust. Geol. Surv. Bull. **130**, 123 (1984)

K. Grey, Stromatolites and other organic remains in the Bangemall Basin. Appendix in Muhling, P.C. and Brakel, A.T., Geology of the Bangemall Group: the evolution of an intracratonic basin. West. Aust. Geol. Surv. Bull. **128**, 221–256 (1985)

K. Grey, Book Review—"Proceedings of the Indo-Soviet Symposium on Stromatolites and Stromatolitic Deposits" by K.S. Valdiya (editor). Precambr. Res. **59**(3/4), 325–327 (1992)

K. Grey, Stromatolites from the Palaeoproterozoic Earaheedy Group, Earaheedy Basin, Western Australia. Alcheringa **18**, 187–218 (1994)

J.P. Grotzinger, Facies and evolution of Precambrian carbonate depositional systems: Emergence of modern platform archetype. SEPM Spec. Publ. **44**, 79–106 (1989)

J.P. Grotzinger, Geochemical model for Proterozoic stromatolite decline. Am. J. Sci. A **290**, 80–103 (1990)

J.P. Grotzinger, D.H. Rothman, An abiotic model for stromatolite morphogenesis. Nature **383**, 423–425 (1996)

G. Gürich, Les spongiostromides di Viséen de la province de Namur. Muséum d'Historie Naturelle de Belgique, mémoires **3/4**, 1–55 (1906)

I.W. Hälbich, D. Lamprecht, W. Altermann, U.E. Horstmann, The carbonate-banded iron formation transition in the Early Proterozoic of South Africa. J. Afr. Earth Sci. **15**(2), 217–236 (1992)

J.D. Hall, *Cryptozoön* (*proliferum*) n.g. and s.p.- Rep. N.Y. State Mus. 36, pl. 6, 1883

H.J. Hofmann, Attributes of stromatolites. Geological Survey Canada, Paper 69/39 (1969), 58pp

H.J. Hofmann, Stromatolites: Characteristics and utility. Earth Sci. Rev. **9**, 339–373 (1973)

H.J. Hofmann, On Aphebian stromatolites and Riphean stromatolite stratigraphy. Precambr. Res. **5**, 175–205 (1977)

H.J. Hofmann, Archean stromatolites as microbial archives, in *Microbial Sediments*, ed. by R.E. Riding, S.M. Awramik (Springer, Berlin, 2000), pp. 315–327

H.J. Hofmann, K. Grey, A. Hickman, R. Thorpe, Origin of 3.45 Ga coniform stromatolites in Warrawoona Group, Western Australia. Geol. Soc. Am. Bull. **111**, 1256–1262 (1999)

P.F. Hoffman, Shallow and deepwater stromatolites in lower Proterozoic platform-to-basin facies change, Great Slave Lake, Canada. Bull. Am. Assoc. Petroleum Geol. **58**(5), 856–867 (1974)

P. Hoffman, Environmental diversity of Middle Precambrian stromatolites, in *Stromatolites*, ed. by M.R. Walter. Developments in Sedimentology, vol. 20 (Elsevier, Amsterdam, 1976), pp. 599–612

R. Höferle, D. Haller, A. Tetzlaff, W. Altermann, The unique assemblage of elongated, coniform stromatolites in the Neoarchean Campbellrand Subgroup, southwestern Kaapvaal Craton, South Africa. Abstracts, 18th Colloq. of Afr. Geol., Graz. J. Afr. Earth Sci. **30**(4), 40 (2000)

R.J. Horodyski, Environmental influences on columnar stromatolite branching patterns: examples from the Middle Proterozoic Belt Supergroup, Glacier National Park, Montana. J. Paleontol. **51**, 661–671 (1977)

E. Kalkowsky, Oolith und Stromatolith im Norddeutschen Bundsandstein. Z. dt. geol. Ges. **60**, 68–125 (1908)

J. Kazmierczak, W. Altermann, Neoarchean biomineralisation by benthic cyanobacteria. Science **298**, 2351 (2002)

J. Kazmierczak, S. Kempe, W. Altermann, Microbial origin of Precambrian carbonates: Lessons from modern analogues, in *The Precambrian Earth: Tempos and Events*, ed. by P.G. Eriksson, W. Altermann, D.R. Nelson, W. Mueller, O. Catuneanu. Developments in Precambrian Geology (Elsevier, Amsterdam, 2004), pp. 545–563

S. Kiyokawa, T. Ito, M. Ikehara, F. Kitajima, Middle Archean volcano-hydrothermal sequence: Bacterial microfossil-bearing 3.2 Ga Dixon Island Formation, coastal Pilbara terrane, Australia. GSA Bull. **118**(1–2), 3–22 (2006)

W.E. Krumbein, Stromatolites – the challenge of a term in space and time. Precambr. Res. **20**, 493–531 (1983)

M.B. Lambert, Stromatolites of the late Archean back River stratovolcano, Slave structural province, Northwest Territories, Canada. Can. J. Earth Sci. **35**(3), 290–301 (1998)

B.W. Logan, Cryptozoon and associated stromatolites from the Recent, Shark Bay, Western Australia. J. Geol. **69**, 517–533 (1961)

B.W. Logan, R. Rezak, R.N. Ginsburg, Classification and environmental significance of algal stromatolites. J. Geol. **72**, 68–83 (1964)

D.R. Lowe, Restricted shallow-water sedimentation of Early Archean stromatolitic and evaporitic strata of the Strelley Pool Chert, Pilbara Block, Western Australia. Precambr. Res. **19**, 239–283 (1983)

D.R. Lowe, M.M. Tice, Tectonic controls on atmospheric, climatic, and biological evolution 3.5–3.4 Ga. Precambr. Res. **158**, 177–197 (2007)

T.R. Mason, V. von Brunn, 3-Gyr-old stromatolites from South Africa. Nature **266**, 47–49 (1977)

D. Mawson, Some South Australian algal limestones in process of formation. Quart. J. Geol. Soc. **85**, 613–623 (1929)

A.D. Miall, *Principles of Sedimentary Basin Analysis* (Springer, New York, 1984), 490pp

S. Moorbath, Dating earliest life. Nature **434**, 155 (2005)

D.R. Nelson, A.F. Trendall, W. Altermann, Chronological correlations between the Pilbara and Kaapvaal cratons. Precambr. Res. **97**(3–4), 165–189 (1999)

N. Noffke, N. Beukes, J. Gutzmer, R. Hazen, Spatial and temporal distribution of microbially induced sedimentary structures: A case study from siliciclastic storm deposits of the 2.9 Ga Witwatersrand Supergroup, South Africa. Precambr. Res. **146**, 35–44 (2006a)

N. Noffke, R.N. Hazen, K.A. Eriksson, E.L. Simpson, A new window into early life: Microbial mats in siliciclastic early Archean tidal flat (3.2 Ga Moodies Group, South Africa). Geology **34**, 253–256 (2006b)

P.E. Playford, Devonian "Great Barrier Reef" of the Canning Basin, Western Australia. AAPG Bull. **64**, 814–840 (1980)

B.R. Pratt, Calcification of cyanobacterial filaments: *Girvanella* and the origin of lower Paleozoic lime mud. Geology **29**, 763–766 (2001)

B.R. Pratt, Calcification of cyanobacterial filaments: *Girvanella* and the origin of lower Paleozoic lime mud— Discussion and reply. Geology **30**, 580 (2002)

W.V. Preiss, The systematics of South Australian Precambrian and Cambrian Stromatolites, Part I. South Aust. R. Soc. Trans. **96**, 67–100 (1972)

W.V. Preiss, The Systematics of South Australian Precambrian and Cambrian Stromatolites, Part II. South Aust. R. Soc. Trans. **97**(2), 91–125 (1973)

W.V. Preiss, The systematics of South Australian Precambrian and Cambrian stromatolites, Part III. South Aust. R. Soc. Trans. **98**, 105–208 (1974)

M.E. Raaben, A.K. Sinha, Classification of stromatolites: in K.S. Valdiya (ed.) Proceedings of the Indo-Soviet Symposium on Stromatolites and Stromatolitic Deposits. Himal. Geol. **13**, 215–227 (1989)

M.E. Raaben, A.K. Sinha, M. Sharma, *Precambrian Stromatolites of India and Russia* (a catalogue of Type-Form-Genera) (Birbal Sahni Institute of Palaeobotany, Army Printing Press, 2001), 125pp

R. Riding, The term stromatolite: towards an essential definition. Lethaia **32**, 321–330 (1999)

R.E. Riding, S.M. Awramik (eds.), *Microbial Sediments* (Springer, Berlin, 2000), 331pp

J. Schieber, P. Bose, P.G. Eriksson, S. Banerjee, S. Sarkar, W. Altermann, O. Catuneanu (eds.), *Atlas of Microbial Mat Features Preserved within the Siliciclastic Rock Record*. Atlases in Geosciences, vol. 2 (Elsevier, Amsterdam, 2007), 311p

J.W. Schopf (ed.), *Earth's Earliest Biosphere: Its Origin and Evolution* (Princeton University Press, Princeton, 1983), 543pp

J.W. Schopf, Earth's earliest biosphere: Status of the hunt, in *The Precambrian Earth: Tempos and Events*, ed. by P.G. Eriksson, W. Altermann, D.R. Nelson, W. Mueller, O. Catuneanu. Developments in Precambrian Geology (Elsevier, Amsterdam, 2004), pp. 516–539

J.W. Schopf, Microfossils of the early Archean Apex chert: New evidence of the antiquity of life. Science **260**, 640–646 (1993)

J.W. Schopf, Fossil evidence of Archaean life. Philos. Trans. R. Soc. Lond. B **361**, 869–885 (2006)

J.W. Schopf, C. Klein (eds.), *The Proterozoic Biosphere: A Multidisciplinary Study* (Cambridge University Press, Cambridge, 1992), 1348pp

J.W. Schopf, A.B. Kudryavtsev, A.D. Czaja, A.B. Tripathi, Evidence of Archean life: Stromatolites and microfossils. Precambr. Res. **158**, 141–155 (2007a)

J.W. Schopf, M.R. Walter, C. Ruiji, Earliest evidence of life on Earth. Precambr. Res. **158**, 139–140 (2007b)

J.W. Schopf, Yu.K. Sovietov, Microfossils in Conophyton from the Soviet Union and their bearing on Precambrian Biostratigraphy. Science **193**, 143–146 (1976)

S.N. Serebryakov, M.A. Semikhatov, Riphean and Recent stromatolites: a comparison. Am. J. Sci. **274**(6), 556–574 (1974)

B.M. Simonson, K.E. Carney, Roll-Up Structures: Evidence of in situ Microbial Mats in Late Archean Deep Shelf Environments. Palaios **14**, 13–24 (1999)

D.Y. Sumner, Decimeter-thick encrustations of calcite and aragonite on the sea-floor and implications for Neoarchean and Neoproterozoic ocean chemistry, in *Precambrian Sedimentary Environments: A Modern Approach to Ancient Depositional Systems*, ed. by W. Altermann, P.L. Corcoran. I.A.S. Spec. Publ., vol. 33 (Blackwell, Oxford, 2002), pp. 107–122

K. Sugitani, K. Grey, A. Allwood, T. Nagaoka, K. Mimura, M. Minami, C.P. Marshall, M.J. Van Kranendonk, M.R. Walter, Diverse microstructures from Archean chert from the Mount Goldsworthy – Mount Grant area, Pilbara Craton, Western Australia: Microfossils, dubiofossils, or pseudofossils? Precambr. Res. **158**, 228–262 (2007)

M.M. Tice, D.R. Lowe, The origin of carbonaceous matter in pre-3.0 Ga greenstone terrains: A review and new evidence from the 3.42 Ga Buck Reef Chert. Earth Sci. Rev. **76**, 259–300 (2006)

J.F. Truswell, K.A. Eriksson, Stromatolitic associations and their palaeo-environmental significance: A reappraisal of a lower Proterozoic locality from the northern Cape Province, South Africa. Sediment. Geol. **10**, 1–23 (1973)

Y. Ueno, Y. Isozaki, H. Yurimoto, S. Maruyama, Carbon isotopic signatures of individual Archean microfossils(?) from Western Australia. Int. Geol. Rev. **40**, 196–212 (2001)

Y. Ueno, Y. Isozaki, K.J. McNamara, Coccoid-like microstructures in a 3.0 Ga Chert from Western Australia. Int. Geol. Rev. **48**, 78–88 (2006a)

Y. Ueno, K. Yamada, N. Yoshida, S. Maruyama, Y. Isozaki, Evidence from fluid inclusions for microbial methanogenesis in the early Archean era. Nature **440**(23), 516–519 (2006b)

M.J. Van Kranendonk, Volcanic degassing, hydrothermal circulation and the flourishing of early life on Earth: A review of the evidence from c. 3490-3240 Ma rocks of the Pilbara Supergroup, Pilbara Craton, Western Australia. Earth Sci. Rev. **74**, 197–240 (2006)

M.J. Van Kranendonk, A.H. Hickman, I.R. Williams, W. Nijman, Archaean geology of the East Pilbara Granite-Greenstone Terrane Western Australia—a field guide. Geological Survey of Western Australia, Record 2001/9, Perth, 2001, 134pp

M.A. Van Zuilen, M. Chaussidon, C. Rollion-Bard, B. Marty, Carbonaceous cherts of the Barberton Greenstone Belt, South Africa: Isotopic, chemical and structural characteristics of individual microstructures. Geochim. Cosmochim. Acta **71**(3), 655–669 (2007)

M.M. Walsh, Microfossils and possible microfossils from the early Archean Onverwacht Group, Barberton Mountain Land, South Africa. Precambr. Res. **54**, 271–293 (1992)

M.R. Walter, Stromatolites and the biostratigraphy of the Australian Precambrian and Cambrian. Paleont. Assoc. Lond. Spec. Pap. **11**, 190 (1972)

M.R. Walter, Introduction, in *Stromatolites*, ed. by M.R. Walter. Developments in Sedimentology, vol. 20 (Elsevier, Amsterdam, 1976a), pp. 1–3

M.R. Walter (ed.), *Stromatolites*. Developments in Sedimentology, vol. 20 (Elsevier, Amsterdam, 1976b), 790pp

M.R. Walter, J.P. Grotzinger, J.W. Schopf, Proterozoic stromatolites, in *The Proterozoic Biosphere*, ed. by J.W. Schopf, C. Klein (Cambridge University Press, New York, 1992), pp. 253–260

M.R. Walter, J. Bauld, T.D. Brock, Microbiology and morphogenesis of columnar stromatolites (Conophyton, Vacerrilla) from hot springs in Yellowstone National Park, in *Stromatolites*, ed. by M.R. Walter. Developments in Sedimentology, vol. 20 (Elsevier, Amsterdam, 1976), pp. 273–310

D.T. Wright, W. Altermann, Microfacies development in Late Archaean stromatolites and oolites of the Campbellrand Subgroup, South Africa, in Carbonate Platform Systems. Components and interactions, ed. by E. Insalco, P.W. Skelton, T.J. Palmer. Geol. Soc. London, Spec. Publ., vol. 178 (2000), pp. 51–70

# Biogenerated Rock Structures

## W.E. Krumbein

Originally published in the journal Space Science Reviews, Volume 135, Nos 1–4.
DOI: 10.1007/s11214-007-9289-9 © Springer Science+Business Media B.V. 2007

**Abstract** Earth as a planet under firm control of life processes since more than 3 Ga has evolved global biogeochemical cycles, biogeomorphogenetic processes and structures also called plates or global tectonics including global climate and movement of water masses. These processes have deep impact on the shape and thickness of continental land masses as well as the chemistry and mineralogy of the crust and upper mantle. Biogenerated rock structures in this sense can be visualised through the analysis of sedimentary rock structures exhibiting e.g., biogenerated stromatolites, onkolites, oolites or cementing structures of sandstones which clearly preserve biochemical processes and biophysical structures. Further the chemical composition including the segregation of mineral layers, ore deposits, sedimentary and metamorphic rocks and granites hint to sun powered energy storage (gas, gas hydrates, hydrocarbons, coal) and tectonic processes initiated or at least modified through the enormous input of external energy through reduced carbon and iron compounds. One can state with considerable reliability that a planet under control of life must exhibit rock chemistry, mineralogy and structures typical for the impact of life on the geodynamic cycles. This includes the idea of top-down geotectonics instead of bottom-up processes.

**Keywords** Biogeomorphogenesis · Biogeochemistry · Geotectonics · Oolites · Onkolites · Platonics stromatolites

## 1 Introduction

Imagine a scenario where intelligent alien life forms, through interstellar travel, approach an unknown planet in search of life. First, they would screen for unusual modulations of the electro-magnetic field, i.e. intelligent background noise or pollution. Second, after determining the planet's composition—gaseous, liquid, or solid—the research team would take a closer look for plant, animal, or other kinds of morphotypes; their analysis would include differentiating between physical and mineralogical products. If the team decides to land—or

W.E. Krumbein (✉)
Geomicrobiology, ICBM, Carl von Ossietzky Universität Oldenburg, 16111 Oldenburg, Germany
e-mail: wek@uni-oldenburg.de

perhaps before landing—they would conduct a more detailed search that might yield living bacteria, micro-algae, fungi or alike.

Let us assume, however, that the planet under examination is dead! No TV, no radio, neither plant nor animal, not even a single living bacterium can be detected with all available techniques.

Will the alien research team now mark the planet as uninhabitable? Rather no. Planets designated uninhabitable may have been habitable in the past or could become habitable in the future, depending on the evolution and circumstances of the star around which the planet revolves. Therefore the team will most probably turn to classical approaches of planetary science (perhaps also to geophysiology). The team will examine rock samples for fossils, petrified traces of ancient life. Let us assume the planet under investigation is Terra in the Solar system. Let us further assume a planetoid or comet impact occurred 3 billion years after Terra's accretion. The impact was so severe, that all established life was abruptly extinguished. Such scenarios are well known as "snow ball" world or "impact" derived extinctions in our own scientific jargon.

The question is: If such an event happened at the end of the Precambrian, or even 700 million years ago, will the research team from afar be able to unravel the planet's life history? Will they find evidence of the biofilm, microbial mats, or the incredible, active photosynthesis cycles and rotations of biologically animated carbon atoms by analyzing the rocks remaining in the planetary crust since the catastrophe? How would this be achieved? Three (or more steps) of investigation would be promising.

- The search for physical fossils (body fossils) in rock materials
- The search for chemical biosignatures in rock materials
- The search for physical (mechanical, structural) biosignatures in rock materials.

In this article we will only briefly touch on the first two possible approaches and focus mainly on the third one, namely physical signatures of unequivocal biologically generated structures, morphologies, or minerals. Last but not least, global structures or platonics will be discussed as unequivocal signs of former or extant life. This will be derived from the ratio of reduced and oxidized compounds in the solid crust of the planet as well as from the resulting morphology.

## 2  Fossils

The geological and paleontological literature is full of attempts to identify and to describe microorganisms in ancient sediments. Many of the structures described and published turned out to be artifacts, erroneous outcome of microscopy and misinterpretation of simple morphologies. The same is true for the multitude of reports on bacteria or other structures detected in meteorites of Martian origin or from farther away. Actually, one can cast doubt on many fossils reported for the first two billion years of life on Earth.

## 3  Chemical Signatures

Following the old idea of dissymmetry—forwarded by Pasteur and Curie and fully developed by Jasper and Vernadsky in the last decade of the nineteenth and first decades of the twentieth centuries (Levit et al. 1999)—many researchers have looked into the dissymmetry of stable isotopes, especially those of a high biological migration potential, namely carbon, sulphur, and oxygen (van Zuilen 2007; Ono 2007). During metabolic activity, these isotopes are fractionated frequently in a way that deviates from normal physical–chemical conditions. Often the lighter isotope is enriched in cell material. This enrichment may be preserved in the dead cell material or organic compounds derived from living systems. Many if not all these signatures, however, have turned out to be ambiguous and could be explained by special climatic conditions or other processes not related to life. Thus, it turned out to be useful to study structures and physical conditions in addition to the search for fossil microbes and their isotope-fractionating activities.

## 4 Biogenerated Rock Signatures

Physical or physical–chemical signatures of life are multiple and absolutely convincing in the case of macroorganisms. Some classical examples are the depression and—at a certain distance—the pile of spaghetti-like excretions produced by *Arenicola marina* or the worm-like tubes of *Rhizocorallium*. Many other examples from the macroscopic end of the mesoscopic physical world could be given. One of the truly enigmatic biosignatures is the *Microcodium* complex in fossil soils. Another very debated set of biosignatures visible to the naked eye are the root and fungal network signatures in agate samples (coloured and banded silica or chert concretions sold in many mineral shops all over the world). The biological remains are repeatedly embedded in cavities in relatively cool silica exhalations.

However, in our initial scenario, life on Earth was abruptly extinguished before multicellular organisms established themselves. Therefore the exploration team must turn to more hidden and cryptic structures and signatures.

## 5 Microbial Growth Structures (Including Kerogen and Carbonate Deposits)

The first reports on such structures apparently, and most astonishingly, are based on observations from the Germanic Trias around the Hercynian Mountains in the centre of Germany. The wandering scholar and Medicus Philipp von Hohenheimb, who named himself Paracelsus (around 1548), reported some interesting structures in a clear mountain creek. He saw a slime forming on some granite pebbles and concluded that this slime transformed into calcium carbonate, cementing the pebbles. He even confirmed his field observations by laboratory experiments in cucurbites (Erlenmeyer flask-like vessels). Thus, under his eyes, biological activity transformed loose pebbles into a carbonate-cemented conglomerate. Many generations of geoscientists, however, claimed that living matter was not necessary to produce such rock structures. Some years later another Medicus (Brueckmann 1721) described some concentric structures previously known as roe stones or frog eggs; these were also depicted by Hooke (1665) as "Kettering stone". Brueckmann first named them oolites and regarded them as made ex vivo. It took another 125 years until Ludwig and Theobald (1852) described the formation of these oolite grains within algal carpets in the thermal waters of Bad Nauheim and demonstrated that the grains, which formed in the algal carpet, were washed away and redeposited downriver upon the decay and elimination of the algal cell material. Kalkowsky (1908) visited the same localities in the Harz Mountains and named for the first time the whole ensemble of these biosedimentary structures. He described very precisely what he called stromatolites (rocks made of individual stromatoids) and oolites (rocks made of individual ooids and ooid bags). Following Ludwig and Theobald (1852), he attributed all these very conspicuous and characteristic sedimentary rock structures to microorganisms.

**Fig. 1** A sandy beach transforms in beachrock (*upper left*) and stromatolite and onkolite (*lower part*)

From then on research on stromatolites and stromatolitic structures, ooids and oolithic structures expanded tremendously. However, especially for ooids and oolites, the discussion went on whether or not biological processes play a role in the formation of these characteristic structures. There is now almost general agreement, however, about the statement of Carl von Linné that the petrifacts in carbonate rocks are not generated in the rock. On the contrary, the carbonate rocks have been generated by the organisms found within them ("omnis calx ex vivo"). Today we know that microorganisms, instead of the embedded macrofossils and their carbonate skeletons, generate most carbonate rocks on earth.

An elegant experimental approach was described by Brehm et al. (2003). They convincingly showed that a complex community of one cyanobacterial species, one diatom species and about a dozen species of heterotroph bacteria organise themselves in a special way to produce at first organic spheres, which later are filled with diatoms and cyanobacteria, subsequently calcifying into ooids. This community is very specific and well organised. Bacteria arrange themselves in spheres, form an organic coated balloon filled with water and surrounded by water, not unlike the macroscopic cyanobacterial spheres of Mares Egg (Oregon) and some Karelian lakes (Russia). These, however, have never been described as calcifying. Thus Paracelsus (1976), Brueckmann (1721), Ludwig and Theobald (1852), and last but not least Kalkowsky (1908) have laid the foundation for the detection, study and char-

**Fig. 2** Step one: mat formation and disintegration according to nutrient and pore water supply

acterisation of biosedimentary signatures derived from subaquatic biofilms or stromatolitic microbial mats with their astonishing individual morphologies, which as a whole can only be produced by living organisms. The structures, however, usually contain no traces of organisms and one individual lamina (stromatoid) of any stromatolite may represent 100 years of microbial mat formation in annual cycles of production and decay (Krumbein et al. 1977; Walter 1976). The controversial discussion on ooids and oolites need not be repeated here. The literature is vast (Dahanayake et al. 1985). One example of in situ formation of biosedimentary structures is given from the coast of the Gulf of Aqaba (Figs. 1, 2, 3 and 4).

## 6 What then Is Important for Our Alien Planet Exploration Team?

Usually biogenic stromatolites and biogenic oolites (still questioned by many authors according to origin or place of formation—biogenic/abiogenic? planktonic/benthonic?) are described from limestone environments. Many siliciclastic structures, however, have been described as well (Figs. 5, 6 and 7).

**Fig. 3** Sand dollar like structures
emerge and stabilise thin layers
of sand in irregular patterns

Examples of ferruginous or phosphatic stromatolites and oolites also exist (Dahanayake and Krumbein 1986); many stromatolite-like structures may be abiogenic as well (Krumbein 1983).

Stromatolites and oolites are usually attributed to cyanobacteria. However, many stromatolites have been identified as being generated by fungi or filamentous nonphotosynthetic bacteria (Krumbein 1983; Gerdes and Krumbein 1987). In exceptional cases, stromaolitic structures also emerge from macroorganisms (Krumbein, 1963, 1983).

Furthermore, the study of recent microbial mats focuses too much on the organisms themselves and less on their envelopes, often called extracellular polymeric substances (EPS). As a matter of fact, EPS are more resistant to decay then the organisms embedded in them. The mechanical stability of EPS is astonishing, comparable to spider web material in the sense that it remains unmatched by any synthetic product.

Finally there is a lack of detailed comparative studies of the contrast between sedimentary rock structures, derived from transport of grains and deposition, and the completely different situation of "Aufwuchs" structures. This term defines the growth of sedimentary rocks from "bottom up" instead of from "top down". Thus, in reality, many sedimentary rocks are not really sediments. They represent biological growth, into which grains are incidentally embedded and captured. Such structures differ considerably from

**Fig. 4** Upon anaerobic total decay of the cyanobacteria biofilm, Onkolites calcify

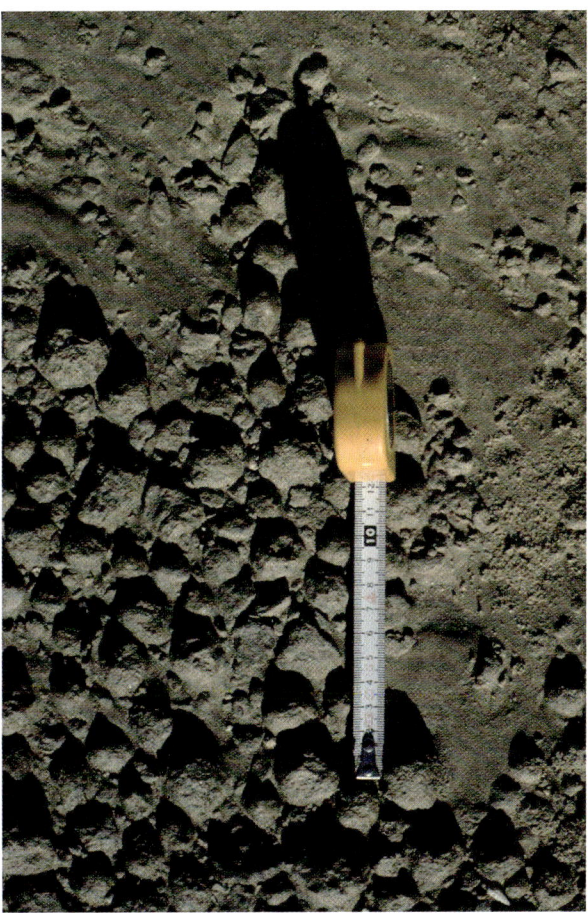

a sedimentation; i.e., a purely physical deposition by gravity (Wachendörfer et al. 1994a, 1994b). In Fig. 6, dark angular quartz grains are floating freely within the extracellular polymeric substances (EPS) produced by a complex microbial community. The EPS were produced in excess amounts for many physiological and ecological reasons. These substances are more rigid and pressure resistant than many minerals, and possess porosities or special cementing minerals that may be generated only in very late stages of sedimentary diagenesis (sometimes only when heated above 100°C and after undergoing some Maillard reaction transformations). Thus very peculiar rock structures are formed in carbonates and sandstones; these, in turn, are regarded as signals of biological growth even when totally transformed into granites. Vernadsky (1929), and more recently many others (Arp et al. 2001; Anderson, 1984, 2006; Krumbein, 1983, 1996; Krumbein et al. 2006; Rosing et al. 2006) discussed these views of a biologically structured crust model.

The most ancient documents of life on Earth are, apparently, the biosedimentary impacts and traces of biofilms or microbial mats. A biofilm, if subaquatic, can be regarded as 99% solid water stabilised and immobilised by large amounts of extracellular compounds (EPS) excreted by few microorganisms (bacteria, cyanobacteria, algae, or fungi). These films could really be called a tissue and exhibit even more specialized structures than animal or plant tissue studied in much more detail (e.g., Walter 1976; Brehm et al. 2006).

**Fig. 5** First demonstration of a biofilm generated sedimentary structure (*Flora Danica*, 1813)

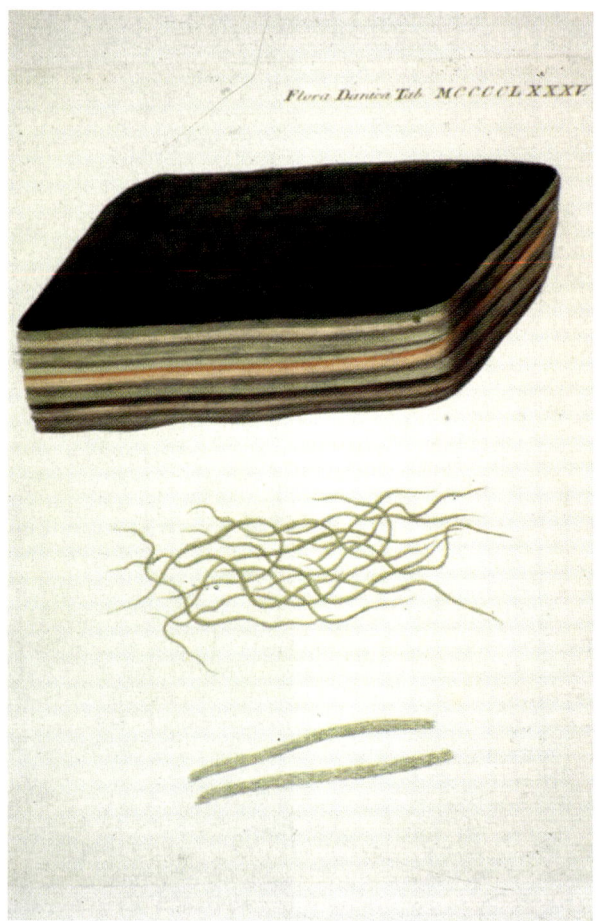

In contrast, a subaerial biofilm can be regarded as highly structured 99% cellular or extracellular material (EPS) surviving and spreading with a minimum supply of water. Dark zones in, e.g., Fig. 8 are spaces not calcified because EPS was characteristically not calcifying in this part of the tissue-like build-up of the cyanobacterium *Scytonema* which forms in the irregular (sporadic) intertidal areas of alpine lakes.

Both types of biofilms and, in addition, microbial networks (biodictyon) within rocks and rock cavities may produce minerals and morphological patterns that differ considerably from purely physically produced structures and morphotypes (Fig. 8). The morphological and topological difference between a physical deposit (varvite) and a biosedimentary deposit (stromatolite) is often difficult to trace. Some authors suggested the term "stromatoloid" when typically laminated structures are observed without unequivocal evidence of microbiota as a source. Furthermore, biofilm-generated ooids and oolites, onkoids and onkolites were described and could also be assigned different names, which would then clearly indicate the biogenic or physical origin of the morphotype in question.

When regarding fossil and present-day biospheres, it is astonishing that for more than 80% of the existence of life on Earth (3.7 Ga) these communities were thriving in a biosphere stabilised for Eons. Only 0.5 Ga ago, complex communities of macroorganisms emerged and associated symbiotically with biofilms to form grasslands, trees, and forests or coral reefs to

**Fig. 6** Sand grains without any contact will get fossilised swimming in carbonate cement

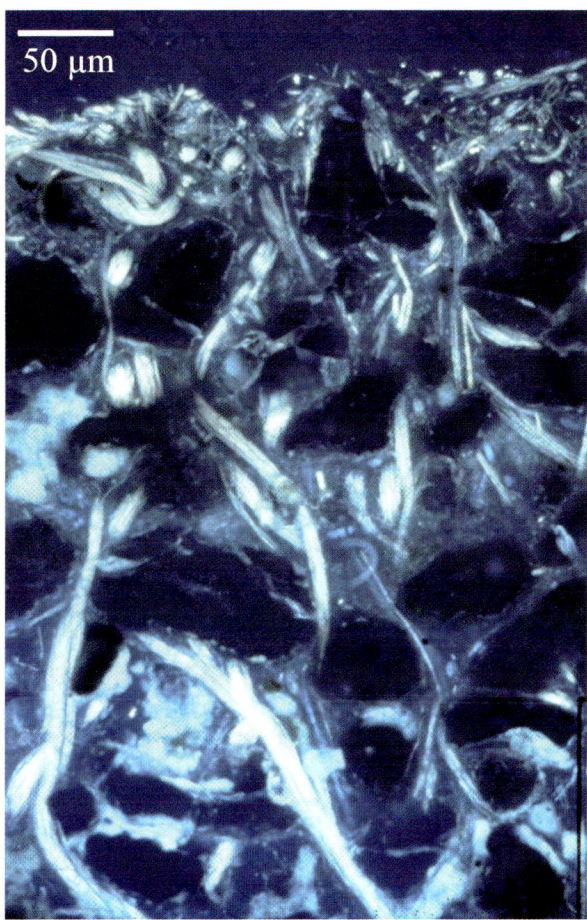

50 µm

name a few examples of these symbiotic mesoscopic success stories. Naturally, this would include rocks or former soils enclosing remains of Coca-Cola bottles.

Subaquatic, subterranean (rock inherent), and subaerial biofilms need to be considered separately. In a global sense biofilms can be regarded as a kind of a super tissue, and para-histology could be applied to these peculiar structures (Levit et al. 1999; Wachendörfer et al. 1994b). The terms biofilm, microbial mat and stromatolite—which describe complex multilayered biofilms creating, designating, and destroying rocks—were only coined in the twentieth century. The first scientific descriptions of such communities and their rock-creating potential stem from the *Flora Danica*, initiated in 1761 by G.C. Oeder. The first polychrome picture of a microbial mat or stromatolite was published in 1813 in issue 25 of the *Flora Danica* (Fig. 5). *Oscillatoria*, and later more correctly *Microcoleus chthonoplastes*, was named soil and rock forming because it was observed that leathery biofilms created by this microorganism could build islands that rose above sea level; not unlike coral reefs built by atoll islands in tropical seas. Considering the evolution of biosedimentology, it is worthwhile to note that studies on stromatolitic structures were reported almost 50 years before Ehrenberg and Darwin published their monographs on coral reef growth as a source of biosedimentary rock structures.

**Fig. 7** Balls of silt and clay are formed by biofilm growth. No microbes left in internal structure

Geosciences unfortunately concentrated on macroorganisms in terms of sedimentology and stratigraphy for more than 150 years. The early work of the Danish (Odense) group, of Ehrenberg (1854) and of Kalkowsky (1908) remained unnoticed until Krumbein (1983), Walter (1976), and others attracted attention to biosedimentary rocks made by biofilms. Consequently, the notion of benthic Oolites, onkolites and stromatolites being constructed by the same microbiota and stromatolitic biofilms was studied and acknowledged. Mud cracks, table cloth folding, Tepee structures were matched with their biosedimentary counterparts such as Petee, Elephant skin and complex biolaminated negative or positive cones and stylolites (Friedman and Krumbein 1985; Gerdes and Krumbein 1987; Krumbein et al. 2004).

Microbial-induced sedimentary structures (MISS) were introduced into the literature by N. Noffke, a student of Seilacher, Gerdes and Krumbein (Noffke et al. 2001). The pattern of structures and individual morphotypes were more clearly defined and made visible in recent and ancient sedimentary systems. Terminology, nomenclature and the history of research into biofilms were sketched out more precisely. Special attention should be given, however, to typical patterns and morphological characteristics, which may enable a clear distinction between purely physical structures and structures that stem from microbially

**Fig. 8** Laminated structure created by *Rivularia* (Attersee, Austria) lacking fossil microbes

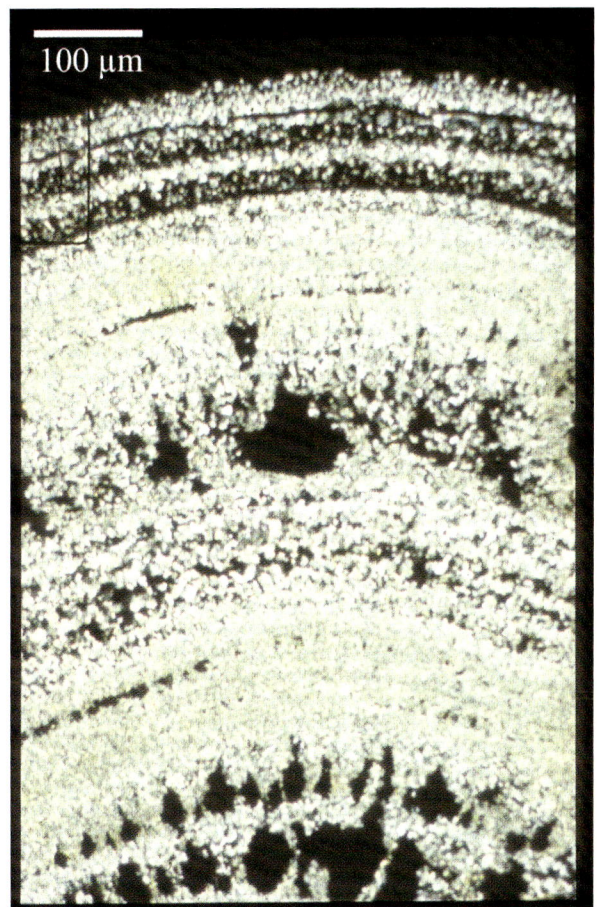

100 μm

induced sedimentary structures (MISS) and microbially altered rock surfaces (MARS). Both MISS and MARS may be measured using different approaches and then transformed into search algorithms for the detection of ancient and extant biogenic rock formation and rock destruction patterns on a purely mathematical measurement base (Kempe et al. 2005; Rodenacker et al. 2003).

## 7 Global Biogenic Structures and Processes

We now return to the alien exploration team. The team went to a new solar system immediately after a first glance at the albedo, the cloud systems in the atmosphere and some glances at the distribution of land and water. Naturally they are more advanced in recognizing the impact of living matter than, say, Lomonosov, Kant, Herder or Vernadsky (to name a few). Thus, the team derived from atmosphere composition, cloud pattern, oceanic currents and the size and distribution of land masses, that the planet was transformed by living matter. The report for the books and the mother planet thus could have been:

> The third planet of a small yellow star in galaxy WEK3.14.37 is (or was) controlled and organised by life processes. Living matter exists (existed) in sufficient amounts

to stabilise planet surface temperature and conditions. The migration of atoms under control of living matter embraces approximately 7% of the planetary mass. Unfortunately, the migration of atoms and structural evolution of morphologies stagnated at a level exhibiting lack of intelligent "noosphere" type developments. Life support is (was) good. However, control of the moon and orbiting around the sun, as well as organised use of the solar energy source are (were) rudimentary and changes in this situation seem improbable (are impossible). On the other hand, the turnover of large amounts of mineral and rock in well-equilibrated mixtures of aqueous and solid portions of the surface layers via carbon-based living matter is well documented. The manipulation of large portions of the crust and water masses exists, but is (was) not selective and flexible enough to guarantee long-term survival. The planet may lose (lost) and is thus not eligible for the list of living planets. We suggest the category habitable.

## 8 Summary

Platonics, i.e. the biologically controlled organisation of large masses of solid matter in a "top down" geotectonic way powered by sun energy (Anderson 1984, 2006; Krumbein 1983; Krumbein et al. 2006; Krumbein and Schellnhuber, 1990, 1992; Rosing et al. 2006) are an outcome of biological migration of atoms through living matter. These processes are powered by external (sun) energy in a top down way, feeding chemical and physical (mechanical and morphological) differences into the upper 7–10% of the planet in order to guarantee energy, nutrient and temperature equilibrium in a global biogeomorphological sense. This combined with the action of living matter guarantees an extension of the window of life on planet Earth and its continued habitability. This way biosedimentary structures extend into global bio-platonic morphologies, which can be analysed as biologically driven processes and patterns from long-distance mapping of the planet's surface. Platonics was derived by Anderson (2006) from plates, tectonics and perhaps even from cratons, all three representing ambiguous terms of present-day Earth system science. The complex composition and evolution (physically and chemically) of large land masses erroneously, termed plates, as contrasted with huge bodies of water and their conveyor functions of temperature and nutrients cannot be explained by simple mechanic functions (implied by the words "plate" or "plate tectonics"). All sediments and sedimentary structures of planets bearing life will be characteristically modified and altered by the impact of living matter. The vertical and horizontal movements and turnover processes of huge amounts of minerals and rocks in life-dependent cycles of geological time dimensions could be called the "breath of Earth", or could be compared to a body with bones, tissue, vessels and nerves as a metaphor to describe the living Earth. Such descriptions were very popular from the Renaissance until the Enlightenment (sixteenth to eighteenth centuries). These cyclic and sometimes cataclysmic changes driven by living matter are detectable from far in space. The migration of atoms, ions and minerals in space and time via living matter involves the uppermost 80–130 km of the planet or 7–10% of its mass. Geophysiology is a more living description for biosignatures on sediments and continents and their fate in space and time then the presently broadly used term earth system science (Krumbein, 1983, 1993b, 1990, 1996).

## References

D.L. Anderson, The Earth as a planet: paradigms and paradoxes. Science **223**, 347–354 (1984)

D.L. Anderson (2006). http://www.gps.caltech.edu/~dla/

G. Arp, A. Reimer, J. Reitner, Photosynthesis-induced biofilm calcification and calcium concentrations in Phanerozoic oceans. Science **292**, 1701–1704 (2001)

U. Brehm, W.E. Krumbein, K.A. Palinska, Microbial spheres: a novel cyanobacterial-diatom symbiosis. Naturwissenschaften **90**, 136–140 (2003)

U. Brehm, W.E. Krumbein, K.A. Palinska, Biomicrospheres generate ooids in the laboratory. Geomicrobiol. J. **23**, 545–550 (2006)

F.E. Brueckmann, Specimen physicum exhibens historiam naturalem, oolithi seu ovariorum piscium & concharum in Saxa Mutatorum, Helmestadii, Salomoni & Schnorrii (1721), 21p

K. Dahanayake, W.E. Krumbein, Ultrastructure of a microbial mat-generated phosphorite. Miner. Depos. **20**, 260–265 (1986)

K. Dahanayake, G. Gerdes, W.E. Krumbein, Stromatolites, oncolites and oolites biogenically formed in situ. Naturwissenschaften. **72**, 513–518 (1985)

C.G. Ehrenberg, *Mikrogeologie* (Voss, Leipzig, 1854), 374p

Friedman, W.E. Krumbein (Eds.), *Hypersaline Ecosystems. The Gavish Sabkha* (Springer, Berlin, 1985), 484pp

G. Gerdes, W.E. Krumbein, *Biolaminated Deposits* (Springer, Berlin, 1987), 183p

R. Hooke, *Micrographia, or Some Physiological Descriptions of Minute Bodies Made by Magnifying Glasses, with Observations and Inquiries thereupon* (Martyn & Allestry, London, 1665), 246p

E. Kalkowsky, Oolith und Stromatolith im Norddeutschen Buntsandstein. Zeitschrift der deutschen Geologischen Gesellschaft **60**, 86–125 (1908)

A. Kempe, U. Brehm, W. Bunk, A.A. Gorbushina, F. Jamitzky, K. Rodenacker, R.W. Stark, W.E. Krumbein, W.M. Heckl, EGU 2005. Geophys. Res. Abstr. **7**, 06655 (2005)

W.E. Krumbein, Über Riffbildung von *Placunopsis ostracina* im Muschelkalk von Tiefenstockheim (Marktbreit) in Unterfranken. Abh. Naturwiss. Ver. Würzburg **4**, 1–15 (1963)

W.E. Krumbein, Stromatolites – The challenge of a term in space and time. Precambr. Res. **20**, 493–531 (1983)

W.E. Krumbein, Der Atem Cäsars. Mitt. Geol.-Paläont. Inst. Univ. Hamburg **69**, 267–301 (1990)

W.E. Krumbein, Geophysiologie, Klima und Biogeomorphogenese. Eine späte Würdigung der Physikotheologie des jungen Immanuel Kant. in Ruprecht-Karsl. Uni. Heidelberg, Hrsg.: Klima: Studium Generale 1992/93, Heidelberg, 1993b, pp. 141–163

W.E. Krumbein, Geophysiology and parahistology of the interactions of organisms with the environment. Marine Ecol. **17**, 1–21 (1996)

W.E. Krumbein, H.J. Schellnhuber, Geophysiology of carbonates as a function of bioplanets, in *Facets of Modern Biogeochemistry*, ed. by A.V. Ittekott, S. Kempe, W. Michaelis, A. Spitzy (Springer, Berlin, 1990), pp. 5–22

W.E. Krumbein, H.-J. Schellnhuber, Geophysiology of mineral deposits – a model for a biological driving force of global changes through Earth history. Terra Nova **4**, 351–362 (1992)

W.E. Krumbein, Y. Cohen, M. Shilo, Solar Lake (Sinai) 4. Stromatolitic cyanobacterial mats. Limnol. Oceanogr. **22**, 635–656 (1977)

W.E. Krumbein, A.A. Gorbushina, E. Holtkamp-Tacken, Hypersaline microbial systems of Sabkhas examples of the thrive of life for survival. Astrobiology **4**, 450–459 (2004)

W.E. Krumbein, W. v. Bloh, S. Franck, H.-J. Schellnhuber, Bacteria rule the world – a survey of planetary tectonics and life. 1st European Planetary Science Congress, Berlin, Abstracts, 2006, p. 244

G. Levit, A.A. Gorbushina, W.E. Krumbein, Geophysiology and Parahistology of benthic microbial mats with special reference to the dissymmetry principle of Pasteur-Curie-Vernadskij, in *Marine cyanobacteria*, Bulletin de l'Institut Océanographique, Monaco, special issue 19, 1999, pp. 175–196

R. Ludwig, G. Theobald, Über die Mitwirkung der Pflanzen bei der Ablagerung des kohlensauren Kalkes. Annalen der Physik und Chemie **87**, 91–107 (1852)

N. Noffke, G. Gerdes, T. Klenke, W.E. Krumbein, Microbially induced sedimentary structures – a new category within the classification of primary sedimentary structures. J. Sediment. Res. **71**, 650–656 (2001)

Ono, Space Sci. Rev. (2007, this issue). doi:10.1007/s11214-007-9267-2

T.B. Paracelsus, in *(1493–1541) Werke V. Pansophische, magische und gabalische Schriften*, ed. by W.-E. Peuckert (Schwabe & Co., Basel, 1976)

K. Rodenacker, B. Hausner, A.A. Gorbushina, Quantification and spatial relationship of microorganisms in subaquatic and subaerial biofilms, in *Fossil and Recent Biofilms*, ed. by W.E. Krumbein, D.M. Paterson, G.A. Zavarzin (Kluwer, Dordrecht, 2003), pp. 387–399

M. Rosing, D.K. Bird, N.H. Sleep, W. Glassley, F. Albarede, The rise of continets – an essay on the geologic consequences of photosynthesis. PALAEO **232**, 199–213 (2006)

V.I. Vernadsky, *La Biosphére* (Alkan, Paris, 1929), 232pp

Wachendörfer, W.E. Krumbein, H.J. Schellnhuber, Bacteriogenic Porosity of marine sediments – A case of
     biomorphogenesis of sedimentary rocks. in *Biostabilization of Sediments*, ed. by W.E. Krumbein, D.M.
     Paterson, L.J. Stal (BIS Oldenburg, 1994a), pp. 203–220
Wachendörfer, H. Riege, W.E. Krumbein, Parahistological sediment in thin sections. in *Biostabilization of
     Sediments*, ed. by W.E. Krumbein, D.M. Paterson, L.J. Stal (BIS Oldenburg, 1994b), pp. 257–277
M. Walter, *Stromatolites* (Elsevier, Amsterdam, 1976), 790p
van Zuilen, Space Sci. Rev. (2007, this issue). doi:10.1007/s11214-007-9268-1

# Morphological Biosignatures in Early Terrestrial and Extraterrestrial Materials

Frances Westall

Originally published in the journal Space Science Reviews, Volume 135, Nos 1–4.
DOI: 10.1007/s11214-008-9354-z © Springer Science+Business Media B.V. 2008

**Abstract** Biosignatures in early terrestrial rocks are highly relevant in the search for traces of life on Mars because the early geological environments of the two planets were, in many respects, similar and, thus, the potential habitats for early life forms were similar. However, the identification and interpretation of biosignatures in ancient terrestrial rocks has proven contentious over the last few years. Recently, new investigations using very detailed field studies combined with highly sophisticated analytical techniques have begun to document a large range of biosignatures in Early Archaean rocks. Early life on Earth was diversified, widespread and relatively evolved, but its traces are generally, but not always, small and subtle. In this contribution I use a few examples of morphological biosignatures from the Early-Mid Archaean to demonstrate their variety in terms of size and type: macroscopic stromatolites from the 3.443 Ga Strelley Pool Chert, Pilbara; a meso-microscopic microbial mat from the 3.333 Ga Josefsdal Chert, Barberton; microscopic microbial colonies and a biofilm from the 3.446 Ga Kitty's Gap Chert, Pilbara; and microscopic microbial corrosion pits in the glassy rinds of 3.22–3.48 Ga pillow lavas from Barberton. Some macroscopic and microscopic structures may be identifiable in an *in situ* robotic mission to Mars and *in situ* methods of organic molecule detection may be able to reveal organic traces of life. However, it is concluded that it will probably be necessary to return suitably chosen Martian rocks to Earth for the reliable identification of signs of life, since multiple observational and analytical methods will be necessary, especially if Martian life is significantly different from terrestrial life.

**Keywords** Early Archaean · Mars · Biosignatures · Stromatolites · Fossil microbial mats · Fossil microbial colonies

## 1 Introduction

The evolution of the early Earth as a habitat for microbial life has been recently reviewed by Westall and Southam (2006), Southam et al. (2007) and Southam and Westall (2007). The

F. Westall (✉)
Centre de Biophysique Moléculaire, CNRS, Rue Charles Sadron, 45071 Orléans cedex 02, France
e-mail: westall@cnrs-orleans.fr

early Earth was probably habitable as soon as the ocean temperatures descended to about 80°C, the temperature thought to be conducive to the association of the prebiotic molecules that formed the building bricks of life. Indirect evidence suggests that an ocean existed at about 4.4 Ga (Wilde et al. 2001) but the earliest documented traces of life occur in rocks from Greenland, South Africa and Australia that formed one billion years after the formation of the Earth. Highly metamorphosed, 3.7–3.8 Ga old rocks from Greenland contain carbon having isotopic compositions similar to those produced by microbial fractionation processes (Mojzsis et al. 1996; Rosing 1999; Rosing and Frei 2004). However, the interpretation of these data as being of biogenic origin is complicated by the fact that the rocks have been subjected to younger contamination and the fact that abiogenic metasomatic processes could have also produced a similar isotopic composition (Westall and Folk 2003; van Zuilen et al. 2002, 2003). Very well-preserved sediments aged 3.3–3.5 Ga and affected by only very moderate prehnite-pumpellyite to lower greenschist metamorphism occur in the Pilbara in Australia and from Barberton in South Africa. These sediments do contain reliable morphological, isotopic and geochemical traces of early life.

The biosignature-hosting Early-Mid Archaean rocks and sediments are good analogues for the kinds of rocks that could have formed on early Mars, when it was "warmer and wetter" (Westall 2005a). Basalts are the most common rock type and the early martian sediments would have been volcaniclastic sands, silts and muds deposited in shallow to deeper water (below wave base) environments in standing bodies of water. Although Mars may not have had a global ocean, it certainly had sufficient water for there to have been a small sea/small ocean covering its northern hemisphere, as well as water-filled volcanic and impact craters (Clifford and Parker 2001; Bibring et al. 2006; Poulet et al. 2005). Having the same prebiotic organic carbon ingredients, other essential elements (HNOPS) and nutrients, liquid water, and energy sources, early Mars was as habitable as the early Earth and had the same types of habitable environments (McKay et al. 1992; Schock 1997; Southam et al. 2007). For example, on the early Earth, the surfaces of volcanic grains and vitreous rinds on lavas provided an inexhaustible source of chemical energy (plus nutrients) for chemolithotrophic microorganisms (Furnes et al. 2004, 2007; Westall et al. 2006a; Southam et al. 2007), whereas sediment/rock surfaces in the photic zone hosted anaerobic photosynthetic microorganisms that constructed tabular microbial mats and three dimensional stromatolites-like edifices reaching macroscopic dimensions in certain locations (Walter 1983; Byerly et al. 1986; Hofmann et al. 1999; Allwood et al. 2006, 2007; Westall et al. 2006a, 2006b). These microorganisms left traces of their existence preserved in lithified sediments. The traces include (1) morphological remnants, such as mineral casts or moulds of the physical structures of the microorganisms, minerals precipitated as a result of the activity of microorganisms, and corrosion features caused by microbial activity, (2) the carbonaceous molecules (albeit, degraded) that made up the original organism, and (3) the isotopic ratios of important biogenic elements, such as carbon, sulphur and possibly nitrogen. All these types of biosignatures are subject to further degradation, first of all during early diagenesis (this includes the activities of heterotrophic microorganisms on the original microbial materials), then during subsequent metamorphic processes throughout geological time, which can completely obliterate the original traces (but not always; cf. Schiffbauer et al. 2007; Bernard et al. 2007).

## 2 Morphological Biosignatures

Although this chapter will concentrate on the morphological biosignatures of early terrestrial life, in almost all cases the interpretations of biogenicity relies on analysis of a combination of biosignatures (morphological, chemical and isotopic), as well as on the local (micro-scale) habitability characteristics of the rock host (Westall 2005b). Note that the degree of reliability of biogenicity is enhanced by the complementary evidence provided by a variety of types of biosignatures (Westall and Southam 2006).

Morphological biosignatures are one of the most visually immediate signatures of life. They include the body fossils of microorganisms, their colonies and biofilms or mats (including stromatolites). Fossilisation of microorganisms occurs when mineral ions in solution complex to certain functional groups in the cell walls of microorganisms (living or dead) and to their exopolymeric substances. The minerals polymerise and thus form a crust around the organic structure, sometimes trapping the degraded remnants of the organic structures (Ferris et al. 1986; Westall et al. 1995; Cady and Farmer 1996; Phoenix et al. 2000; Benning et al. 2002; Toporski et al. 2002; Konhauser et al. 2003; Rancourt et al. 2005). Once embedded in a lithified matrix, the mineralised fossil remains of the microorganisms have the potential of being further preserved, depending upon the post-burial history of the sediment/rock. However, not all microorganisms are susceptible to fossilisation and cell wall composition plus the robustness of the individual cells to small changes in environmental conditions can influence the fossilisation potential of a microorganism (Westall 1997).

Chemolithotrophic microorganisms obtain their energy from redox reactions at the surfaces of reactive rocks and minerals. Volcanic rocks and minerals therefore represent sources of energy and also sources of essential elements, such as Mn, P, Mg and trace metals. In order to obtain these elements, the microorganisms secrete organic acids that dissolve the surface of the rock/mineral, leaving cavities that may be lined with carbonaceous matter, such as DNA, resulting from the microbial activity and surrounded by rock that is depleted in certain elements (including Fe, Mn, Ca and Na) (Thorseth et al. 1992, 2001; Alt and Mata 2000, and references therein). These traces can also be lithified and preserved for posterity (e.g. Furnes et al. 2004, 2007; Benzerara et al. 2006).

Finally, minerals can be indirectly precipitated by microorganisms whose metabolic activities can change the immediate local physico-chemical environment. The precipitation of calcium carbonate on microbial surfaces is one such example (Riding 2000). For example, the huge thicknesses of Proterozoic stromatolites, three-dimensional structures created by microbial mats in shallow, platform environments, were largely preserved as calcium carbonate deposits sometimes subsequently silicified, sometimes directly replaced by silica or iron oxide, e.g. the Gunflint Formation of Ontario, Canada (1.9 Ga) (Barghoorn and Tyler 1965; Hofmann and Schopf 1983).

## 2.1 Identifying Morphological Biosignatures in Early Terrestrial Rocks

The recent controversies surrounding the search for morphological traces of past life in some of the oldest terrestrial rocks (Schopf et al. 2002, 2007; Schopf 2006; Brasier et al. 2002, 2004), not to mention those related to the purported traces of life in the martian meteorite ALH84001 (McKay et al. 1996; Becker et al. 1999; Treiman 1998), have highlighted the difficulties involved in such an undertaking. The main problems of such studies are related to: (1) recent contamination and the establishment of the syngenicity of a purported microfossil with the rock encasing it, and (2) establishment of the biogenicity of the purported microfossil. Even with the most sophisticated instrumentation available in the laboratory, this task can be arduous and sometimes simply not conclusive if the potential biosignatures are particularly badly preserved. This situation furthermore underlines the difficulties involved in the search for biosignatures in extraterrestrial materials, whether *in situ* or in returned samples. However, these controversies have been useful in spurring new developments in the understanding of *bona fide* biosignatures. For instance, examples of advances that have been made include a better understanding of the size limits of cellular life (NRC 1999, on Size Limits of Very Small Microorganisms), mixed carbonaceous/mineral abiogenic bacteriomorphs (Garcia-Ruiz et al. 2003; Bittarello and Aquilino 2007), abiogenic production of carbon having isotopic signatures similar to those of microorganisms (van Zuilen et al. 2002).

As mentioned above, the most important aspect in the search for traces of life in ancient and extraterrestrial materials is the necessity of using a multitude of complementary studies in order to reach an acceptable level of reliability in any interpretation of biogenicity of a potentially fossiliferous structure (or structures). Examination of multiple preparations of the same sample can aid distinguishing recent contamination from *bona fide in situ* potential microfossils. Recent contamination tends to occur on the surfaces of natural fractures or within the pores of naturally porous rocks. Examination of freshly prepared fracture surfaces (i.e. not old and potentially contaminated fractures), as well as observation by optical and/or electron microscope methods of the microfossils within a thin section or polished rock surface, can help in the identification of microfossils that are syngenetic to the rock (c.f. Westall and Folk 2003; Westall et al. 2006a, 2006b). Interpretation of syngenicity is also related to a sound understanding of the local context and habitability of the environment: are the macroscopic to microscopic environmental conditions of the host rock conducive to the kinds of microorganisms found within it? For example, structures interpreted as the fossil remains of oxygenic photosynthetic microorganisms were described from a rock that was subsequently identified as a hydrothermal chert vein (Schopf 1993; Brasier et al. 2002). It is evident that hydrothermal veins do not represent a suitable environment for photosynthesising microorganisms, although the latter occur in hot spring pools fed by such conduits, as in Yellowstone hot springs (Madigan et al. 2002). Microbial habitats may be wide-ranging in size, encompassing small areas of tens up to a few hundreds of micrometers in size (for instance a microcavity in a rock or the pore space in water-logged sediments), to much larger environments of the order of hundreds of meters to even kilometers in extent (for example, mudflats, carbonate platforms, the floors of lagoons or lakes) (Gerdes et al. 1993; Southam et al. 2007; Krumbein et al. 1994; Noffke et al. 1997).

A number of criteria have been established for the identification of the biogenicity of a potential microfossil (Buick 1990; Schopf 1993; Westall 1999; Cady et al. 2004; Westall and Southam 2006; Brasier et al. 2006). These criteria were incorporated in the assessment of Westall and Southam (2006) of which a resume is provided below.

Biogenicity criteria are related to both the structural aspects of a cell (i.e., the morphology of a cell, its colony and its biofilm, or mat) as well as to its living processes (i.e., the chemical composition of its constituent parts, the fractionation of various elements, such as C, S, and possibly N). They include: (i) the morphological aspects of individual microorganisms (size, shape, cell division, cell death, cell envelope texture, flexibility); (ii) colonial and biofilm characteristics (association with a number of other organisms of the same species, association with different species in the vicinity (consortium), association with microbially-produced polymer (extracellular polymeric substances, EPS); (iii) evidence of interaction between individual microorganisms and their biofilms with their microenvironment (biolaminations, e.g. Gerdes et al. 1993; Noffke et al. 1997; microbial corrosion, e.g. Fisk et al. 1998; Thorseth et al. 1992, 2001; and microbial mineral precipitation, e.g. Lowenstam and Weiner 1989); and (iv) biogeochemical characteristics including (a) biomarkers molecules (e.g. Summons et al. 1999; Brocks et al. 1999, 2003), (b) bio-elements, such as N, S, and P (e.g. Furnes et al. 2004, 2007), (c) associated heavy metals and REEs (*N.B.* heavy metals and REEs can fix to the surface of microorganisms and their EPS, or can be concentrated by them e.g. Ferris et al. 1988; Mullen et al. 1989; Kamber and Webb 2001; Webb and Kamber 2000), (d) C, S, and possibly N isotopes (e.g. Schidlowski 1988; Lyons et al. 2003).

Problems with biosignatures may arise, such as morphological similarity with non-biogenic bacteriomorphs (e.g. Yushkin 2000; Garcia-Ruiz et al. 2003 (in the latter case, the nonbiogenic bacteriomorph also exhibited a carbon isotopic signal)), contamination of ancient biomolecular signatures by more recent biomolecules (e.g. Brocks et al. 1999), or inorganic processes producing isotopic signatures similar to those produced by microbial fractionation (e.g. van Zuilen et al. 2002, 2003); hence the desirability of using a multiplicity of biosignatures (morphological, chemical and isotopic) to increase the reliability of biogenicity interpretations.

With respect to potential morphological biosignatures on the early Earth (and possibly Mars), the size of the early microorganisms was influenced by the anaerobic environment in which they existed. The anaerobic metabolisms used by early microorganisms were less efficient in terms of energy production than oxygenic metabolisms, such as oxygenic photosynthesis (Des Marais 2000). The amount of biomass produced, for instance, by chemolithotrophic microorganisms that obtain both their carbon and energy from inorganic sources is generally far smaller than that produced by photosynthesisers (anoxygenic or oxygenic) and therefore their physical and chemical expressions are generally very subtle (Westall and Southam 2006; Southam et al. 2007; Southam and Westall 2007) (*N.B.* there are exceptions to this "rule of thumb" and there are chemolithotrophic microorganisms, such as *Thiomargarita*, that are larger than photosynthetic ones). For instance, coccoidal microfossils from a 3.446 Ga chert from the Pilbara range in size from 0.4 to about 0.8 μm form colonies on the orders of tens of microns in size whereas filaments up to tens of microns long are between 0.25–0.3 μm wide (Westall et al. 2006a). On the other hand, Early Archaean microbial mats (including stromatolites) formed on rock/sediment surfaces in the photic zone by probable anaerobic photosynthesisers can reach macroscopic sizes (millimeters to centimeters) (Byerly et al. 1986; Hofmann et al. 1999; Westall 2004; Allwood et al. 2006, 2007). Apart from these large-scale constructions, it is the very subtleness of many of the biosignatures in the Early to Mid Archaean rocks, small size and often low carbon contents, that makes their observation, analysis and interpretation difficult.

## 3 Morphological Biosignatures from the Early-Mid Archaean

Table 1 lists the published reports of morphological microfossils in Early-Mid Archaean formations. It is not the aim of this chapter to evaluate the biogenicity of each of these reports. Most of them have already been the subject of previous reviews (Schopf 1993; Altermann and Kazmierczak 2003; Westall and Southam 2006; and Buick 2007) in which the problems associated with individual studies have been dissected in detail. This review concentrates on few recent multi-disciplinary studies that have greatly improved our understanding of early life in terms of life in its ecosystem. The Strelley Pool Chert stromatolites (3.343 Ga) from the North Pole Dome in the Pilbara are representative of large-scale morphological biosignatures; the Josefsdal Chert, a 3.333 Ga-old biolaminite, contains a meso-microscopic anaerobic photosynthetic microbial mat; the "Kittys Gap Chert" in the 3.446 Ga Coppin Gap Greenstone Belt exhibits microscopic colonies of chemolithotrophic microorganisms on volcanic clast surfaces; lastly, 3.22–3.48 Ga-old pillow lavas from the Barberton

**Table 1** Early Archaean morphological biosignatures

Macroscopic to microscopic laminations

| Biosignature type | Environment | Lithology | Structural details | Interpretation | Formation/Group | Other biosignatures | Reference |
|---|---|---|---|---|---|---|---|
| Domical stromatolites | Carbonate platform, shallow water | Carbonate/silica | Domical stromatolites | Anoxygenic photosynthesisers | Warrawoona, North Pole | | Walter et al. 1980 |
| | | | | (+ others e.g. heterotrophs) | | | Lowe 1980 |
| | | | | | | | Walter 1983 |
| | | | | | | | Nijman et al. 1998 |
| | | | | | | | Westall et al. 2002 |
| | | | | | | | Westall 2003, 2004 |
| | | | | | | | van Kranendonk et al. 2003 |
| | | | Microenvironmental speciation of domical stromatolites | | | REEs | Allwood et al. 2006, 2007, Kamber and Webb 2001 |
| | Shallow water | Silica | | | Fig Tree, Barberton | | Byerly et al. 1986 |
| Tabular biolaminations (stromatolites) | Shallow water | Silica | Macro-microscopic carb. laminations | Anoxygenic photosynthesisers | Onverwacht, Barberton | C isotopes, microfossils | Walsh 1992, 2004; Walsh and Lowe 1985, 1999 |
| | Mud flats/littoral | Carbonate/silica | Macro-microscopic carb. laminations | (+ others e.g. heterotrophs) | Onverwacht, Barberton | C isotopes, microfossils | Westall et al. 2001, 2007 |

**Table 1** (*Continued*)

Macroscopic to microscopic laminations

| Biosignature type | Environment | Lithology | Structural details | Interpretation | Formation/Group | Other biosignatures | Reference |
|---|---|---|---|---|---|---|---|
| | Littoral | | Reworked mats | | Onverwacht, Barberton | C isotopes | Tice and Lowe 2004 |
| | Mud flats | | Microscopic carb. mat | Anoxygenic photosynthesisers (+ others e.g. heterotrophs) | Warrawoona, Coppin Gap Gst. Belt | C isotopes, microfossils | Westall et al. 2006c |
| | Deep water, assoc. hydrothermal | Silica | carb. Laminations | Hyperthermophile mats?? | Sulphur Springs group. N. Pole | C isotopes, microfossils | Duck et al. 2007 |

Microscopic morphological biosignatures

| Biosignature type | Environment | Lithology | Structural details | Interpretation | Formation/Group | Other biosignatures | Reference |
|---|---|---|---|---|---|---|---|
| Microbial/colonial microfossils | Associated with carbon. laminae | Silica | Carbonaceous filaments | Anoxygenic photosynthesisers | Onverwacht, Barberton | C isotopes, other microfossils, biolaminae | Walsh 1992, 2004; Walsh and Lowe 1985, 1999; Westall et al. 2001, 2006b, 2007 |
| | Within sediments (lithotrophic environment) | Silicified volcanic sands/silts | Multi-species colonies of carbonaceous coccoids; rods; filaments on volcanic grains | Chemolithotrophs | Warrawoona, Coppin Gap Gst. Belt | C isotopes, HR-TEM | Westall et al. 2006a |
| | Hydrothermal sediments? | | Carbonaceous filaments | Oxygenic photosynthesisers | Warrawoona, Chinaman Creek | C isotopes | Schopf 1993 |
| | Hydrothermal sediments | Silicified hydrothermal sediments | Carbonaceous filaments | Chemolithotrophs | Warrawoona | C/N ratios | Orberger et al. 2006 |

**Table 1** (*Continued*)

Macroscopic morphological biosignatures

| Biosignature type | Environment | Lithology | Structural details | Interpretation | Formation/Group | Other biosignatures | Reference |
|---|---|---|---|---|---|---|---|
| | Hydrothermal sediments | Silicified hydrothermal sediments | Pyritised filaments | Chemolithotrophs | Sulphur Springs group | | Rasmussen 2000 |
| | Hydrothermal sediments | Silicified hydrothermal sediments | Carbonaceous filaments | Chemolithotrophs | North Pole | C isotopes | Ueno et al. 2001a, 2001b |
| | Associated with hydrothermal sediments | Silicified | Carbonaceous filaments | Hyperthermophiles | Sulphur Springs group, N. Pole | C isotopes | Duck et al. 2007 |
| | Reworked chert clasts | Silicified | Carbonaceous filaments | ? | Strelley Pool Sdst., N. Pole | | Wacey et al. 2006 |

Greenstone belt contain microscopic tubules along fractures interpreted as resulting from microbial corrosion.

## 3.1 The 3.430 Ga Strelley Pool Chert Stromatolites

Laminated sedimentary deposits resembling stromatolites were first described from the Strelley Pool Chert (3.430 Ga) in the North Pole Dome region of the Pilbara by Lowe (1980, 1983). Lowe subsequently refuted a biogenic origin for any stromatolite-like structures older than 3.2 Ga (Lowe 1994). According to Krumbein (1983), stromatolites are the lithified vestiges of two dimensional or three dimensional microbial mats. Modern microbial stromatolites mats are highly complex structures consisting of layered communities of microbes having different metabolic activities depending on the physico-chemical microenvironment of the mat layer in which they are living (Krumbein 1983; Jørgensen et al. 1983; Gerdes and Krumbein 1986; Cohen and Rosenberg 1989). Photosynthetic microorganisms are the primary producers of the mat. In modern mats the uppermost aerated layers are dominated by oxygenic phototrophs, such as cyanobacteria. Underneath these layers, anoxygenic phototrophs, such as *Chloroflexus*, inhabit the anaerobic parts of the mat at depths where photons can still penetrate. Associated with the primary producers are a whole host of aerobic and anaerobic heterotrophic microorganisms that degrade the organic matter of the dead primary producers and occur in a strict vertical hierarchy, depending on the physico-chemical conditions (Costerton et al. 1995). Within the body of the mat, the degraded organic matter typically takes on an alveolar texture that has been termed "kopara" (Défarge et al. 1994; Défarge 1997; Benzerara et al. 2006). Extracellular polymeric substances (EPS) are abundant in these mats and detrital particles are invariably trapped within the body of the mats as a result of these sticky polymers.

The term stromatolite is often associated with vertical microbial mat constructions, although Krumbein (1983) describes two-dimensional microbial mats coating sediment surfaces as tabular or stratiform stromatolites. Resulting from the primary activity of photosynthesisers, such structures are necessarily formed in environments that can be reached by sunlight. Such mats may cover enormous areas, as for instance in the Laguna Fugueroa, Baja California where the whole lagoon floor is coated by photosynthetic microbial mats (Stolz and Margulis 1984). The three-dimensional forms of many ancient and modern stromatolites are due to the relationship between upward growth of a phototropic or phototactic biofilm and mineral accretion normal to the surface (Batchelor et al. 2005).

Detailed biogeochemical studies across the vertical structure of the mats using microelectrodes (Jørgensen et al. 1983) document the changes in solute concentrations, pH, Eh on the scale of tens of microns. *In situ* mineralisation of these microbial mats is probably due to a variety of processes. For instance, an increase in alkalinity caused by the activities of chemoheterotrophic microorganisms using nitrate ammonification and sulphate reduction, plus the release of $Ca^{2+}$ from EPS/sheaths, leads to localised supersaturation in $CaCO_3$ and the precipitation of micrite (Konhauser 2007; Braissant et al. 2007). Cyanobacteria can also play a minor role in the precipitation of micrite (Paerl et al. 2001; Chafetz and Buczynski 1992). The calcium carbonate is precipitated on the degraded organic matter and EPS in the form of micro-crystallites (Défarge et al. 1994, 1997; Verrechia et al. 1995; Braissant et al. 2007), thus contributing to the lithification of the structures.

Early Archaean stromatolites are less common and less well-developed (in terms of thickness of deposit) than in the Late Archaean-Proterozoic periods but occur in both the Pilbara (Australia) and the Barberton (South Africa) Greenstone Belts.

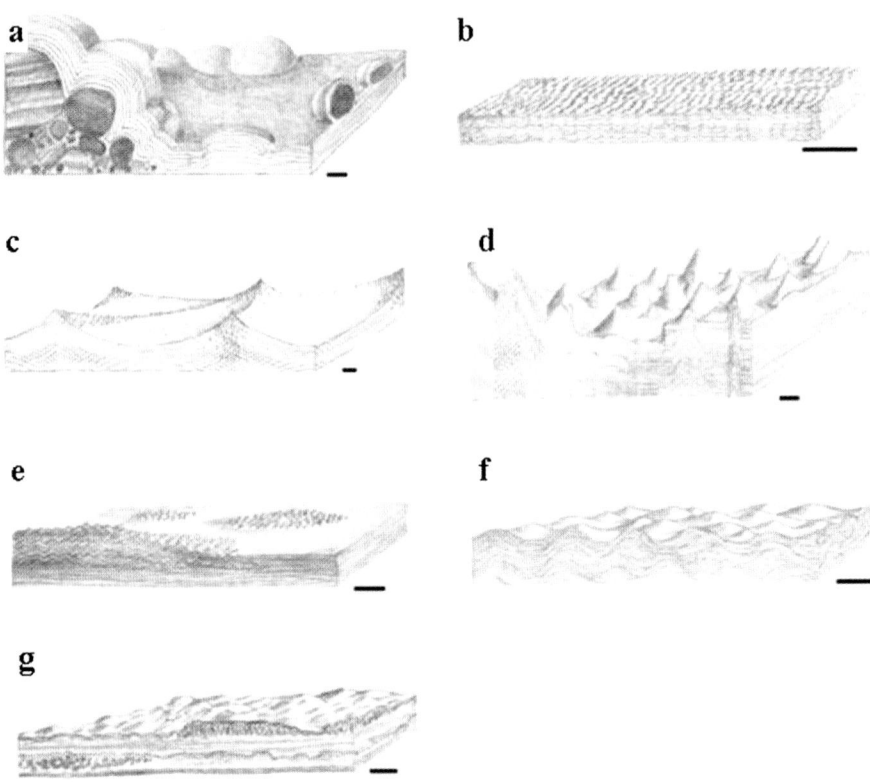

**Fig. 1** Different types of stromatolites morphologies in the 3.443 Ga Strelley Pool Chert, Pilbara (Allwood et al. 2006, 2007). (**a**) Encrusting, domical laminites; (**b**) small crest/conical laminites; (**c**) cuspate swales; (**d**) large complex cones; (**e**) egg-carton laminites, (**f**) wavy laminites; (**g**) iron-rich laminites

Detailed field mapping and lithological studies show that the stromatolites of the Strelley Pool Chert in the North Pole Dome region of the Pilbara developed on a peritidal carbonate platform (Allwood et al. 2006, 2007). Although abiogenic explanations for the origin of these structures were advanced (Lowe 1994; Grotzinger and Rothman 1996; Grotzinger and Knoll 1999; Lindsay et al. 2005), Hofmann et al. (1999) suggested that the complicated domical morphologies characterising the "stromatolites" from the Strelley Pool Chert indicated that they most probably originated from microbial activity. Allwood et al. (2006, 2007) took the morphological characterisation of these stromatolites one step further. After making a field study to place the chert horizon in an evolving geological context, i.e. that of an isolated peritidal carbonate platform on a rocky coastline subjected to transgressive sea level rises, the group undertook detailed mapping of the stromatolites, as well as petrographic investigations, that revealed that different morphological types occurred in very specific micro-environmental locations (Fig. 1; Table 2).

Identification of these structures as biogenic constructions was based on: (1) the occurrence of seven different stromatolite morphotypes in different micro-environments, (2) the syngenicity of the structures with the formation of the host rock, (3) the complexity exhibited by the structures, (4) the occurrence of morphologies inconsistent with purely mechanical deposition, and (5) inconsistency of the palaeoenvironment with hydrothermal precipitation.

**Table 2** Stromatolite morphologies and environmental locations for the Strelley Pool Chert (Allwood et al. 2006, 2007)

| Macroscopic stromatolite structure | Environmental location |
| --- | --- |
| Encrusting/domical laminites | On basal layer, water depth 1 m |
| Small crested/domical | Shallower water depths (infilling embayment) |
| Laminites/flat laminite | |
| Cuspate swales | Intermittently exposed, intertidal to lower supra-tidal carbonates |
| Large complex cones | As above |
| Egg-carton laminites | As above |
| Wavy laminites | Subtidal carbonate facies (wave influence) |
| Iron-rich laminites | Shallow, silica/clastic depositing, partially restricted conditions |

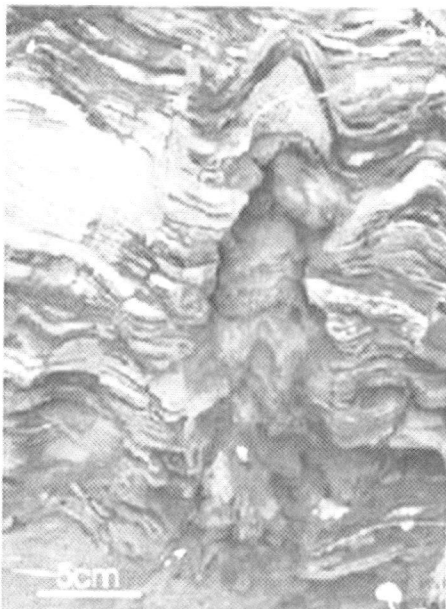
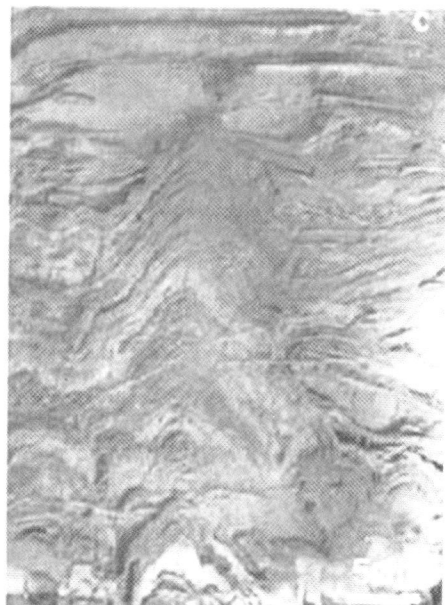

**Fig. 2** Large complex cone stromatolite facies, Strelley Pool Chert (Allwood et al. 2007)

As an example, the conical structures at the Trendall Locality (Fig. 2) show great geometric and textural complexity. The cones are heterogeneously distributed, their laminae are non-isopachous, and they have vertical column alignment similar to that exhibited by modern upward migrating microbial colonies at the sediment-water interface, i.e. photosynthetic microorganisms. In addition, the textural differences between the cones and the layered interspaces suggest that the interspaces grew from mechanical sediment deposition whereas trapping and binding as well as intra mat precipitation characterised the cones. There is additional chemical evidence for the biogenicity of these stromatolites. They exhibit a 250-fold increase in REEs compared to the purely chemical chert precipitates (Kamber and Webb 2001), values that are typical of microbial carbonates (Webb and Kamber 2000). Allwood et al. (2006, 2007) were therefore able to determine the likely biogenicity of these stromato-

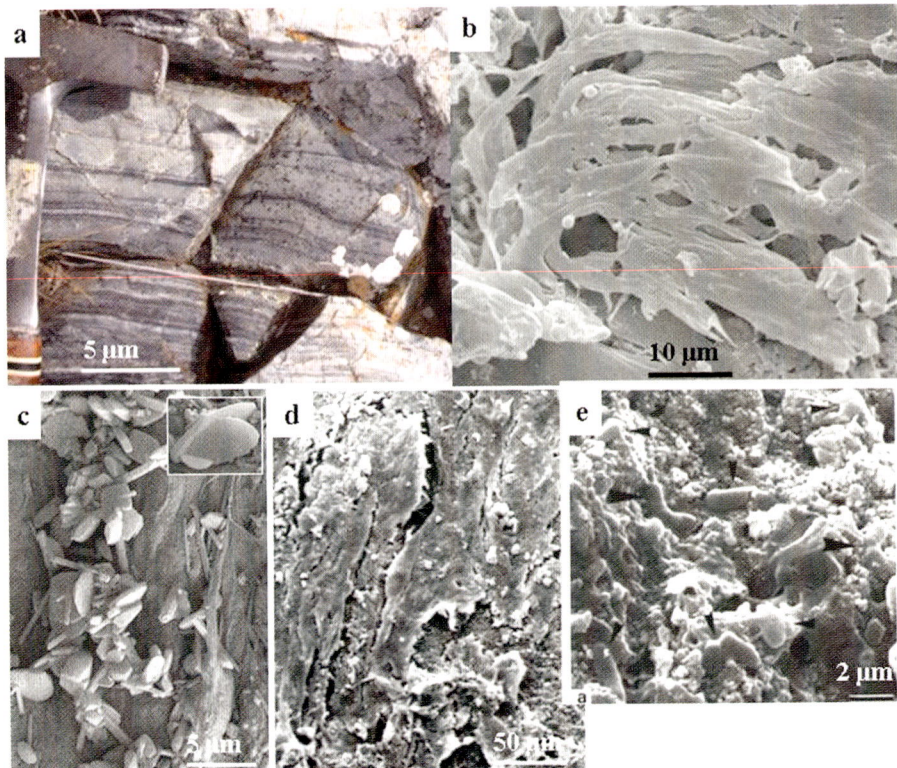

**Fig. 3** Microbial mat from the 3.333 Ga Josefsdal Chert, Barberton (Westall et al. 2001, 2006b, 2007). (**a**) Black and greenish white, biolaminated chert outcrop; (**b**) overturned filaments forming the filamentous mat, (**c**) pseudomorph evaporite minerals encrusting the surface of the mat with pseudomorph gypsum in the inset, (**d**) desiccation cracks in the filamentous mat, (**e**) small colony of rod to vibroid-shaped microorganisms adjacent to the filamentous mat

lites based on detailed field and petrographic studies, as well as geochemical data. Note that no actual fossil microorganisms were described from this site.

## 3.2 The 3.333 Ga-old Josefsdal Chert Microbial Mat

The Josefsdal Chert in the Onverwacht Group of Barberton, South Africa, represents a sequence of silicified volcanic sands and silts that were deposited in a littoral tidal-flat environment (Westall et al. 2006b, 2007). In fact, the finely laminated deposits have been interpreted as biolaminites (Westall et al. 2007; cf. ancient biolaminites have also been described in 3.2 Ga-old clastic sediments from the Pongola Group by Noffke et al. 2006).

The chert consists of alternating black and greenish white laminae. On the top of one of the black laminae and underlying a green layer are the remains of a microbial mat covering an area of approximately 6 mm$^2$ (Fig. 3). The surface of the 1–5 μm thick mat consists of 0.25 μm diameter filaments (tens of micrometers in length) that are embedded in, and coated with, copious quantities of a filmy substance interpreted as EPS. Indications of formation under flowing water are given by the streamlined orientation of the filaments and overturned and mechanically torn portions of the mat. Trapped detrital particles (volcanic spherules, clasts, quartz) occur within the mat as "floating" particles. The topographically highest portions of the mat surface are coated with a suite of evaporite mineral pseudomorphs that include aragonite, gypsum, Mg calcite, and a halide. Thin layers of evaporite minerals are also interspersed within the thickness of different parts of the mat. Together with other indications of desiccation (cracks in the mat), these observations suggest that the mat was episodically exposed to air-drying. Vertical sections through the mat show that only the uppermost surface exhibits perfect structural preservation of the filaments (that were coated with silica). Beneath this surface, the mat exhibits an alveolar texture. Adjacent to the mat is a small colony of rod to vibroid-shaped microfossils 2–3.8 μm in length and about 1 μm in diameter, also embedded in a film-like substance interpreted to be EPS. Carbon isotope analyses of the dark layers are −22.7 to −26.8‰ and Raman spectrometry showed that the carbon had a maturity consistent with the age and metamorphic history of the rock (lower greenschist). Carbon contents of the black layers are up to 0.7% whereas contents of the bulk sample are 0.07%.

Westall et al. (2001, 2006b) interpreted this structure as the remnants of a filamentous microbial mat. As with the example of the Strelley Pool Chert stromatolites, it was necessary to demonstrate (1) the biogenicity of the mat structure and (2) its syngenicity with the formation of the original rock. The biogenicity of the mat was determined on the basis of the combined morphological and chemical characteristics of the structures observed. The size and shape of the individual filaments and rod/vibroid-shaped structures are identical to those known for modern prokaryotes and the filaments demonstrate flexibility. Evidence of cell division is shown within the colony of rods/vibroids. All the microfossils are associated with EPS and exhibit colony, biofilm and mat formation. The carbon isotope data are also consistent with microbial fractionation. The structures demonstrate direct interaction with their environment (intergrowth of filaments with underlying sediment particles at the base of the mat, orientation and overturning under water flow, incorporation of detrital particles, desiccation and evaporite mineral precipitation in response to exposure to the atmosphere). The complexity of the mat composition and structure is inconsistent with an abiogenic origin. Thus, as with the Strelley Pool Chert stromatolites, the heterogeneous complexity of the microbial structures argues for their biogenicity.

Westall et al. (2006b) demonstrated the syngenicity of the microbial mat using the arguments of (1) mat formation on a planar sediment surface (not a fracture surface), (2) filaments embedded *in situ* in early diagenetic quartz crystals, (3) mat interaction with its immediate environment (response to current flow, direct stabilisation of the underlying sediment surface, incorporation of particles coming from the immediate environment), (4) the intermittent desiccation of the mat, and (5) the well-defined vertical structure of the mat and its preservation by a coating of silica. Having established the biogenicity and syngenicity of the microbial mat, Westall et al. (2006b) concluded that, given the littoral environment of formation of the Josefsdal Chert, the filamentous organisms forming the mat were probably anoxygenic microorganisms.

## 3.3 The 3.446 Ga-old Kitty's Gap Chert

The 3.446 Ga "Kitty's Gap Chert" represents volcaniclastic sediments deposited in a tidal channel/mud flat environment that gradually silted up, possibly with a brief period of subaerial exposure (De Vries 2004). A transgressive sequence of fine ash deposited in a low energy environment (i.e. quiet water conditions) tops the tidal deposits (Westall et al. 2006a). Penecontemporaneous silicification rapidly lithified the sediments. The REE signature of the cherts indicates that much of the ambient seawater silica was of hydrothermal origin (Orberger et al. 2006).

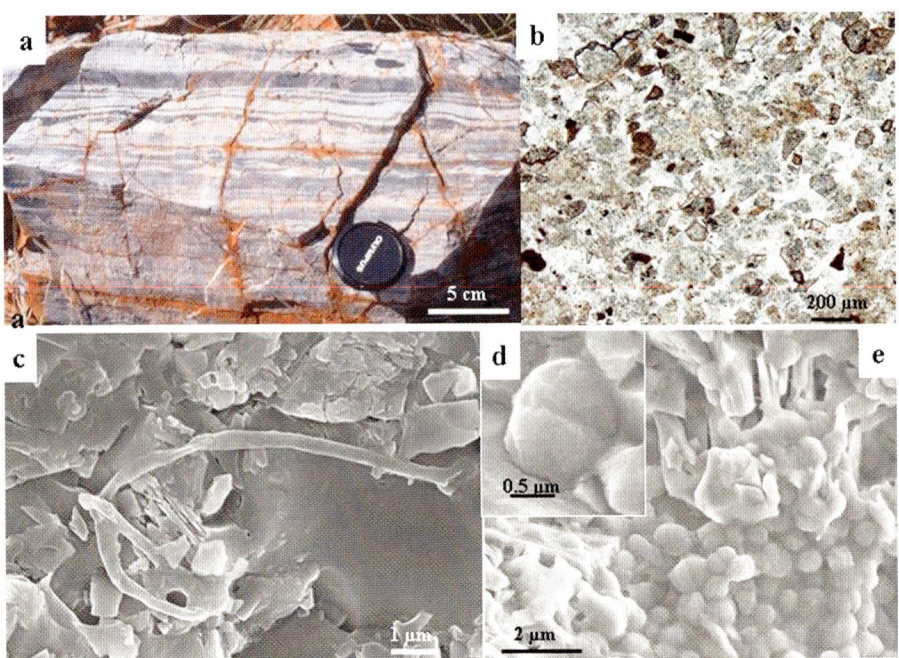

**Fig. 4** Tidal flat volcanic sediments with colonies of microfossils on volcanic clast surfaces and a biofilm on a stable sediment surface, 3.446 Ga Kitty's Gap Chert, Pilbara (Westall et al. 2006a). (**a**) Laminated sediments in outcrop, (**b**) petrographic view of the volcanic clasts, (**c**) microbial filament; (**d, e**) a two-species colony of a coccoids with detail showing cell division

These sediments host a number of potentially habitable environments: stable sediment surfaces representing periods on non-deposition where microbial mats could develop, as well as the pore space of the water-logged volcanic sediments and the surfaces of the volcanic particles. The latter represent ideal substrates for chemolithotrophic microorganisms that, as mentioned above, obtain both their energy and carbon from inorganic sources— the energy from chemical reactions at the surface of the volcanic clasts (that also liberated useful elements as nutrients) and the carbon from the $CO_2$-rich seawater. Westall et al. (2006a) documented small colonies of coccoidal-shaped microstructures on the surfaces of the volcanic clasts and within layers of dust-sized volcanic material (Fig. 4). The colonies consisted of (1) coccoidal structures with bimodal sizes (0.4–0.5 and 0.7–0.8 μm) occurring in closely-packed "carpet" colonies some tens to >100 μm in diameter and consisting of hundreds of individuals, and (2) colonies of individuals that are more loosely associated including coccoids, chains of 3–10 coccoids, rod-shaped structures 0.4 μm in diameter, 0.8 μm in length and filaments 0.3–0.45 μm in diameter and tens of μm in length. The distribution of the different types of colonies was closely related to different sediment layers. One stable sediment surface hosted the development of a very delicate, multispecies biofilm consisting of coccoids, rods, filaments and EPS. Carbon isotope analyses of the individual layers produced values ranging from −25.9‰ to −27.8‰, whereas *in situ* Raman spectrometry showed that the kerogenous material was mature and consistent with the age and metamorphic history of the rock (prehnite-pumpellyite-lower greenschist; Kisch and Nijman 2004). The carbon contents of the individual layers range from 0.01–0.05%. High resolution transmission electron microscopy (HR-TEM) of the (002) lattice fringe of the kerogen indicated that the precursor material was biogenic (cf. Boulmier et al. 1982; Oberlin 1989; Rouzaud and Clinard 2002).

The global environment of deposition, tidal mud flats, was clearly conducive to microbial growth and activity. Interpretations of biogenicity of the microstructures observed were based on the similarity of the morphological characteristics of the microstructures with those of modern microorganisms (size and shape, cell division, evidence of both living and dead cells in the same colony (prior to silicification), filament flexibility, EPS association, colony formation, multispecies colony and biofilm formation). The totality of morphological features described above, the associations of different morphotypes in groups similar to colonies, and their specific distributions related to specific microhabitats are inconsistent with formation by abiogenic processes. Taking into account all the characteristics of these putative microfossils, i.e. their carbonaceous composition and the HR-TEM evidence for precursor biogenic carbon, their isotopic composition and morphological features, Westall et al. (2006a) concluded that these structures probably represent silicified microorganisms. The silicification must have occurred rapidly, as many of the microorganisms were still living when silicified, although some collapsed individuals had obviously lysed before silicification. There is also sedimentological evidence that the volcanic muds were silicified very rapidly during early diagenesis.

Syngenicity was established on the basis of the heterogeneous but micro-scale dependent distribution of the different associations of microfossils, the clear association of certain types of colonies around the edges of volcanic clasts, the formation of a bedding plane parallel biofilm, and the fact that the microfossils are encased in early diagenetic quartz.

Given the close association of colonies of microfossils around the edges of volcanic clasts, Westall et al. (2006a) concluded that they most likely represented anaerobic chemolithotrophic organisms. The delicate biofilm formed on a sediments surface may have represented the initial growth of an anaerobic photosynthetic mat.

3.4 Microbial corrosion on the rinds of 3.22–3.48 Ga-old pillow lavas from Barberton

In contrast to the previous example, where the fossilised remains of microbial colonies and biofilms on volcanic particles and sediment surfaces was described, Furnes et al. (2004, 2007) documented tubular pits in fractures in early-Mid Archaean pillow lavas rinds from the Barberton and Pilbara Greenstone Belts that they interpreted as evidence of microbial activity on the basis of morphological and geochemical evidence. As noted above, the dissolution of volcanic glasses can provide nutrients and essential elements for chemolithotrophic microorganisms but there is some debate as to whether microtunnels, such as those described by Furnes et al. (2004, 2007) are biogenic or abiogenic in origin. In recent basalt rinds, direct evidence of microbes or their remains in corrosion pits in the glassy rinds of younger basalts has been documented, as has microbial mediation of the alteration of the basalt rinds (cf. Torsvik et al. 1998; Benzerara et al. 2006 and references therein).

The purported microbial structures in glassy pillow rims from the Barberton Greenstone belt are mineralised tubes that are up to 200 μm long and between 1–9 μm in width (Fig. 5) (Furnes et al. 2004). The tubes, extending into the volcanic glass from the fractures, are filled with fine-grained titanite. Some of them appear to be segmented and contain sub-spherical shapes 1–9 μm in diameter. Carbon is associated with the walls of the tubes. Carbon isotope measurements on the carbonate-altered glassy rims ($+3.9$ to $-16.4‰$) differ from those of the interior of the pillow lavas ($+0.7$–$6.9‰$). Bacteriomorph structures of similar size and shape from similar-aged pillow lava rinds in the Pilbara were also reported by Furnes et al. (2007). Furnes et al. (2004, 2007) determined the biogenicity of the ancient tubular

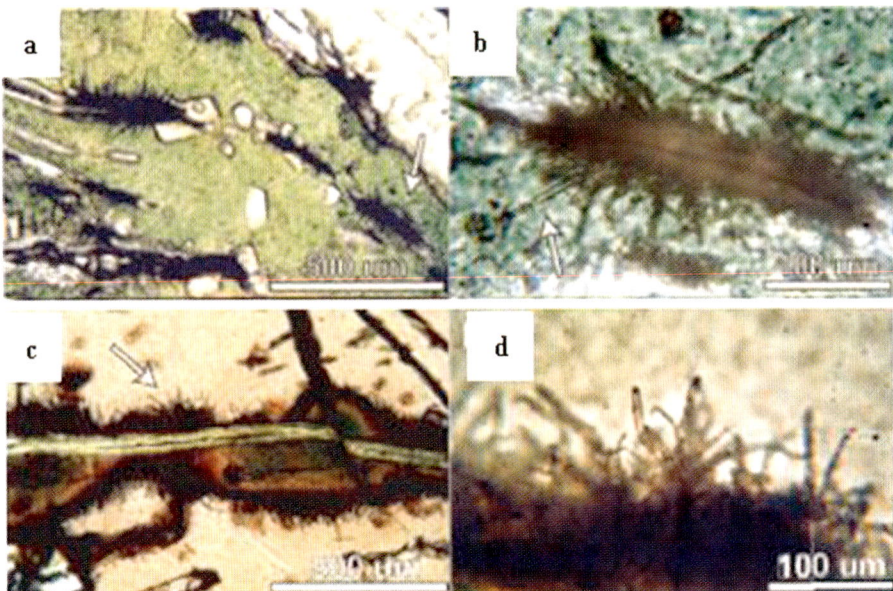

**Fig. 5** Microbial corrosion tubes in the glassy rims of ancient and modern pillow lava rims (Furnes et al. 2004). (**a, b**) corrosion tubes radiating out from a fracture in the rim of 3.22–3.48 Ga pillow lavas from Barberton. (**c, d**) Similar corrosion tubes in modern oceanic pillow lava rims

structures on the basis of their similarities with modern microbially-induced pits and tubes: their shape, size and distribution (location of the tubes with respect to fractures), the concentration of important bio-elements (C, N, P), and different $\delta^{13}C$ signature compared to the interior of the lavas. They demonstrate the syngenicity of the structures by the fact that the latter were influenced by early metamorphic alteration of rocks (precipitation of titanite and disruption caused by the growth of diagenetic chlorite).

## 4 Conclusions and Perspectives for the Search for Martian Life

The above examples document a variety macroscopic to microscopic morphological biosignatures, including the fossilised remains of individual microorganisms, colonies and biofilms, macroscopic microbial constructions (stromatolites) and microbial corrosion features. In all cases, interpretation of the biogenicity of a particular structure or group of structures was strengthened by the use of more than one type of biosignature. However, in certain circumstances, the complexity of the morphological biosignatures, especially in the Strelley Pool Chert stromatolites, the Josefsdal Chert microbial mat and the "Kitty's Gap" Chert colonies and biofilms, is a very strong indicator in itself of biogenicity; likewise the heterogeneity in the distribution of different types of morphological biosignatures with respect to different micro-habitats.

In terms of searching for traces of life *in situ* on Mars, these examples are revealing. In the first place they occur in the types of rocks (basaltic lavas and volcanic sediments) that would be expected on Noachian Mars. Secondly, they were formed at a period only slightly later than that during which the surface of Mars was habitable (from ~4.4 to about 3.8 Ga, Bibring et al. 2006; Poulet et al. 2005). Most of the biogenetic features associated with the

Strelley Pool Chert and the finely laminated texture of the Josefsdal Chert biolaminite would be visible using macroscopic to microscopic instrumentation on a rover, such as the planned Mars Science Laboratory of NASA (launch 2009) or the ExoMars mission of ESA (launch 2013). It will not be possible to identify individual fossil microorganisms, colonies or mats of the kinds found in the Josefsdal and Kitty's Gap Cherts, or the biocorrosion features in pillow lavas using *in situ* microscopy on any of the planned missions. However, although the total quantity of organic carbon in these cherts is low (0.01–0.07%) and the distribution of the carbon is heterogeneous, it may be possible to detect the carbon and to partly analyse its molecular and isotopic composition with a number of instruments on MSL (SAM) and ExoMars (Raman, MOMA, Urey, and MOD). Given the low quantities of organic carbon expected in the Martian rocks, the possibility of a false positive, i.e. the detection of biogenic organics of terrestrial origin, will be of fundamental importance. Ideally, the detection of Martian organics should go hand in hand with detailed studies of their micro to nano-scale distribution and relationship to other phenomena that might be indicative of life, such as morphological fossils. This will be particularly important if the potential martian life is significantly different from terrestrial life. The conclusion is that it will probably be necessary to return carefully chosen rocks from Mars in order to obtain reliable evidence of potential Martian life.

## References

A.C. Allwood, M.R. Walter, B.S. Kamber, I.W. Burch, Nature **441**, 714 (2006)
A.C. Allwood, M.R. Walter, I.W. Burch, B.S. Kamber, Precambrian Res. **158**, 198–227 (2007)
J.C. Alt, P. Mata, Eart Planet. Sci. Lett. **181**, 301–313 (2000)
W. Altermann, J. Kazmierzcak, Res. Microbiol. **154**, 611–617 (2003)
E.S. Barghoorn, S.A. Tyler, Science **147**, 563–577 (1965)
R.V. Batchelor, M.T. Burne, B.I. Henry, T. Slatyer, Physica A **350**, 6–11 (2005)
L. Becker, B. Popp, T. Rust, J.L. Bada, Earth Planet. Sci. Lett. **167**, 71 (1999)
L.G. Benning, V. Phoenix, N. Yee, M.G. Tobin, K.O. Konhauser, B.W. Mountain, *Geochemistry of the Earth's Surface* (2002), p. 259
K. Benzerara, N. Menguy, P. López-Garcia, T.-H. Yoon, J. Kazmierczak, T. Tyliszczak, F. Guyot, G.E. Brown, PNAS **103**, 9440–9445 (2006)
S. Bernard, K. Benzerara, O. Beyssac, N. Menguy, F. Guyot, G.E. Brown Jr., B. Goffé, Earth Planet. Sci. Lett. **262**, 257–272 (2007)
J.P. Bibring et al., Science **312**, 400–404 (2006)
E. Bittarello, D. Aquilino, Eur. J. Mineral. **19**, 345–351 (2007)
J.L. Boulmier, A. Oberlin, J.N. Rouzaud, M. Villey, in *Scanning Electron Microscopy*, ed. by A.M.F. O'Hare (SEM, Chicago, 1982), p. 1523
O. Braissant, A.W. Decho, C. Dupraz, C. Glunk, K.M. Sprzekop, P.T. Visscher, Geobiology **5**, 401–411 (2007)
M.D. Brasier, O.R. Green, A.P. Jephcoat, A.K. Kleppe, M. van Kranendonk, J.F. Lindsay, A. Steele, N. Grassineau, Nature **416**, 76 (2002)
M.D. Brasier, O.R. Green, J.F. Lindsay, A. Steele, Orig. Life Evol. Biosph. **34**, 257 (2004)
M.D. Brasier, N. McLoughlin, O.R. Green, D. Wacey, Philos. Trans. R. Soc. B **361**, 887–902 (2006)
J.J. Brocks, G.A. Logan, R. Buick, R.E. Summons, Science **285**, 1033–1036 (1999)
J.J. Brocks, G.D. Love, C.E. Snape, G.A. Logan, R.E. Summons, R. Buick, Geochim. Cosmochim. Acta **67**, 1521 (2003)
R. Buick, Palaios **5**, 441 (1990)
R. Buick, in *Planets and Life*, ed. by W.T. Sullivan, J.A. Baross (Cambridge University Press, Cambridge, 2007), pp. 237–264
M.M. Byerly, G.R. Walsh, D.L. Lowe, Nature **319**, 489 (1986)
S.L. Cady, J.D. Farmer, in *Evolution of Hydrothermal Ecosystems on Earth (and Mars)*, ed. by G.R. Bock, J.A. Goodie. Ciba Symposium, vol. 202 (Wiley, Chichester, 1996), p. 150
S.L. Cady, J.D. Farmer, J.P. Grotzinger, W.J. Schopf, A. Steele, Astrobiology **3**, 351 (2004)
H.S. Chafetz, C. Buczynski, Palaios **7**, 277–293 (1992)

S.M. Clifford, T.J. Parker, Icarus **154**, 40–79 (2001)

Y. Cohen, E. Rosenberg, *Microbial Mats: Physiological Ecology of Benthic Microbial Communities* (ASM, Washington, 1989)

J.W. Costerton, Z. Lewandowski, D.E. Caldwell, D.R. Korber, H.M. Lappin-Scott, Rev. Microbiol. **49**, 711–745 (1995)

S.T. De Vries, Early Archean sedimentary basins: depositional environment and hydrothermal systems. Examples from the Barberton and Coppin Gap greenstone belts. Geologica Ultraiectina, University of Utrecht (2004), 159 pp

C. Défarge, C. R. Acad. Sci. Paris Sér. IIa **324**, 553–561 (1997)

C. Défarge, J. Trichet, A. Maurin, M. Hucher, Sediment. Geol. **89**, 9–23 (1994)

L.J. Duck, M. Glikson, S.D. Golding, R.E. Webb, Precamb. Res. **154**, 205–220 (2007)

D.J. Des Marais, Science **289**, 1703 (2000)

F.G. Ferris, T.J. Beveridge, W.S. Fyfe, Nature **320**, 609 (1986)

F.G. Ferris, W.S. Fyfe, T.J. Beveridge, Geology **16**, 149–152 (1988)

M.R. Fisk, S.J. Giovannoni, I.H. Thorseth, Science **281**, 978–980 (1998)

H. Furnes, N.R. Banerjee, K. Muehlenbachs, H. Staudigel, M. de Wit, Science **304**, 578–581 (2004)

H. Furnes, N.R. Banerjee, H. Staudigel, K. Muehlenbachs, N. McLoughlin, M. de Wit, M. van Kranendonk, Precambrian Res. **158**, 156–176 (2007)

J.M. Garcia-Ruiz, S.T. Hyde, A.M. Carnerup, A.G. Christy, M.J. van Krankendonk, N.J. Welham, Science **302**, 1194 (2003)

G. Gerdes, W.E. Krumbein, *Biolaminated Deposits* (Springer, Berlin, 1986), 183 pp

G. Gerdes, M. Claes, K. Dunajtschik-Piewak, H. Riege, W.E. Krumbein, H.-E. Reineck, Facies **29**, 61–74 (1993)

J.P. Grotzinger, A.H. Knoll, Annu. Rev. Earth Planet. Sci. **27**, 313–358 (1999)

J.P. Grotzinger, D.H. Rothman, Nature **383**, 423–425 (1996)

H.J. Hofmann, J.W. Schopf, in *Earth's Earliest Biosphere*, ed. by J.W. Schopf (Princeton University Press, Princeton, 1983), pp. 321–362

H.J. Hofmann, K. Grey, A.H. Hickman, R.I. Thorpe, Geol. Soc. Am. Bull. **111**, 1256–1262 (1999)

B.B. Jørgensen, N.P. Revsbech, Y. Cohen, Limnol. Oceanogr. **28**, 1075–1093 (1983)

B.S. Kamber, G.E. Webb, Geochim. Cosmochim. Acta **65**, 2509–2525 (2001)

H.J. Kisch, W. Nijman, in *Field Forum on Processes on the early Earth*, ed. by W.U. Reimold, A. Hofmann, Kaapvaal Craton, South Africa, 2001 (University of Witwatersrand, Johannesburg, 2004), p. 47

K.O. Konhauser, *Introduction to Geomicrobiology* (Blackwell, Oxford, 2007)

K.O. Konhauser, B. Jones, A.-L. Reysenbach, R.W. Renaut, Can. J. Earth Sci. **40**, 1713 (2003)

W.E. Krumbein, Precambrian Res. **20**, 493–531 (1983)

W.E. Krumbein, D.M. Paterson, L.J. Stal (eds.), *Biostabilization of Sediments: Oldenburg.* Biblioteks und Informationssystem der Carl von Ossietzky Universität (1994), 526 pp

J.F. Lindsay, M.D. Brasier, N. McLoughlin, O.R. Green, M. Fogel, A. Steele, S.A. Mertzman, Precambrian Res. **143**, 1–22 (2005)

D.R. Lowe, Nature **284**, 441–443 (1980)

D.R. Lowe, Precambrian Res. **19**, 239–283 (1983)

D.R. Lowe, Geology **22**, 287–390 (1994)

H.A. Lowenstam, S. Weiner, *On Biomineralisation* (Oxford University Press, Oxford, 1989)

T.W. Lyons, C.L. Zhang, C.S. Romanek, Chem. Geol. **195**, 1 (2003)

J.M. Madigan, M.T. Martinko, J. Parker, *Brock Biology of Microorganisms*, 10th edn. (Prentice-Hall, New Jersey, 2002)

C.P. McKay, R.L. Mancinelli, C.R. Stoker, R.A. Wharton, in *Mars*, ed. by Kieffer et al. (University of Arizona Press, Tucson, 1992), pp. 1234–1245

D.S. McKay, E.K. Gibson, K.L. Thomas-Keprta, H. Vali, C.S. Romanek, S.J. Clemett, X.D.F. Chillier, C.R. Maedling, R.N. Zare, Science **273**, 924–930 (1996)

S.J. Mojzsis, G. Arrhenius, K.D. McKeegan, T.M. Harrison, A.P. Nutman, C.R.L. Friend, Nature **384**, 55–59 (1996)

M.D. Mullen, D.C. Wolf, F.G. Ferris, T.J. Beveridge, C.A. Flemming, G.W. Bailey, Appl. Environ. Microbiol. **55**, 3143–3149 (1989)

National Research Council, in *Size Limits of Very Small Microorganisms: Proceedings of a Workshop* (Space Studies Board, National Academies Press, Washington, 1999)

W. Nijman, B.A. Willigers, A. Krikke, Precambrian Res. **88**, 83 (1998)

N. Noffke, G. Gerdes, T. Klenke, W.E. Bein, Sediment. Geol. **110**, 1–6 (1997)

N. Noffke, K.A. Eriksson, R.M. Hazen, E.L. Simpson, Geology **34**, 253–256 (2006)

A. Oberlin, in *Chemistry and Physics of Carbon*, vol. 22, ed. by P.A. Thrower (Marcel Dekker, New York, 1989), pp. 1–143

B. Orberger, V. Rouchon, F. Westall, S.T. de Vries, C. Wagner, D.L. Pinti, in *Processes on the Early Earth*, ed by W.U. Reimold, R. Gibson. Geol. Soc. Amer. Spec. Paper, vol. 405 (2006), p. 133

H.W. Paerl, T.F. Steppe, R.P. Reid, Environ. Microbiol. **3**, 123–130 (2001)

V.R. Phoenix, D.G. Adams, K.O. Konhauser, Chem. Geol. **169**, 329 (2000)

F. Poulet, J.-P. Bibring, J.F. Mustard, A. Gendrin, N. Mangold, Y. Langevin, R.E. Arvidson, B. Gondet, C. Gomez, Nature **438**, 623–627 (2005)

D.G. Rancourt, P.-J. Thibault, D. Mavrocordatos, G. Lamarche, Geochim. Cosmochim. Acta **69**, 553 (2005)

B. Rasmussen, Nature **405**, 676–679 (2000)

R. Riding, Sedimentology **47**, 179–214 (2000)

M.T. Rosing, Science **283**, 674–676 (1999)

M.T. Rosing, R. Frei, Earth Planet. Sci. Lett. **217**, 237–244 (2004)

J.N. Rouzaud, C. Clinard, Fuel Process. Technol. **77–78**, 229–235 (2002)

M. Schidlowski, Nature **333**, 313–318 (1988)

J.D. Schiffbauer, L. Yin, R.J. Bodnar, A.J. Kaufman, F. Meng, J. Hu, B. Shen, X. Yuan, H. Bao, S. Xiao, Astrobiology **7**, 684–704 (2007)

E.L. Schock, J. Geophys. Res. **102**, 23687–23694 (1997)

J.W. Schopf, Science **260**, 646 (1993)

J.W. Schopf, Philos. Trans. Roy. Soc. B. **361**, 869–885 (2006)

J.W. Schopf, A.B. Kudryavtsev, D.G. Agresti, T.J. Wdowiak, A.D. Czaja, Nature **416**, 73 (2002)

J.W. Schopf, A.B. Kudryavtsev, A.D. Czaja, A.B. Tripathi, Precambrian Res. **158**, 141–155 (2007)

G. Southam, F. Westall, in *Treatise on Geophysics*, ed. by T. Spohn. Planets and Moons, vol. 10 (Elsevier, Amsterdam, 2007), pp. 421–438

G. Southam, L. Rothschilde, F. Westall, Space Sci. Rev. **7**, 34 (2007)

J.F. Stolz, L. Margulis, Orig. Life **14**, 671–679 (1984)

R.E. Summons, L.L. Jahnke, J.M. Hope, J.H. Logan, Nature **400**, 554 (1999)

Y. Ueno, S. Maruyama, Y. Isozaki, H. Yurimoto, in *Geochemistry and the Origin of Life*, ed. by S. Nakashsima, S. Maruyama, A. Brack, B.F. Windley (Univ. Acad. Press, Tokyo, 2001a), pp. 203–236

Y. Ueno, Y. Isozaki, H. Yurimoto, S. Maruyama, Int. Geol. Rev. **43**, 196–212 (2001b)

I. Thorseth, H. Furnes, M. Heldal., Geochim. Cosmochim. Acta **56**, 845–850 (1992)

I.H. Thorseth, T. Torsvik, V. Torsvik, F.L. Daae, R.B. Pedersen, Earth Planet. Sci. Lett. **194**, 31–37 (2001)

M. Tice, D.R. Lowe, Nature **431**, 549–552 (2004)

J.K.W. Toporski, F. Westall, K.A. Thomas-Keprta, A. Steele, D.S. Mckay, Astrobiology **2**, 1 (2002)

T. Torsvik, H. Furnes, K. Muehlenbachs, I.H. Thorseth, O. Tumyr, Earth Planet. Sci. Lett. **162**, 165–176 (1998)

A.H. Treiman, Meteorit. Planet. Sci. **33**, 753 (1998)

M.J. van Kranendonk, G.E. Webb, B.S. Kamber, Geobiology **1**, 91–108 (2003)

M. van Zuilen, A. Lepland, G. Arrhenius, Nature **418**, 627 (2002)

M. van Zuilen, A. Lepland, J. Teranes, F. Finarelli, M. Wahlen, G. Arrhenius, Precambrian Res. **126**, 331 (2003)

E.P. Verrechia, P. Feytet, K.E. Verrechia, J.L. Dumont, J. Sedimentol. Res. **65**, 690–700 (1995)

D. Wacey, N. McLoughlin, O.R. Green, J. Parnell, C.A. Stoakes, M.D. Brasier, Int. J. Astrobiol. **6**, 1–10 (2006)

M.M. Walsh, Precamb. Res. **54**, 271–293 (1992)

M.M. Walsh, D.R. Lowe, Nature **284**, 443–445 (1985)

M.M. Walsh, D.R. Lowe, in *Geologic Evolution of the Barberton Greenstone Belt, South Africa*, ed. by D.R. Lowe, G.R. Byerly, Geol. Soc. Am. Spec. Paper, vol. 329 (1999), pp. 115–132

M.M. Walsh, Astrobiology **4**, 429–437 (2004)

M.R. Walter, in *Earth's Earliest Biosphere*, ed. by J.W. Schopf (Princeton University Press, Princeton, 1983), pp. 187–213

M.R Walter, R. Buick, J.S.R. Dunlop, Nature **284**, 443–445 (1980)

G.E. Webb, B.S. Kamber, Geochim. Cosmochim. Acta **64**, 1557–1565 (2000)

F. Westall, in *Astronomical and Biochemical Origins and the Search for Life in the Universe*, ed. by C.B. Cosmovici, S. Bowyer, D. Werthimer (Editrici Compositori, Bologna, 1997), pp. 491–504

J. Westall, J. Geophys. Res. Planets **104**, 16437–16451 (1999)

F. Westall, Palevol **2**, 485–501 (2003)

F. Westall, in *Astrobiology: Future Perspectives*, ed. by P. Ehrenfreund (Kluwer, Dordrecht, 2004), p. 287

F. Westall, in *Water on Mars and Life*, ed. by T. Tokano. Advances in Astrobiology and Biogeophysics (2005a), p. 45

F. Westall, Science **434**, 366 (2005b)

F. Westall, R.L. Folk, Precambrian Res. **126**, 313 (2003)

F. Westall, G. Southam, in *Archean Geodynamics and Environments*, ed. by K. Benn et al. AGU Geophys. Monogr., vol. 164 (2006), p. 283

F. Westall, L. Boni, M.E. Guerzoni, Palaeontology **38**, 495 (1995)

F. Westall, M.J. De Wit, J. Dann, S. Van Der Gaast, C. De Ronde, D. Gerneke, Precambrian Res. **106**, 94–112 (2001)

F. Westall, A. Brack, B. Barbier, M. Bertrand, A. Chabin, in *Exo/Astrobiology, Proceedings Second European Workshop on Exo/Astrobiology*, Graz, 2002, ed. by H. Lacoste, vol. SP-518 (ESA, Noordwijk, 2002), pp. 131–136.

F. Westall, S.T. de Vries, W. Nijman, V. Rouchon, B. Orberger, V. Pearson, J. Watson, A. Verchovsky, I. Wright, J.-N. Rouzaud, D. Marchesini, S. Anne, Geol. Soc. Am. Spec. Pap. **405**, 105 (2006a)

F. Westall, C.E.J. de Ronde, G. Southam, N. Grassineau, M. Colas, C. Cockell, H. Lammer, Philos. Trans. R. Soc. B. **361**, 1857 (2006b)

F. Westall, S.T. de Vries, W. Nijman, V. Rouchon, B. Orberger, V. Pearson, J. Watson, A. Verchovsky, I. Wright, J.-N. Rouzaud, D. Marchesini, S. Anne, in *Processes on the Early Earth*, ed. by W.U. Reimold, R. Gibson, vol. 405 (Geol. Soc. Am. Spec. Publ., 2006c), pp. 105–131

F. Westall, G. Gerdes, A. Hofmann, in *European Astrobiology Network Association Meeting*, October, Türku, Finland, Abst. (2007)

S.A. Wilde, J.W. Valley, W.H. Peck, C.M. Graham, Nature **409**, 175–178 (2001)

N.P. Yushkin, in *Instruments, Methods, and Missions for Astrobiology* III, ed. by R.B. Hoover. Proceedings of SPIE, vol. 4137. International Society for Optical Engineering (2000), pp. 22–35

# Biosignatures: Molecular Biosignatures

# Gene Transfer and the Reconstruction of Life's Early History from Genomic Data

**J. Peter Gogarten · Gregory Fournier ·
Olga Zhaxybayeva**

Originally published in the journal Space Science Reviews, Volume 135, Nos 1–4.
DOI: 10.1007/s11214-007-9253-8 © Springer Science+Business Media B.V. 2007

**Abstract** The metaphor of the unique and strictly bifurcating tree of life, suggested by
Charles Darwin, needs to be replaced (or at least amended) to reflect and include processes
that lead to the merging of and communication between independent lines of descent. Gene
histories include and reflect processes such as gene transfer, symbioses and lineage fusion.
No single molecule can serve as a proxy for the tree of life. Individual gene histories can
be reconstructed from the growing molecular databases containing sequence and structural
information. With some simplifications these gene histories can be represented by furcating
trees; however, merging these gene histories into web-like organismal histories, including
the transfer of metabolic pathways and cell biological innovations from now-extinct lineages, has yet to be accomplished. Because of these difficulties in interpreting the record
retained in molecular sequences, correlations with biochemical fossils and with the geological record need to be interpreted with caution. Advances to detect and pinpoint transfer
events promise to untangle at least a few of the intertwined histories of individual genes
within organisms and trace them to the organismal ancestors. Furthermore, analysis of the
shape of molecular phylogenetic trees may point towards organismal radiations that might
reflect early mass extinction events that occurred on a planetary scale.

**Keywords** Tree of life · Horizontal gene transfer · Late heavy bombardment · Coalescence

## 1 Trees, Forests, Webs: How to Depict Evolutionary History

### 1.1 Darwin's Coral of Life

Bifurcating trees have long been used to depict evolutionary history. The earliest depictions
were family trees, tracing the ancestry of individuals. It is noteworthy that these family

J.P. Gogarten (✉) · G. Fournier
Department of Molecular and Cell Biology, University of Connecticut, Storrs, CT 06269-3125, USA
e-mail: gogarten@uconn.edu

O. Zhaxybayeva
Department of Biochemistry and Molecular Biology, Dalhousie University, Halifax, NS B3H 1X5,
Canada

trees or pedigrees often have their root in the extant individual, and that the tree grows more branches as one moves back into the past (two parents, four grandparents, eight great-grandparents, etc.). In contrast, species trees trace back the history of groups of organisms to common ancestral groups. For example, all animals are thought to have evolved from a single-celled choanoflagellate ancestor (Lang et al. 2002; Philippe et al. 2004). Lamarck (1815), in his work on the classification of invertebrates, was the first to consider and depict species evolution as a bifurcating, tree-like process. Darwin recognized that species evolve, with natural and sexual selection causing a parent species to diverge into two different new species. As a logical consequence, he concluded that all living organisms could be traced back to a single ancestor (Darwin 1859). Darwin frequently used the term "tree of life", but he also pointed out that the term "coral of life" would be more appropriate. Like the tree of life, the base of the branches of coral is made out of extinct, dead organisms, whereas in botanical trees living cells are found throughout the tree (Darwin 1836–1844). The metaphor of a coral of life is also appropriate because in many corals the thin layer of living organisms sits on the massive, richly connected skeletons formed by their ancestors (compare Fig. 1).

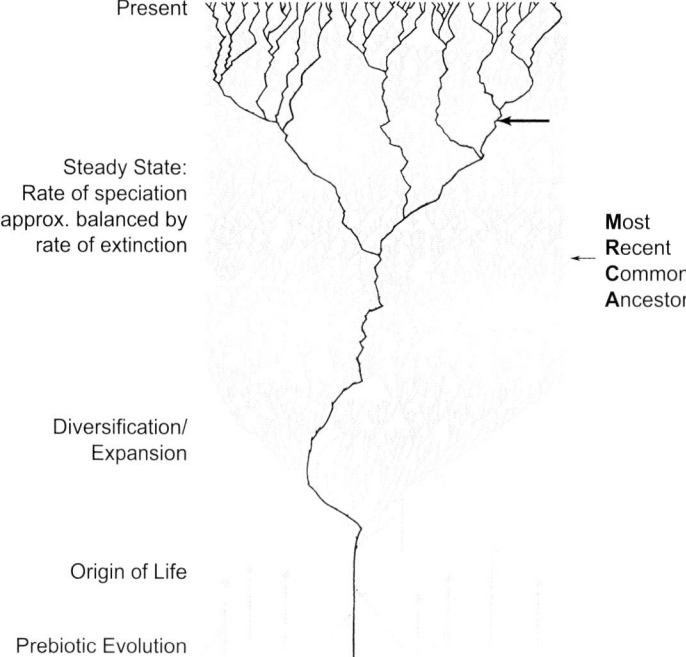

**Fig. 1** Schematic representation of the "coral of life". In his notebook, Charles Darwin suggested that the term "coral of life" might be preferable to the term tree of life (Darwin 1836–1844), because the layer of living, extant organisms rests on dead branches. In this diagram lineages with extant representatives are drawn in black, whereas extinct lineages are given in gray. The depicted scenario assumes that life originated from one or several prebiotic chemical processes, and that after its origin life adapted to different ecological niches available on the early Earth. Later on evolution represents an approximate balance between extinction and speciation; the evolution of extant lineages can be described by a coalescence process (Zhaxybayeva and Gogarten 2004). Note that the most recent common ancestor (MRCA) of all organisms only came into existence some time after the origin of life, and that many other, now-extinct lineages existed at the same time as well. The *horizontal black arrow* exemplifies that some genes that evolved early in life's history might have been transferred from now extinct lineages into extant lineages

## 1.2 Intertwined Trees and Coalescence

### 1.2.1 Exchange Groups, Species

The biological species concept describes species as groups of organism that can create offspring (Mayr 1942). As a consequence of recombination, within a species molecular phylogenies will not be congruent, and innovations can be shared within the species. In case of prokaryotes, procreation is independent of recombination, and genetic exchange is not limited within a species. Species boundaries, especially within prokaryotes, are fuzzy because some genes transfer across them (Gogarten and Townsend 2005; Hanage et al. 2005). This finding is not restricted to prokaryotes (Arnold 2006), but among eukaryotes these processes are mainly observed in recently diverged species (e.g., Grant et al. 2004).

The exchange of genetic material between divergent organisms greatly accelerates evolution (Jain et al. 2003). Horizontal gene transfer (HGT) can transfer an invention or an improvement made in one part of the tree of life to other lineages. Microorganisms need not reinvent a metabolic pathway, they can acquire it from other organism. It therefore is not surprising that genes encoding enzymes involved in metabolism are frequently found transferred between very divergent organisms (see Sect. 3.2.1).

However, HGT has not been so rampant as to create a continuum of phenotypes. A strain of *Escherichia coli* is clearly recognized by its pheno- and genotype as belonging to this species, which in turn is clearly placed within the gamma proteobacteria, which in turn are clearly identified as bacteria. This clear identification as *E.coli* is made even though the comparison of genomes of three *E. coli* strains revealed that less than 40% of the common shared gene pool was represented in all three genomes (Welch et al. 2002). Despite this enormous within-species diversity with respect to genome content, the measure traditionally used to define species boundaries in prokaryotes (70% DNA reassociation after melting the DNA mixture from two organisms) is in good agreement with percent SSU rRNA identity (about 98%) (Rossello-Mora and Amann 2001) and average nucleotide identity in those genes that are present in both organisms (more than 94%) (Konstantinidis and Tiedje 2005).

Currently, three extreme models are discussed to explain the cohesion within species (Gevers et al. 2005), and probably each of these is close to reality in some instances: First, in species with little recombination an advantageous mutation that spreads through a population will carry the whole genome along, thereby leading to populations that only recently diverged from their Most Recent Common Ancestor (MRCA); second, in species with a high rate of recombination, the biological species concept might apply, provided that within-species recombination is more frequent than gene flow between species; and third, species might appear coherent, because they separated a long time ago, and the intermediate forms went extinct.

### 1.2.2 Pedigrees and Species Trees

The trees of family history depicting the ancestry of specific individuals are all deeply intertwined through recombination and sexual procreation. Entwined together, the family trees of all the members of a species form the lineage depicting the descent of the entire species. Somewhere within this lineage there exists an individual from which all living members of a species can claim descent. In sexual interbreeding populations, the time to the most recent of these common ancestor is surprisingly short, because the total number of ancestors for each individual grows geometrically (at least initially) as one moves back in time. For example, the most recent common ancestor of all humans was estimated to have lived just a few thousand years ago, even under the assumption of mating occurring mainly in geographically determined subpopulations (Rohde et al. 2004). Note that the most recent common ancestor defined in the sense of a pedigree has not necessarily contributed any genes to all its descendants' genomes.

### 1.2.3 Gene Trees and Species Trees

In some regards things are simplified, if one adopts a gene-centered view of evolution in which organisms are considered the vessels constructed by the genes in order to be propagated into future generations (Dawkins 1976). Gene trees, in contrast to pedigrees, bifurcate in the same way as species trees—i.e., the individual gene usually is derived from only one immediate ancestor and it can have multiple descendents. Genomic recombination, gene transfer and lineage sorting can cause differences between trees defined by speciation and trees defined by the descent of a particular gene (Page and Holmes 1998; Felsenstein 2003). The most recent common ancestor of two individuals belonging to different species or of two alleles present in a population might have lived some time before the two species actually separated. In small populations with frequent recombination between individuals this time difference usually can be ignored because the time required for lineage sorting is negligible compared to the time separating two speciation events. However, in case of large populations fixation due to genetic drift requires more time on average; in populations with infrequent recombination between genomes, the time required for lineage sorting might no longer be negligible in a geographically dispersed population. Both of these factors, large population size and low recombination frequency, are common in prokaryotic populations, where procreation is not linked to recombination and effective population sizes are estimated to be larger than $10^8$ individuals (Lynch and Conery 2003; Lynch 2006). Longer time intervals for lineage sorting also result when different coexisting alleles provide distinct adaptive advantages.

The inverse problem—i.e. the gene ancestor being more recent than the species ancestor—occurs perhaps more frequently. Species boundaries are not impermeable, and often organisms from recently diverged species can and do interbreed (Arnold 2006). Thus, the most recent common ancestor of two individuals from different species might be more recent or more ancient than the speciation event that separates the two species. Although gene transfer across species boundaries occurs perhaps more frequently in prokaryotes than in eukaryotes, the differences between gene and species trees that are due to lineage sorting and gene flow occur in eukaryotes as well as prokaryotes.

### 1.2.4 Most Recent Common Ancestors in Gene and Species Trees

In sexual eukaryotic populations, whole genomes line up and can undergo homologous recombination. The linkage between different genes breaks down and each individual gene (or gene fragment) has its individual history. Uniparentally inherited genes in the recombining human population provide a useful illustration of individual gene histories being different from one another. The non-recombining part of the Y-chromosome coalesces to a common ancestor that lived only about 50,000 years ago (Thomson et al. 2000; Underhill et al. 2000), while the mitochondrial genomes in humans (inherited via the female lineage) have a common ancestor that existed about 200,000 years before present (Cann et al. 1987; Vigilant et al. 1991). Y-chromosome Adam never met mitochondrial Eve, and at the time the ancestral genes existed, many other mitochondrial genomes and Y-chromosomes co-existed in the same population. However, these other genes did not make it into today's human population. Y chromosome Adam and mitochondrial Eve were not alone, and many of their human contemporaries likely contributed other genes to the now existing gene pool; however, because of recombination, these other gene lineages cannot be traced as effectively as in the case of mitochondria and Y chromosomes.

### 1.2.5  Prokaryotic Evolution and the Trees of Life

In case of prokaryotes the situation appears even more complicated. Recombination is not restricted to members of the same species, since genes can be transferred between divergent organisms. Furthermore, the evolutionary lineages of prokaryotic genomes are surrounded by a halo of mobile genetic elements that only sometimes reside in genomes, but perhaps more often reside in phages, viruses and plasmids. Many genes found in microbial genomes were acquired from this cloud of transitory genes, and often these genes do not persist in the genome for long periods of time (Daubin et al. 2003; Gogarten and Townsend 2005; Hsiao et al. 2005).

This pool of horizontally transferred genes is characterized by nucleotide (high A and T content) and codon usage biases. These biases might reflect the mutation bias (mutations from GC pairs to AT pairs occur more frequently than vice versa), and indicate that these genes are under weak purifying selection only. Most of the genes transferred between prokaryotes belong to this transitory, weakly selected pool of genes; however, all types of genes, including ribosomal RNA and protein coding genes, are known to have been transferred (Gogarten et al. 2002; Lawrence and Hendrickson 2003) (see Sect. 3 for more discussion).

In view of gene transfer between divergent species, it is not surprising to find that different genes have different histories. For example, the tyrosyl-tRNA synthetases of animals and fungi group with their haloarchaeal homologs (Huang et al. 2005), the ATP synthases of the Deinococcaceae (a group of bacteria, with bacterial cell wall, bacterial membranes and bacterial ribosomes) group with homologs from archaea (Olendzenski et al. 2000; Lapierre et al. 2006). Clearly, every gene has its own phylogeny. Ignoring intragenic recombination, the history of any individual gene can be described by a bifurcating tree. Yet these trees are different from one another, and in particular, the most recent common ancestors of individual genes, even if they are present in all extant organisms, might have existed at different times and in different lineages (see Fig. 1 and Zhaxybayeva and Gogarten (2004)). Relationships between organismal lineages may generally take the form of a bifurcating tree. However, organismal evolution certainly will include some reticulations. For example, eukaryogenesis (the emergence of the eukaryotic cell) involved at least the symbiosis between an archaea related host cell and a bacterial endosymbiont that became the ancestor of today's mitochondria (Gogarten et al. 1989; Margulis 1995; Martin and Koonin 2006). More importantly, while we might be able to define an organismal most recent common ancestor, we need to remain aware that most recent common ancestors of individual genes probably did not exist together inside the organismal most recent common ancestor.

The extension of population genetics principles to the evolution of microbial species does not depend on horizontal exchange during the early evolution of life being more frequent than today (Woese 1998): even the low rates of exchange between divergent organisms observed today are sufficient justification. However, if rates of genetic exchange were higher than they are today, this would argue even more strongly for the application of population genetics to the early evolution of life.

## 1.3 The Shape of the Tree of Life

### 1.3.1 Forces and Processes that Lead to Bush-Like Phylogenies

Many evolutionary and analytical processes shape phylogenetic trees. This shape can be described through "lineages through time" plots that give the number of lineages existing over time (Raup 1985). In many instances it was observed that the number of lineages increases in an exponential fashion over time (Martin et al. 2004). A rapid succession of speciation events leads to a bush-like phylogeny, also known by the term radiation (Rokas and Carroll 2006). Other processes that lead to phylogenies with similar appearance are extinction events and the under-sampling of present species. Species radiations can follow a large ecological change that provides for many empty ecological niches that can be occupied by subpopulations that subsequently evolve into distinct species. These ecological changes can be caused by biology [e.g., disruptions due to invasive species (Gurevitch and Padilla 2004)] or by geological or astronomical processes [e.g., mass extinctions caused by meteorite impacts (Raup 1989; Gogarten-Boekels et al. 1995; Nisbet and Sleep 2001)].

### 1.3.2 Coalescence and Long, Empty, Basal Branches

Coalescence is the process of tracing lineages backwards in time to their common ancestors (Felsenstein 2003). Every two extant lineages coalesce to their most recent common ancestor. Eventually, all lineages coalesce to one lineage. The shape of the resulting tree depends on whether all lineages or only extant ones are considered. For example, animal phylogenies frequently include extinct organisms known from the fossil record. In contrast, molecular phylogenies are based on molecules from extant species only, with the possible exception of a few species that went extinct recently, but whose DNA has survived to the present day [e.g., the wooly mammoth or the Neanderthal humans (Gibbons 2005; Willerslev and Cooper 2005; Donoghue and Spigelman 2006)].

Considering only lineages with extant representatives greatly impacts the shape of the resulting phylogeny. A simple steady-state model in which speciation is balanced by extinction (compare Fig. 1) results in a faster than exponential growth in lineages through time plots (Zhaxybayeva and Gogarten 2004). The branching pattern in the phylogenies derived from this simple model is described by the Kingman coalescence (Felsenstein 2003), in which the coalescence of the last two lineages to their common ancestor occupies on average about half the time of the total coalescence process. Under the assumption of a steady state between extinction and speciation, for every group defined by shared ancestry, on average the two deepest branches are expected to cover half the time the group is in existence (Zhaxybayeva and Gogarten 2004). In the case of some molecules and for some features of the trees of life this expectation is met, for example, the ancestors of the archaeal and bacterial domain frequently are connected to their common ancestor by long branches (Gogarten-Boekels et al. 1995). Deviations from this expectation point towards processes that deviate from the simple steady-state model (see Sect. 1.3.3).

### 1.3.3 The Bacterial Radiation

One of the deviations from the expectation of long basal branches occurs in the early evolution of the bacterial domain. Bergey's manual, the main reference for bacterial and archaeal taxonomy recognizes 24 different bacterial phyla (Garrity et al. 2004) (see Fig. 2). Many

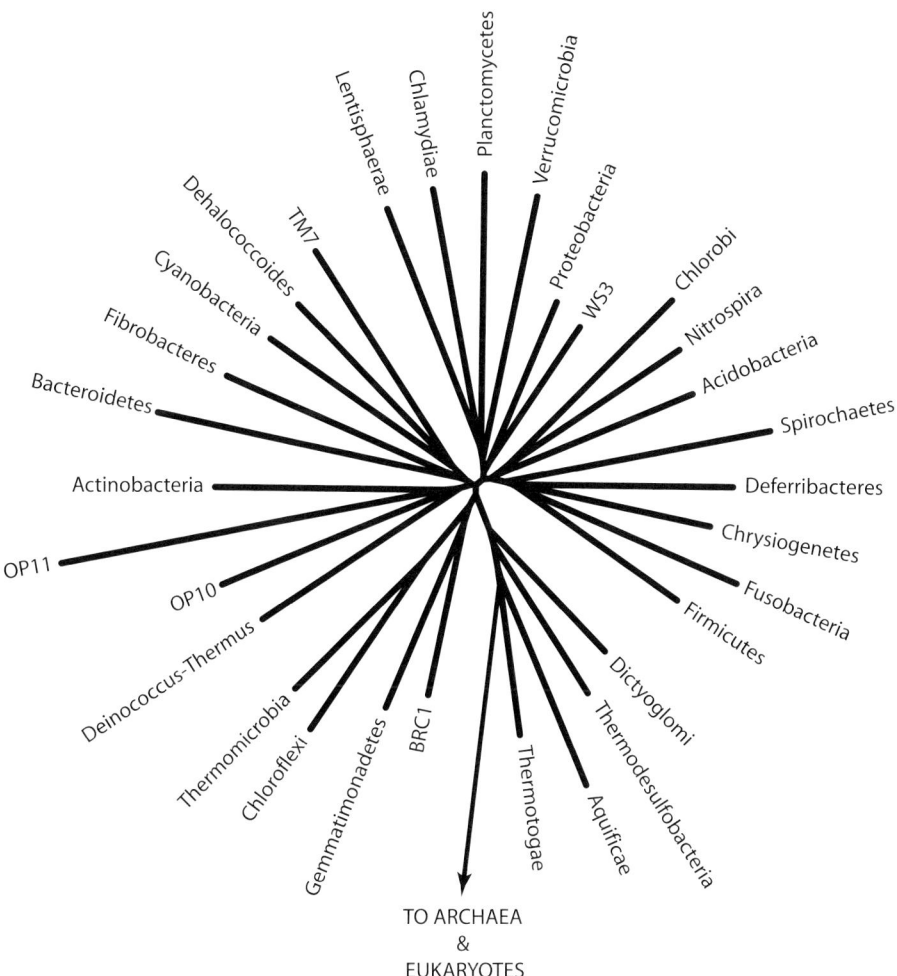

**Fig. 2** Unrooted phylogenetic tree for 31 bacterial phyla listed in the Ribosomal Databank (RDP) version 9.44 (Cole et al. 2005). Each phylum is represented by one branch. The alignment and tree were obtained from the RDP website (http://rdp.cme.msu.edu/). Note the deviation in the tree shape from "long branches leading to MRCA" observed under simple birth-death models (Zhaxybayeva and Gogarten 2004). The shape of the shown tree suggest an actual radiation (a rapid succession of bifurcations) at the base of the bacterial domain

additional bacterial phyla are recognized based on amplification of genes encoding ribosomal RNA isolated from the environment (e.g., Ley et al. 2006). Many of these bacterial phyla include only organisms that have not been cultured at present (Schloss and Handelsman 2004). All of these ribosomal RNA defined phyla (see Sect. 2) diverged over a very small phylogenetic distance very close to the root of the bacterial domain (Fig. 2). At present the branching order between the bacterial phyla has to be considered as unresolved. The reasons for this are two-fold: First, the outgroup that can be used to place the domain ancestor relative to the bacterial phyla (either archaea or the eukaryotic nucleocytoplasm) is connected to the bacteria by a very long branch; the placement of the end of this branch inside the bacterial domain is difficult, and might be subject to artifacts. Second, the branches separating the different phyla are short; even omitting the outgroup the relationships cannot be resolved with confidence. These findings suggest that a real radiation occurred at the base of the bacterial domain. One reason for this radiation to occur might have been a mass extinction that was triggered by the late heavy bombardment (Gomes et al. 2005). Only organisms that before the bombardment had adapted to deep-surface environments had a chance to survive this violent episode in Earth's history. After the end of the late heavy bombardment these survivors would have found many ecological niches closer to Earth's surface into which to adapt (Raup 1989; Gogarten-Boekels et al. 1995; Nisbet and Sleep 2001).

## 2 The Tree of Life According to Ribosomal RNA

### 2.1 Attempts to Create a Prokaryotic Taxonomy

Traditionally, single-celled organisms without a nucleus were placed into a group called monera (Haeckel 1866) or prokaryotes (Stanier and Van Niel 1962). Classification of single-celled organisms within this group was initially based on morphology (e.g., cell shape), physiology and biochemistry (e.g., fermentation type, and temperature ranges for growth) [see more on the history of prokaryotic taxonomy in Rossello-Mora and Amann (2001); Olendzenski et al. (2004); Sapp (2005)].

The study of ribosomal RNA (rRNA) molecules by Carl Woese, George Fox and colleagues (Woese and Fox 1977) launched a new era in approaches to phylogeny and provided a measure by which all organisms could be compared. rRNA molecules are ubiquitous in distribution, and perform the same function in all organisms. Furthermore, rRNAs contain both highly conserved and highly variable regions that can be compared between organisms of very different degrees of relationship (Woese 1987). The ability to amplify rDNA (rRNA encoding DNA) outside the living organism allows analyses not only from cultured organisms, but also from environmental samples (e.g. Ley et al. 2006; Sogin et al. 2006). With the isolation of environmental rDNA sequences that did not match any cultured organisms came the realization that the diversity of the prokaryotes had been vastly underestimated (Schloss and Handelsman 2004; Sogin et al. 2006). It is generally believed that, depending on environment sampled, only between 0.001% and 15% of existing prokaryotic diversity has been cultured (Amann et al. 1995). Sequencing of rDNA has become a standard procedure to characterize cultures and environmental samples. On December 6, 2006, more than 286,000 aligned sequences were available in the ribosomal RNA databank (Cole et al. 2005); this wealth of sequence information facilitates phylogenetic placement of organisms characterized by only small stretches of rDNA sequences (Sogin et al. 2006).

### 2.2 The Ribosomal RNA Based Tree of Life

Ribosomal RNA (rRNA) has been used to reconstruct the tree of life. Figure 3 summarizes some of the features frequently associated with the rRNA phylogeny. The ribosomal RNA-based tree of life clearly distinguishes three domains: Bacteria, Archaea, and Eukaryotes (Woese et al. 1990).

**Fig. 3** Diagram of the small subunit ribosomal RNA-based tree of life. According to this tree, all life can be divided into three monophyletic groups, or domains (Bacteria, Archaea and Eukaryotes). The tree is strictly bifurcating, shows no reticulations, and only extant lineages are depicted (since 16S rRNA genes are sequenced from contemporary organisms). The tree is based on a single molecular phylogeny. All lineages on the tree trace back to a most recent common ancestor (MRCA). (The tree used as a backbone was photographed in the Point Pleasant Park in Halifax, Nova Scotia)

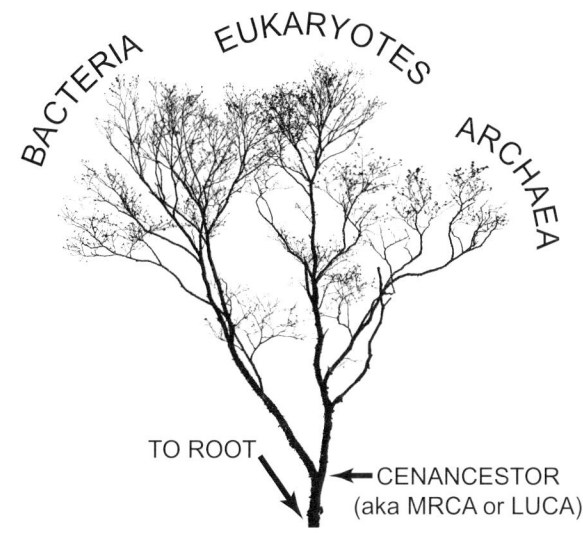

### 2.2.1 The Root of the Tree of Life

The tree of life calculated from rRNA sequences has to be considered unrooted. The placement of the root in the branch leading to the bacteria was not determined using rRNA, rather the placement of the root is based on the study of other evolutionary constrained molecules (ATPase, elongation factors, signal recognition particles) that underwent gene duplications early in life's history (Gogarten et al. 1989; Iwabe et al. 1989; Gogarten and Taiz 1992; Brown and Doolittle 1995; Gribaldo and Cammarano 1998); an overview on the current state of the sometimes controversial and ongoing debate on where to place the root in the tree of life is included in Zhaxybayeva et al. (2005).

The root of a molecular phylogeny refers to the deepest (oldest) bifurcation in the tree (and not to the bottom of the trunk). The organism represented by this deepest bifurcation is also known as the Most Recent Common Ancestor (MRCA) of all life, as cenancestor (Fitch and Upper 1987), or as the Last Universal Common Ancestor (LUCA). This organism existed at a rather long distance from the origin of life. Due to gene transfer, this organism—or population of organisms—almost certainly did not harbor all of the MRCAs of the different molecular systems found in all of today's organisms (see Sect. 1.2.4); however, this organism probably possessed membranes used in chemiosmotic coupling (Gogarten and Taiz 1992; Pereto et al. 2004), its genes were arranged on one or more chromosomes, and it used a complete genetic code encoding 20 amino acids (Anantharaman et al. 2002; Delaye et al. 2005).

### 2.2.2 Ribosomal RNA and Horizontal Gene Transfer

Like other molecular phylogenies, the small subunit rRNA based tree of life is depicted as a strictly bifurcating tree (e.g., Woese et al. 1990). However, some organisms are known to harbor multiple, sometimes rather divergent, rRNA operons (Mylvaganam and Dennis 1992; Yap et al. 1999; Acinas et al. 2004) and there are indications that ribosomal RNA operons can be transferred between different species, and that subsequent to the transfer recombination between the divergent copies can occur (Dennis et al. 1998; Yap et al. 1999;

Gogarten et al. 2002; Boucher et al. 2004). Although these transfer and recombination events can make it difficult to identify organisms accurately (Morandi et al. 2005), most of these events took place between closely related organisms. Since homologous recombination requires short stretches of identical sequences (Shen and Huang 1986), the conserved regions of rRNA may be more prone to undergo recombination than protein-coding genes. In the latter, the redundancy of the genetic code permits synonymous substitutions, allowing for rapid divergence on the DNA sequence level that prevents homologous recombination (Gogarten et al. 2002). These recombination events are a feature that the rRNA may share with the genome as a whole: both are mosaics in which the different parts can have different histories.

### 2.2.3 *Other Molecular Markers and Phylogentic Reconstruction Artifacts*

Regardless of the difficulties, ribosomal RNA has become the gold standard for microbial taxonomy. As more molecular data became available, different gene trees were reconstructed and compared to rRNA trees. Some were in agreement with rRNA trees; others were not. One explanation for the incongruence of phylogenetic trees constructed using different markers is HGT; however, another important consideration is artifacts of phylogenetic reconstruction. The study of deep branching eukaryotic lineages reveals that one should not lose sight of the fact that the small subunit (SSU) rRNA-based tree of life at best depicts the evolutionary history of a single molecule. In some lineages this molecule evolved faster than in others. Many phylogenetic reconstruction algorithms tend to group long branches together, even if they are not specifically related (Felsenstein 1978). For example, the microsporidia, considered a deep-branching lineage based on SSU rRNA (Vossbrinck et al. 1987), have been recognized as a more recently emerging, rapidly evolving relatives of the fungi (Embley and Hirt 1998).

Branches in a molecular phylogeny are scaled with their lengths proportional to the amount of sequence change that occurred along a branch. Usually these branches are not scaled with respect to time. Based on the analyses of protein coding genes, the tree of eukaryotic evolution appears more bush-like than is the case in rRNA-based phylogenies. The clear distinction between crown group and deep branching eukaryotes appears to have been a particularity of ribosomal RNA (Simpson and Roger 2004).

## 3  Detecting HGT and Measuring Its Extent

Several types of phylogenetic and non-phylogenetic methods for detection of HGT events are being developed and constantly improved. Due to varying underlying assumptions, different methods detect HGTs at different phylogenetic distances and of different age, and therefore often return non-overlapping sets of HGT candidates (Ragan 2001a; Lawrence and Ochman 2002; Ragan 2002). In addition, all methods are imperfect and suffer from high rates of false positives and false negatives (Cortez et al. 2005).

### 3.1  Methods to Detect HGT

### 3.1.1 *Approach #1: Surrogate Methods*

Since a horizontally transferred gene comes from a different genomic background, its nucleotide sequence can contain signatures of its previous "home" genome. One group of HGT detection methods uses either atypical nucleotide composition (Lawrence and Ochman 1997) and/or atypical codon usage patterns (Lawrence and Ochman 1998) to infer which genes in a genome are instances of HGT. Because these methods do not rely on phylogenetic reconstruction, they are sometimes called surrogate methods (Ragan 2001b). Because genes "ameliorate" (i.e., adapt to the signatures of its new genome) quickly (Lawrence and Ochman 1997), these methods are applicable to detection of very recent transfers only. While easily applicable to completely sequence genomes, these methods were criticized for returning high rates of false positives and negatives (Koski et al. 2001; Azad and Lawrence 2005; Cortez et al. 2005). An application of a compositional approach to 116 available genomes revealed that the number of recently transferred genes ranges from 0.5% in pea aphid endocellular symbiont *Buchnera* sp. APS to 25.2% in anaerobic methane-producing archaeon *Methanosarcina acetivorans* C2A (Nakamura et al. 2004).

Other surrogate methods are applicable only to very closely related organisms. Extent of HGT among the closely related organisms can be judged through a comparison of gene content of their genomes. For example, three sequenced *Escherichia coli* genomes share only 39.2% genes, while each *E. coli* genome separately has a substantial proportion of genes absent from two other strains (585 genes in non-pathogenic *E. coli* K12, 1,623 genes in uropathogenic *E. coli* CFT073 and 1,346 genes in enterohaemorrhagic *E. coli* O157:H7) (Welch et al. 2002).

### 3.1.2 Approach #2: Unusual Phyletic Patterns

Another way to assess whether a gene could have been transferred is to do a BLAST (Altschul et al. 1997) (or any other similarity or clustering algorithm) search of a sequence database (such as NCBI's non-redundant database) to find homologs to the query gene, define a gene family using this information and to look at the taxonomic distribution of members of the gene family (so called phyletic patterns). A significant top-scoring BLAST hit itself may suggest the most similar sequence in the database; this has been used to get a rough estimate of number of horizontally transferred genes in a genomes [e.g., bacterium *Thermotoga maritima* genome was proposed to have 24% of horizontally transferred genes from Archaea based on the top-scoring BLAST hits (Nelson et al. 1999)]. However, a top-scoring BLAST hit might not represent a sequence grouping with the query sequence, if a phylogenetic tree is reconstructed (Koski and Golding 2001), and therefore this is not a reliable approach for HGT detection. Phyletic patterns, however, can be further used to infer whether the patchy distribution of a gene is most parsimoniously explained by HGTs or by gains and losses (Snel et al. 2002; Kunin and Ouzounis 2003; Mirkin et al. 2003). The outcome of the inferences depends on a value of "HGT penalty" (a ratio between HGT events and gene losses), which is not known, but has to be estimated or set a priori, and different studies disagree on what value to use. A recent attempt to apply this type of approach to 165 microbial genomes resulted in an inference of ~40,000 horizontal gene transfers, ~90,000 gene losses and over ~600,000 vertical transfers in all analyzed gene families (Kunin and Ouzounis 2003). While the numbers given here may be interpreted as showing only a limited number of HGTs among the 165 genomes, one should not forget that those estimates do not consider HGTs resulting in orthologous replacement, which could constitute a substantial part of a genome (see approach #3).

### 3.1.3 Approach #3: Phylogenetic Incongruence

These methods rely on reconstruction of phylogenetic trees for sets of orthologous genes and comparison of them to each other, assuming that trees with unexpected (i.e., topologically incongruent) phylogenetic histories are results of horizontal gene transfer. These methods typically use the expected phylogenetic history (organismal tree) as a reference tree for the comparison. One of the earliest such analyses came from comparison of gene families from *Aquifex aeolicus* genome to rRNA tree, with the conclusion that one gets "different phylogenetic placements based on what genes are used" (Pennisi 1998). Later, 205 gene families from 13 gamma-proteobacteria were compared to their concatenated phylogeny (Lerat et al. 2003) and 22,432 gene families from 144 prokaryotic genomes were compared to the supertree constructed from compatible bipartitions (Beiko et al. 2005). Choices of reference trees (as a proxy of organismal trees) include rRNA trees, genome trees, trees derived from concatenation of selected datasets, or trees (possibly only partially resolved) supported by a plurality of sets of orthologous genes (consensus trees or supertrees). Ideally, if the organismal tree is not known, all possible tree topologies should be tried as a reference tree [examples of such methodologies to analyze four and five genomes are in (Zhaxybayeva and Gogarten 2002; Zhaxybayeva et al. 2004a))]. However, this approach is computationally impossible for large-scale analyses, due the vast number of possible tree topologies. As an alternative, the trees to be analyzed could be broken into smaller pieces (e.g., bipartitions or quartets). There is a significantly smaller number of possible bipartitions/quartets than trees for a given number of analyzed genomes, and therefore all possibilities can be evaluated (giving rise to bipartition and quartet decomposition analyses) (Zhaxybayeva et al. 2004b, 2006). For example, analysis of 1,128 gene families in 10 cyanobacterial genomes resulted in 685 gene families with phylogenetic trees incongruent with a reference tree supported by a plurality of gene families (Zhaxybayeva et al. 2006), and hence providing candidates for HGTs.

One drawback of phylogenetic approaches (aside from artifacts of phylogenetic reconstruction, which are not discussed here) is that HGTs between neighboring taxa on the reference tree are invisible for these methods, because these transfers do not result in a change of tree topology. Another drawback is that a weak phylogenetic signal in a set of orthologous genes often results in an unresolved (or unsupported) tree topology. The latter topologies cannot be used to delineate HGT events, but they also should not be used [although unfortunately they are sometimes used, e.g. Snel et al. (2002)] as evidence for absence of HGT. The third drawback is that often a choice of reference tree may bias the results of HGT quantification. This is particularly a problem when a reference tree is obtained as a plurality tree (or supertree) from the same sets of genes that are subject to HGT detection in the study. The underlying assumption is that the number of HGTs should be minimized, and that the plurality of genes therefore reflects organismal evolution and not a reoccurring pattern of HGT.

## 3.2  Biological Consequences of HGT

### 3.2.1  Types of Transferred Genes

Jain et al. (1999) proposed that informational genes should be less likely to be transferred in comparison to operational genes because the former are part of highly interactive molecular assemblies. Although genome-wide analyses indicate that all types of genes are among inferred HGTs, certain functional categories are over- or under-represented among horizontally transferred genes as compared to their genome-wide distribution. In HGTs detected by a compositional method, Nakamura et al. (2004) reported bias of HGTs in functional categories of cell surface, DNA binding and pathogenicity related genes. In HGTs detected by a phylogenetic method, Beiko et al. (2005) reported over-representation of "energy metabolism" and "mobile and extrachromosomal element functions" among genes with discordant bipartitions in their phylogenies (i.e., among HGT candidates). However, transfer distance may need to be taken into account when types of transferred genes are evaluated. For example, in genome-wide analyses of HGT in cyanobacteria, no bias toward a particular functional category is found for HGTs inferred to occur within cyanobacteria, but an excess of metabolic genes and decrease of informational genes is observed in transfers inferred to occur between cyanobacteria and other phyla (Zhaxybayeva et al. 2006).

### 3.2.2 How Tangled is the Web of Life?

Several recent large-scale genome analyses attempted to estimate HGT across all available sequenced genomes. Beiko et al. (2005) analyzed gene families from 144 prokaryotic genomes. They compared 95,194 strongly supported bipartitions from 22,432 gene family trees and found that 86.6% (82,473) of bipartitions are in agreement with bipartitions of a reference supertree (constructed from all highly supported bipartitions). Ge et al. (2005) analyzed 297 gene families from gene clusters of 40 microbial genomes and compared gene family phylogenies with a genome tree, applying an additional criteria to increase stringency. This resulted in 33 HGT events in 11.1% of 297 analyzed gene families. In the latter study, the investigators limited themselves to analyses of very strictly defined (ubiquitous) core genes, hence severely underestimating the number of HGT events. Analyzing larger number of gene families shows that in cyanobacteria as many as 61% of gene families may be affected by HGT (Zhaxybayeva et al. 2006). It should also be noted that all these analyses are based on phylogenetic approach (see Sect. 3.1.3) and hence ignore the pool of transitory genes (see Sect. 1.2.4), therefore underestimating the amount of gene transfer.

Nevertheless, attempts are still being made to resolve the tree of life using the vast amount of available genomic information. Recently Ciccarelli et al. (2006) performed an analysis of 191 genomes where they aimed to find genes with "indisputable orthology" and ended up with only 31 such genes (mainly ribosomal proteins). The joint analysis of these genes results in a phylogeny that has some resolution; however, the tree's backbone is poorly resolved ($<80\%$), perhaps due to insufficient phylogenetic information and hence does not provide a clearly resolved strictly bifurcating tree. Dagan and Martin (2006) referred to this tree as "the tree of 1 percent" since only 1% of the genes were determined to fit the null hypothesis of a single strictly bifurcating tree of life. It is not sufficient to describe evolutionary relationships among only 1% of the genes and to claim that it represents an evolutionary history of the organisms.

### 3.2.3 Horizontal Gene Transfer and Genetic Life Rafts

As one moves down the tree of life, the branches tend to become sparse. Rather than suggesting that this reveals early life as less diverse, this might be a natural consequence of the large amount of extinction that has occurred over the history of life on Earth. Random models of extinction and speciation show that such sparse "deep branches" in phylogenies are expected, even when actual diversity remains constant over time (see Sect. 1.3). This observation has profound consequences in light of the large amount of horizontal gene transfer seen across today's extant lineages.

There is every reason to believe that HGT was at least as prevalent in the distant past as it is today (Woese 1998). However, as we move backward in time, it becomes less likely that the participants in any transfer event have any surviving descendants. Consequently, the more ancient the transfer event, the less likely that the donor lineage will still be in existence, even if the recipient has living descendants.

Furthermore, the more ancient the transfer event, the more divergent an extinct donor is likely to have been from currently existing lineages. In many cases, these genes may be the only surviving biological signature from entire clades of extinct organisms, a "genetic life-raft". If these genes have any detectable homologs in existing species, the transferred gene will appear more deeply branching in a phylogeny of the gene family, even if the organismal tree shows the recipient lineage as highly derived. For example, the pyrrolysine aminoacyl-tRNA synthetase gene (PylS) is found only within a single family of derived Euryarchaea (the Methanosarcinales) and in one bacterial species (*Desulfitobacterium hafniense*) (Srinivasan et al. 2002). However, a phylogenetic tree of related aminoacyl-tRNA synthetases roots PylS more deeply than the MRCA of Bacteria and Archaea. This gene may well represent an ancient transfer from an extinct lineage. Since the amino acid pyrrolysine is only used at a single position within a specific set of enzymes unique to methanogenesis (the metabolic process of reducing single-carbon compounds to methane for energy production) that have no identified homologs (mono-, di-, and trimethylamine methyltransferases) (Ibba and Soll 2002), this may also suggest that this system must have originated outside the lineages that lead to extant methanogens. Evolution of a system to use an entirely novel amino acid would almost certainly require considerable positive selection, requiring (and resulting in) far more uses for Pyl. However, transfer of even a single advantageous gene requiring Pyl, along with the Pyl incorporation machinery, would result in its retention in the recipient lineage.

Perhaps even more interestingly, if there are no surviving homologs at all (as is the case with the methylamine methyltransferases), these genes will be listed as "orphans", and often may be uncharacterized simply for lack of comparative evidence from other more well-studied genes. Several pathways in methanogenesis contain many such orphans (Fournier, unpublished). As these pathways are found exclusively within a derived group of the Euryarchaea, the life-raft model provides a succinct explanation of their presence, which would otherwise have to be explained by either very rapid evolution, or ancestral presence with selective loss in other lineages.

Geological signatures and biochemical fossils (e.g., Schidlowski et al. 1983; Brocks et al. 2003) can be correlated to metabolic processes and physiological properties. However, in mapping these traits on to the tree of life (House et al. 2003), gene transfer, and in particular gene transfer form extinct lineages needs to be taken into consideration. A metabolic pathway might have existed a long time before it was transferred into an extant lineage; the geological signature or chemical fossil should not considered automatic proof that the lineage that today carries this trait was also responsible for creating the signature. Careful consideration of molecular phylogenies, including ancient gene duplications and gene transfers events, promises to yield a better understanding of the order in which metabolic pathways emerged.

## 4 Concluding Remarks

Advances in genome sequencing have revolutionized our understanding of microbial evolution. Reconstructing the evolutionary history of organisms turned out to be more difficult than anticipated two decades ago. Gene transfer and unequal substitution rates create artifacts that initially were not recognized.

Extending population genetic principles to larger taxonomic units helps to unravel evolutionary histories not only of the organisms, but also of the key pathways that allowed early organisms to survive on early Earth. The emerging picture of life's history less resembles a tree, but a coral formed by a tangle of lineages with many interconnections generated through the transfer of genes, or through the fusion of independent lines of descent. Most of the organisms that form the bulk of this "coral of life" do not have extant representatives. However, some these extinct lineages transferred genes to lineages still alive today.

**Acknowledgements**    Work in JPG's lab was supported the NSF (MCB-0237197), the NASA Applied Information Systems Research (NNG04GP90G) and NASA Exobiology Programs (NNX07AK15G). OZ is supported through a CIHR Postdoctoral Fellowship and is an honorary Killam Postdoctoral Fellow at Dalhousie University.

# References

S.G. Acinas, L.A. Marcelino, V. Klepac-Ceraj, M.F. Polz, J. Bacteriol. **186**(9), 2629–2635 (2004)

S.F. Altschul, T.L. Madden, A.A. Schaffer, J. Zhang, Z. Zhang, W. Miller, D.J. Lipman, Nucleic Acids Res. **25**(17), 3389–3402 (1997)

R.I. Amann, W. Ludwig, K.H. Schleifer, Microbiol. Rev. **59**(1), 143–169 (1995)

V. Anantharaman, E.V. Koonin, L. Aravind, Nucleic Acids Res. **30**(7), 1427–1464 (2002)

M. Arnold, *Evolution through Genetic Exchange* (Oxford University Press, Oxford, 2006)

R.K. Azad, J.G. Lawrence, PLoS Comput. Biol. **1**(6), e56 (2005)

R.G. Beiko, T.J. Harlow, M.A. Ragan, Proc. Natl. Acad. Sci. USA **102**(40), 14332–14337 (2005)

Y. Boucher, C.J. Douady, A.K. Sharma, M. Kamekura, W.F. Doolittle, J. Bacteriol. **186**(12), 3980–3990 (2004)

J.J. Brocks, R. Buick, R.E. Summons, G.A. Logan, Geochimica Cosmochimica Acta **67**(22), 4321–4335 (2003)

J.R. Brown, W.F. Doolittle, Proc. Natl. Acad. Sci. USA **92**(7), 2441–2445 (1995)

R.L. Cann, M. Stoneking, A.C. Wilson, Nature **325**(6099), 31–36 (1987)

F.D. Ciccarelli, T. Doerks, C. von Mering, C.J. Creevey, B. Snel, P. Bork, Science **311**(5765), 1283–1287 (2006)

J.R. Cole, B. Chai, R.J. Farris, Q. Wang, S.A. Kulam, D.M. McGarrell, G.M. Garrity, J.M. Tiedje, Nucl. Acids Res. **33**(Suppl. 1), D294–D296 (2005)

D.Q. Cortez, A. Lazcano, A. Becerra, In Silico Biol. **5**(5–6), 581–592 (2005)

T. Dagan, W. Martin, Genome Biol. **7**(10), 118 (2006)

C. Darwin, *Charles Darwin's Notebooks, 1836–1844, Transcription Published in 1987* (Cornell University Press, Ithaca, 1836–1844)

C. Darwin, *On the Origin of Species by Means of Natural Selection, or the Preservation of Favoured Races in the Struggle for Life* (John Murray, London, 1859)

V. Daubin, E. Lerat, G. Perriere, Genome Biol. **4**(9), R57 (2003)

R. Dawkins, *The Selfish Gene* (Oxford University Press, Oxford, 1976)

L. Delaye, A. Becerra, A. Lazcano, Orig. Life Evol. Biosphere **35**(6), 537–554 (2005)

P.P. Dennis, S. Ziesche, S. Mylvaganam, J. Bacteriol. **180**(18), 4804–4813 (1998)

H. Donoghue, M. Spigelman, Proc. R. Soc. B **273**(1587), 641–642 (2006)

T.M. Embley, R.P. Hirt, Curr. Opin. Genet. Dev. **8**(6), 624–629 (1998)

J. Felsenstein, Syst. Zool. **27**, 401–410 (1978)

J. Felsenstein, *Inferring Phylogenies* (Sinauer, Sunderland, 2003)

W.M. Fitch, K. Upper, Cold Spring Harb. Symp. Quant. Biol. **52**, 759–767 (1987)

G.M. Garrity, J.A. Bell, T.G. Lilburn, *Bergey's Taxonomic Outline* (Springer, New York, 2004). http://dx.doi.org/10.1007/bergeysoutline200310

F. Ge, L.-S. Wang, J. Kim, PLoS Biol. **3**(10), e316 (2005)

D. Gevers, F.M. Cohan, J.G. Lawrence, B.G. Spratt, T. Coenye, E.J. Feil, E. Stackebrandt, Y. Van de Peer, P. Vandamme, F.L. Thompson, J. Swings, Nat. Rev. Microbiol. **3**(9), 733–739 (2005)

A. Gibbons, Science **310**(5756), 1889 (2005)

J.P. Gogarten, W.F. Doolittle, J.G. Lawrence, Mol. Biol. Evol. **19**(12), 2226–2238 (2002)

J.P. Gogarten, H. Kibak, P. Dittrich, L. Taiz, E.J. Bowman, B.J. Bowman, M.F. Manolson, R.J. Poole, T. Date, T. Oshima et al., Proc. Natl. Acad. Sci. USA **86**(17), 6661–6665 (1989)

J.P. Gogarten, L. Taiz, Photosynth. Res. **33**, 137–146 (1992)

J.P. Gogarten, J.P. Townsend, Nat. Rev. Microbiol. **3**(9), 679–687 (2005)

M. Gogarten-Boekels, E. Hilario, J.P. Gogarten, Orig. Life Evol. Biosphere **25**(1–3), 251–264 (1995)

R. Gomes, H.F. Levison, K. Tsiganis, A. Morbidelli, Nature **435**(7041), 466–469 (2005)

P.R. Grant, B.R. Grant, J.A. Markert, L.F. Keller, K. Petren, Evolution **58**(7 ), 1588–1599 (2004)

S. Gribaldo, P. Cammarano, J. Mol. Evol. **47**(5), 508–516 (1998)

J. Gurevitch, D.K. Padilla, Trends Ecol. Evol. **19**(9), 470–474 (2004)
E. Haeckel, *Generelle Morphologie der Organismen: Allgemeine Grundzüge der organischen Formen-Wissenschaft mechanisch begründet durch die von Charles Darvin reformierte Descendenz-Theorie* (Georg Riemer, Berlin, 1866)
W.P. Hanage, C. Fraser, B.G. Spratt, BMC Biol. **3**(1), 6 (2005)
C.H. House, B. Runnegar, S.T. Fitz-Gibbon, Geobiology **1**(1), 15–26 (2003)
W.W. Hsiao, K. Ung, D. Aeschliman, J. Bryan, B.B. Finlay, F.S. Brinkman, PLoS Genet. **1**(5), e62 (2005)
J. Huang, Y. Xu, J.P. Gogarten, Mol. Biol. Evol. **22**(11), 2142–2146 (2005)
M. Ibba, D. Soll, Curr. Biol. **12**(13), R464–R466 (2002)
N. Iwabe, K. Kuma, M. Hasegawa, S. Osawa, T. Miyata, Proc. Natl. Acad. Sci. USA **86**(23), 9355–9359 (1989)
R. Jain, M.C. Rivera, J.A. Lake, Proc. Natl. Acad. Sci. USA **96**(7), 3801–3806 (1999)
R. Jain, M.C. Rivera, J.E. Moore, J.A. Lake, Mol. Biol. Evol. **20**(10), 1598–1602 (2003)
K.T. Konstantinidis, J.M. Tiedje, Proc. Natl. Acad. Sci. USA **102**(7), 2567–2572 (2005)
L.B. Koski, G.B. Golding, J. Mol. Evol. **52**(6), 540–542 (2001)
L.B. Koski, R.A. Morton, G.B. Golding, Mol. Biol. Evol. **18**(3), 404–412 (2001)
V. Kunin, C.A. Ouzounis, Bioinformatics **19**(11), 1412–1416 (2003)
J.-B. Lamarck, *Histoire naturelle des animaux sans vertebras* (Paris, 1815)
B.F. Lang, C. O'Kelly, T. Nerad, M.W. Gray, G. Burger, Curr. Biol. **12**(20), 1773–1778 (2002)
P. Lapierre, R. Shial, J.P. Gogarten, Syst. Appl. Microbiol. **29**(1), 15–23 (2006)
J.G. Lawrence, H. Hendrickson, Mol. Microbiol. **50**(3), 739–749 (2003)
J.G. Lawrence, H. Ochman, J. Mol. Evol. **44**(4), 383–397 (1997)
J.G. Lawrence, H. Ochman, Proc. Natl. Acad. Sci. USA **95**(16), 9413–9417 (1998)
J.G. Lawrence, H. Ochman, Trends Microbiol. **10**(1), 1–4 (2002)
E. Lerat, V. Daubin, N.A. Moran, PLoS Biol. **1**(1), E19 (2003)
R.E. Ley, J.K. Harris, J. Wilcox, J.R. Spear, S.R. Miller, B.M. Bebout, J.A. Maresca, D.A. Bryant, M.L. Sogin, N.R. Pace, Appl. Environ. Microbiol. **72**(5), 3685–3695 (2006)
M. Lynch, Annu. Rev. Microbiol. **60**, 327–349 (2006)
M. Lynch, J.S. Conery, Science **302**(5649), 1401–1404 (2003)
L. Margulis, *Symbiosis in Cell Evolution: Microbial Communities in the Archean and Proterozoic Eons* (Freeman, 1995)
A.P. Martin, E.K. Costello, A.F. Meyer, D.R. Nemergut, S.K. Schmidt, Evol. Int. J. Org. Evol. **58**(5), 946–955 (2004)
W. Martin, E.V. Koonin, Nature **440**(7080), 41–45 (2006)
E. Mayr, *Systematics and the Origin of Species* (Columbia Univ. Press, New York, 1942)
B.G. Mirkin, T.I. Fenner, M.Y. Galperin, E.V. Koonin, BMC Evol. Biol. **3**(1), 2 (2003)
A. Morandi, O. Zhaxybayeva, J.P. Gogarten, J. Graf, J. Bacteriol. **187**(18), 6561–6564 (2005)
S. Mylvaganam, P.P. Dennis, Genetics **130**(3), 399–410 (1992)
Y. Nakamura, T. Itoh, H. Matsuda, T. Gojobori, Nat. Genet. **36**(7), 760–766 (2004)
K.E. Nelson, R.A. Clayton, S.R. Gill, M.L. Gwinn, R.J. Dodson, D.H. Haft, E.K. Hickey, J.D. Peterson, W.C. Nelson, K.A. Ketchum, L. McDonald, T.R. Utterback, J.A. Malek, K.D. Linher, M.M. Garrett, A.M. Stewart, M.D. Cotton, M.S. Pratt, C.A. Phillips, D. Richardson, J. Heidelberg, G.G. Sutton, R.D. Fleischmann, J.A. Eisen, O. White, S.L. Salzberg, H.O. Smith, J.C. Venter, C.M. Fraser, Nature **399**(6734), 323–329 (1999)
E.G. Nisbet, N.H. Sleep, Nature **409**(6823), 1083–1091 (2001)
L.O. Olendzenski, L. Liu, O. Zhaxybayeva, R. Murphey, D.G. Shin, J.P. Gogarten, J. Mol. Evol. **51**(6), 587–599 (2000)
L.O. Olendzenski, O. Zhaxybayeva, J.P. Gogarten, in *Microbial Genomes*, ed. by C.M. Fraser, T. Read, K. Nelson (Humana Press, 2004), pp. 143–154
R.D.M. Page, E.C. Holmes, *Molecular Evolution: A Phylogenetic Approach* (Blackwell, 1998)
E. Pennisi, Science **280**(5364), 672–674 (1998)
J. Pereto, P. Lopez-Garcia, D. Moreira, Trends Biochem. Sci. **29**(9), 469–477 (2004)
H. Philippe, E.A. Snell, E. Bapteste, P. Lopez, P.W. Holland, D. Casane, Mol. Biol. Evol. **21**(9), 1740–1752 (2004)
M.A. Ragan, Curr. Opin. Genet. Dev. **11**(6), 620–626 (2001a)
M.A. Ragan, FEMS Microbiol. Lett. **201**(2), 187–191 (2001b)
M.A. Ragan, Trends in Microbiol. **10**(Suppl. 1), 4 (2002)
D.M. Raup, Paleobiology **11**(1), 42–52 (1985)
D.M. Raup, Philos. Trans. R. Soc. Lond. B. Biol. Sci. **325**, 421–431 (1989); discussion 431–435
D.L. Rohde, S. Olson, J.T. Chang, Nature **431**(7008), 562–566 (2004)
A. Rokas, S.B. Carroll, PLoS Biol. **4**(11), e352 (2006)

R. Rossello-Mora, R. Amann, FEMS Microbiol. Rev. **25**(1), 39–67 (2001)

J. Sapp, Microbiol. Mol. Biol. Rev. **69**(2), 292–305 (2005)

M. Schidlowski, J.M. Hayes, I.R. Kaplan, in *Earth's Earliest Biosphere*, ed. by J.W. Schopf (Princeton University Press, Princeton, 1983), pp. 149–187.

P.D. Schloss, J. Handelsman, Microbiol. Mol. Biol. Rev. **68**(4), 686–691 (2004)

P. Shen, H.V. Huang, Genetics **112**(3), 441–457 (1986)

A.G. Simpson, A.J. Roger, Curr. Biol. **14**(17), R693–R696 (2004)

B. Snel, P. Bork, M.A. Huynen, Genome Res. **12**(1), 17–25 (2002)

M.L. Sogin, H.G. Morrison, J.A. Huber, D.M. Welch, S.M. Huse, P.R. Neal, J.M. Arrieta, G.J. Herndl, Proc. Natl. Acad. Sci. USA **103**(32), 12115–12120 (2006)

G. Srinivasan, C.M. James, J.A. Krzycki, Science **296**(5572), 1459–1462 (2002)

R.Y. Stanier, C.B. Van Niel, Arch. Mikrobiol. **42**, 17–35 (1962)

R. Thomson, J.K. Pritchard, P. Shen, P.J. Oefner, M.W. Feldman, Proc. Natl. Acad. Sci. USA **97**(13), 7360–7365 (2000)

P.A. Underhill, P. Shen, A.A. Lin, L. Jin, G. Passarino, W.H. Yang, E. Kauffman, B. Bonne-Tamir, J. Bertranpetit, P. Francalacci, M. Ibrahim, T. Jenkins, J.R. Kidd, S.Q. Mehdi, M.T. Seielstad, R.S. Wells, A. Piazza, R.W. Davis, M.W. Feldman, L.L. Cavalli-Sforza, P.J. Oefner, Nat. Genet. **26**(3), 358–361 (2000)

L. Vigilant, M. Stoneking, H. Harpending, K. Hawkes, A.C. Wilson, Science **253**(5027), 1503–1507 (1991)

C.R. Vossbrinck, J.V. Maddox, S. Friedman, B.A. Debrunner-Vossbrinck, C.R. Woese, Nature **326**(6111), 411–414 (1987)

R.A. Welch, V. Burland, G. Plunkett, 3rd, P. Redford, P. Roesch, D. Rasko, E.L. Buckles, S.R. Liou, A. Boutin, J. Hackett, D. Stroud, G.F. Mayhew, D.J. Rose, S. Zhou, D.C. Schwartz, N.T. Perna, H.L. Mobley, M.S. Donnenberg, F.R. Blattner, Proc. Natl. Acad. Sci. USA **99**(26), 17020–17024 (2002)

E. Willerslev, A. Cooper, Proc. R. Soc. B: Biol. Sci. **272**(1558), 3–16 (2005)

C. Woese, Proc. Natl. Acad. Sci. USA **95**(12), 6854–6859 (1998)

C. Woese, O. Kandler, M. Wheelis, Proc. Natl. Acad. Sci. USA **87**, 4576–4579 (1990)

C.R. Woese, Microbiol. Rev. **51**(2), 221–271 (1987)

C.R. Woese, G.E. Fox, Proc. Natl. Acad. Sci. USA **74**(11), 5088–5090 (1977)

W.H. Yap, Z. Zhang, Y. Wang, J. Bacteriol. **181**(17), 5201–5209 (1999)

O. Zhaxybayeva, J. Gogarten, BMC Genomics **3**, 4 (2002)

O. Zhaxybayeva, J.P. Gogarten, Trends Genet. **20**(4), 182–187 (2004)

O. Zhaxybayeva, J.P. Gogarten, R.L. Charlebois, W.F. Doolittle, R.T. Papke, Genome Res. **16**(9), 1099–1108 (2006)

O. Zhaxybayeva, L. Hamel, J. Raymond, J. Gogarten, Genome Biol. **5**(3), R20 (2004a)

O. Zhaxybayeva, P. Lapierre, J.P. Gogarten, Trends Genet. **20**(5), 254–260 (2004b)

O. Zhaxybayeva, P. Lapierre, J.P. Gogarten, Protoplasma **227**(1), 53–64 (2005)

# Molecular Biosignatures

**Roger E. Summons · Pierre Albrecht · Gene McDonald ·**
**J. Michael Moldowan**

Originally published in the journal Space Science Reviews, Volume 135, Nos 1–4.
DOI: 10.1007/s11214-007-9256-5 © Springer Science+Business Media B.V. 2007

**Abstract** Life, as we know it, is based on carbon chemistry operating in an aqueous environment. Living organisms process chemicals, make copies of themselves, are autonomous and evolve in concert with the environment. All these characteristics are driven by, and operate through, carbon chemistry. The carbon chemistry of living systems is an exact branch of science and we have detailed knowledge of the basic metabolic and reproductive machinery of living organisms. We can recognise the residual biochemicals long after life has expired and otherwise lost most life-defining features. Carbon chemistry provides a tool for identifying extant and extinct life on Earth and, potentially, throughout the Universe. In recognizing that certain distinctive compounds isolable from living systems had related fossil derivatives, organic geochemists coined the term biological marker compound or biomarker (e.g. Eglinton et al. in Science 145:263–264, 1964) to describe them. In this terminology, biomarkers are metabolites or biochemicals by which we can identify particular kinds of living organisms as well as the molecular fossil derivatives by which we identify defunct counterparts. The terms biomarker and molecular biosignature are synonymous.

A defining characteristic of terrestrial life is its metabolic versatility and adaptability and it is reasonable to expect that this is universal. Different physiologies operate for carbon acquisition, the garnering of energy and the storage and processing of information. As

R.E. Summons (✉)
Dept. Earth, Atmospheric and Planetary Sciences, Massachusetts Institute of Technology,
77 Massachusetts Avenue E34-246, Cambridge, MA 02139, USA
e-mail: rsummons@mit.edu

P. Albrecht
Institut de Chimie, Université Louis Pasteur, CNRS-UMR 7177, ECPM, 25 rue Becquerel,
67087 Strasbourg Cedex 2, France

G. McDonald
Dept. of Chemistry and Biochemistry, University of Texas at Austin, 1 University Station A5300,
Austin, TX 78712, USA

J.M. Moldowan
Department of Geological & Environmental Sciences, Stanford University, Stanford, CA 94305-2115,
USA

well as having a range of metabolisms, organisms build biomass suited to specific physical environments, habitats and their ecological imperatives. This overall 'metabolic diversity' manifests itself in an enormous variety of accompanying product molecules (i.e. natural products). The whole field of organic chemistry grew from their study and now provides tools to link metabolism (i.e. physiology) to the occurrence of biomarkers specific to, and diagnostic for, particular kinds of metabolism.

Another characteristic of living things, also likely to be pervasive, is that an enormous diversity of large molecules are built from a relatively small subset of universal precursors. These include the four bases of DNA, 20 amino acids of proteins and two kinds of lipid building blocks. Third, life exploits the specificity inherent in the spatial, that is, the three-dimensional qualities of organic chemicals (stereochemistry). These characteristics then lead to some readily identifiable and measurable generic attributes that would be diagnostic as biosignatures.

Measurable attributes of molecular biosignatures include:

- Enantiomeric excess
- Diastereoisomeric preference
- Structural isomer preference
- Repeating constitutional sub-units or atomic ratios
- Systematic isotopic ordering at molecular and intramolecular levels
- Uneven distribution patterns or clusters (e.g. C-number, concentration, $\delta^{13}C$) of structurally related compounds.

In this paper we address details of the chemical and biosynthetic basis for these features, which largely arise as a consequence of construction from small, recurring sub-units. We also address how these attributes might become altered during diagenesis and planetary processing. Finally, we discuss the instrumental techniques and further developments needed to detect them.

**Keywords** Biomarkers · Lipids · Isomerism · Chirality · Life-detection · Diagnostic molecules

## 1 Introduction and Operating Principles

Earthly life is based on carbon chemistry operating in an aqueous environment. In this section, we address the potential for detecting extra-terrestrial life using molecular organic chemistry although we also acknowledge that other kinds of chemistry may constitute 'life'.

All organisms on Earth process chemicals, make copies of themselves, are autonomous and evolve in concert with the environment. These characteristics are driven by, and operate through, carbon chemistry which is an exactly known branch of science. Carbon chemistry affords a robust tool for identifying extant and extinct life on today's Earth, in the sedimentary record of the early Earth and, potentially, throughout the Universe.

In recognizing that certain distinctive compounds isolable from living systems had related fossil derivatives, organic geochemists coined the term *biological marker compound* or *biomarker* (Eglinton et al. 1964; Eglinton and Calvin 1967) to describe them. Biomarkers are defined here as metabolites or biochemicals by which we can identify particular kinds of living organisms and their defining physiologies together with the molecular fossil derivatives through which we identify defunct counterparts. Thus biological marker, biomarker and molecular biosignature are used interchangeably.

Amino acids, being simple and essential components common to all known life, have received the most attention in the quest for effective molecular biosignatures (e.g. Kminek et al. 2000; Bada et al. 2007). This focus may have diverted attention from other kinds of molecules with structural features that could indicate not only biogenicity but specific biological provenance as well. Nucleic acids, proteins, carbohydrates and intermediary metabolites are also essential components of life and are obviously potential molecular biosignatures, especially for extant life. On the other hand, these compound classes are rapidly recycled by other living systems and are not known for their ability to survive in the external environment over geological timescales (Nagy 1982). Lipids and structural biopolymers are biologically essential compound classes which are known for their wide variety of diagnostic structural attributes as well as for their overall recalcitrance under harsh environmental conditions (Engel and Macko 1993). Some lipids are known to be stable, when buried in the subsurface and isolated from oxygen, over billion year timescales (Eglinton et al. 1964; Summons et al. 1988; Brocks and Summons 2003). Lipids are the biomarkers most commonly used to study and reconstruct terrestrial ecosystems. Accordingly, we have chosen to focus much of this discussion on the generic molecular qualities of lipids and lipidic biopolymers in the knowledge that other classes of organic compounds might be more appropriate targets in some circumstances (e.g. DNA in the case of extant life).

Coal seams, crude oil and dispersed macromolecular organic matter (kerogen), comprise extraordinary and complex mixtures of fossil organic compounds. Crude oil, in particular, is mostly of lipid derivation and can be stable in the subsurface for tens of millions to billions of years (Peters et al. 2005). In the course of analysing petroleum and seeking to understand its composition and origins from kerogen, organic and petroleum geochemists have constructed a mature branch of the biogeosciences and this naturally embraces analysis of environmental and paleoenvironmental samples containing organic matter. Organic geochemical knowledge is used extensively in exploration for and production of petroleum resources (e.g. Peters et al. 2005) and is also a powerful tool used in Earth history reconstruction (e.g. Eglinton et al. 1964; Eglinton and Calvin 1967; Summons and Walter 1990). In seeking to understand the fossilisation of organic matter and the formation of petroleum, organic geochemists have established that the transition from metabolite to molecular fossil can be accompanied by specific structural and stereochemical changes. Rather than just providing unwanted complexity, these chemical changes encode a historical record of the fossilisation process. Known as *diagenesis*, the fossilisation process results in conversion of functioning biochemicals to molecular fossils. Chemical changes that take place during diagenesis can be diagnostic for issues such as the redox state, water chemistry and temperature history of the sedimentary environment.

In summary, molecular biosignatures are organic compounds used to identify specific kinds of Earthly organisms and physiologies and to track these in the sedimentary rock record. This suggests that biomarkers will provide particularly useful tools by which we might recognise and understand alien, carbon-based life. In this section, we set out to describe the attributes of terrestrial biomarker molecules as observed in modern and fossilised organic matter and to compare these attributes with those characteristic of abiological organic matter. Identification of several or all of the 'biological' characteristics in extraterrestrial organic matter would betray a biosynthetic origin.

Some of the key attributes of biomarkers include:

- Enantiomeric excess
- Diastereoisomeric preference
- Structural isomer preference
- Repeating constitutional sub-units or atomic ratios

- Systematic isotopic ordering at molecular and intramolecular levels
- Uneven distribution patterns or clusters (e.g. C-number, concentration, $\delta^{13}$C) of structurally related compounds.

These attributes are defined and elaborated upon in the following sections.

## 2 Attributes of Molecular Biosignatures in Terrestrial Organic Matter

### 2.1 Enantiomeric Excess

As introduced earlier, stereochemistry is one of the most fundamental chemical attributes of biogenic organic compounds. A *chiral* molecule is one that can exist as two non-identical mirror image structures known as enantiomers (Eliel et al. 1994; Barron 2007, this volume). *Chirality* results from the fact that a saturated carbon atom may have all four available valencies satisfied by different substituents (*a chiral centre*). Transposition of two substituents of a chiral centre results in another molecule which differs from the original in an analogous manner to right-handed and left-handed gloves. This is illustrated for the amino acid alanine in Fig. 1.

Most biologically formed chiral compounds (compounds with a chiral centre) are synthesised exclusively as one or another enantiomer. The high fidelity with which enantiomers are produced by biological systems is due to the fact that the molecules are put together by reactions catalysed by enzymes, proteins which are themselves made of exclusively L-amino acids in a lock-and-key manner. Chirality in organic compounds confers the physical property of optical activity (Barron 2007). Different organisms may synthesise the same chiral compound in different enantiomeric forms. Once an organism dies, and its biochemicals

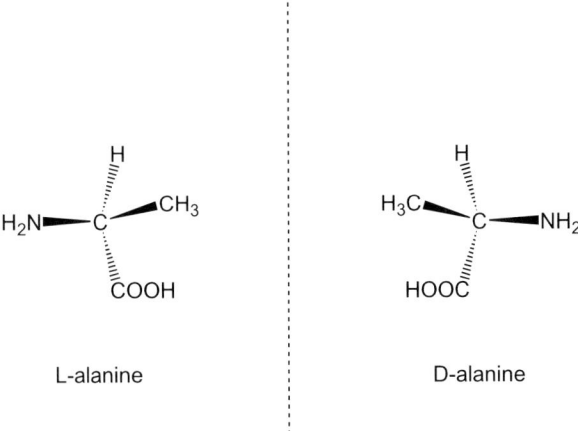

**Fig. 1** The D- and L- enantiomers of alanine are mirror-image structures that cannot be superimposed. They can be distinguished by the direction in which they rotate a beam of plane-polarised light. Equal mixtures of the two enantiomers are known as racemic mixtures and do not rotate the polarised light because the effect of the D-enantiomer is exactly cancelled by its L-counterpart. Unequal mixtures are said to be optically active and a compound comprising 100% of the D-enantiomer gives the maximum rotation in one direction while 100% of the L-enantiomer gives maximum rotation in the opposite direction. Most terrestrial organisms synthesise and use $\alpha$-amino acids exclusively in the L-form (*homochirality*) although there are some bacteria with the capacity to make and use D-amino acids, especially in toxins

are released into the environment, their chiral purity (and optical activity) may or may not persist. The degree to which chiral purity persists depends upon on the relative chemical stability of bonds in the vicinity of the chiral centre. Various natural chemical processes can lead to *racemization*, that is, the formation of mixtures of the two enantiomers. While this may result in loss or corruption of a biological signature, the rate at which it happens can also have a practical application. The best known example of this is the dating of fossil organic matter using the degree of amino acid racemization. It is also recognised that chiral agents other than enzymes exist, including the recent discovery that magnetochiral anisotropy can induce synthesis of compounds which have one enantiomer in excess of its counterpart (Rikken and Raupach 2000). Nevertheless, biology is still considered the most likely source of compounds that are produced purely or predominantly as one enantiomer.

Enantiomeric excess can be detected in a number of ways. Direct observation of optical activity requires a significant amount of material in crystal or a solution form but can also be used to measure the specific amount of rotation $[\alpha]_d$ characteristic of any particular optically active compound. Biochemical detection is also possible although biochemical methodologies are generally specific to individual compounds or compound types. Indirect measurement through a chromatographic procedure, for example, gas chromatography (GC) or gas chromatography-mass spectrometry (GC–MS) on a chiral stationary phase, has wide applicability and is very sensitive. In fact, it was high resolution GC that ultimately led to the observation of a slight enantiomeric excess in some meteoritic amino acids (Cronin and Pizzarello 1997).

## 2.2 Diastereoisomeric Preference

Chirality is just one of several types of isomerism (*stereoisomerism*) that stem from the spatial arrangement of atoms in a molecule. In a molecule with a single chiral centre, such as an amino acid, inversion of that centre leads to the alternate enantiomer. In a molecule with more than one chiral centre, inversion of one centre leads to the formation of an epimer while inversion of all centres is required to form the enantiomer, or non-superimposable mirror image of the original. The isomers that are not related as mirror images are diastereomers or diastereoisomers and, unlike enantiomers, have different physical and chemical properties. This is illustrated for tartaric acid in Fig. 2.

Just as biologically produced amino acids (single chiral centre) generally occur preferentially as one enantiomer, natural products with multiple chiral centres or other kinds of molecular asymmetry, are preferentially synthesised as a unique diastereomer. Organic acids and simple sugars provide examples of this kind of stereochemical selectivity. More complex

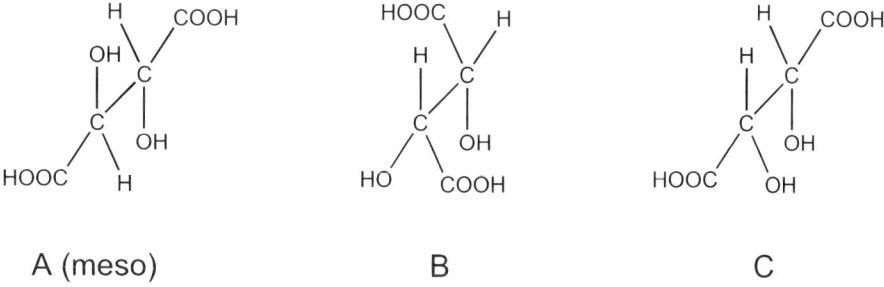

**Fig. 2** Stereoisomerism in tartaric acid. **B** and **C** are enantiomers. **A, B** and **A, C** are pairs of diastereomers

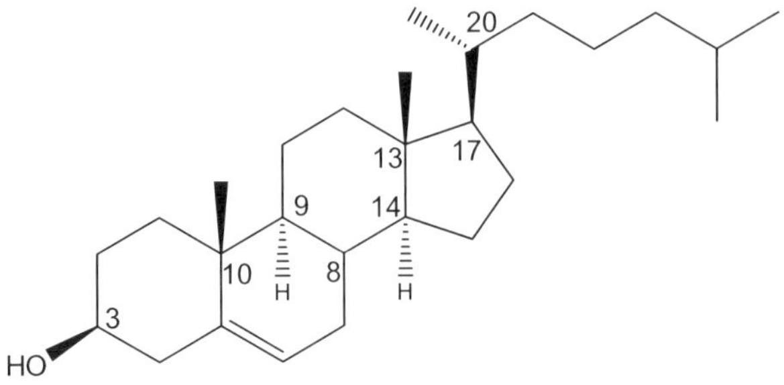

**Fig. 3** Structure of cholesterol with its eight asymmetric carbon atoms identified with their position number. Theoretically, this compound could exist in as many as 256 ($2^8$) possible stereoisomers and yet biosynthesis produces only the one illustrated (Peters et al. 2005)

natural products can have numerous sites of possible asymmetry but, with few exceptions, these compounds are also biosynthesised as a single stereoisomer. This is illustrated in Fig. 3 with reference to the ubiquitous sterol cholesterol.

Because diastereomers are chemically and physically distinct chemicals, there are many different ways to detect them singularly or in mixtures. Again GC or liquid chromatography (LC) are ideal means to separate such compounds while mass spectrometry (MS) is generally the most sensitive and informative detection method (see the following).

## 2.3 Structural Isomer Preference

The foregoing examples illustrate, in a somewhat limited way, two types of (biologically controlled) stereoselectivity in organic compounds. The propensity of carbon compounds to exist with multiple ring systems and unsaturations opens the way for an enormous variety of possible structures for each combination of $C_pH_qN_rO_sP_tS_u$. These isomers are *structural* isomers, otherwise known as *constitutional* isomers (Eliel et al. 1994), that is, compounds that are different in the connectivity of their chemical bonds. Despite the potential for variety, we observe that naturally synthesised biochemicals fall into patterns and the number of known compounds comprises but a small subset of what is chemically feasible. Moreover, the natural product may be the thermodynamically least favoured structure within a set of possible isomers.

Structural isomers are readily separated using chromatography. In many, but not all cases, their mass spectra are also distinctive. As with other forms of isomerism, combinatorial approaches such as GC–MS and LC–MS provide the most sensitive and diagnostic analyses.

## 2.4 Repeating Constitutional Sub-Units or Atomic Ratios

Virtually all natural products are constructed from a limited number of simpler sub-units, the best known examples being proteins and nucleic acids (McKay 2007). Lipids are significantly different in that there are only two basic building blocks. Lipids are polymers of either acetyl or isopentenyldiphosphate precursors (Eigenbrode 2007). The final products lack a hydrolysable chemical functionality, such as a peptide or glycosidic linkage, at the point where sub-units join and, unlike other proteins and nucleic acids, lipids cannot be depolymerised.

Polymers of acetate (acetogenic lipids) include fatty acids, wax esters, alkenones and long-chain hydrocarbons and generally occur as series of compounds differing by two methylene (i.e. $C_2H_4$ or $-CH_2-CH_2-$) units. The classic examples are the fatty acids which, when esterified to glycerol, are essential components of membrane lipid bi-layers of Bacteria and Eukarya. The most common fatty acids are all-acetate products and thus have

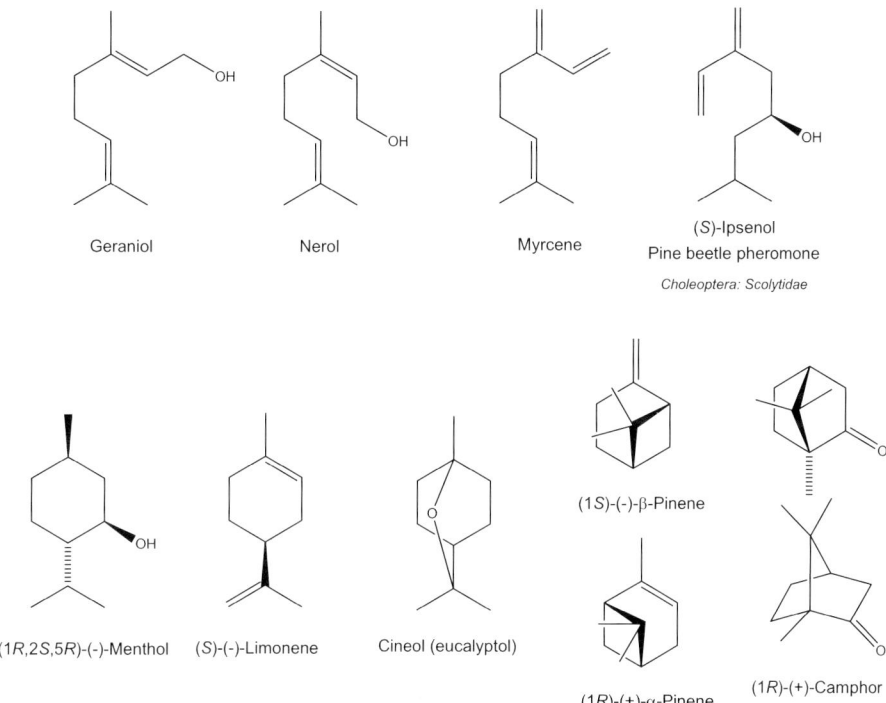

**Fig. 4** Structures of a group of plant- and animal-derived $C_{10}$ natural products (monoterpenes). The diphosphate ester of geraniol, itself formed by dimerization of $\Delta^3$-isopentenyldiphosphate, is the biochemical precursor of the other structures. Limonene, myrcene, $\alpha$-pinene and $\beta$-pinene are just a few of the biologically generated structural isomers with a molecular formula $C_{10}H_{16}$. Nerol, geraniol, ipsenol and cineol are isomers with a molecular formula $C_{10}H_{18}O$. There would be many hundreds of possibilities for non-biological isomers with the same formulae

even carbon numbers (e.g. $C_{14}$, $C_{16}$, $C_{18}$ and $C_{20}$). Odd carbon-numbered members, generally synthesised from a non-acetyl starter, exist but are subordinate. Extension of fatty acid precursors to higher carbon numbers proceeds with addition of further acetate units. Terminating and modifying reactions such as desaturation, reduction or decarboxylation yield common intermediate molecular weight series such as the plant and algal waxes comprising even-numbered alcohols (e.g. $C_{26}$, $C_{28}$, $C_{30}$, $C_{32}$) and odd-numbered hydrocarbons (e.g. $C_{25}$, $C_{27}$, $C_{29}$, $C_{31}$). Long acetogenic chains are also incorporated into structural biopolymers such as algaenans (Kadouri et al. 1988; Tegelaar et al. 1989; Gelin et al. 1996).

Terpenoids, polymers of $\Delta^3$-isopentenyldiphosphate, have somewhat more complex origins (Nes and McKean 1977; Flesch and Rohmer 1988; Schwender et al. 1996; Disch et al. 1998; Lichtenthaler 1999; Rohmer 2003). The multiplicity of isoprenoid biosynthetic pathways, their distribution across different phylogenetic groups, their requirement, or otherwise, for molecular oxygen, and the types of post-synthesis modification are generally held to have evolutionary significance (e.g. Nes and McKean 1977; Ourisson et al. 1987; Ourisson and Albrecht 1992). The result is the widespread occurrence of isoprene-based lipids known as polyisoprenoids (also called terpenoids). Some structures derived from the linkage of two isoprene units are shown in Fig. 4.

Further complexity arises through the linkage processes which yield 'regular' head–tail compounds such as phytol or 'irregular' tail–tail or head–head products such as squalene

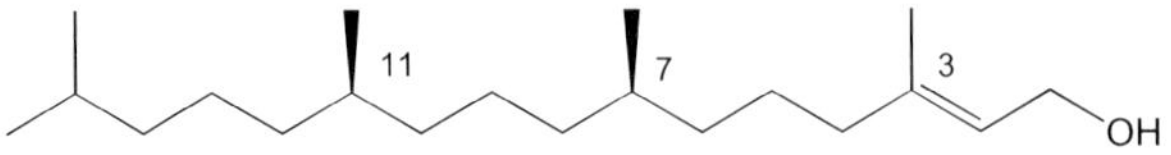

**Fig. 5** Structure of the diterpenoid phytol composed of four head-tail linked $C_5$ ('isoprene') units. Also note that phytol has two sites of asymmetry and a double bond each of which could deliver additional isomers if they were produced in other than natural circumstances. Phytol occurs uniquely as the E-3, 7R, 11R, 15-tetramethylhexadecene-2-enol structure (see Peters et al. 2005 and references therein for further explanation of terminology)

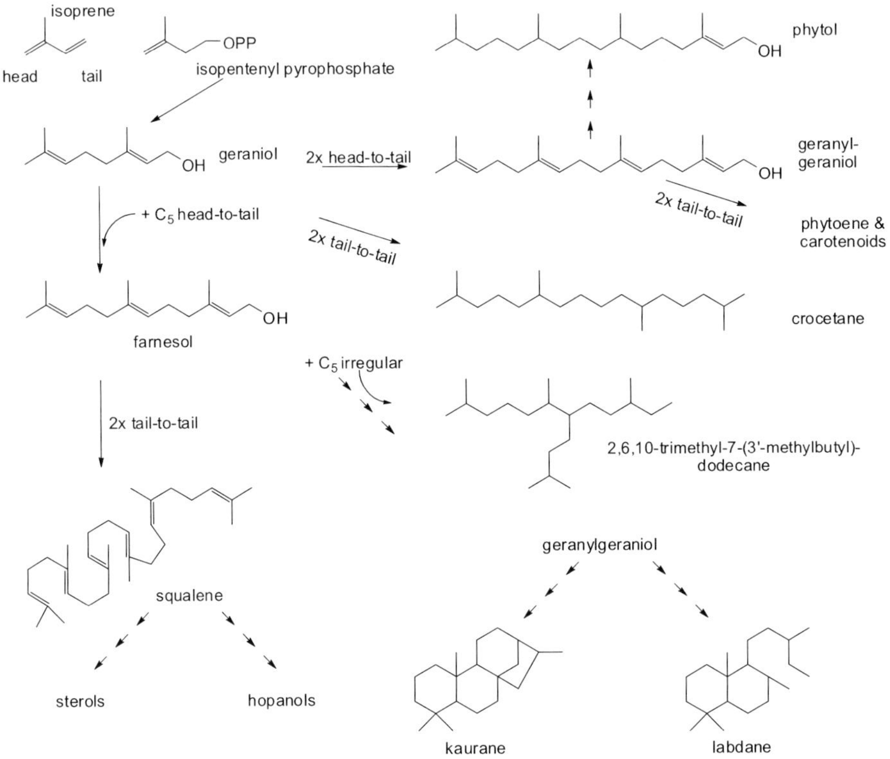

**Fig. 6** Biosynthetic origins and structures of some regular, irregular and cyclic $C_{20}$ (diterpenoid) and $C_{30}$ (triterpenopid) and $C_{40}$ (tetraterpenoid) hydrocarbons that have been identified in bitumen and which illustrate a variety of biosynthetic patterns based on repeating $C_5$ sub-units after Hayes (2001). Crocetane, 2,6,10-trimethyl-7-(3'-methylbutyl)-dodecane and squalene are irregularly branched compounds while phytol, labdane and kaurane are 'regular' and constructed from four head-tail linked isoprene units. These compounds also illustrate how different structures can be diagnostic for specific physiologies (phytol and farnesol for photosynthesis; crocetane for anaerobic oxidation of methane) or specific organisms (2,6,10-trimethyl-7-(3-methylbutyl)-dodecane for diatoms; labdane and kaurane for conifers)

or biphytane, respectively (Figs. 5 and 6). Finally, polyisoprenoids are prone to folding and ring-formation affording the extensive range of cyclic terpenoids (sterols, bacterio-hopanepolyols, carotenoids, tocopherols, cyclic biphytanes, etc.) which are essential components of all life. Although there are many and varied individual structures known, biochemistry invariably produces each one as a unique constitutional and stereochemical isomer. Cholesterol (Fig. 3) and phytol (Fig. 5) are typical of this phenomenon.

In summary, lipids display the same kind of stereochemical and structural 'purity' found in all biologically synthesised compounds. An additional consequence of their biosynthesis from smaller sub-units is the occurrence of clusters of compounds that differ by $C_2H_4$ units (acetogenic lipids) or $C_5H_{10}$ units (polyisoprenoids). In a typical sample of fossil lipids, for example, one would find a predominance of even-carbon numbered fatty acids, odd carbon-numbered hydrocarbons, $C_{15}$, $C_{20}$ and $C_{25}$ acyclic isoprenoids, $C_{20}$ and $C_{30}$ cyclic terpenoids including steroids, and $C_{40}$ carotenoids (Eigenbrode 2007). Subsets of these traits are even identifiable in highly derived products such as petroleum where $n$-alkanes may exhibit weak odd over even or even over odd carbon number preferences. Isoprenoids occur in clusters of compounds around $C_{15}$, $C_{20}$ and $C_{30}$. An interesting and distinctive 'biosignature' occurs as a consequence of the branching pattern of isoprenoids and becomes evident through thermolytic cracking of their precursors in petroleum. This results in very low relative concentrations of $C_{12}$, $C_{17}$ and $C_{22}$ acyclic isoprenoids and aryl isoprenoids (e.g. Summons and Powell 1987; Kissin 1998) compared to their corresponding pseudohomologues.

## 2.5 Systematic Isotopic Ordering at Molecular and Intramolecular Levels

The building blocks of natural products are well known for their isotopic inhomogeneity. Acetate provides one of the best examples because it shows significant differences in the $^{13}C$ contents of its methyl and carboxyl carbons (DeNiro and Epstein 1977). The most overt consequences are isotopic ordering in fatty acids (Monson and Hayes 1982) and a major isotopic difference between acetogenic and polyisoprenoid lipids (Schouten et al. 1998). The latter are more enriched in $^{13}C$, often by significant amounts (e.g. Hayes 1993) but there are a number of exceptions (Summons et al. 1994; Hayes 2001; Schouten et al. 1998). In a single organism, the differences between acetogenic and polyisoprenoid lipids depend on how many of the polyisoprenoid carbon atoms arise from acetate versus carbohydrate metabolism.

## 2.6 Uneven Distribution Patterns or Clusters (e.g. C-Number, Concentration, $\delta^{13}C$) of Structurally Related Compounds

This molecular biosignature is an outcome of the 'repeating sub-unit' origin of many biochemicals. A second contributory factor is the tendency of functional biochemicals such as lipids to show discrete molecular weight ranges (e.g. $C_{14}$–$C_{20}$ fatty acids, hydrocarbons centred around $C_{17}$ and $C_{25}$–$C_{33}$, $C_{26}$–$C_{31}$ sterols, $C_{30}$ triterpenoids and $C_{35}$ bacteriohopanepolyols, etc.). Third, most samples of biologically produced organic matter come from organisms living in complex ecosystems. The volatile components of a microbial mat, for example, will show compound classes with carbon numbers distributed roughly as described earlier. Moreover, the $C_{25}$–$C_{30}$ fraction might contain more material than the $C_{15}$–$C_{20}$ fraction. This is in stark contrast to distributions recorded for meteorite (Sephton et al. 2000) and other (McCollum 2003) abiological volatiles where relative abundances drop as carbon number increases. $^{13}C$ contents may also vary widely and according to the physiology of the source organisms. Meteoritic $^{13}C$ distributions, in stark contrast, show patterns diagnostic for synthesis by addition of single carbon atoms or cracking at single C–C bonds (e.g. Sephton and Gilmour 2001; Sephton et al. 2000).

The molecular size distribution of natural products is difficult to assess accurately because our conventional analytical window is skewed in favour of volatile molecules or those that can be made volatile through derivatisation. What we can be sure of is that high molecular weight natural products show the same degree of stereochemical and structural purity as their low molecular weight counterparts. The degree to which these features are preserved or lost during and subsequent to fossilisation is reviewed below in Sects. 5 and 6.

## 3 Prebiotic and Abiotic Organic Compounds

The use of structural, isotopic and stereochemical criteria for assessing the biological origins of extra-terrestrial organic matter hinges on whether or not they are common to all biochemistries, Earthly and alien. Similarly, the worth of criteria to characterise abiogenic organic matter depends on their *absence* in all biochemistries. While natural products chemistry has given us criteria to discern features that are common to compounds of biological origin, two types of studies have contributed to the identification of attributes of abiological organic materials. These are simulations of planetary and hydrothermal chemistries and analyses of carbonaceous meteorites.

### 3.1 Prebiotic Syntheses

Experimental simulations of prebiotic chemistry, pioneered by Miller (1953), opened an area of research in which laboratory models of planetary environments were used to develop insights into the abiotic organic chemistry of early Earth and other Solar System bodies. Most of these studies focused on electron, photon and ion irradiations of gases. The energy sources were intended to simulate the effects of electrical and coronal discharges, solar ultraviolet radiation, cosmic rays and magnetospheric particles on planetary atmospheres. Gas compositions covered a wide range of experimental redox states from reducing mixtures of $CH_4$–$N_2$–$H_2$ to mimic Titan (e.g. Thompson et al. 1991) to neutral mixtures of $CO_2$–$N_2$–$H_2O$ to mimic a presumed prebiotic Earth (e.g. Stribling and Miller 1987). These experiments showed that the reducing power of the gas mixtures exerted a fundamental control on the product yield and composition.

    Since Earth and Mars lack an accessible geological record from which to reconstruct atmospheric compositions operating in their first 700 million years, considerable uncertainty surrounds the prebiotic organic chemistry that may have prevailed initially. Thus, the nature and abundance of organic matter that may have been produced on both planets remains highly reliant on the behaviour of models. Nonetheless, some useful generalisations emerge from the simulation experiments. If the atmospheric compositions of early Mars and Earth were alike, similar products would have formed by processes common to both planets. As Mars lies at greater distance from the Sun, planetary organic production rates would have been lower due to solar energy flux. Experiments performed in reducing atmospheres generally show that product abundances decrease with increasing carbon number and that virtually all possible isomers in a given class of compounds are formed at low carbon numbers. This appears to be true for syntheses of hydrocarbons, amino acids, carboxylic acids and dicarboxylic acids (Lemmon 1970; Zeitman et al. 1974). The resulting amino acids and dicarboxylic acids also lack enantiomeric excess (Ring et al. 1972; Zeitman et al. 1974). In neutral to oxidizing gas compositions fewer products are formed, and these are invariably of low molecular weight (Cronin and Chang 1993; Stribling and Miller 1987).

    In addition to atmospheric models for prebiotic organosynthesis, considerable attention has been paid to hydrothermal systems as sites for thermochemical organosynthesis and emergence of life on early Earth and Mars (Shock, 1992, 1996, 1997). In this approach, the redox state and temperature of fluids in metastable equilibrium with mineral assemblages have been modelled to determine their contents of organic compounds. Amino acids, carboxylic acids, alcohols, aldehydes, ketones and hydrocarbons are among the compounds predicted to form. The relative abundances of straight chain homologs are calculated to vary according to redox and temperature with strongly reducing fluids favouring the higher over lower homologs. Lack of thermodynamic data for the multitude of possible branched chain isomers limits the predictions of relative abundances to straight chain species. When CO is the carbon source in experimental tests of the metastable equilibrium theory, normal hydrocarbons, carboxylic acids and alcohols form as predicted, but these products appear not to be synthesised when $CO_2$ is employed (McCollum et al. 1999; McCollum and Simoneit 1999). In a recent series of experiments, polyaromatic hydrocarbons (PAH) have been shown to form during the thermal decomposition of siderite (iron carbonate) in water at 300°C (McCollom 2003). Although there is new information on the isotopic outcomes of such syntheses (McCollom and Seewald 2006), more experimental tests of the hydrothermal synthesis model are needed to verify its predictions and identify characteristic patterns of molecular structural variation among its products.

## 3.2 Delivery of Exogenous Organic Matter

The cratered surfaces of virtually all solid bodies in the Solar System underscore the potential role of meteor impacts in delivering exogenous material to planetary surfaces (Chyba and Sagan 1992). How much cometary and asteroidal organic matter survived collisions on Mars and Earth is poorly constrained, but the bulk of surviving material is thought to have entered as interplanetary dust particles (IDP) in the size range of 10–50 μm (Anders 1989; Flynn 1966). Like volcanic dust and atmospheric aerosols, IDP will have been incorporated in a planet's regolith and sediments. C, H, N, S abundances have been measured in IDPs, and a significant fraction of them contains assemblages of sub-micron-sized organic carbon grains, some of which are highly deuterium enriched (McKeegan et al. 1985; Zinner 1988). These grains bear some resemblance to the acid insoluble carbon component of carbonaceous meteorites (see the following) and may be related to the CHON grains of Comet Halley (Fomenkova et al. 1994). Little is known about the organic compounds present in IDP beyond the identification of a range of polycyclic aromatic hydrocarbons (Clemett et al. 1993). By default, carbonaceous meteorites probably provide the best natural model, not just for exogenous organic matter on Mars (e.g. Biemann et al. 1976), but for abiotic organic matter in general.

## 3.3 Organic Chemistry of Carbonaceous Meteorites

Carbonaceous meteorites contain condensates from our solar nebula and other stars, mineral phases formed in liquid water (in CI and CM types) and diverse organic components. Apparently, prebiotic evolution occurred early on some parent bodies of carbonaceous meteorites, but fell short of the origin of life. Thus, the organic chemistry of meteorites provides a useful model of the abiotic organic matter that may occur on Mars, Europa and other extraterrestrial environments. Since these bodies also exhibit spectral properties that resemble those of the abundant C-type main belt asteroids (Zellner and Bowell 1977), the proximity of the asteroid belt to Mars ensures that the planet's surface contains carbonaceous meteoritic debris and IDP of asteroidal origins.

Before the fall of the Murchison meteorite (CM type) in 1969, terrestrial contamination obscured the organic chemistry of carbonaceous meteorites. In preparing to study lunar samples returned that same year, new analytical methods were devised to minimise the impact of terrestrial contamination and a new generation of meteorite analyses were applied to the Murchison samples. These investigations uncovered an extensive inventory of individual compounds whose structures, stable isotopic compositions and stereochemistry reveal a rich organic chemistry (Sephton and Botta 2007; Cronin and Chang 1993; Pizzarello 2006; Sephton 2002; Weber and Pizzarello 2006).

The 2% total carbon in the Murchison meteorite exists in a variety of forms. Water- and solvent-soluble compounds comprise about 0.2 and 0.3 mole fraction of the carbon and this varies from sample to sample. The bulk of the carbon, ranging in mole fraction from 0.6 to 0.8, exists as high molecular weight, solvent- and acid-insoluble components whose detailed properties remain poorly constrained. This material appears to be a mixture of sub-micron carbonaceous grains resembling the CHON particles detected in the coma of Comet Halley (Fomenkova et al. 1994). The insoluble fraction also contains minor amounts of extrasolar condensates in the form of nanometer-sized diamonds, silicon carbide and graphite grains (Anders and Zinner 1993). Carbonates formed during the epoch of liquid water make up the remaining 0.02 to 0.1 mole fraction of carbon.

The classes of meteoritic organic compounds that have familiar biochemical counterparts include amino acids, fatty acids, purines, pyrimidines and sugars (G. Cooper, personal communication). In addition there are alcohols, aldehydes amides, amines, mono- and di-carboxylic acids, aliphatic and aromatic non-polar hydrocarbons, polar hydrocarbons as heterocyclic aromatics, hydroxy acids, ketones, phosphonic and sulfonic acids, sulfides (S. Pizzarello, personal communication) and ethers (G. Cooper, personal communication). Concentrations of compound classes vary widely from less than ten ppm (amines) to tens of ppm (amino acids) to hundreds of ppm (carboxylic acids) (Cronin et al. 1988). Chromatographic analyses of virtually all classes of acyclic compounds reveal complex mixtures which include homologous series of compounds up to $C_{12}$ in some cases (carboxylic acids).

The structural variety and molecular weights of organic compounds show three distinct types of correlation. Data for amino acids and hydroxy acids are chosen to illustrate these features. The amino acids are of particular interest because of their centrality to terrestrial biochemistry, their wide range of possible structures and the availability of relevant meteoritic data. Moreover, future analyses of extra-terrestrial organic matter will target amino acids because an alien biochemistry is likely to evolve catalytic polymers consisting of similar multi-functional building blocks.

1. Molecular abundances decrease with increasing carbon number.
   As illustrated by the $\alpha$-methyl branched and straight chain series of $\alpha$-amino acids, plots of their concentrations (log nmole/g) versus carbon numbers yield linear correlations with declining slopes of about $-7$ (Cronin et al. 1988). These trends suggest growth of amino acid carbon skeletons was by single carbon additions.
2. Abundances of branched chain isomers exceed those of the straight chain.
   For example, for compounds containing the same number of carbon atoms, the concentration of the $\alpha$-methyl and the $\beta$-methyl branched isomers each surpasses that of the straight chain $\alpha$-isomer.
3. Complete constitutional diversity prevails at the lower carbon numbers.
   This last pattern is illustrated by Table 1, which compares the numbers of known isomers of the acyclic monoamino monocarboxylic acids and the acyclic monohydroxy monocarboxylic acids with the numbers of theoretically possible structures. Overall, 57 amino acids occur among the 159 possible $C_2$ to $C_7$ isomers. Analytical sensitivity and diminishing abundances at higher molecular weights limit further identifications. Remarkably, almost all constitutional isomers from $C_2$ to $C_7$ occur in the $\alpha$-amino and $\alpha$-hydroxy acids; from $C_2$ to $C_5$ nearly all isomers of $\beta$- and $\gamma$-substituted acids appear. Murchison also contains suites of structurally diverse acyclic monoamino- and monohydroxy-substituted dicarboxylic acids (Cronin et al. 1993). In contrast, life on Earth employs only 20 protein $\alpha$-amino acids and all of these have an $\alpha$-H. Biochemical counterparts account for fewer than $1/3$ of the 33 amino acids listed in the second column of Table 1.

**Table 1** Comparison of numbers of acyclic monoamino monocarboxylic acids and monohydroxy, monocarboxylic acids of the Murchison meteorite with the number of possible *theoretical isomers* (after Cronin et al. 1993). Compounds with other substituent groups are omitted and $\alpha$, $\beta$, etc. denotes which carbon atom carries the amino or hydroxy substituent

| C-atoms | $\alpha$ $\beta$ $\gamma$ $\delta$ $\varepsilon$ $\zeta$<br>HOOC–C–C–C–C–C–C<br>X = NH$_2$ or **OH** | | | | | | |
|---|---|---|---|---|---|---|---|
| | $\alpha$ | $\beta$ | $\gamma$ | $\delta$ | $\varepsilon$ | $\zeta$ | Unknown |
| 2 | **1**, 1, **1** | – | – | – | – | – | – |
| 3 | **1**, 1, **1** | **1**, 1, **1** | – | – | – | – | – |
| 4 | **2**, 2, **2** | **2**, 2, **2** | **1**, 1, **1** | – | – | – | – |
| 5 | **3**, 3, **3** | **6**, 6, **3** | **3**, 3, **3** | **1**, 1, **0** | – | – | **1** |
| 6 | **8**, 8, **8** | *12*, 3, **1** | *11*, 4, **0** | **4**, 2, **0** | **1**, 1, **0** | – | **2** |
| 7 | *18*, 18, **12** | *29*, 0, **0** | *29*, 0, **0** | *20*, 0, **0** | **5**, 0, **0** | **1**, 0, **0** | **2** |

Structural isomer diversity also occurs among cyclic meteoritic compounds. A plethora of polycyclic aromatic hydrocarbons up to 750 dalton have been found in Murchison samples, along with a multitude of $C_{15}$ to $C_{30}$ branched alkyl-substituted mono-, di- and tricyclic alkanes (Cronin and Chang 1993). Becker et al. (2000) also reported an extensive suite of high molecular weight $C_{72}$ to $C_{270}$ fullerenes. These polycyclic hydrocarbons together with the alkyl phosphonic and sulfonic acids comprise the most thermally stable compounds in the Murchison inventory; they and the acid insoluble material are the likeliest meteoritic components to survive delivery to and diagenesis on Mars.

In contrast to biosynthetic pathways, which manifest themselves in a pattern of limited structural variation within specific compound classes, abiotic synthetic pathways yield distinctly different arrays arising from structural diversity across a wide array of meteoritic compound classes. The three patterns of variation in molecular structure and abundance with increasing carbon number suggest synthetic routes entailing small free radical initiators and intermediates (Cronin and Chang 1993). Such pathways tend to discriminate against high molecular weight compounds and favour formation of all possible constitutional isomers at lower carbon numbers by more or less random synthesis. Qualitatively similar patterns occur in both carbonaceous meteorites and laboratory prebiotic syntheses (see earlier discussion). Most importantly, none of the three patterns is exhibited by the classes of compounds used in living systems. These considerations lend confidence to the proposal that such patterns can distinguish between biosynthetic and abiotic organic matter.

## 3.4 Chirality of Amino Acids in the Murchison Meteorite

Syntheses of organic compounds do not produce chiral products in the absence of a chiral agent, whether it be an energy source (e.g., circularly polarized light), a surface substrate such as quartz crystals (Hazen et al. 2001), reactant (Bonner 1998) or catalyst (Weber and Pizzarello 2006). The absence of chirality in products of prebiotic evolution experiments strengthened the presumption that naturally occurring abiotic organic compounds would lack chirality. Indeed, early analysts of Murchison amino acids declared them to be racemic (e.g., Kvenvolden et al. 1970). Later reports of L-enantiomeric excesses in several meteoritic counterparts of protein amino acids (Engel and Nagy 1982;

Engel et al. 1990) met stiff resistance (Bada et al. 1983). In 1997, Cronin and Pizzarello published work that favoured modest L-enantiomeric excesses of 2 to 9%. They avoided the pitfalls of contamination by making measurements on 2-amino-2,3-dimethylpentanoic acid, $\alpha$-methyl norvaline and isovaline. All three compounds are $\alpha$-methyl substituted; the first two have no known biological counterparts and the third has a restricted distribution in fungal antibiotics. Using an $^{15}$N isotopic approach, Engel and Macko (1997) produced evidence for L-excess in alanine and glutamic acid that complemented earlier work using $^{13}$C (Engel et al. 1990) and showing that their observed enantiomeric excesses could not have been due to terrestrial contamination. Additional examples of L-enantiomeric excess in $\alpha$-methyl-$\alpha$-amino acids have been published (Pizzarello and Cronin 2000) although the latter authors observed that the alanine in their samples of the Murray and Murchison meteorites was essentially racemic.

Bailey et al. (1998) suggested that the observed enantiomeric excesses could have been induced by circularly polarised (CP) light arising from dust scattering in regions of high mass star formation. These sources occur more widely than do supernova remnants or pulsars that were first proposed by Rubenstein et al. (1983) as sources of CP synchrotron radiation. Regardless of production mechanism, enantiomeric excesses (Cronin and Pizzarello 1997; Engel and Macko 1997) testify to the reality of a naturally occurring abiotic chiral process. Future researchers will have to exercise caution in using chirality as a unique criterion for assigning biological origins to extra-terrestrial compounds.

## 3.5 Isotopic Compositions of C, H and N in Meteoritic Organic Matter

Although bulk analyses of the isotopic compositions of C, H and N of the Murchison meteorite hold no surprises, detailed examination of organic components reveals anomalies that call for unusual production mechanisms and environments of origin. The organic matter generally contains deuterium enrichments that start above the upper limits of the terrestrial range ($\delta$D $+100‰$) and extend to $+2,500‰$ approaching those of interstellar molecules. Values of $\delta^{13}$C range from $-18$ to $+40$. Results are displayed in Fig. 7 for organic components on which simultaneous measurements of $\delta^{13}$C and $\delta$D are available.

The heavy isotope enrichments in C and H vary widely from component to component and extend well beyond the limits of terrestrial experience. This is also true for $\delta^{15}N$ values, which range from $+18$ to more than $+98‰$ (Cronin and Chang 1993). Variations are large even within compound classes. Such variability suggests different reaction pathways, wide ranging formation temperatures, sampling of isotopically dissimilar reactant reservoirs or some combination of these and other factors. Temperature effects accord with the view that meteoritic organic matter originated over a wide range of low temperatures: Reactions in interstellar gas and on grains at temperatures as low as $-263°$C (Zinner 1988) coupled with aqueous processing in parent bodies up to and beyond $100°$C (Cronin and Chang 1993) are capable of yielding a range of large isotopic fractionations. The presence of interstellar molecules and stellar condensates assures sampling of distinctive nucleosynthetic isotopic reservoirs (Anders and Zinner 1993).

In contrast, biochemicals typically do not exhibit wide isotopic variability within or among compound classes; nor do the products of laboratory prebiotic syntheses (Chang et al. 1983; Kung et al. 1979). Fossil organic matter preserved in sediments of increasing age exhibits a relatively narrow range of carbon isotopic compositions for both extractable (Summons 1992) and insoluble fractions (Hayes et al. 1983; Strauss et al. 1992). Samples from some select environments such as the vicinity of methane seeps, exhibits a much greater heterogeneity in $\delta^{13}$C values. The $+20$ to $+35‰$ difference ($\Delta\delta$) between the organic carbon and the isotopically heavier coeval carbonate (representing the source reservoir) stands

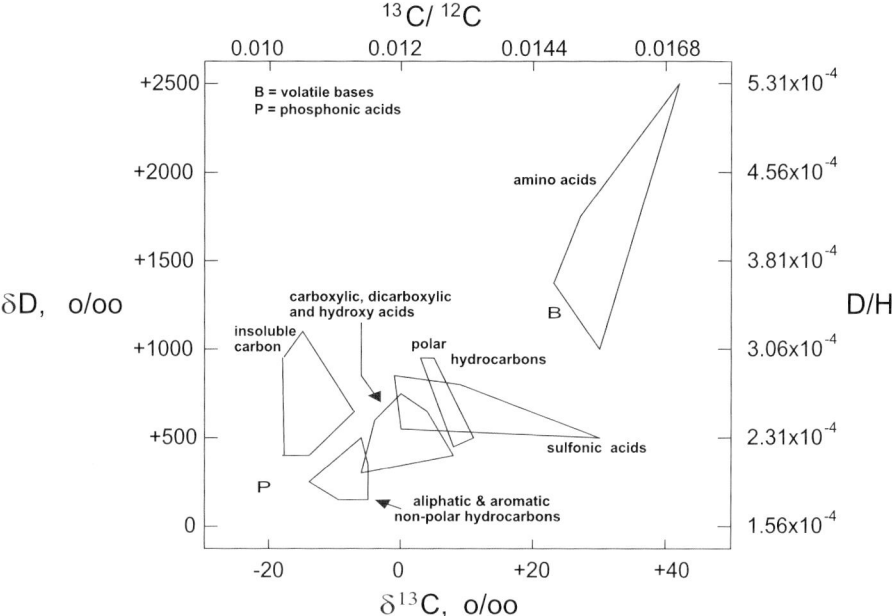

**Fig. 7**  C and H isotopic composition of organic components in Murchison meteorite (data from Cronin and Chang 1993; Cooper et al. 1997)

as an isotopic biomarker. On Earth, both biochemistry and planetary chemistry use a single dominant isotopic reservoir for each of the elements C, H and N; and the temperatures over which significant isotopic fractionations can occur under kinetic control are relatively restricted. Life on Earth operates over a relatively narrow temperature range ($\sim$0 to 120°C). Although some abiotic syntheses may occur at comparably low temperatures, reactions at higher temperatures and energetic processes like electrical discharges and meteor shocks would yield isotopic fractionations smaller than those of biochemistry.

Murchison organics preserved in ancient sediments on Mars, however, might well yield a narrow range of compositions analogous to that of fossil life on Earth. As diagenesis destroys the meteoritic amino acid and carboxylic acid components, the thermally stable insoluble carbon (kerogen), hydrocarbons and sulfonic and phosphonic acids would remain as relics. The $\delta^{13}C$ of carbonate in Murchison is about +42‰; therefore, depending on the proportions of the surviving components, the $\Delta\delta$ could range from +20 to +60‰, overlapping the terrestrial biomarker values. As with chirality, isotopic criteria alone must be used cautiously for distinguishing biological from non-biological materials. When used together as criteria, however, the combination of molecular structural patterns, chirality and isotopic compositions provides a powerful tool for distinguishing abiotic organic matter from compounds produced by living systems.

Compounds extracted from returned Martian samples can be subjected to intermolecular and intramolecular isotopic analyses. Measurements on individual Murchison compounds show each carboxylic acid containing $\sim$16‰ more $^{13}C$ than the light hydrocarbon of corresponding carbon number. In both homologous series, $\delta^{13}C$ values decline in parallel fashion as carbon number increases (Yuen et al. 1984). Intramolecular measurements on acetic acid indicate that the carboxyl group is relatively enriched in $^{13}C$ by 15 to 30‰ (Yuen et al. 1984). These results suggest synthesis of carboxylic acids by stepwise addition of the elements of

$CO_2$ to hydrocarbons. Cooper et al. (1997) also carried out intramolecular carbon, hydrogen and sulfur isotopic measurements on individual Murchison sulfonic acids, while an interesting correlation between $\delta^{13}C$ and H/C ratios for meteoritic PAH is evidently diagnostic for their synthesis under kinetic control in the interstellar medium (Naraoka et al. 2000). Most recently, Sephton et al. (2004) measured the molecular patterns of PAH released by stepped hydropyrolysis of Murchison kerogen and found that the refractory component probably consists of a network dominated by at least five- or six-ring PAH units cross-linked together.

Measurements at similar levels of molecular detail may prove invaluable in establishing sources and production mechanisms; they should certainly be included in the isotopic arsenal for future studies of extra-terrestrial organic matter. Extension of combined gas chromatography-isotope ratio mass spectrometry (GC-IRMS) to include all of the biogenic elements—C, H, O, N, S, with simultaneous multi-element measurements—would provide an extremely powerful tool for elucidating patterns of molecular structure and isotopic composition (Hinrichs et al. 2001). Use of chiral column substrates for gas chromatographic separations adds a further dimension.

## 4  Reduction Processes and the Preservation of Organic Matter

Bioproductivity and degree of preservation are key factors controlling the ultimate accumulation of organic matter from decaying organisms in sedimentary environments. On one hand, productivity is driven by environmental parameters (e.g. light level, nutrient supply, temperature, etc.). Preservation, on the other hand, is greatly enhanced under oxygen-depleted conditions, which protect lipids against remineralisation due to abiotic oxidation or aerobic biodegradation (Tissot and Welte 1984; Peters et al. 2005). In addition, efficient biogeochemical processes, involving sulfurisation, defunctionalisation and/or reduction of labile functionalities from lipids have been proposed to explain the transformations undergone by the organic matter deposited under reducing conditions. The latter processes are thought to be the major route for conversion of (unstable) unsaturated compounds to their related saturated counterparts. Prominent biomarkers preserved in this way include acyclic isoprenoids, carotenoids (e.g. Schaeflé et al. 1977; Adam et al., 1993, 1998; Sinninghe Damsté and Koopmans 1997; Hebting et al. 2006) and bacteriohopanepolyols (Ourisson and Albrecht 1992; Sinninghe Damsté et al. 1995). Although the cited examples indicate that reduction processes are commonly involved in the transformation of organic matter, little is known on the specifics of the mechanism(s) with the exception of thermal reduction and reactions with hydrogen sulfide (Hebting et al. 2006). In the former case, reduction operates by hydrogen disproportionation during late diagenesis and in the oil window and proceeds by homolytic cleavage of carbon–heteroatom or carbon–carbon bonds, followed by quenching of the formed radicals (e.g. Lewan 1997). In recent sediments, however, reduction reactions clearly result from different mechanisms which are not yet fully understood and referred to using the generic term 'hydrogenation' in the literature. Early reduction processes could be biologically induced, being therefore operative during the earliest stage of organic matter diagenesis, i.e. in the water column or in the first centimetres of the sediment, where active populations of micro-organisms are present (e.g. Wakeham 1989). Abiotic reduction reactions are also likely involved (Schneckenburger et al. 1998; Hebting et al. 2006) and could operate within the water column, as well as the sediments, provided that the redox potential is low enough, for example, in the presence of abundant hydrogen sulfide. There are several reports showing that partly reduced lipids, in particular carotenoids, occur in recent or very immature sediments (e.g. Watts and Maxwell 1979; Boon et al. 1981; De Castro 1994; Sinninghe Damsté and Koopmans 1997; Schaeffer et al., 1997a, 1997b) and hydrogen sulfide seems to be a key player as the agent of reduction (Hebting et al. 2006).

## 4.1 Other Reactions with Sulfur

As indicated, sulfur chemistry has been shown to exert a major influence on the fate of terrestrial organic matter. This is an important issue because the degree to which biosignatures are preserved, or lost, has only recently been rationalised through knowledge of the chemistry of sulfur-bound intermediates. These reactions lead to the formation of organic sulfur compounds (OSC) that are less susceptible to bacterial degradation and thus have a higher preservation potential (e.g. Sinninghe Damsté and de Leeuw 1990). The availability of Fe in the form of ferric oxide or oxyhydroxide (Canfield 1989) is also important as it is thought to react faster with reduced S species than organic matter (Gransch and Posthuma 1974). Sulfur-rich organics are, therefore, usually deposited in Fe-depleted systems such as stratified basins where anoxic conditions can establish and the supply of Fe is less than sulfide production.

Apart from coal deposits that may, or may not, have sulfur involved in their formation, the most important reservoir of sedimentary organic matter are rocks that were initially deposited in anoxic, marine environments. These are favourable for micro-organisms involved in the reduction of sulfate to reduced species such as hydrogen sulfide, elemental sulfur or polysulfides. The sulfur content of organic matter deposited in such environments may reach 20%. Sulfurisation of organic matter results in the formation of low molecular weight constituents such as thiophenes, sulfides and thiols as well as sulfur cross-linked macromolecules (Sinninghe Damsté and de Leeuw 1990 and references therein). Most of what is now known was uncovered in the last decade and includes full structural identifications of numerous low molecular weight organo-sulfur compounds and sulfur-rich macromolecular component. Selective chemical degradation procedures and reagents (e.g. Raney nickel, nickel boride, $LiAlH_4$ or $CH_3Li/CH_3I$) release sub-units as low molecular weight components amenable to conventional geochemical analyses. These studies have shown that sulfur-rich macromolecules are derived from highly functionalised precursors and that the locus of bound sulfur is generally the primary chemical functionalities (Valisolalao et al. 1984; Sinninghe Damsté and de Leeuw 1990; Kohnen et al. 1991a, 1991b, 1993; Adam et al. 1992; Schaeffer et al. 1995). This indicates that functionalities—in particular double bonds, hydroxyl and carbonyl groups—are crucial for the incorporation of sulfur and also that sulfurisation occurs relatively early during diagenesis and under very benign, low-temperature and near-neutral pH, conditions. Sulfurisation even occurs in the water column as demonstrated by the widespread occurrence of organo-sulfur compounds in recent sediments ($<1,000$ yr; e.g. Barbé et al. 1990; Wakeham et al. 1995; Poinsot et al. 1998; Kok et al. 2000). The fate of many lipids that are not trapped as sulfur adducts during the early stages of diagenesis probably involves progressive aromatisation (e.g. Murray et al. 1997).

## 5 Processes Leading to Loss or Alteration of Molecular Biosignatures

### 5.1 Oxidation as a Secondary Alteration Process of Sedimentary Organic Matter

When exposed to atmospheric conditions, sedimentary organic matter undergoes various alteration processes, such as water washing, evaporation, biodegradation and abiotic oxidation. Organic matter is rapidly destroyed in the presence of molecular oxygen and this is mostly mediated by the action of micro-organisms (e.g. Engel and Macko 1993; Hedges and Keil 1995). Organic matter destruction can also be initiated by light and trace oxidants in the atmosphere (ozone, peroxides; e.g. Cooper and Zika 1983; Cooper et al. 1989; Aneja et al. 1994). Although oxidation itself is a widespread process, insufficient attention has been paid to abiotic oxidation as an independent phenomenon leading to a loss of biomarker integrity. Results gathered by the Opportunity rover team at Meridiani Planum, demonstrate an abundance of Fe(III) and sulfate minerals that formed from liquid water (Squyres et al. 2004). The presence of these minerals suggests oxidizing conditions prevailed during their formation and that organic compounds are unlikely to be preserved within them.

Although there is considerable interest in determining the relative recalcitrance of specific biochemicals under oxic and anoxic conditions during sedimentation (e.g. Teece et al. 1998), somewhat different issues are involved when fossil organic matter is exhumed from buried material. The focus of the remainder of this section concerns the fate of organics which survived early diagenesis to become preserved in rocks as kerogen or bitumen.

## 5.2 Bulk Analysis of Oxidized Sedimentary Organic Matter

The relative proportions of the extractable fractions from shales or bitumens (saturates, aromatic hydrocarbons, macromolecular asphaltenes and resins) depends drastically on the severity of oxidative alteration. Indeed, a preferential removal of aromatic hydrocarbons compared to saturated hydrocarbons is often observed in weathered shales (Leythaeuser 1973; Clayton and Swetland 1978), possibly as a consequence of water solubility, and in altered bitumen samples (e.g. Wang et al. 1994; Charrié-Duhaut et al. 2000). In contrast, proportions of asphaltenes in weathered samples increase relative to low molecular weight fractions with values reaching up to 90% (cf. Wang et al. 1994). Analysis of weathered shale also reveals an important loss of total organic carbon (e.g. Petsch et al. 2000). The formation of organic material, insoluble in organic solvents, is also observed in severely oxidized samples (Charrié-Duhaut et al. 2000).

Infrared (IR) spectroscopic analyses of severely altered bitumen samples reveals the appearance of oxygen functionalities (e.g. Speight 1984). In non-degraded bitumens, the aliphatic and aromatic signals prevail whereas absorptions corresponding to oxygenated functions are low. After oxidative alteration, new IR bands corresponding to hydroxyls and carboxyls appear, suggesting the presence of constituents such as phenols, alcohols, mono- or polycarboxylic acids. Ketones, aldehydes, esters, sulfones, sulfoxides and sulfonates can also be detected by IR spectroscopy, the corresponding bands becoming predominant in samples more severely affected by degradation.

The effect of oxidative alteration on organic sulfur compounds can be measured by XANES (X-ray and near edge) spectroscopy. Thus, the relative sulfur content of bitumen samples (thiophenic, alkyl or aryl sulfur; Brown et al. 1992) estimated by the intensity of the S L-edge signals has been shown to decrease with increasing alteration, but the most degraded samples contain oxidized organic sulfur species (sulfoxides, sulfones, sulfonates and sulfates).

Carbon isotopic composition does not seem to be significantly affected by oxidative alteration (Connan et al. 1992; Sarret et al. 1999). However, a slight enrichment in $^{13}C$ is observed in severely weathered shales (e.g. Leythaeuser 1973). In contrast to carbon, hydrogen/deuterium isotopic compositions are strongly affected by the alteration processes (Charrié-Duhaut et al. 2000). The observed increase in deuterium content with increasing alteration may be explained by H–D exchange reactions with rainwater or percolating groundwater. Variations in $\delta D$ values parallel the increase in oxygen content, indicating that incorporation of oxygen from the atmosphere as hydroxyl, carboxyl or carbonyl is likely part of such a process.

## 5.3  Molecular Composition of Oxidized Sedimentary Organic Matter

It has been clearly shown that some of the saturated and aromatic hydrocarbons disappearing during alteration are transformed into oxygenated compounds. Indeed, analysis of the polar fractions (ketones, sulfones, carboxylic acids) from altered bitumens led to the identification of oxygen-containing compounds clearly resulting from the oxidation of typical petroleum components (Charrié-Duhaut et al. 2000). For example, it could be shown that the benzylic positions of aromatic compounds are the most sensitive to oxidation. Thus, benzylic ketones related to mono- and triaromatic steroids, ring $D$ monoaromatic 8,14-secohopanoids have been identified (Charrié-Duhaut et al. 2000). Similarly, carboxylic acids related to steroids, tricyclic terpenoids, hopanoids and benzothiophenes could also be detected (Charrié-Duhaut et al. 2000). Sulfones and sulfoxides, which are the result of the oxidation of steroid, terpenoid or hopanoid sulfides are also present in altered bitumens (Strausz et al. 1990; Schouten et al. 1993; Charrié-Duhaut et al. 2000). Maleimides (1H-pyrrole-2,5-diones) are examples of diagenetic products formed through oxidation although they provide useful geochemical information (Grice et al. 1996).

The hypothesis that these compounds are, at least in part, the result of abiotic oxidation processes is supported by the fact that oxygen incorporation generally occurred without diastereomeric discrimination.

## 5.4  Oxidation of Sedimentary Organic Matter: Simulation Experiments

Most of the oxygenated products formed in simulation experiments involving typical aromatic constituents of mature petroleum (triaromatic steroids, alkylated benzothiophenes, alkylaromatics), oxygen and light are similar to the benzylic ketones or acids detected in naturally oxidized bitumens. This also supports the hypothesis that abiotic oxidation processes are operative.

The photo-oxidation of aromatic compounds leads principally to the formation of oxygenated compounds (alcohols and ketones) resulting from attack at benzylic positions. Degradation of the carbon skeletons (C–C bond cleavage reactions) affecting either alkyl chains or aromatic polycyclic systems has also been demonstrated (Erhardt and Petrick 1984; Fox and Olive 1979; Charrié-Duhaut et al. 2000). Similarly, alkylbenzothiophenes were rapidly transformed into the corresponding benzylic ketones and alkylated carboxybenzothiophenes formed by oxidative cleavage at benzylic positions (e.g. Andersson and Bobinger 1996). However, the major degradation pathway involves the opening of the thiophenic ring and leads mainly to alkylated 2-thiobenzoic acids or 2-sulfobenzoic acids and the related phenols (Andersson and Bobinger 1996; Charrié-Duhaut et al. 1999).

Non-aromatic sulfur compounds, such as thianes or thiolanes, are even more sensitive to photo-oxidation than benzo-and dibenzothiophenes. Thus, thiolanes are easily oxidized to sulfoxides and sulfones. In a further stage, opening of the thiolane ring—leading in particular to the formation of sulfonic acids and ketones—is also observed (Takata et al. 1985; Charrié-Duhaut et al. 1999).

The possibility that non-aromatic ketones may also be formed by abiotic oxidation is supported by simulated (photo)oxidation of saturated hydrocarbons ($n$-alkanes, isoprenoid alkanes, hopanes) in the presence of a photosensitiser (Ehrhardt and Petrick 1985; Rontani and Giusti, 1987, 1989; Giuliano et al. 1997; Tritz et al. 1999) which led, notably, to the formation of the corresponding ketones.

Thus, simulation experiments involving petroleum lipids, oxygen and light suggest that abiotic oxidation processes significantly contribute to the degradation of petroleum macromolecular structures, generally considered as highly refractory to atmospheric alteration. Such processes favour the oxidative release of sub-units as oxygenated low-molecular weight components by cleavage of C–C bonds at the benzylic positions of alkylaromatic subunits or by degradation of the (poly)aromatic skeletons. Oxidative cleavage reactions affecting, in particular, benzothiophenic sub-units in petroleum macromolecules may also contribute to the formation of hydrophilic functionalities such as phenols, sulfonic acids together with carboxylic acids. As a consequence, macromolecular organic matter may progressively be transformed into more water-soluble, humic-like, substances.

## 5.5 Oxidants and Abiotic Oxidation Mechanisms for Destruction of Petroleum in Natural Environments

Weathering of exposed petroleum has been intensively studied and serves as a model for the oxidation of carbonaceous matter with size ranges from small organics to macromolecular material. Among the oxidants naturally occurring in nature, and responsible for the formation of oxygenated compounds detected in weathered petroleum, triplet oxygen is the most abundant and is known to participate in radical-type autoxidation reactions (Ingold 1969; Voronenkov et al. 1970; Korcek et al. 1972; Sheldon and Kochi 1981). It affects more likely benzylic or allylic positions as well as tertiary alkyl positions. Such a process may result in the formation of oxygenated compounds such as acids, alcohols, ketones and epoxides. Thus, this type of reaction can account for the formation of aromatic ketones or acids (oxygen functionality at the benzylic position) identified in weathered petroleum. Such processes may also induce cleavage and may contribute to the degradation of high molecular weight constituents of the bitumens, leading to the release of low molecular weight oxygenated components. Light may favour the oxidation of the organic matter from petroleum, and the reaction products, such as benzylic ketones, may even accelerate the process by inducing formation of free radicals necessary to initiate oxidation by triplet oxygen.

Besides triplet oxygen, other oxidants, such as singlet oxygen, ozone, hydroxyl or peroxy radicals, superoxides, nitrogen oxides and hydrogen peroxide (Cooper et al. 1989; Aneja et al. 1994) may also be involved in the alteration of hydrocarbons. Some, although present in low concentrations, are ubiquitous (e.g. hydrogen peroxide, nitrogen oxides). Other highly reactive species are transient (e.g. hydroxyl radicals, singlet oxygen). All may occur, in nature, from photochemical reactions involving photosensitisers (Cooper et al. 1989) such as humic substances, aromatic hydrocarbons, benzylic ketones and petroporphyrins. Relatively stable oxidants such as hydrogen peroxide, formed by photochemical reactions in nature can be transported in meteoric water. Alternatively, oxidants may be formed by photochemical reactions between sensitisers already present in bitumens and react directly with the bitumen. In such a situation, unstable, highly reactive oxidants such as hydroxyl radicals or singlet oxygen may also participate in the oxidation process.

## 6 The Effects of Thermal Cracking and Ionising Radiation

Over very long time scales (i.e. $>10^8-10^9$ years), thermal cracking and ionising radiation are major risk factors for preservation of molecular biosignatures in both terrestrial and extra-terrestrial samples. Evidence exists for widespread distribution of Archean-aged, organic rich sediments but overwhelmingly these show metamorphic grades known to preclude preservation of biomarkers (e.g. Hayes et al. 1983). Hydrocarbon generation and migration were evidently widespread in Archean times but the remnants consist mainly of pyrobitumen, minute hydrocarbon bearing inclusions in quartz and other minerals and bitumen rims on radioactive mineral grains (Buick et al. 1998; Dutkiewicz et al. 1998). The deleterious effects of ionising radiation are profoundly demonstrated in uraniferous samples of the Alum Shale (ca. 500 Ma). Despite having conventional indications of low thermal maturity, and abundant biological markers in other parts of the sedimentary sequence, high uranium samples are virtually devoid of these compounds (Dahl et al. 1988).

Most samples of Archean organic matter so far examined have vanishingly low contents of molecular biosignatures, and those with the best evidence for indigenous biomarkers are still somewhat tenuous (Brocks et al. 1999; 2003a, 2003b, 2003c, 2003d). Aromatic hydrocarbons released by hydropyrolysis of highly mature Archaean kerogens are the least ambiguous remains of Archaean microbes detected so far (Brocks et al. 2003a).

## 7 Methods of Chemical Analysis

### 7.1 Analytical Technology

With the possible exception of chirality, the molecular attributes described here require that the analytical approach can separate and accurately identify the components of complex mixtures. Sample availability for terrestrial applications always tends to pose limitations and this issue is extreme for extra-terrestrial samples. Consequently, analytical sensitivity is paramount. Gas chromatography–mass spectrometry (GC–MS) provides the optimal combination of isomer separation and compound identification for known as well as unknown molecules. Accordingly, for lending flexibility, adaptability, sensitivity, combinatorial techniques such as GC–MS, GC–MS–MS, GC–GC–MS, LC–MS, Pyrolysis–GC–MS and GC–IRMS remain at the forefront of the analytical methods used in organic geochemical and biogeochemical applications. Recent improvements in mass analysers for GC–MS (Time of Flight (TOF) and Q-TOF and Ion Traps) ensure that mass spectrometry will retain its status as the most useful analytical tool. Chromatography is also seeing continuous improvement, the emergence of 'Comprehensive GC' being one such example (Marriott and Kinghorn 1997; Frysinger et al. 2002). Avoidance of sample pre-treatments (e.g. solvent extraction) is desirable for avoiding the pitfalls of contamination. Thermal extraction of volatile organic components directly onto a GC has many attractive features for microanalysis (Hayes and Biemann 1968; Mahaffy 2007).

Spatial resolution is the key attribute of secondary ion mass spectrometry (SIMS) and laser ablation approaches ($L^2MS$) which allow us to examine specific objects and inhomogeneities in compound distribution (Kovalenko et al. 1992). Combined with TOF mass analysers, these also allow exceptional mass resolution and very fast data acquisition. Although these techniques do not resolve isomer distributions, they provide valuable means for screening minute samples. SIMS with $^{13}C$ analysis using an ion microprobe is increasingly used to examine isotopic heterogeneity within microscopic objects, including living cells (Orphan et al. 2002). Other spectroscopic techniques (e.g. Laser Raman, fluorescence spectroscopy; X-ray absorption spectroscopy i.e., XANES/EXAFS) are non-destructive but lack the facility to unravel complex compound mixtures. However, they are likely to be extremely valuable when used in a screening sense.

### 7.2 Chemical and Pyrolytic Degradation of Macromolecular Organic Matter

Macromolecular organic matter, including notably kerogen and coal, represents the quantitatively most important type of fossil organic matter in the geosphere (Tissot and Welte 1984).

Different biogeochemical processes are thought to be involved in the formation of kerogen and related macromolecular organic matter and the last decade has seen major revisions in earlier models for 'geopolymerization' of small to medium biochemicals, as discussed earlier. Analytical pyrolysis led to the discovery of insoluble, non-hydrolysable aliphatic biopolymers in extant organisms (higher plants, algae) and fossil organic matter. This led to the concept of 'selective preservation' whereby a significant proportion of the kerogen comprises biosynthetic polymers such as algeanans (Tegelaar et al. 1989). Macromolecular organic matter is also the most abundant form of carbon in meteorites and pyrolytic methods have also been used to study this (Sephton et al., 2000, 2001, 2004).

Knowledge of the chemical composition of kerogen and related macromolecules has also been strongly improved by studies involving the use of chemical reagents which selectively cleave sub-units covalently bound within the macromolecular network. These use oxidative, reductive, hydrolytic and specific bond (e.g. ether) cleavage reactions and give information not only on the nature of sub-units bound within kerogen, but also on the mode binding of sub-units (i.e. esters, amides, sulfide and polysulfide bonds, ethers and carbon–carbon bonds linked to aromatic moieties). Labelling experiments using deuterated reagents (e.g. $LiAlD_4$) give additional information about sites and types of chemical linkages. Strong oxidizing reagents ($KMnO_4$ or $RuO_4$) are also employed with sub-units being released as carboxylic acids and this has been useful for degradation of highly aromatic geomacromolecules. Finally, the value of chemical degradation techniques can be enhanced when several are used in a sequential manner. With respect to the analysis of extra-terrestrial organic matter, chemical degradation approaches are likely to require significant amounts of material.

Hydropyrolysis (rapid pyrolysis in a high pressure hydrogen stream) is a new and potentially powerful tool for conversion of macromolecular organic matters into identifiable components while retaining a high degree of stereochemical integrity (Murray et al. 1998; Love et al. 1995; Sephton et al. 2004).

## 8 Strategies to Strengthen Criteria Technologically

We need much-improved knowledge of recalcitrant biochemicals (those that are resistant to geological degradation) that are produced by bacteria and archaea. Many new organisms are known only from their nucleic acid sequences. Expansive surveys of the lipid and other biomarker compositions of these 'new' organisms is critically important.

There needs to be development and testing of immunological or other tagging methods for recognition of specific organic biomarkers that might be more sensitive than mass spectrometry.

We need better understanding of organic compound distributions produced by abiogenic synthesis, especially with regard to aromatic hydrocarbons (including PAH) and alkanes. Simulations of hydrothermal organosyntheses under widely different conditions are needed to better understand the richness of structures produced in this way as well as their isotopic characteristics. We also need more information on mechanisms and pathways of abiotic oxidation that lead to loss of organic biosignatures.

Successful exploitation of molecular biosignatures would be advanced by marked improvements to the sensitivity of existing mass spectrometers, both commercial and experimental. Attention should be paid to achieving incremental improvements in all aspects of GC–MS, LC–MS and capillary electrophoresis (CE) instrumentation including thermal desorption inlet systems, laser tools for volatilisation and pyrolysis of organic matter at microscopic spatial scales, microbore and multi-dimensional chromatographic systems, SIMS, laser and electron impact ion sources, mass analysers and detector systems to increase signal to noise performance. For example, coupling a TOF analyser to an electron impact source yielded 20- to 50-fold improvements in both sensitivity and mass resolution compared to the commonly used benchtop quadrupole GC–MS systems.

The specificity of chemical structure information can be greatly enhanced through simultaneous multi-element isotope analyses. Hence, structural and multi-element isotopic data collected in tandem will greatly expand the information that can be decoded from molecular biosignatures.

## 9  Mars—The Challenges

Very small sample sizes for returned materials, and likely low organic carbon contents, dictate that we should continue to improve and optimise the sensitivity of our analytical instruments and techniques.

Micro-sample handling and the avoidance of contamination will continue to be critical issues for analysis of extra-terrestrial materials. Evaluation of contamination sources and vectors will also require detailed attention.

The Martian environmental context suggests that surface and near-surface samples will have been exposed to chemical reactions that are difficult to predict. Alteration by cosmic rays, UV irradiation, peroxides and array of non-aqueous chemical oxidations over very long timescales are just some of the possibilities through which molecular biosignatures could be corrupted or lost entirely.

Recent reports of studies of Mars analogue soils (Navarro-Gonzalez et al., 2006, 2003) propose that the reactive nature of Mars surface soils presents obstacles to preservation and detection using instruments employing thermal volatilisation GC–MS since organics could be subject to oxidation during the sample heating. This research merits further scrutiny (Biemann 2007) since chromatographic instruments hold the most promise for identifying non-Earth-centric molecular biosignatures as described herein (Mahaffy 2007; for an alternative extraction protocol see Buch et al. 2006 and Bada et al. 2007). The potential instability of organic compounds is such an environment over geological timescales adds further risk (Sumner 2004; Aubrey et al. 2006). These topics require deeper understanding if we are to adequately address the present and past habitability of Mars.

**Acknowledgements**   We are deeply indebted the Sherwood Chang for his contributions to the sections on meteoritic organic matter. The manuscript was improved significantly by the comments of two anonymous reviewers and discussions by the National Research Council Committee on the Astrobiology Strategy for the Exploration of Mars during the preparation of their 2007 report to the Space Studies Board of the U.S. National Academies. RES is supported by grants from the NASA Exobiology program and the NASA Astrobiology Institute.

## References

P. Adam, B. Mycke, J.C. Schmid, P. Albrecht, Energy Fuels **6**, 553–559 (1992)
P. Adam, J.C. Schmid, B. Mycke, C. Strazielle, J. Connan, A. Huc, A. Riva, P. Albrecht, Geochimica Cosmochimica Acta **57**, 3395–3419 (1993)
P. Adam, E. Philippe, P. Albrecht, Geochimica Cosmochimica Acta **62**, 265–271 (1998)
E. Anders, Nature **342**, 255–257 (1989)
E. Anders, E. Zinner, Meteoritics **28**, 490–514 (1993)
J.T. Andersson, S. Bobinger, Polycycl. Aromat. Compd. **11**, 145–151 (1996)
V.P. Aneja, M. Das, D.S. Kim, B.E. Hartsell, Israel J. Chem. **34**, 387–401 (1994)

A. Aubrey, H.J. Cleaves, J.H. Chalmers, A.M. Skelley, R.A. Mathies, F.J. Grunthaner, P. Ehrenfreund, J.L. Bada, Geology **34**, 357–360 (2006)
J.L. Bada, J.R. Cronin, M.-S. Ho, K.A. Kvenvolden, J.G. Lawless, S.L. Miller, S. Oro, J. Steinberg, Nature **301**, 494–497 (1983)
J.L. Bada et al., Space Sci. Rev. (2007, this issue). doi:10.1007/s11214-007-9213-3
J. Bailey, A. Chrysostomou, J.H. Hough, T.M. Gledhill, A. McCall, F. Menard, M. Tamura, Science **281**, 672–674 (1998)
A. Barbé, J.O. Grimalt, J.J. Pueyo, J. Albaigés, in *Advances in Organic Geochemistry*, ed. by B. Durand, F. Behar (Pergamon Press, Oxford, 1990), pp. 815–828
L. Barron, Space Sci. Rev. (2007, this issue). doi:10.1007/s11214-007-9254-7
L. Becker, R.J. Poreda, T.E. Bunch, Proc. Natl. Acad. Sci. USA **97**, 2979–2983 (2000)
K. Biemann, Proc. Natl. Acad. Sci. USA **104**, 10310–10313 (2007)
K. Biemann, J. Oro, P. ToulminIII, L.E. Orgel, A.O. Nier, D.M. Anderson, P.G. Simmonds, D. Flory, A.V. Diaz, D.R. Rushneck, J.A. Biller, Science **194**, 72–76 (1976)
W. Bonner, *Homochirality and Life. D-Amino Acids in Sequences of Secreted Peptides of Multicellular Organisms* (Birkhaeuser, Basel, 1998) pp. 159–188
J.J. Boon, H. Hines, A.L. Burlingame, J. Klok, W.I.C. Rijpstra, J.W. de Leeuw, K.E. Edmunds, G. Eglinton, in *Advances in Organic Geochemistry* ed. by M. Bjoroy et al. (Wiley, Chichester, 1981), pp. 207–227
J.J. Brocks, G.A. Logan, R. Buick, R.E. Summons, Science **285**, 1033–1036 (1999)
J.J. Brocks, R.E. Summons, in *Treatise on Geochemistry*, ed. by D.H.a.K.K.T. Heinrich (Pergamon, Oxford, 2003), pp. 63–115
J.J. Brocks, G.D. Love, C.E. Snape, G.A. Logan, R.E. Summons, R. Buick, Geochimica Cosmochimica Acta **67**, 1521–1530 (2003a)
J.J. Brocks, R.E. Summons, G.A. Logan, R. Buick, Geochimica Cosmochimica Acta **67**, 4289–4319 (2003b)
J.J. Brocks, R.E. Summons, G.A. Logan, R. Buick, Geochimica Cosmochimica Acta **67**, 4321–4335 (2003c)
J.J. Brocks, R.E. Summons, R. Buick, G.A. Logan, Org. Geochem. **34**, 1161–1175 (2003d)
J.R. Brown, M. Kasrai, G.M. Bancroft, K.H. Tan, J.M. Chen, Fuel **71**, 649–653 (1992)
A. Buch, D.P. Glavin, R. Sternberg, C. Szopa, C. Rodier, R. Navarro-Gonzalez, F. Raulin, M. Cabane, P.R. Mahaffy, Planet. Space Sci. **54**, 1592–1599 (2006)
R. Buick, B. Rasmussen, B. Krapez, AAPG Bull. **82**, 50–69 (1998)
D.E. Canfield, Geochimica Cosmochimica Acta **53**, 619–632 (1989)
S. Chang, D. Des Marais, R. Mack, S.L. Miller, G.E. Strathearn, in *Earth's Earliest Biosphere*, ed. by J.W. Schopf (Princeton University Press, Princeton, 1983), pp. 53–92
A. Charrié-Duhaut, S. Lemoine, P. Adam, J. Connan, P. Albrecht, J. Conf. Abstr. **4**, 572 (1999)
A. Charrié-Duhaut, S. Lemoine, P. Adam, J. Connan, P. Albrecht, Org. Geochem. **31**, 977–1003 (2000)
C.F. Chyba, C. Sagan, Nature **355**, 125–132 (1992)
J.L. Clayton, P.J. Swetland, Geochimica Cosmochimica Acta **42**, 305–312 (1978)
S. Clemett, C. Maechling, R. Zare, P. Swan, R. Walker, Science **262**, 721–725 (1993)
J. Connan, A. Nissenbaum, D. Dessort, Geochimica Cosmochimica Acta **56**, 2743–2759 (1992)
G.W. Cooper, M.H. Thiemens, T.L. Jackson, S. Chang, Science **277**, 1072–1074 (1997)
W.J. Cooper, R.G. Zika, Science **220**, 711–712 (1983)
W.J. Cooper, R.G. Zika, R.G. Petasne, A.M. Fisher, in *Aquatic Humic Substances*, ed. by I.H. Suffet P. MacCarthy. ACS Symposium Series, vol. 219 (American Chemical Society, Washington, 1989) pp. 333–362
J.R. Cronin, S. Pizzarello, D.P. Cruikshank, in *Meteorites and the Early Solar System*, ed. by J.F. Kerridge M.S. Matthews (University of Arizona Press, Tucson, 1988) pp. 819–857
J.R. Cronin, S. Chang, in *The Chemistry of Life's Origin*, ed. by J.M. Greenberg et al. (Kluwer, Netherlands, 1993) pp. 209–258
J.R. Cronin, S. Pizzarello, S. Epstein, R.V. Krishnamurthy, Geochimica Cosmochimica Acta **57**, 4745–4752 (1993)
J.R. Cronin, S. Pizzarello, Science **275**, 951–955 (1997)
J. Dahl, R. Hallberg, I.R. Kaplan, Org. Geochem. **12**, 559–571 (1988)
I.M. De Castro, Lipides de microorganismes et de sédiments actuels. PhD thesis, Université Louis Pasteur, Strasbourg, France, 1994
M.J. DeNiro, S. Epstein, Science **197**, 261–263 (1977)
A. Disch, J. Schwender, C. Müller, H.K. Lichtenthaler, M. Rohmer, Biochem. J. **333**, 381–388 (1998)
A. Dutkiewicz, B. Rasmussen, R. Buick, Nature **395**, 885–888 (1998)
G. Eglinton, P.M. Scott, T. Besky, A.L. Burlingame, M. Calvin, Science **145**, 263–264 (1964)
G. Eglinton, M. Calvin, Sci. Am. **261**, 32–43 (1967)
Eigenbrode, Space Sci. Rev. (2007, this issue). doi:10.1007/s11214-007-9252-9
M. Erhardt, G. Petrick, Marine Chem. **15**, 47–58 (1984)

M. Ehrhardt, G. Petrick, Marine Chem. **16**, 227–238 (1985)

E.L. Eliel, S.H. Wilen, L.N. Mander, *Stereochemistry of Organic Compounds* (Wiley, New York, 1994) 1267 pp

M.H. Engel, B. Nagy, Nature **296**, 837–840 (1982)

M.H. Engel, S.A. Macko, eds., *Org. Geochem. Principles and Applications* (Plenum Press, New York, 1993)

M.H. Engel, S.A. Macko, J.A. Silfer, Nature **348**, 47–49 (1990)

M.H. Engel, S.A. Macko, Nature **389**, 265–268 (1997)

G. Flesch, M. Rohmer, Eur. J. Biochem. **175**, 405–411 (1988)

G. Flynn, Earth, Moon, Planets **72**, 469–474 (1966)

M.N. Fomenkova, S. Chang, L.M. Mukhin, Geochimica Cosmochimica Acta **58**, 4503–4512 (1994)

M.A. Fox, S. Olive, Science **205**, 582–583 (1979)

G.S. Frysinger, R.B. Gaines, C.M. Reddy, Environ. Forensics **3**, 27–34 (2002)

F. Gelin, I. Boogers, A.A.M. Noordeloos, J.S. Sinninghe Damsté, P. Hatcher, J.W. de Leeuw, Geochimica Cosmochimica Acta **60**, 1275–1280 (1996)

M. Giuliano, F. El Anba-Luro, P. Doumenq, G. Mille, J.F. Rontani, J. Photochem. Photobiol. A: Chem. **102**, 127–132 (1997)

J.A. Gransch, J. Posthuma, in *Advances in Organic Geochemistry*, ed. by B. Tissot, F. Bienner (Editions Technip, Paris, 1974), pp. 727–739

K. Grice, R. Gibbison, J.E. Atkinson, L. Schwark, C.B. Eckardt, J.R. Maxwell, Geochimica Cosmochimica Acta **60**, 3913–3924 (1996)

J.M. Hayes, Marine Geol. **113**, 111–125 (1993)

J.M. Hayes, in J.W. Valley, D.R. Cole eds., Stable isotopic geochemistry, Rev. Mineral. Geochem. **43**, 225–277 (2001)

J.M. Hayes, K. Biemann, Geochimica Cosmochimica Acta **32**, 239–267 (1968)

J.M. Hayes, I.R. Kaplan, K.W. Wedeking, in *Earth's Earliest Biosphere; Its Origin and Evolution*, ed. by J.W. Schopf (Princeton University Press, Princeton, 1983) pp. 93–134

R.M. Hazen, T.R. Filley, G.A. Goodfriend, Proc. Nat. Acad. Sci. USA **98**, 5487–5490 (2001)

Y. Hebting, P. Schaeffer, A. Behrens, P. Adam, G. Schmitt, P. Schneckenburger, S.M. Bernasconi, P. Albrecht, Science **312**, 1627–1631 (2006)

K.-U. Hinrichs, G. Eglinton, M.H. Engel, R.E. Summons, Geochem. Geophys. Geosyst. **1** (2001). 2001GC000142

J.I. Hedges, R.G. Keil, Marine Chem. **49**, 81–115 (1995)

K.U. Ingold, Acc. Chem. Res. **2**, 1–9 (1969)

A. Kadouri, S. Derenne, C. Largeau, E. Casadevall, C. Berkaloff, Phytochemistry **27**, 551–557 (1988)

Y.V. Kissin, Org. Geochem. **29**, 947–962 (1998)

G. Kminek, J.L. Bada, O. Botta, D.P. Glavin, F. Grunthaner, Planet. Space Sci. **48**, 1087–1091 (2000)

M.E.L. Kohnen, J.S. Sinninghe Damsté, A.C. Kock-van Dalen, J.W. de Leeuw, Geochimica Cosmochimica Acta **55**, 1375–1394 (1991a)

M.E.L. Kohnen, J.S. Sinninghe Damsté, J.W. de Leeuw, Nature **349**, 775–778 (1991b)

M.E.L. Kohnen, J.S. Sinninghe Damsté, M. Baas, A.C. Kock-van-Dalen, J.W. de Leeuw, Geochimica Cosmochimica Acta **57**, 2515–2528 (1993)

M.D. Kok, W.I.C. Rijpstra, L. Robertson, J.K. Volkman, J.S. Sinninghe Damsté, Geochimica Cosmochimica Acta **64**, 1425–1436 (2000)

S. Korcek, J.H.B. Chenier, J.A. Howard, K.U. Ingold, Can. J. Chem. **50**, 2285–2297 (1972)

L.J. Kovalenko, C.R. Maechling, S.J. Clemett, J.M. Philippoz, R.N. Zare, Analyt. Chem. **64**, 682–690 (1992)

C.C. Kung, R. Hayatsu, M.H. Studier, R.N. Clayton, Earth Planet. Sci. Lett. **38**, 421–435 (1979)

K. Kvenvolden, J. Lawless, K. Pering, E. Peterson, J. Flores, C. Ponnamperuma, I.R. Kaplan, C. Moore, Nature **228**, 623–626 (1970)

R.M. Lemmon, Chem. Rev. **70**, 95–109 (1970)

M.D. Lewan, Geochimica Cosmochimica Acta **61**, 3691–3723 (1997)

D. Leythaeuser, Geochimica Cosmochimica Acta **37**, 113–120 (1973)

H.K. Lichtenthaler, Annu. Rev. Plant Physiol. Plant Mol. Biol. **50**, 47–65 (1999)

G.D. Love, C.E. Snape, A.D. Carr, R.C. Houghton, Org. Geochem. **23**, 981–986 (1995)

Mahaffy, Space Sci. Rev. (2007, this issue). doi:10.1007/s11214-007-9223-1

P.J. Marriott, R.M. Kinghorn, Anal. Chem. **69**, 2582–2588 (1997)

T. McCollum, Geochimica Cosmochimica Acta **67**, 311–317 (2003)

T. McCollum, G. Ritter, B.R.T. Simoneit, Orig. Life Evol. Biosphere **29**, 153–166 (1999)

T.M. McCollom, J.S. Seewald, Earth Planet. Sci. Lett. **243**, 74–84 (2006)

T. McCollum, B.R.T. Simoneit, Orig. Life Evol. Biosphere **29**, 167–186 (1999)

McKay, Space Sci. Rev. (2007, this issue). doi:10.1007/s11214-007-9229-8

K.D. McKeegan, R.M. Walker, E. Zinner, Geochimica Cosmochimica Acta **49**, 1971–1987 (1985)

S.L. Miller, Science **117**, 527–528 (1953)

K.D. Monson, J.M. Hayes, Geochimica Cosmochimica Acta **46**, 139–149 (1982)

A.P. Murray, I.B. Sosrowidjojo, R. Alexander, R.I. Kagi, C.M. Norgate, R.E. Summons, Geochimica Cosmochimica Acta **61**, 1261–1276 (1997)

I.P. Murray, G.D. Love, C.E. Snape, N.J.L. Bailey, Org. Geochem. **29**, 1487–1505 (1998)

B. Nagy, Naturwissenschaften **69**, 301–310 (1982)

H. Naraoka, A. Shimoyama, K. Harada, Earth Planet. Sci. Lett. **184**, 1–7 (2000)

R. Navarro-Gonzalez, F.A. Rainey, P. Molina, D.R. Bagaley, B.J. Hollen, J. de la Rosa, A.M. Small, R.C. Quinn, F.J. Grunthaner, L. Caceres, B. Gomez-Silva, C.P. McKay, Science **302**, 1018–1021 (2003)

R. Navarro-Gonzalez, K.F. Navarro, J. de la Rosa, E. Iniguez, P. Molina, L.D. Miranda, P. Morales, E. Cienfuegos, P. Coll, F. Raulin, R. Amils, C.P. McKay, Proc. Nat. Acad. Sci. USA **103**, 16089–16094 (2006)

W.R. Nes, M.L. McKean, *Biochemistry of Steroids and Other Isopentenoids* (University Park Press, Baltimore, 1977) 690 pp

G. Ourisson, P. Albrecht, Acc. Chem. Res. **25**, 398–402 (1992)

G. Ourisson, M. Rohmer, K. Poralla, Annu. Rev. Microbiol. **41**, 301–333 (1987)

V.J. Orphan, C.H. House, K.-U. Hinrichs, K.D. McKeegan, E.F. Delong, Proc. Nat. Acad. Sci. USA **99**, 7663–7668 (2002)

K.E. Peters, C.C. Walters, J.M. Moldowan, *The Biomarker Guide*, 2nd edn. (Cambridge University Press, Cambridge, 2005), Parts 1 and 2, 1155 pp

S.T. Petsch, R.A. Berner, T.I. Eglinton, Org. Geochem. **31**, 475–487 (2000)

S. Pizzarello, J.R. Cronin, Geochimica Cosmochimica Acta **64**, 329–338 (2000)

S. Pizzarello, Acc. Chem. Res. **39**, 231–237 (2006)

J. Poinsot, P. Schneckenburger, P. Adam, P. Schaeffer, J.M. Trendel, A. Riva, P. Albrecht, Geochimica Cosmochimica Acta **62**, 805–814 (1998)

G.L.J.A. Rikken, E. Raupach, Nature **405**, 932–934 (2000)

D. Ring, Y. Wolman, N. Friedmann, S.L. Miller, Proc. Nat. Acad. Sci. USA **69**, 765–769 (1972)

M. Rohmer, Pure Appl. Chem. **75**, 375–387 (2003)

J.F. Rontani, G. Giusti, J. Photochem. Photobiol. A: Chem. **40**, 107–120 (1987)

J.F. Rontani, G. Giusti, J. Photochem. Photobiol. A: Chem. **46**, 357–365 (1989)

E. Rubenstein, W.A. Bonner, H.P. Noyes, G.S. Brown, Nature **300**, 118 (1983)

G. Sarret, J. Connan, M. Kasrai, G.M. Bancroft, A. Charrié-Duhaut, S. Lemoine, P. Adam, P. Albrecht, L. Eybert-Bérard, Geochimica Cosmochimica Acta **63**, 3767–3779 (1999)

P. Schaeffer, C. Reiss, P. Albrecht, Org. Geochem. **23**, 567–581 (1995)

P. Schaeffer, P. Adam, P. Wehrung, P. Albrecht, Tetrahedron Lett. **48**, 8413–8416 (1997a)

P. Schaeffer, P. Adam, P. Wehrung, P. Albrecht, in *Proc. 18th International Meeting on Organic Geochemistry 1997*, Maastricht, the Netherlands, Abstracts Part I (1997b), pp. 57–58

J. Schaeflé, B. Ludwig, P. Albrecht, G. Ourisson, Tetrahedron Lett. **41**, 3673–3676 (1977)

P. Schneckenburger, P. Adam, P. Albrecht, Tetrahedron Lett. **39**, 447–450 (1998)

S. Schouten, M.E.L. Kohnen, J.S. Sinninghe Damsté, J.W. de Leeuw, Org. Geochem. **23**, 129–138 (1993)

S. Schouten, W.C.M. Klein Breteler, P. Blokker, X. Schogt, W.I.C. Rijpstra, K. Grice, M. Baas, J.S. Sinninghe Damsté, Geochimica Cosmochimica Acta **62**, 1397–1406 (1998)

J. Schwender, M. Seeman, H.K. Lichtenthaler, M. Rohmer, Biochem. J. **316**, 73–80 (1996)

M.A. Sephton, Botta, Space Sci. Rev. (2007, this issue). doi:10.1007/s11214-007-9171-9

M.A. Sephton, I. Gilmour, Mass Spectrom. Rev. **20**, 111–120 (2001)

M.A. Sephton, C.T. Pillinger, I. Gilmour, Geochimica Cosmochimica Acta **64**, 321–328 (2000)

M.A. Sephton, C.T. Pillinger, I. Gilmour, Planet. Space Sci. **47**, 181–187 (2001)

M.A. Sephton, Nat. Prod. Res. **19**, 292–311 (2002)

M.A. Sephton, G.D. Love, J.S. Watson, A.B. Verchovsky, I.P. Wright, C.E. Snape, I. Gilmour, Geochimica Cosmoshimica Acta **68**, 1385–1393 (2004)

R.A. Sheldon, J.K. Kochi, *Metal-Catalyzed Oxidations of Organic Compounds* (Academic, New York, 1981)

E.L. Shock, Orig. Life Evol. Bios. **22**, 67–108 (1992)

E.L. Shock, Ciba Found. Symp. **202**, 40–60 (1996)

E.L. Shock, J. Geophys. Res. **102**(23), 687–694 (1997)

J.S. Sinninghe Damsté, M.P. Koopmans, Pure Appl. Chem. **69**, 2067–2074 (1997)

J.S. Sinninghe Damsté, A.C.T. van Duin, D. Hollander, M.E.L. Kohnen, J.W. de Leeuw, Geochimica Cosmochimica Acta **59**, 5141–5147 (1995)

J.S. Sinninghe Damsté, J.W. de Leeuw, in *Advances in Organic Geochemistry 1989*, ed. by B. Durand, F. Behar (Pergamon, Oxford, 1990), pp. 1077–1101

J.G. Speight, in *Caractérisation des Huiles Lourdes et des Résidus Pétroliers*, ed. by B.P. Tissot. International Symposium, Lyon (Technip, Paris, 1984), pp. 32–41

S.W. Squyres, J.P. Grotzinger, R.E. Arvidson, J.F. Bell 3rd, W. Calvin, P.R. Christensen, B.C. Clark, J.A. Crisp, W.H. Farrand, K.E. Herkenhoff, J.R. Johnson, G. Klingelhofer, A.H. Knoll, H.Y. McSween Jr., R.V. Morris, J.W. Rice Jr., R. Rieder, L.A. Soderblom, Science **306**, 1709–1714 (2004)
H. Strauss, D.J. Des Marais, J.M. Hayes, R.E. Summons, in *The Proterozoic Biosphere*, ed. by J.W. Schopf, C. Klein (Cambridge University Press, New York, 1992), pp. 117–127
O.P. Strausz, E.M. Lown, J.D. Payzant, in *Geochem. of Sulfur in Fossil Fuels*, ed. by W.L. Orr, C.M. White ACS Symposium Series, vol. 429 (American Chemical Society, Washington, 1990), pp. 368–396
R. Stribling, S.L. Miller, Orig. Life **17**, 261–273 (1987)
R.E. Summons, T.G. Powell, Geochimica Cosmochimica Acta **51**, 557–566 (1987)
R.E. Summons, T.G. Powell, C.J. Boreham, Geochimica Cosmochimica Acta **52**, 1747–1763 (1988)
R.E. Summons, M.R. Walter, Am. J. Sci. **290-A**, 212–244 (1990)
R.E. Summons, in *The Proterozoic Biosphere*, ed. by J.W. Schopf, C. Klein (Cambridge University Press, New York, 1992), pp. 101–115
R.E. Summons, L.L. Jahnke, Z. Roksandic, Geochimica Cosmochimica Acta **58**, 2853–2863 (1994)
D.Y. Sumner, J. Geophys. Res. – Planets **109**(12), 12007 (2004)
T. Takata, K. Ishibashi, W. Ando, Tetrahedron Lett. **26**, 4609–4612 (1985)
M.A. Teece, J.M. Getliff, J.W. Leftly, R.J. Parkes, J.R. Maxwell, Org. Geochem. **29**, 863–880 (1998)
E.W. Tegelaar, J.W. de Leeuw, S. Derenne, C. Largeau, Geochimica Cosmochimica Acta **53**, 3103–3106 (1989)
W.R. Thompson, T.J. Henry, J.M. Schwartz, B.N. Khare, C. Sagan, Icarus **90**, 57–73 (1991)
B.P. Tissot, D.H. Welte, *Petroleum Formation and Occurrence* (Springer, New York, 1984), 699 p.
J.P. Tritz, D. Herrmann, P. Bisseret, J. Connan, M. Rohmer, Org. Geochem. **30**, 499–514 (1999)
J. Valisolalao, N. Perakis, B. Chappe, P. Albrecht, Tetrahedron Lett. **25**, 1183–1186 (1984)
V.V. Voronenkov, A.N. Vinogradov, V.A. Belyaev, Russ. Chem. Rev. **39**, 944–952 (1970)
S. Wakeham, Nature **342**, 787–790 (1989)
S.G. Wakeham, J.S. Sinninghe Damsté, M.E.L. Kohnen, J.W. de Leeuw, Geochimica Cosmochimica Acta **59**, 521–533 (1995)
Z. Wang, M. Fingas, G. Sergy, Environ. Sci. Technol. **28**, 1733–1746 (1994)
C.D. Watts, J.R. Maxwell, Geochimica Cosmochimica Acta **41**, 493–497 (1979)
A.L. Weber, S. Pizzarello, Proc. Nat. Acad. Sci. USA **103**, 12713–12707 (2006)
G. Yuen, N. Blair, D.J. Des Marais, S. Chang, Nature **307**, 252–254 (1984)
B. Zeitman, S. Chang, J.G. Lawless, Nature **251**, 42–43 (1974)
B. Zellner, E. Bowell, in *Comets–Asteroids–Meteorites*, ed. by A.H. Delsemme (University of Toledo Press, Toledo, 1977), pp. 185–195
E. Zinner, in *Meteorites and the Early Solar System*, ed. by J.F. Kerridge, M.S. Matthews (University of Arizona Press, Tucson, 1988), pp. 956–983

# Fossil Lipids for Life-Detection: A Case Study from the Early Earth Record

Jennifer L. Eigenbrode

Originally published in the journal Space Science Reviews, Volume 135, Nos 1–4.
DOI: 10.1007/s11214-007-9252-9 © Springer Science+Business Media B.V. 2007

**Abstract** The geological preservation of lipids from the cell membranes of organisms bestows a precious record of ancient life, especially for the Precambrian eon (>542 million years ago) when Earth life was largely microbial. All organisms produce lipids that, if the lipids survive oxidative degradation, become molecular fossils entrained with information on biological diversity, environmental conditions, and post-depositional alteration history. As with most biosignatures, the molecular fossil record that is indigenous (of the same place) and syngenetic (of the same age) to host rocks can be compromised by the introduction from and reaction with foreign or younger materials (e.g., petroleum or endolithic life). Deciphering the resulting complex pool of organic signals requires tests for the provenance of molecular fossils and the overall quality of the geobiological record itself. This paper reviews the basis for the very existence of a molecular fossil record from lipid biochemistry to mechanisms of organic-matter preservation and geochemical alteration. A systematic approach to resolving the provenance of molecular fossils and historical qualities of the record is presented in a case study of an early Earth record. This example demonstrates the value of geological context and the integration of independent geobiological parameters, which are critical to the detection and understanding of the ecological processes responsible for records of life.

**Keywords** Lipids · Molecular fossils · Organic matter · Preservation · Alteration · Extraterrestrial life · Microbial ecology

## 1 Introduction

Microbial life has existed on Earth for more than three billion years (e.g., Shen et al. 2001; Rasmussen 2000, 2005; Tice and Lowe 2006). Even today microbes have claimed

A manuscript prepared for ISSI Workshop on Strategies for Life Detection, Space Science Reviews, Space Science Series of International Space Science Institute.

J.L. Eigenbrode (✉)
Geophysical Laboratory, Carnegie Institution of Washington, 5251 Broad Branch Rd. NW, Washington, DC 20015, USA
e-mail: jeigenbrode@ciw.edu

nearly every extreme environmental niche observed on Earth (cf. Horikoshi and Grant 1998; Seckbach 1999; Amils et al. 2007). Given life's tenacity and resilience, perhaps life exists or has existed elsewhere. Detecting and understanding ancient records of life on early Earth has been a confounding challenge because most biosignatures afford ambiguous interpretations of a biosphere having an unfamiliar biogeochemistry and history. Detecting extraterrestrial life (i.e., non-Earth life) will likely present a similar challenge. The Precambrian geobiological record of microbial life allows us to develop tests for ancient life that will potentially be applicable to extraterrestrial exploration.

The most direct biosignatures we have for early Earth life are molecular fossils, organic compounds that were biosynthesized by once-living organisms (cf. Summons et al. 2007; Eglinton and Calvin 1967). The majority of molecular fossils found in sedimentary organic matter and petroleum of ancient geological samples are derived from lipids of cell membranes (Peters et al. 2005). The recalcitrant hydrocarbon backbone of lipids is responsible for the high preservation potential of lipid-derived molecular fossils relative to those of other biomolecules.

Petroleum geochemical (cf. Tissot and Welte 1984; Hunt 1996; Chilingar et al. 2005), environmental (e.g., Brassell et al. 1986; Summons and Powell 1987; Freeman et al. 1994; Farrimond et al. 2000) and biological (e.g., Rohmer et al. 1979; Ourisson et al. 1987a; Aiello et al. 1999; Blumenberg et al. 2004; Volkman 2005; Ladygina et al. 2006) studies demonstrate that lipid compositions and distributions reflect biological- and environment-specific characteristics of modern and ancient ecosystems, because both genetics and environment influence lipid production by source organisms. Since biological materials, which include lipids, undergo in situ chemical and thermal reactions dictated by the geobiological and/or geological conditions in water, sediments, and rock, the structures and distributions of molecular fossils often reflect post-depositional conditions and alteration. Thus, the quality of the organic geochemical record directly bears on the record's interpretive significance as potential evidence of life. Both must be assessed. The case study described here provides an example of an effective approach to interpreting and evaluating the quality of molecular records from early Earth and elsewhere.

## 2 From Organisms to Molecular Fossils

### 2.1 Lipid Biochemistry and Diversity

#### 2.1.1 Lipid Membrane Formation and Functions

All living organisms on Earth have membranes that separate cellular materials from their ambient environment. Membranes are mainly composed of oils, fats, and waxes (i.e. the lipids) and proteins. Like soap micelles, lipid membranes are chemical and physical barriers due to the self-association of hydrophilic head groups and hydrophobic hydrocarbon chains (Fig. 1) (cf. Dowham and Bogdanov 2002). As such, mono- and bi-layer cellular membranes likely arose from the self-assembly of prebiotic amphiphiles such as fatty acids (Deamer 1997; Mahajan et al. 2003).

In addition to providing a barrier and structural integrity, lipid membranes host proteins and act as a dynamic medium that directly affects cellular processes, such as the regulation of metabolites across membranes and other protein–lipid interactions (Dowham and Bogdanov 2002; Konings et al. 2002; Vorob'eva 2004). Consequently, organisms produce assorted lipid molecules and membrane structures to fulfill their functional roles, particularly at the environment–cell interface.

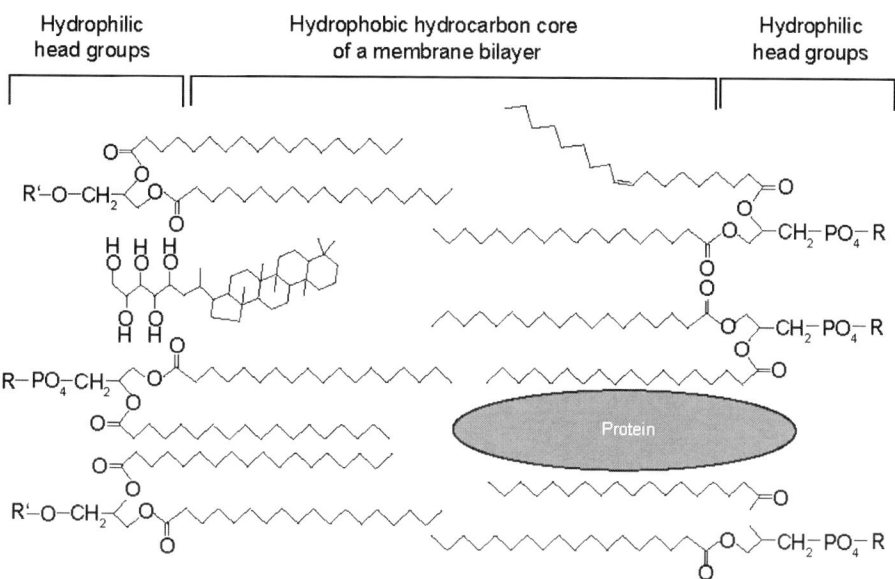

**Fig. 1** Organization of a cell membrane, lipid bilayer. Each lipid is composed of a hydrophilic head group and hydrophobic portion that makes up the hydrocarbon core. Typical head groups of phospholipids include R = H, ethanoamine, choline, serine, and glycerol. Sugars attach to glycolipids (R′) or to phospholipids forming phosphoglycolipids

### 2.1.2 Membrane Fluidity and Adaptations to Extreme Environments

The diversity of lipid constituents in organisms today reflect the adaptability and flexibility of membranes required for organisms to persist under the broad range of environmental conditions found on Earth (Dowham 1997). In particular, regulating the fluidity of the hydrocarbon core in membranes, and thus, membrane integrity and permeability (Melchior 1982), is critical to life in extreme environments. Permeability passively increases with temperature; however, cells actively enhance permeability by increasing the content of unsaturated (i.e., less hydrogen content by the presence of double- and triple-bonded carbons), branched or short alkyl chains. Conversely, membrane rigidity and impermeability, which are necessary under high-temperature or osmotic stress (imposed by salinity, acidity, alkalinity, toxins, etc.), require less fluidity and/or greater membrane thickness and lipid-packing density. These membrane qualities are achieved by (1) a high degree of saturation (greater H-C content), (2) extended alkyl-chain length, (3) addition of five- and six-carbon rings to hydrocarbon chains, and/or (4) increased abundance of membrane spanning lipids (e.g. biphytanyl diglycerol diether). Fluidity, permeability, and volume of the hydrocarbon core are also dependent on the composition of the polar head group, which dictates the molecular shape of lipids (Dowham and Bogdanov 2002).

Other lipid adaptations include ether linkages connecting hydrocarbon tails to hydrophilic head groups and ancillary lipid-membrane constituents. Ether linkages are more resilient to extreme environmental conditions compared to ester linkages (Gelin et al. 1996; Schouten et al. 1998; van de Vossenberg et al. 1998). Carotenoids, which absorb light-energy, are a good example of ancillary lipids for they provide photo-oxidative protection while serving other cellular functions (Krinsky 1989; Lesser 2006). Further review of extremophile lipid-membrane adaptations is beyond the scope of this paper. Dowham and Bogdanov (2002), van de Vossenberg et al. (2000), Morgan-Kiss et al. (2006), and Chintalapati et al. (2004) provide more detailed discussion of this topic.

### 2.1.3 Lipid Taxonomic Specificity

Although the molecular diversity of lipids strongly reflect physiological adaptation to environmental conditions, genetics also play a key role in the observed lipid diversity, because genes encode for enzymes used in lipid biosynthesis. Lipid structures exhibit varying degrees of taxonomic specificity, which is attributed to gene-controlled biochemistry, but complicated by environmental distribution of organisms and physiological response. In some cases, biological factors or environmental conditions can interfere with the expression of lipid-metabolism genes. Consequently, the observed lipid diversity expressed by lipid-metabolism genes is different than lipid diversity observed in samples. Ultimately, both genetic composition and physiological factors influence lipid composition expressed by organisms in culture and environments.

On the grandest of scales, each of the three domains of life, Archaea, Bacteria, and Eukarya, show some general lipid distinctions. For instance, only the Archaea produce acyclic isoprenoids that are composed of phytanyl chains linked to various phosphatic and glycerol head groups by ether bonds (Fig. 2). In contrast, many bacteria produce pentacyclic triterpenoids, namely bacteriohopanepolyols from which many hopanoid molecular fossils are derived (Rohmer et al. 1984; Ourisson et al. 1987b; Farrimond et al. 1998) and distributed in sediments (Sinninghe Damsté et al. 1995; Innes et al. 1998; Farrimond et al. 2000; Talbot et al. 2003). Neither eukaryotes nor archaea are known to produce bacteriohopanepolyols. Instead, most eukaryotes and few bacteria produce tetracyclic triterpenoids called sterols (e.g., cholesterol in animals; (Bode et al. 2003; Pearson et al. 2003) and references therein), which are preserved as sterenes, steranes, and aromatic steroids in the geological record (Volkman 2005; Brocks and Summons 2005; Summons et al. 2006). In addition to these triterpenoids, bacteria and eukaryotes produce fatty acids having alkyl chains and ester linkages.

Some lipid structures or molecular patterns appear to be limited to explicit taxonomic groups. For instance, several fatty acid compositions are unique to particular bacteria (e.g., 10-methyl hexadecanoic acids from *Geobacter metallireducens*; Lovley et al. 1993) and are often used for their identification (Tornabene et al. 1980; Kaneda 1991; Ladygina et al. 2006). Other examples include the branched alkene, botyococcene (Fig. 2), produced by the green alga *Botryococcus braunii* (McKirdy et al. 1986; Metzger and Largeau 2005), unusual steroids in sponges (Aiello et al. 1999; Blumenberg et al. 2002) and numerous di- and triterpenoids (e.g., cadinanes and $\beta$-amyrin, lupeol, taraxerol, and friedelanol, respectively) from higher plants (Peters et al. 2005; Simoneit 2005). Of particular value in early Earth records are hopanoid fossils of bacteriohopanepolyols that have a methyl group at the C-2 and C-3 position. These are distinguishing lipids of some oxygen-producing cyanobacteria (Summons et al. 1999) and aerobic methanotrophic bacteria (Summons and Jahnke 1992), respectively. Fossil carotenoids are also important in reconstructing early Earth ecology (Summons and Powell 1986; Requejo et al. 1992; Brocks et al. 2004) for they have taxonomic specificity among phototrophic organisms (Krinsky 1994; Sinninghe Damsté and Koopmans 1997; Vershinin 1999).

Only recently have genomic analyses of modern organisms been used to evaluate the phylogenetic evolution of lipid synthesis (e.g., Pearson et al. 2003; Benveniste 2004; Peretó et al. 2004; Fischer et al. 2005). In particular, Pearson and colleagues have targeted phylogenetic relationships among hopanoids and steroids (Pearson et al. 2003;

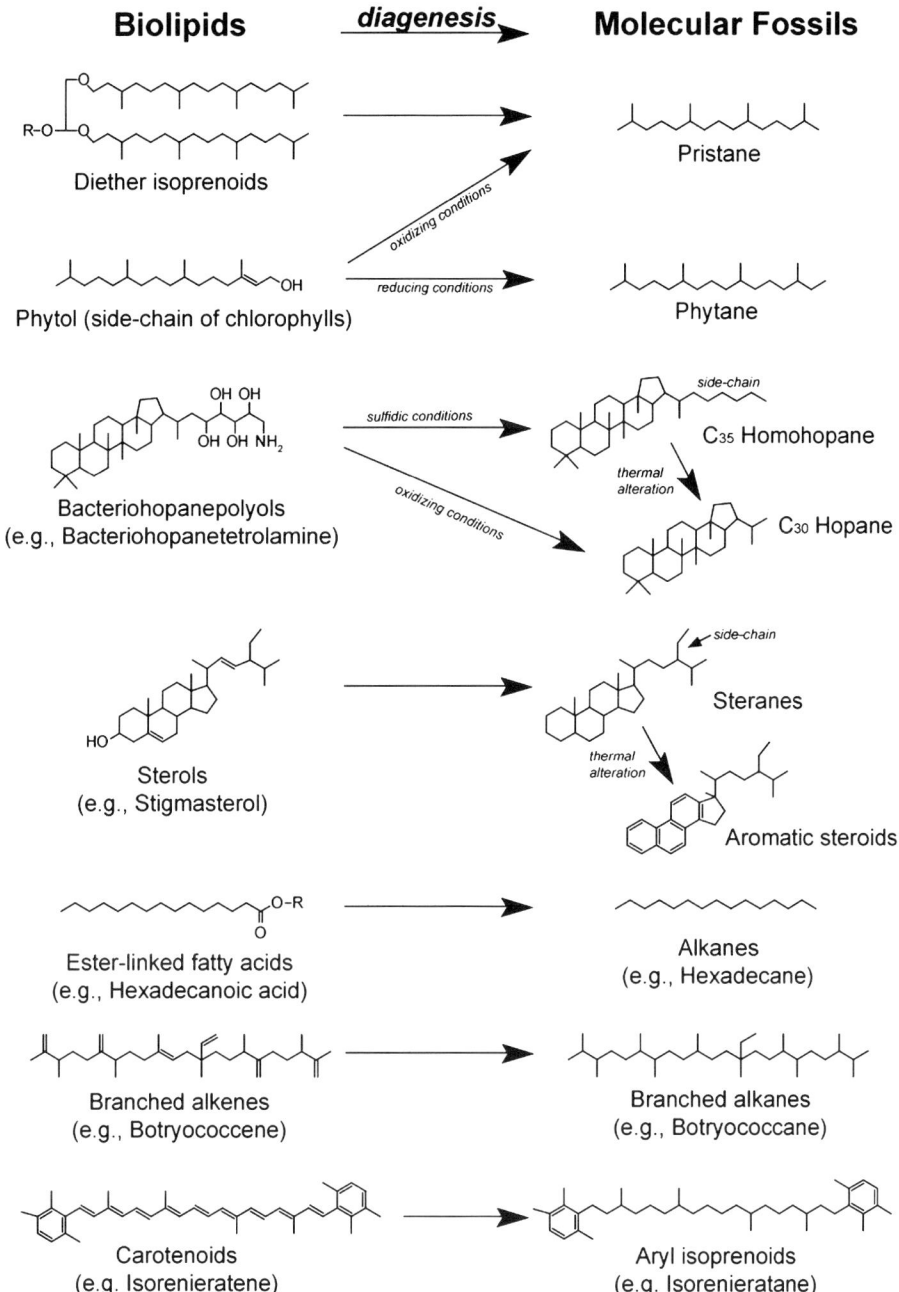

**Fig. 2** Biolipids commonly found in organisms and their geochemical derivatives (molecular fossils) after diagenesis and thermal alteration. *R* denotes various polar head groups

Brocks and Pearson 2005) building upon early suggestions of lipid evolution among organisms (Rohmer et al. 1979; Bloch 1994, 1983). Future phylogenetic studies may help constrain the relative importance of physiological adaptation to environment and biochemical heritage in the molecular diversity in the lipids of organisms.

### 2.1.4 Lipid-Specific Isotopic Compositions

Diversity among lipids is not limited to molecular structure. The enzymes used for lipid synthesis also exert an isotopic effect on the stable isotopic content of the lipids produced relative to that of the source materials. In particularly, carbon isotopic compositions (expressed by conventional notation: $\delta^{13}C$, $\permil = 10^3$ ($^{13}R_{sample}/^{13}R_{standard} - 1$), and $^{13}R = ^{13}C/^{12}C$) reflect both the isotopic composition of carbon substrates and metabolism for carbon assimilation (Hayes 1993, 2001). Moreover, $\delta^{13}C$ values have revealed pathways of microbial carbon cycling on ecological scales (Freeman et al. 1990; van der Meer et al. 2000; Pancost and Sinninghe Damsté 2003). Most of the carbon isotopic effect comes from the initial primary production of biomass by autotrophs, which fix $CO_2$ by reducing it to organic carbon. Autotrophs exhibit substantial diversity in their expressed isotopic effects (Zerkle et al. 2005), whereas heterotrophs, which consume organic molecules, exert little fractionation (typically a 0–3‰ increase in biomass $\delta^{13}C$ value relative to the source organics). Although the $\delta^{13}C$ values of molecules are most often studied, compound-specific isotopic analysis (CSIA) has also proven useful in understanding the biochemical and ecological flow of hydrogen and nitrogen (Sessions et al. 1999; Sauer et al. 2001; Montoya 2003).

### 2.1.5 Modern Lipids as Analogues for Early Earth and Extraterrestrial Life

The functional nature of lipids, the environmental triggers (and inhibitors) of lipid biosynthesis, and the genetic controls on lipid biosynthesis are all in rudimentary exploration. Thus, the power of fossil lipid compositions as ancient records of environmental adaptation and taxonomic diversity is limited to inferences that assume present relationships are representative of past relationships. A common inverse approach to the interpretation of the molecular fossil record relies on observed relationships among molecular, biogeochemical, and environmental proxies in geological records. For example, pristane/phytane and 28,30-dinorhopane abundance tend to correlate to redox conditions and clay abundance of depositional environments (Peters et al. 2005 and references therein). Molecular fossil interpretations are strengthened by the convergence of more than one approach and the recognition of interpretive limitations, particularly for early Earth studies in which the antiquity and potential uniqueness of the biology present a challenge. For the same reasons, biological interpretations of extraterrestrial molecular records warrant even more caution.

Despite limitations in relating modern microbial physiology and biochemistry to ancient and extraterrestrial life, modern life-detection studies provide strategic direction and interpretive constraints for early Earth and extraterrestrial investigations (e.g., Karl et al. 1999; Marion et al. 2003; Amundsen et al. 2004; Miteva et al. 2004; Blackhurst et al. 2005; Kelley et al. 2005; Eigenbrode et al. 2006; Fogel and Team 2006; Souza-Egipsy et al. 2006) and abiogenic hydrocarbon investigations (e.g., Sephton 2002; Sherwood Lollar et al. 2002; McCollom 2003; McDonald and Storrie-Lombardi 2006; Morrill et al. 2006). Distinguishing features of life come in the form of orderly patterns that range from molecular (Summons et al. 2007) to ecosystem scales (cf. Begon et al. 2005). These patterns reflect biocomplexity, which transpires from the selective nature of life to tap energy and nutrient resources, to concentrate substrates necessary for vitality and growth, and to create specific biomolecules via biosynthetic pathways. Patterns can result from geochemical processes but they are seldom as pervasive and multifaceted as those arising from living processes (Lazcano 2007). Thus, the key to detecting and understanding records of life is to test for biocomplexity at multiple scales using multiple proxies. The structural and isotopic composition and distribution of molecular fossils in the geological record provide a principal set of proxies for this purpose.

## 2.2 Molecular Fossil Preservation and Alteration

### 2.2.1 Diagenesis

Not all sedimentary rocks retain a molecular fossil record. The presence and quality of the molecular fossil record depends on many factors that begin with the preservation of organic carbon during early diagenesis and is ultimately converted to carbon dioxide. In order to preserve organic matter for the fossil record it must be thermodynamically inert to its surroundings (Mayer 2004). Molecular fossil preservation is aided by rapid incorporation of organics into macromolecules (biopolymers, humics, or kerogen) or sorption by minerals, whether the molecules are chemically bound, physically encased, or both. Summons et al. (2007) review the mechanisms by which organic molecules are preserved.

With regards to the Precambrian organic matter, preservation must have been strongly aided by the absence or low levels of oxygen (Berner 1992; Hedges and Keil 1995) and the presence of sulfide (Hebting et al. 2006). Both conditions are consistent with the hydrothermally driven biogeochemistry of that time (Holland 1984, 2006) and the notably low concentrations of oxygen in the atmosphere and ocean, especially before 2.4 billion years ago when the Earth's atmosphere was anoxic (Pavlov and Kasting 2002; Farquhar and Wing 2003). The prevalence of reducing conditions probably influenced both selective preservation of lipids and possibly biopolymers (Tegelaar et al. 1989; Derenne and Largeau 2001; de Leeuw et al. 2006) through the in situ polymerization of re-calcitrant sedimentary aliphatics into macromolecules, as they do today (e.g., Briggs 1999; Gupta et al. 2007). Cross-linking of organic molecules by sulfur bonds (Sinninghe Damsté et al. 1989; Hartgers et al. 1997; Hebting et al. 2006), rather than oxygen bonds as in modern soil and marine humification processes (Gatellier et al. 1993; Hedges et al. 2000, 2001; Riboulleau et al. 2001; Versteegh et al. 2004; Hertkorn et al. 2006), probably dictated diagenetic condensation reactions at this time.

In addition to reaction mechanisms, occlusion of organics in minerals may have also been an important protection mechanism for Precambrian molecular fossils. Early diagenetic chemical precipitation of silica, carbonates, sulfides, and iron oxides was a hallmark sedimentary process during the Precambrian, especially in the Archean (3.8 to 2.5 billion years ago). Encapsulation of organic matter in inert minerals may have retarded chemical and biological oxidation. This mechanism is supported by observations of kerogenous microfossils in Precambrian cherts and carbonates, but not reactive iron oxides (see chapters by Javaux, Altermann, and Westall this volume), and molecular fossils observed in pyrite (Mycke et al. 1988).

Phyllosilicate clay minerals (Mayer 1994; Ransom et al. 1997; Ding and Henrichs 2002; Kennedy et al. 2002) and detrital mineral grains (e.g. quartz and feldspar) (e.g., Dutkiewicz et al. 2006) may have also contributed to the protection of Precambrian organic matter during both diagenesis and subsequent thermal alteration. Although the exact mechanisms are not understood, surface area, mineral chemistry, and geochemical history appear to be important. Evaporites may have behaved similarly, particularly halogenated salts (Pasteris et al. 2006), though organic preservation by sulfate minerals has also been suggested (but see,

Sumner 2004; Aubrey et al. 2006). Minerals may have been important in protecting organic material from photo-oxidation in the Archean, before there was an ozone layer in Earth's atmosphere. Certainly, further research is necessary to understand fully the roles reaction mechanisms and minerals play in protecting organics during diagenesis and over long geological time scales.

### 2.2.2 Thermal Influences on the Molecular Record

The end product of diagenesis is macromolecular organic matter. Through the loss of superficial hydrophilic functional groups (e.g. –OH and –COOH), this organic matter eventually becomes the solvent-insoluble fossil carbon called kerogen. The heterogeneous structure of kerogen reflects assorted biological sources as well as post-depositional processing. Molecular fossils are chemically bound or occluded in the kerogen matrix.

As temperatures and pressures increase during burial, kerogens become thermodynamically unstable and diagenesis gives way to catagenesis and metagenesis, in which waxy and gaseous hydrocarbons are released (cracked) from the kerogen, respectively (Hunt 1996). Geothermal gradients impose instability over long periods with deep burial (Peters et al. 2005), or over short periods due to molten intrusions and volcanic extrusions (Farrimond et al. 1996), or by hydrothermal (Simoneit et al. 2004) and meteorite-impact events (Parnell et al. 2005; Eglinton et al. 2006). Chemically, kerogen cracking results from the redistribution of hydrogen (disproportionation) during cyclization and aromatization reactions within the macromolecular structure. The hydrogen facilitates C–S, C–N, C–O, and C–C bond cleavage of alkyl chains, isoprenoid, and triterpenoid constituents, giving rise to petroleum containing molecular fossils. If the initial concentration of kerogen in a rock is greater than $\sim 0.5\%$, hydrocarbons may sufficiently saturate the rock for oil expulsion; otherwise the hydrocarbons may remain in the rock as bitumen (Mackenzie and Quigley 1988), which can later be cracked to generate gas at higher temperatures.

Expelled oil and gas migrate into adjacent strata or to the surface. In the process, expelled petroleum can mix with other hydrocarbons and thermally stable kerogens, which may act as absorbents (Luthy et al. 1997; Cornelissen et al. 2005) under some subsurface conditions. In rare circumstances, hydrocarbons may thermally or radiolytically condense into macromolecular material called pyrobitumen (Curiale 1986; Gray et al. 1998; Melezhik et al. 1999), which may form in the source rock before migration (e.g., Rasmussen 2005). Sedimentary organic matter can also be thermally transformed by hydrothermal fluids (e.g., Kawka and Simoneit 1994; Simoneit et al. 2004).

In addition to petroleum generation, geothermal stress causes molecular structures to lose their biological configurations due to thermodynamic instability. The resulting geological configurations are commonly used to assess petroleum-generation potential or geothermal history (see Summons et al. 2007). A recent comprehensive review of the thermal maturity parameters for hydrocarbons and kerogens is found in Peters et al. (2005) and Price (1993).

### 2.3 Molecular Fossils as Records of Life

Unlike the bulk of the hydrocarbons in bitumen, molecular fossils exhibit structural patterns and distributions diagnostic of biosynthetic pathways (Summons et al. 2007). Many nonspecific alkanes result from catagenesis, which gives no odd or even carbon-chain length preference (Peters et al. 2005) even though they originally were of biological origin. Experimental studies show that the absence of a chain-length pattern in $n$-alkanes is also characteristic of abiogenic synthesis (McCollom and Seewald 2001; McCollom 2003). Polycyclic aromatic hydrocarbons (PAHs) can be derived from aromatized diterpenoids, triterpenoids, and carbohydrates either geologically (Peters et al. 2005) or via incomplete combustion (Simoneit 2002); however, PAHs are also abundant in interstellar organics (Sephton 2002; Ehrenfreund and Sephton 2006). Since many hydrocarbons can have biogenic, thermogenic, or abiogenic sources, organic geochemists reserve the terms "molecular fossil" and "bio-marker" for organic compounds that have an unambiguous link to living or once living organisms. Compound-specific isotopic compositions of hydrocarbons may be helpful in identifying terrestrial (Sherwood Lollar et al. 2005; Morrill et al. 2006) and extraterrestrial (McKeegan and Leshin 2001; Alexander et al. 2007, in prep; Cody et al. 2007, submitted) abiogenic sources. Summons et al. (2007) review the principles and limitations of organic compounds as biomarkers.

## 3 Caveats to Molecular Fossil Studies

### 3.1 The Importance of Syngenicity and Indigeneity

The usefulness of any biosignature is contingent upon an inherent relationship to host rocks with respect to both time (syngenicity) and place (indigeneity) (Fig. 3). Syngenicity (Schopf and Walter 1983), as it is used herein, refers to the concurrent deposition to early diagenetic origin of molecular fossils, kerogen, and host rock. Demonstrating indigeneity and syngenicity are difficult tasks but critical to all interpretations that stem from molecular fossil records, or any other biosignature. Testing for relationships between molecular fossils and host rocks is paramount when the molecular fossils are present in trace quantities ($<100$ ppb compound in rock) as in exotic samples, such as Archean rocks (e.g., Brocks et al. 2003), geological materials from some extreme environments, meteorites (e.g., Naraoka et al. 2000) and other extraterrestrial samples. In such cases, the molecular record is highly susceptible to adulteration. Other complications in understanding the organic-rock relationship arise if sedimentary rocks include detrital material from eroded rocks and ice, e.g., glacial tillites and diamictites, having kerogen and/or bitumen from different geological sources.

### 3.2 Adulteration of Authentic Records

The introduction of non-indigenous hydrocarbons containing molecular fossils can migrate into geological materials naturally via oil and groundwater migration or anthropogenically through storage, handling, and processing that alter the composition of bitumen (Fig. 3). Moreover, indigenous endolithic organisms in surface rocks (e.g., Johnston and Vestal 1991; Bell 1993; Eigenbrode et al. 2006) and in the subsurface (Parkes et al. 1994; Boivin-Jahns et al. 1996; Whitman et al. 1998; Takai et al. 2003; Horsfield et al. 2006) may selectively biodegrade hydrocarbons, as is known to occur in petroleum reservoirs (Connan 1984; Larter et al. 2006). Indigenous endolithic organisms may also leave biomolecular traces behind (Coolen et al. 2001; Petsch et al. 2001; Petsch et al. 2003) though the quantitative and qualitative effect on molecular composition is probably highly variable. The net effect of these geobiological processes is mixing and overprinting of the indigenous, syngenetic biomarkers. In such cases, establishing the provenance of molecular fossils is imperative to paleobiological interpretations.

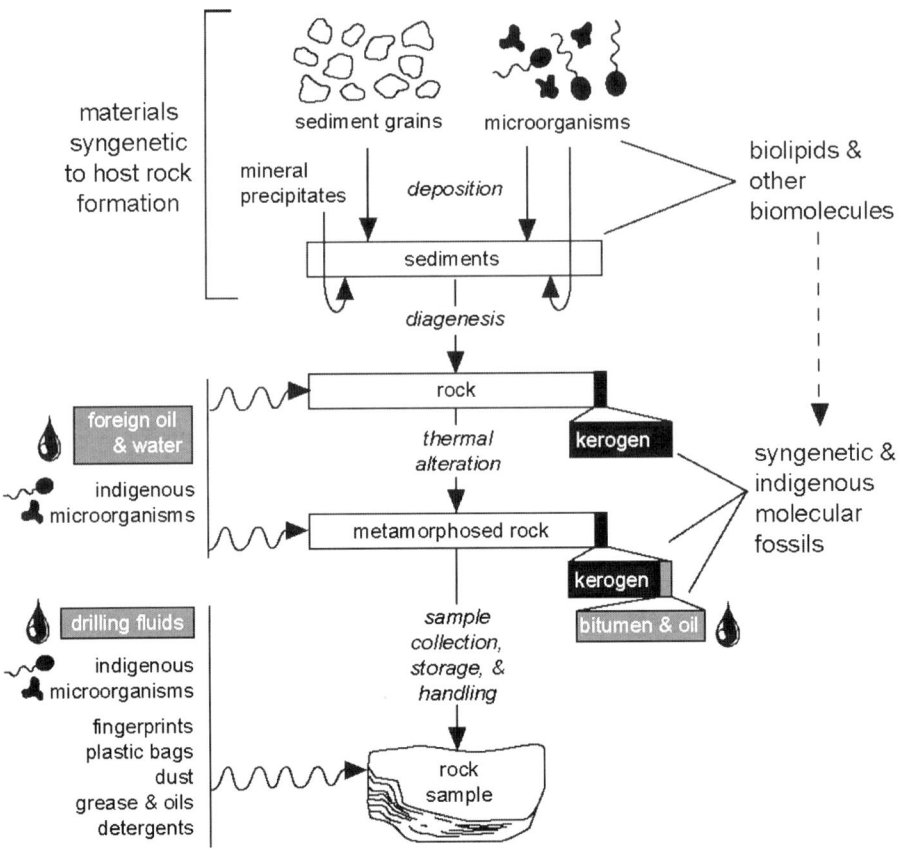

**Fig. 3** Schematic showing the possible origins of molecular fossils in rock samples and timing of their introduction relative to deposition. The only molecular fossil source of both syngenetic and indigenous nature is microorganisms active during deposition and early diagenesis

## 3.3  Approaches to Establishing the Provenance of Molecular Fossils

Organic geochemists take several approaches to test for the presence of non-indigenous molecular fossils that are not indigenous to the host rock; all are adapted from principles of petroleum geochemistry. Many studies have targeted organic-rich black shales, which are common source rocks for molecular-fossil rich oil and are rather impermeable compared to other sediments; thus, organic geochemists often assume black shales retain an indigenous signature. Even so, all studies must evaluate the consistency of thermal maturity levels of biomarkers, macromolecular organic carbon (kerogen), and host rock metamorphic grade and regional geothermal history. Another common practice is to compare the molecular fossils present to those reported in rocks of similar age (cf. Moldowan and Jacobson 2000) (e.g., Summons et al. 1988; Brocks et al. 1999; Summons et al. 1999; Logan et al. 2001; Greenwood et al. 2004). For the thermal maturity and age tests, a positive correlation, however, does not "prove" a relationship, but a negative correlation may be strong compelling evidence for a lack of a relationship. In some cases, a positive thermal-maturity test result may be consistent with syngenicity to host rocks, but further assessment in the context of the regional geological history is also necessary to build confidence in that interpretation (e.g., Brocks et al. 2003; Eigenbrode et al. 2007b).

The third and most reliable approach to testing for foreign molecular fossils involves the correlation among various molecular fossil parameters and other geochemical and geological parameters (Seifert and Moldowan 1978). This approach is commonly applied in the petroleum industry for correlation to petroleum source-rocks and oils and for identifying mixtures (Peters et al. 2005). The premise of this approach is that an inherent relationship among the parameters will be reflected by the patterns among the signatures retained in host rocks. Unfortunately, extreme thermal history and poor preservation of many early Earth rock sections, common of the early Earth record, limit molecular fossil preservation and variability in sample qualities, which are necessary for correlations. However, well-preserved early Earth records do exist. An example of this approach applied to an early Earth record follows.

## 4 An Early Earth Molecular Fossil Case Study

The case study described here demonstrates how molecular fossils derived from lipids can be used to detect life and to assess indigeneity and syngenicity. As described in Sect. 2, structural diversity among lipids observed in environmental samples is controlled by both genetics and physiological factors, which are both influenced by environment. Thus, patterns among molecules and patterns among molecular composition and environmental parameters are characteristics of life that we can search for in ancient records containing molecular fossils. Large datasets composed of a multitude data types (e.g., molecular fossil, isotopic, and geological data) can be effectively reduced to meaningful observations using multivariate analysis, such as principal component analysis (PCA), which resolves major dataset qualities by identifying combinations of linear correlations among variables.

In this case study PCA was employed to statistically identify relationships among bulk geochemical and molecular fossil parameters. PCA resolves multiple components by correlating variables in a dataset (Hill and Lewicki 2006). Background information of the rock section, its relevance to Earth history, and the biogeochemical implications of the molecular fossil data are reported elsewhere (Eigenbrode et al. 2007a, 2007b) and largely irrelevant to this report. As a general rule, data for each parameter must exhibit variability for the results to be meaningful (Hill and Lewicki 2006). Statistically significant correlations, as determined by $p$-values (discussed below), are expected if the parameters are source and/or process related. Therefore, if the molecular fossils are biological signatures from the time of deposition-to-early diagenesis, then the molecular parameters should correlate to bulk geochemical parameters that record conditions of shared environmental and ecological significance.

### 4.1 Principal Component Analysis

#### 4.1.1 PCA Variables and Constraints

PCA is employed with a large dataset having limited cases (15 rock samples) and many variables. In preliminary analyses, some parameters—such as total organic carbon (wt.% TOC), total sulfur (wt.% TS), sample location, lithology, and depositional environment—were investigated, but either showed insufficient variation for correlation, too many outliers, or simply no statistically meaningful correlations [e.g., wt.% TOC and $\delta^{13}C$ of kerogen in Eigenbrode and Freeman (2006)]; several parameters were dropped from subsequent PCA tests. In this report, relationships revealed by PCA are more closely evaluated using scatterplots and environmental facies are used to discern non-statistical patterns within them.

Two bulk geochemical parameters, dolomite abundance (%Dol) and carbon isotopic composition of kerogen ($\delta^{13}$C), are included in the analysis, because they show substantial variability within the dataset (0–99% and −50 to −28‰, respectively) and they most likely record conditions at the time of rock depositional-to-early diagenesis. Dolomite abundance mainly reflects original carbonate precipitation in depositional environments, though dolomite abundance can be diluted by early diagenetic precipitation of other geochemical cements (e.g., $SiO_2$, chert) or altered by vein minerals, including carbonates. Samples used for this study did not contain visually obvious, abundant veins. With regards to kerogens in general, intensive investigations show that kerogens with thermally sensitive H/C > 0.15 have $\delta^{13}$C that may be enriched in $^{13}$C by 2–2.5‰ (Des Marais 2001 and references therein). The samples studied herein have H/C of 0.15–0.30, consistent with sub-greenschist metamorphism (175–280°C). The isotopic alteration associated with this thermal maturity level is relatively insignificant for the dataset given the unusually broad, 40‰ range in $\delta^{13}$C values. On the contrary, the range in values reflects changes in the isotopic contributions from microbial communities due to ecological change (Eigenbrode and Freeman 2006). Dolomite abundance and kerogen $\delta^{13}$C values are the cornerstones of the statistical test for molecular fossil provenance.

Molecular fossil parameters from the saturated hydrocarbon fraction of rock extracts can be determined using chromatographic peak areas by gas chromatography–mass spectrometry (GCMS) analysis or metastable-reaction-monitoring (MRM)-GCMS (Table 1). MRM-GCMS allows for highly specific molecular detection at low-resolution by using dual magnetic sectors to focus on parent-ions with specific fragment ions. As a result, co-elution of steranes and hopanes, which plagues standard GCMS methods, is overcome (see Summons 1987 and Peters et al. 2005 for further review). To simplify PCA tests, only molecular fossil parameters included were those that generally show strong biological source signatures in Phanerozoic petroleum studies, as opposed to robust thermal maturity signals.

Molecular fossil parameters that are strongly influenced by source are also sensitive to thermal alteration to varying degrees (Peters et al. 2005) and some thermal maturity parameters are sensitive to lithology (e.g., Radke et al. 1982; Moldowan et al. 1986; Schulz et al. 2001). Phanerozoic studies show that samples that have undergone similar maturation tend to maintain source relationships (cf. Peters et al. 2005). Thus, the molecular fossil parameters used in this study may still reflect variations in biological sources, despite thermal alteration and lithology-dependent geochemical processes. The impact of thermal maturity required an independent investigation (Eigenbrode 2004), which will not be discussed here, except to summarize that the degree of thermal maturity as indicated by certain biomarker maturity ratios was consistent with geochemical and geological parameters and that the biomarker composition was consistent with other early Earth organic materials.

## 4.2 PCA Results for Molecular Fossil Provenance

PCA for the nine parameters listed in Table 1 revealed three components (S1, S2, and S3; Fig. 4) that together account for 85% of the variability in the dataset (Eigenbrode 2004). Each component was calculated from linear correlations among the variables, whereas loadings for each of the nine variables are correlation coefficients for the variable and the component. Groups of variables that exhibit high loadings for a component (loadings approaching ±1) may be expressing linear correlations. The PCA results in Fig. 4 suggest three variable groupings relate to each component: (S1) $\delta^{13}$C$_{ker}$, 3$\beta$MHI, Pr/Ph, HHI, and BNH/C$_{30}$H,

**Table 1** Variables used in principal component analysis

| Variables | Symbol | Source Indications | References |
|---|---|---|---|
| Dolomite abundance | %Dol | bulk geochemistry & sedimentology of depositional environment | |
| Stable carbon isotopic composition of kerogen | $\delta^{13}C$ | isotopic fractionations imposed by ecosystems during deposition and diagenesis | (Hayes et al. 1983; Freeman et al. 1990; Hayes et al. 1999) |
| $2\alpha$-Methylhopane index | $2\alpha$MHI | (1) cyanobacterial lipid contributions during deposition; (2) high values linked to carbonate environments | (Summons et al. 1999 and references therein) |
| $3\beta$-Methylhopane index | $3\beta$MHI | Aerobic methanotrophic bacterial contributions | (Rohmer et al. 1984; Neunlist and Rohmer 1985; Zundel and Rohmer 1985a; Zundel and Rohmer 1985b; Summons and Jahnke 1992; Cvejic et al. 2000) |
| 28,30-dinorhopane abundance relative to $C_{30}$-hopane | DNH/$C_{30}$H | (1) high values linked to reducing conditions; (2) chemoautotrophic bacterial lipid contributions (?) | (Noble et al. 1985; Mello et al. 1988; Schoell et al. 1992) |
| Pristane/phytane | Pr/Ph | (1) derived from photosynthetic chlorophylls; (2) commonly indicates relative redox conditions during deposition and diagenesis; (3) archaeal et al. lipid contributions | (Didyk et al. 1978; ten Haven et al. 1987; Freeman 1989; Freeman et al. 1990; Rowland 1990; Hughes et al. 1995) |
| Homohopane index | HHI | High values linked to carbonate environments and sulfides during diagenesis | (Sinninghe Damsté et al. 1995; Koster et al. 1997) |
| 30-Norhopane abundance relative to $C_{30}$-hopane | 30norH/$C_{30}$H | high values linked to carbonate environments | (Moldowan et al. 1991; Peters et al. 2005) |
| $C_{27}$ sterane abundance | %$C_{27}$St | subset of steroid contributions from eukaryotes | (Volkman 2005) |

Biomarker ratios were determined using uncorrected peak areas from $M^+ \rightarrow [M^+\text{-x}]$ chromatograms from MRM-GCMS analyses, with the exception to ratios using pristane and phytane, which were determined using 85 Da from GC/MS scans and corrected using response factors. Parameters are defined as: DNH/$C_{30}$H: $C_{28}$ dinorhopane/$C_{30}$ $17\alpha$-hopane ratio, Pr/Ph: pristane/phytane ratio corrected using response factors, $3\beta$MHI: $3\beta$-methyl hopanes index = $C_{31}$ $3\beta$-methyl hopane/($C_{30}$ $17\alpha$-hopane + $C_{31}$ $3\beta$-methyl hopane), $\delta^{13}C_{ker}$: Kerogen carbon-isotope values (‰) = $1000 \times (R_{kerogen} - R_{PDB\ std}/R_{PDB\ std})$ where $R = {}^{12}C/{}^{13}C$, HHI: $C_{35}$-homohopane index = $C_{35}$ homohopanes/sum $C_{31-35}$ homohopanes, $2\alpha$MHI: $2\alpha$-methyl hopanes index = $C_{31}$ $2\alpha$-methyl hopane/($C_{30}$ $17\alpha$-hopane + $C_{31}$ $2\alpha$-methyl hopane), 30norH/$C_{30}$H: 30-norhopane/$C_{30}$ $17\alpha$-hopane ratio, %Dol: estimated dolomite abundance = (wt.% total carbon − wt.%TOC) × 7.6, %$C_{27}$St: $C_{27}$ sterane abundance = sum of $C_{27}$ steranes/sum of $C_{27-30}$ steranes

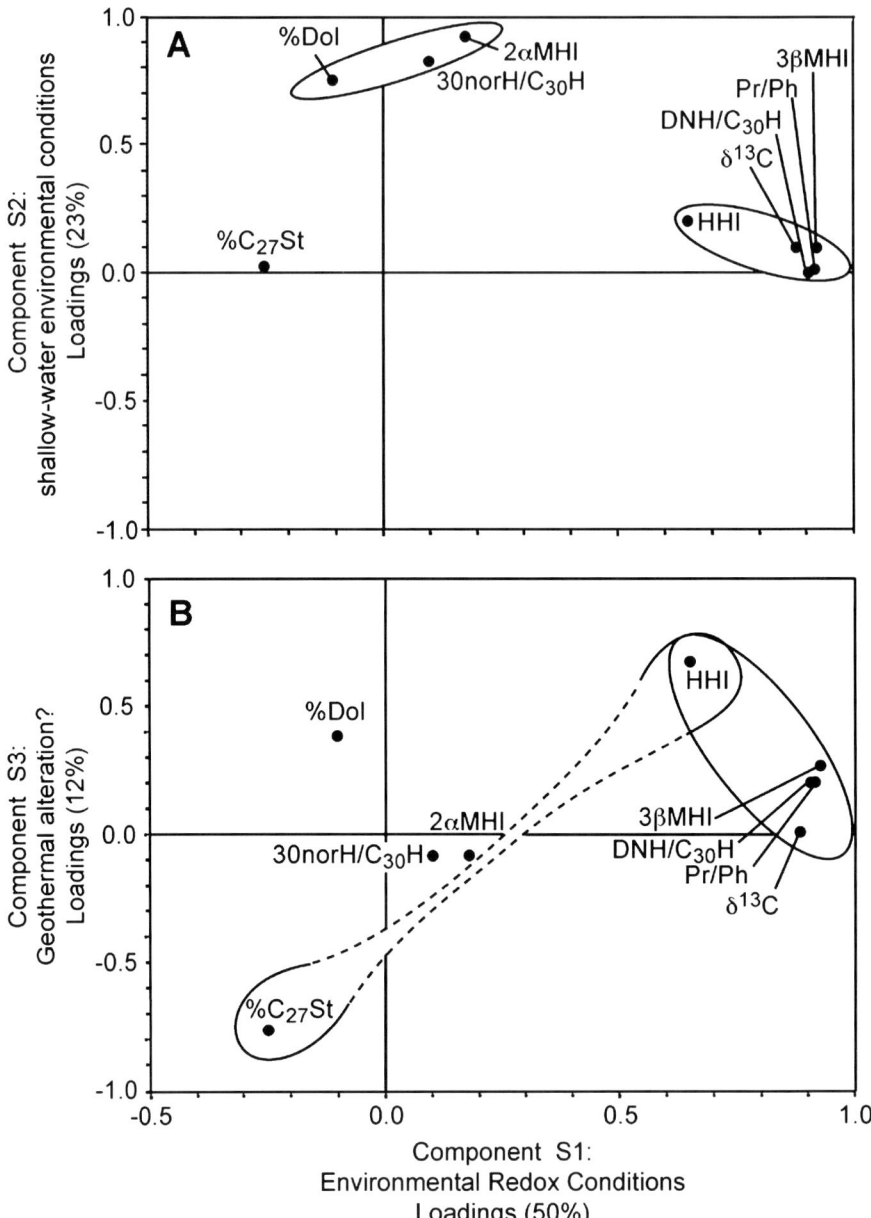

**Fig. 4** Principal component analysis results are plotted as loadings for three extracted components resolved from nine variables (Eigenbrode 2004). Percentages indicate the amount of variability in the dataset attributed to each component. Variables showing relationships to components are *circled*. (**a**) Component S1 groups 5 redox sensitive parameters ($3\beta$MHI, DNH/C30H, Pr/Ph, $\delta^{13}$C, and HHI; see Table 1), which all have positive, high S1 loadings. Component S2 groups dolomite abundance (%Dol), $2\alpha$-methylhopane index ($2\alpha$MHI), and 30-norhopane relative abundance (30norH/C$_{30}$H), which also have positive, high S2 loadings. (**b**) Component S3 groups C$_{27}$ sterane abundance (%C$_{27}$St) and the homohopanes index (HHI) that have negative and positive high S3 loading, respectively. All statistical analyses were conducted using Statistica software (StatSoft, Inc., Tulsa, Oklahoma) with varimax rotation. Pair-wise deletion and mean substitution are used to contend with missing data though similar results were obtained for both methods

(S2) $2\alpha$MHI, 30norH/$C_{30}$H, and %Dol, and (S3) %$C_{27}$St and HHI. All loadings that define the groups are positive, with exception to %$C_{27}$St, which negatively correlates to all components, particularly S3.

Principal components are quantitative; however, they may reflect a shared qualitative property that can be interpreted from variables. In this case, component S1 is interpreted as environmental redox conditions due to the positive, linear correlation among five redox sensitive parameters ($3\beta$MHI, DNH/C30H, Pr/Ph, $\delta^{13}$C, and HHI). Component S2 resolved from strong positive relationship among dolomite abundance (%Dol), $2\alpha$-methylhopane index ($2\alpha$MHI), and 30-norhopane relative abundance (30norH/$C_{30}$H) is consistent with signature for phototrophic activity associated with relatively shallow-water environments. Component S3 suggests a negative correlation between $C_{27}$ sterane abundance (%$C_{27}$St) and the homohopanes index (HHI) possibly due to side-chain cleavage of C28–C30 steranes and >C32–C35 homohopanes during geothermal alteration.

### 4.2.1 Indigeneity Tests

Components S1 and S2 of the PCA results (Fig. 4a) suggest that the hopane and short-chained acyclic isoprenoid biomarkers are not from foreign petroleum. Even if the distribution of molecular constituents of non-indigenous petroleum were influenced by lithology during migration or thermal maturation, the isoprenoid biomarkers would not also be distributed based on kerogen isotopic compositions. Lithology is not the primary control of kerogen $\delta^{13}$C composition—if it were related, dolomite abundance and $\delta^{13}$C would plot together, correlating to a single component. On the contrary, the sample suite contains black carbonaceous shales that have $-50‰$ to $-40‰$ $\delta^{13}$C values, which overlap the $-44‰$ to $-28‰$ range for rocks richer in dolomite (<10 wt.%). Eigenbrode and Freeman (2006) showed that the isotopic variability is related to changes in biological inputs associated with shallow versus deeper water facies. Invalidation of the hypothesis that hopanes and acyclic isoprenoids have a non-indigenous source is further exemplified by the difference in association of the $2\alpha$-methylhopane and $3\beta$-methylhopane indices. Both parameters normalized methyl hopane abundances to the abundance of the common $17\alpha$-$C_{30}$-hopane, yet the methyl hopanes have methyl groups that differ by only one carbon position on the first ring. Thus, the $2\alpha$- and $3\beta$-methyl hopanes likely behaved similarly during post-depositional alteration.

In contrast, the PCA results are not conclusive for tests of sterane indigeneity (but see Sect. 4.4). Figure 4b shows that $C_{27}$ sterane abundance, the sole sterane parameter of the PCA test, does not correlate to components involving host-rock geochemistry or composition variables. Instead, $C_{27}$ sterane abundance and the homohopanes index, weakly related to component S3, suggest a sterane-hopane relationship. Whether or not this connection highlights a syngenetic link or shared post-depositional alteration history is unclear with just the PCA results.

### 4.2.2 Syngenicity Test

The conclusion that hopanes, acyclic isoprenoids, and possibly steranes are indigenous to host rocks does not imply syngenicity. A separate test is necessary to exclude the presence of molecular fossils from microorganisms that lived in the rock (endolithic) after diagenesis (Fig. 3).

Petsch et al. (2001) provided a fitting example of endolithic microbiota metabolizing ancient kerogen and incorporating the carbon-isotopic signature of the kerogen into their own biomass while producing lipid precursors of molecular fossils. The phospholipid fatty acids observed by Petsch et al. were biological, thermally immature molecular biomarkers. In contrast, the molecular fossils of the early Earth case study have stereochemical and distribution features diagnostic of high thermal maturity, which is consistent with the maturity of other hydrocarbons, kerogen, rock metamorphic grade and regional geothermal history (Brocks et al. 1999, 2003; Eigenbrode 2004). Thus, a post-metamorphism, endolithic source for the hopanes, acyclic isoprenoids, and steranes is very unlikely. Notably, this conclusion is dependent on background knowledge of the terrestrial biological sources and thermal alteration pathways of molecular fossils, as well as regional geological context. This type of background information may not always be available in early Earth studies.

In some cases, the degree of recycling of ancient kerogen by endolithic biota may also depend on the substrates and habitat provided by the host rocks. If so, a microbial lipid-rock relationship could result. However, the most parsimonious explanations for the PCA results which rely on biological and environmental context (discussed in the following) are inconsistent with endolithic sources for molecular fossils. An extraterrestrial study would likely lack these constraints.

## 4.3 Environmental and Ecological Signatures Revealed by PCA

The indigeneity and syngenicity of the hopanes, acyclic isoprenoids, and steranes is supported by constraints developed through PCA. Statistical tests can also help resolve (1) the paleobiological interactions among microbes and with environments and (2) the strictly geological processes responsible for the observed relationships. To this end, the relationships revealed by PCA were investigated using scatterplots with data coded for environmental facies, determined from sedimentological and stratigraphic investigations (Eigenbrode and Freeman 2006). The statistical significance of relationships was also determined, where $p$-values less than $10^{-3}$ were regarded as high statistical significance (Hill and Lewicki 2006).

The simplest relationship revealed by PCA is illustrated by the positive linear correlation of dolomite abundance to $2\alpha$-methylhopane index (Fig. 5). Correlation with the relative

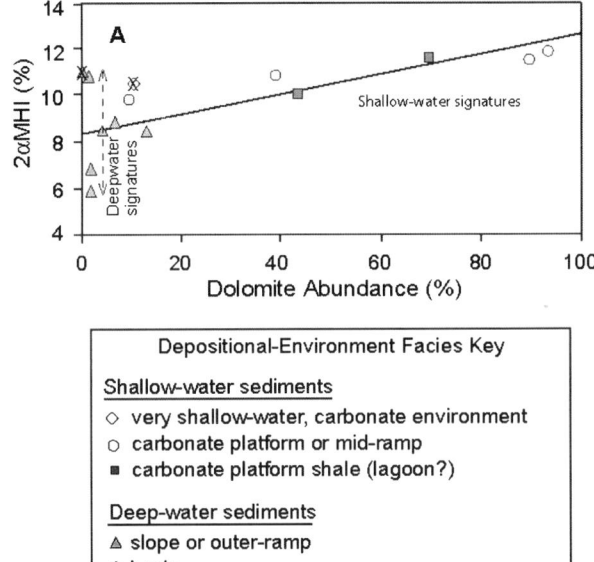

**Fig. 5** $2\alpha$-methylhopane index ($2\alpha$MHI) versus dolomite abundance (%Dol) are plotted according to environmental facies and show a positive correlation for non-silicified samples ($r^2 = 0.53$, $p = 5.7 \times 10^{-2}$). Silicified samples are marked with an "$X$". Shallow-water derived dolomite-rich samples tightly fit the regression, whereas deepwater samples show greater scatter. Analytical uncertainties are within the symbol size (modified from Eigenbrode 2004)

**Fig. 6** $3\beta$-methylhopane index ($3\beta$MHI), pristine/phytane (Pr/Ph), and the homohopanes index (HHI) versus kerogen-$\delta^{13}$C values are plotted according to depositional-environment facies (see Fig. 4 for key). (**a**) $\delta^{13}$C and $3\beta$-methylhopane index show negative correlation ($r^2 = 0.85$ and $p = 3.3 \times 10^{-4}$). (**b**) $\delta^{13}$C and pristane/phytane show separate positive correlations for each core (SV1 $r^2 = 0.92$; RHDH2a $r^2 = 0.81$, $p = 7.3 \times 10^{-4}$; WRL1 $r^2 = 0.95$, $p = 7.9 \times 10^{-4}$). (**c**) $\delta^{13}$C very weakly correlates to the homohopanes index ($r^2 = 0.51$, $p = 1.1 \times 10^{-2}$). All plots suggest facies relationships. Analytical uncertainties calculated from the range of values determined from standards run multiple times under the same conditions are noted with *bars* or are within symbol size (modified from Eigenbrode 2004)

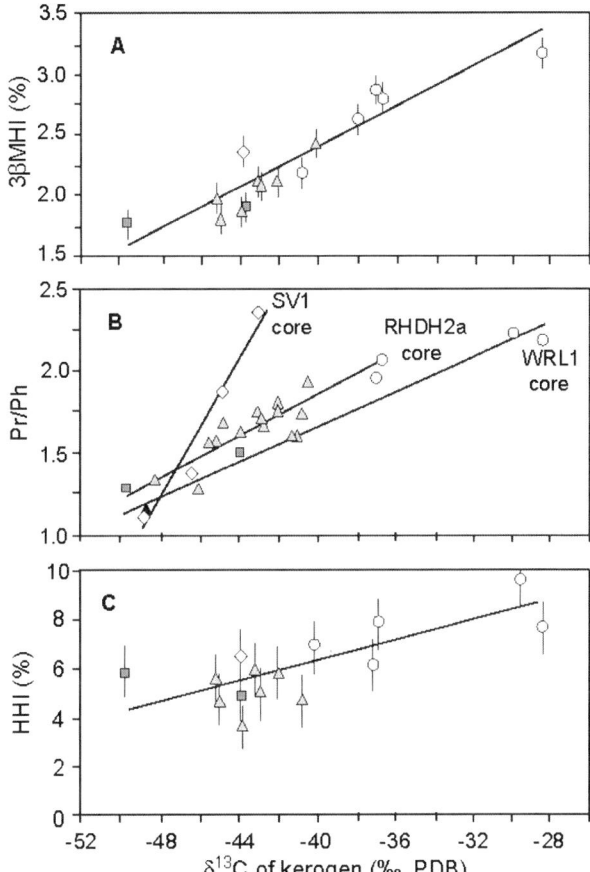

30-norhopane abundance (30nor/C30H) is not shown, but both relationships are statistically significant ($p$-values $< 10^{-2}$). Cyanobacteria are a likely source of $2\alpha$-methyl hopanes (Summons et al. 1999), whereas the specific sources of 30-norhopanes are not known. Still, the commonly observed 30-norhopane association with carbonate rocks (Peters et al. 2005) is consistent with photosynthetic sources, such as cyanobacteria, living in shallow (photic zone) environments. Hence, component S2 is an environmental signature reflecting photosynthetic activity specific to shallow-water environments.

A more complex, independent set of relationships are indicated by the variability linked to component S1, which most likely reflects the sensitivity of microorganisms and non-biological diagenetic processes to environmental redox conditions. The strongest relationship observed in the case study is between $3\beta$-methylhopane index and $\delta^{13}$C of kerogens. The $3\beta$-methylhopane index likely records inputs from aerobic respiring bacteria, particular those that can metabolize methane (Summons and Jahnke 1992; Cvcjic et al. 2000), inferring the biological consumption of environmental molecular oxygen and methane. Kerogen-$\delta^{13}$C differs from the $3\beta$-methylhopane index in that it records the isotopic fractionation imposed by a microbial community. In this case, an increase in $\delta^{13}$C indicates an increase in the importance of respiration by heterotrophs relative to the recycling $^{13}$C-depleted substrates, including methane, particularly in shallow-water environments (Eigenbrode and Freeman

2006). Like the $3\beta$-methylhopane index, kerogen-$\delta^{13}C$ record indicates the changes in the biological consumption of oxidants, organic reductants, and methane available in environments. The correlation between $3\beta$-methylhopane index and kerogen-$\delta^{13}C$ is an ecological signature reflecting the interactions among different microbes and the environment. With regards to the other parameters of S1, Phanerozoic studies demonstrate that pristane/phytane, the homohopanes index, and dinorhopane/$C_{30}$-hopane are primarily associated with redox-dependent reactions during deposition and early diagenesis, though they are also highly susceptible to thermal alteration that can obscure environmental signals (see references in Table 1). Still, all of the parameters point towards redox as an explanation for component S1.

4.4 Geochemical Signatures

In contrast, component S3 (Fig. 4) shows opposing correlations to $C_{27}$ sterane abundance and the homohopanes index that are difficult to explain from environmental and ecological perspectives (Eigenbrode 2004). The most parsimonious explanation for their weak, negative correlation (Fig. 7) is that both variables express thermal alteration, a geochemical process. The weak correlation partially reflects greater analytical error of the homohopanes index. Petroleum studies show that side-chain cleavage of $C_{31-35}$ homohopanes occurs during thermal maturation (Peters and Moldowan 1991), resulting in a lower homohopanes index. Perhaps a similar alteration occurs for the side-chain (i.e., the C-24-alkyl group) of $C_{28-30}$ steranes, resulting in greater $C_{27}$ sterane abundance, though this process has not been noted in the literature to the author's knowledge. If side-chain cleavage is responsible for variation in $C_{27}$ sterane abundance, then component S3 must represent conditions of thermal alteration by which both homohopanes and steranes were degraded.

How then, do we reconcile the relationship of the homohopanes index to both components S1 and S3? There are two possibilities. Notably, the homohopanes index is highest in shallow-water facies and lower in deeper water facies (Fig. 7). A connection between the two environmental facies groups and dolomite abundance is not clear cut (Fig. 5); however, there is a lax relationship. High homohopanes index is common in thermally immature bitumens and oils from Phanerozoic carbonate rocks and attributed to early diagenetic sulfur quenching under reducing conditions (Sinninghe Damsté et al. 1995; Koster et al. 1997). Perhaps the homohopanes index in the bitumens of the early Earth record is retaining a relict signature of a similar process, which would explain the mutual association of homohopanes index to other redox sensitive molecular and isotopic parameters and with thermal maturity. Alternatively, thermal maturity may have completely obscured an early diagenetic signature and is itself dependent on lithology, though lithologic dependence on thermal degradation of homohopanes index has not been shown by

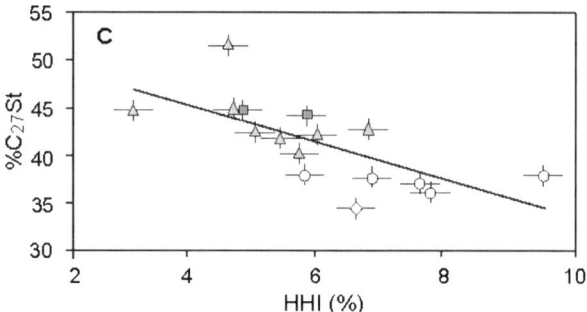

**Fig. 7** $C_{27}$ sterane abundance (%$C_{27}$St) versus the homohopane index (HHI) are plotted according to environmental facies (see Fig. 4 for key) and show a negative, linear correlation ($r^2 = 0.45$, $p = 2.2 \times 10^{-2}$) and a facies pattern (modified from Eigenbrode 2004)

other investigations. Weak correlation to other S1 variables, such as $\delta^{13}C$ (Fig. 5c), may simple be an artifact of the strong $\delta^{13}C$-depositional facies relationship reported by Eigenbrode and Freeman (2006). Both scenarios place the homohopanes (and steranes by association) in the host rocks at the time of early diagenesis or before geothermal maxima. Unless there is an undiscovered connection between side-chain cleavage and early diagenesis that is consistent with sulfur quenching of homohopanes, the variability in homohopanes index probably reflects lithology-dependent, thermal maturation. Based on similar arguments, dinorhopane/$C_{30}$-hopane values may be expressing a similar lithology-dependent maturation signature.

## 4.5 Case Study Summary

The early Earth molecular fossil study described here demonstrates the powerful utility of statistical analyses, such as principal component analysis, for identifying, qualifying, and quantifying meaningful relationships among a diverse pool of parameters. Although the presence of certain biomarkers indicated biological contributions from certain organisms (e.g., $3\beta$-methylhopanes from aerobic methanotrophs) and products from particular biosynthetic pathways (e.g., sterol synthesis), and inferred likely physiologies (e.g. $2\alpha$-methylhopane for oxygenic photosynthesis), PCA revealed (1) that the variability among the data was from redox conditions, phototrophy, and thermal maturity, and (2) that some molecular fossil parameters correlate to bulk geochemical and lithologic variables, providing strong negative result for tests for molecular fossil sources that are neither indigenous nor syngenetic. Hence, this latter result supports preservation of indigenous and syngenetic signals, even without background knowledge of biological and thermal constraints. On the contrary, these multidimensional set of patterns also strongly support the indigeneity and syngenicity of the fossil lipids, kerogen, and host rocks. Biological and geological context was necessary to explain the multi-variable relationships observed. Together the patterns expressed by the fossil lipids, geochemical parameters, and environmental facies are signatures of ancient environment conditions, and ecological and geochemical alteration processes. Similar analyses combining a wide range of biogeochemical, morphological, and geological parameters are likely to help constrain the timing, place, origin, and alteration history of other records and provide deeper insight into the nature of early Earth ecosystems.

## 5 Implications for Life-Detection on Early Earth and Elsewhere

Lipid membranes are characteristics expected for cellular, extraterrestrial life, based on their nature and function. If extraterrestrial life exists or existed, its biochemistry would reflect adaptation to environmental conditions, extreme by Earth standards, just as Earth biota does today. This adaptation includes their lipid biochemistry, which has high preservation potential if protected from strong oxidation. In the case of Mars the strong chemical and photooxidation potential on the rocky surface may largely preclude organic matter preservation (Sumner 2004; Navarro-Gonzalez et al. 2006; Souza-Egipsy et al. 2006). However—based on lessons learned from the habitability of extreme environments on Earth—extraterrestrial microbes may have adapted to this extreme niche. Conditions of the Martian subsurface are poorly constrained, thus assessing the potential for organic-carbon preservation and habitability is difficult. In particular, it is not clear if micro-scale conditions within minerals, such as clays, provided sufficient organic-carbon protection from oxidation. Icy materials, on the other hand, may thermodynamically enhance organic carbon preservation. Psychrophilic organisms dwelling in cold-climate rock, ice, and permafrost (Bolter 2004; Chintalapati et al.

2004; Castello and Rogers 2005 and references therein; Morgan-Kiss et al. 2006) and in deep glacial ice (Karl et al. 1999; Price 2000; Sheridan et al. 2003) are promising analogues for extraterrestrial life (Des Marais 2006). In all four cases, organic carbon (and fossil lipid) abundance will be limited by production and preservation. An ideal alternative is to focus on life detection at hydrothermal systems, particularly in icy regions. Earth analogues indicate geochemical support for prolific chemolithotrophic life (Reysenbach and Cady 2001; Sogin et al. 2006), and both organic and inorganic biosignature preservation in hydrothermal systems. Although hydrothermal systems have not been identified yet on Mars, the presence of mixed sulfate and iron oxide sediments on Mars (Zolotov and Shock 2005) supports the possibility that hydrothermal vents once existed. In any case, sensitive chromatography and mass spectrometry techniques that achieve high-chromatographic resolution of organic compounds (Mahaffy 2007) will be necessary to characterize the molecular structure of organic carbon.

In the event that organic carbon or other potential biological record is found, a scientific approach that integrates molecular, biogeochemical, geological, and morphological observations at various scales may reveal the system that produced the organic record. Unlike on Earth, where natural hydrocarbon contamination of organic records is a major concern, on Mars, migration of martian hydrocarbons into samples may still provide a biosignature, though it could lead to more specific interpretive complications. Endeavors on Earth and elsewhere must be wary of contamination during sample recovery and handling. Paleo-Earth and Mars-analogue studies provide a means for developing strategies to minimize that contamination for extraterrestrial missions focused on organic detection (e.g., NASA's Mars Science Laboratory, ESA's ExoMars, and possibly NASA's Astrobiology Field Laboratory). If enough variation exists in the extraterrestrial datasets, statistical analyses might discern environmental and ecological signatures from records of geochemical processes and from the introduction of spacecraft materials. Together with structural patterns within molecules (Summons et al. 2007), statistics might also provide an unbiased measure of confidence for observed patterns and correlations, which could reflect biocomplexity, the ultimate signature of life.

**Acknowledgements**    The case study presented herein was part of J.L.E.'s Ph.D. research at Penn State University and was supported by NSF funds (grants EAR-00-73831 and EAR-00-80267) awarded to Katherine H. Freeman of the Department of Geosciences at Penn State, with additional support from NASA Astrobiology Institute, Penn State Astrobiology Research Center (PSARC; NAI Cooperative Agreement NNA04CC06A). Molecular analyses were conducted in collaboration with Roger E. Summons, Massachusetts Institute of Technology, with instrument support from NASA Exobiology. J.L.E is grateful for additional fellowship support from the Pennsylvania Space Grant Consortium (NASA grant NGT5-40090). The author thanks Matt Hurtgen, Mark Barley, Brian Krapez, and many others for their field and laboratory assistance.

This review was conducted with fellowship support for J.L.E. from (NAI) Cooperative Agreements NNA04CC09A (Carnegie Institution of Washington) and the Geophysical Laboratory, Carnegie Institution of Washington. The author thanks Marilyn Fogel, Shuhei Ono, and Roger Summons for constructive reviews of this work.

# References

A. Aiello, E. Fattorusso, M. Menna, Steroids **64**, 687 (1999)
C. Alexander, M.F. Fogel, G. Cody (2007, in prep.)
R. Amils, C. Ellis-Evans, H.G. Hinghofer-Szalkay (eds.), *Life in Extreme Environments* (Springer, New York, 2007), p. 370
H.E.F. Amundsen, A. Steele, M. Fogel, J. Kihle, M. Schweizer, J. Toporski, A.H. Treiman, LPI Contrib. No. 1197 (2004)

A. Aubrey, H.J. Cleaves, J.H. Chalmers, A.M. Skelley, R.A. Mathies, F.J. Grunthaner et al., Geology **34**, 357 (2006)

R.A. Bell, J. Phycol. **29**, 133 (1993)

M. Begon, C.R. Townsend, J.L. Harper, *Ecology: From individuals to ecosystems* (Blackwell Publishing, Oxford, 2005), p. 752

P. Benveniste, Annu. Rev. Plant Biol. **55**, 429 (2004)

R.A. Berner, Am. J. Sci. **282**, 451 (1992)

R.L. Blackhurst, M.J. Genge, A.T. Kearsley, M.M. Grady, J. Geophys. Res. – Planets **110** (2005)

K. Bloch, in *Blondes in Venetian Paintings, the Nine-Banded Armadillo, and Other Essays in Biochemistry*, ed. by K. Block (Yale Univ. Press, New Haven, 1994), p. 14

K.E. Bloch, Critical Rev. Biochem. **14**, 47 (1983)

M. Blumenberg, V. Thiel, T. Pape, W. Michaelis, Naturwissenschaften **89**, 415 (2002)

M. Blumenberg, R. Seifert, J. Reitner, T. Pape, W. Michaelis, Proc. Natl. Acad. Sci. USA **101**, 11111 (2004)

H.B. Bode, B. Zeggel, B. Silakowski, S.C. Wenzel, H. Reichenbach, R. Muller, Mol. Microbiol. **47**, 471 (2003)

V. Boivin-Jahns, R. Ruimy, A. Bianchi, S. Daumas, R. Christen, Appl. Environ. Microbiol. **62**, 3405 (1996)

M. Bolter, Cell. Mol. Biol. **50**, 563 (2004)

S.C. Brassell, G. Eglinton, F.J. Mo, Org. Geochem. **10**, 927 (1986)

D.E.G. Briggs, Phil. Trans. R. Soc., Lond. B **354**, 7 (1999)

J.J. Brocks, G.A. Logan, R. Buick, R.E. Summons, Science **285**, 1033 (1999)

J.J. Brocks, R. Buick, G.A. Logan, R.E. Summons, Geochim. Cosmochim. Acta **67**, 4289 (2003)

J.J. Brocks, G.D. Love, R. Summons, G.A. Logan, Geochim. Cosmochim. Acta **68**, A796 (2004)

J.J. Brocks, A. Pearson, in *Molecular Geomicrobiology*, ed. by J. Banfield, J. Cervini-Silva, K. Nealson (Mineralogical Society of America, Washington, 2005), p. 233

J.J. Brocks, R. Summons, in *Treatise on Geochemistry*, vol. 8: Biogeochemistry, ed. by H.D. Holland, K.K. Turekian (Elsevier, New York, 2005), p. 63

J.D. Castello, S.O. Rogers (eds.), *Life in Ancient Ice* (Princeton University Press, Princeton, 2005), p. 307

G.V. Chilingar, L. Buryakovsky, N.A. Eremenko, M.V. Gorfunkel, *Geology and Geochemistry of Oil and Gas* (Elsevier, New York, 2005), p. 390

S. Chintalapati, M.D. Kiran, S. Shivaji, Cell. Mol. Biol. **50**, 631 (2004)

G. Cody, C. Alexander, M.F. Fogel, Nature (2007, submitted)

J. Connan, in *Advances in Petroleum Geochemistry*, ed. by J. Brooks, D.H. Welte (Academic, London, 1984), p. 299

M.J.L. Coolen, H. Cypionka, A.M. Sass, H. Sass, J. Overmann, Science **296**, 2407 (2001)

G. Cornelissen, O. Gustafsson, T.D. Bucheli, M.T.O. Jonker, A.A. Koelmans, P.C.M. Van Noort, Environ. Sci. Technol. **39**, 6881 (2005)

J.A. Curiale, Org. Geochem. **10**, 559 (1986)

J.H. Cvejic, L. Bodrossy, K.L. Kovacs, M. Rohmer, FEMS Microbiol. Lett. **182**, 361 (2000)

J.W. de Leeuw, G.J.M. Versteegh, P.F. van Bergen, Plant Ecol. **182**, 209 (2006)

D. Deamer, Microbiol. Mol. Biol. Rev. **61**, 239 (1997)

S. Derenne, C. Largeau, Soil Sci. **166**, 833 (2001)

D.J. Des Marais, in *Stable Isotope Geochemistry*, ed. by J.W. Valley, D.R. Cole (Mineralogical Society of America, Washington, 2001), p. 555

D.J. Des Marais, Astrobiology **6**, 168 (2004)

B.M. Didyk, B.R.T. Simoneit, S.C. Brassell, G. Eglinton, Nature **272**, 216 (1978)

X.L. Ding, S.M. Henrichs, Marine Chem. **77**, 225 (2002)

W. Dowham, Annu. Rev. Biochem. **66**, 199 (1997)

W. Dowham, M. Bogdanov, in *Biochemistry of Lipids, Lipoproteins and Membranes*, ed. by D.E. Vance, J.E. Vance (Elsevier, New York, 2002), p. 1

A. Dutkiewicz, H. Volk, S.C. George, J. Ridley, R. Buick, Geology **34**, 437 (2006)

G. Eglinton, M. Calvin, Sci. Am. **261**, 32 (1967)

L.B. Eglinton, D. Lim, G. Slater, G.R. Osinski, J.K. Whelan, M. Douglas, Org. Geochem. **37**, 688 (2006)

P. Ehrenfreund, M.A. Sephton, Faraday Discuss. **133**, 277 (2006)

J. Eigenbrode, Ph.D. Dissertation, The Pennsylvania State University, University Park, 2004

J. Eigenbrode, A. Steele, M. Schweizer, M. Fries, M.L. Fogel, V. Starke et al., Astrobiology **6**, 220 (2006)

J. Eigenbrode, K.H. Freeman, R.E. Summons, Earth Planet. Sci. Lett. (2007a, in preparation)

J. Eigenbrode, R.E. Summons, K.H. Freeman, Geochim. Cosmochim. Acta (2007b, in preparation)

J.L. Eigenbrode, K.H. Freeman, PNAS **103**, 15759 (2006)

J. Farquhar, B.A. Wing, Earth Planet. Sci. Lett. **213**, 1 (2003)

P. Farrimond, J.C. Bevan, A.N. Bishop, Org. Geochem. **25**, 149 (1996)

P. Farrimond, G.E. Fox, H.E. Innes, I.P. Miskin, I.M. Head, Anc. Biomol. **2**, 147 (1998)

P. Farrimond, I.M. Head, H.E. Innes, Geochim. Cosmochim. Acta **64**, 2985 (2000)

W.W. Fischer, R.E. Summons, A. Pearson, Geobiology **3**, 33 (2005)

M. Fogel, A. Team, Astrobiology **6**, 148 (2006)

K.H. Freeman, Geol. Soc. Am. – Abstr. Programs **21** (1989)

K.H. Freeman, J.M. Hayes, J.-M. Trendel, P. Albrecht, Nature **343**, 254 (1990)

K.H. Freeman, S.G. Wakeham, J.M. Hayes, Org. Geochem. **21**, 629 (1994)

J. Gatellier, J.W. Deleeuw, J.S. Sinninghe Damsté, S. Derenne, C. Largeau, P. Metzger, Geochim. Cosmochim. Acta **57**, 2053 (1993)

F. Gelin, I. Boogers, A.A.M. Noordeloos, J.S. Sinninghe Damsté, P.G. Hatcher, J.W. de Leeuw, Geochim. Cosmochim. Acta **60**, 1275 (1996)

G.J. Gray, S.R. Lawrence, K. Kenyon, C. Cornford, J. Geol. Soc. Lond. **155**, 39 (1998)

P.F. Greenwood, K. Arouri, G.A. Logan, R.E. Summons, Org. Geochem. **35**, 331 (2004)

N.S. Gupta, R. Michels, D.E.G. Briggs, M.E. Collinson, R.P. Evershed, R.D. Pancost, Org. Geochem. **38**, 28 (2007)

W.A. Hartgers, J.F. Lopez, J.S. Sinninghe Damsté, C. Reiss, J.R. Maxwell, J.O. Grimalt, Geochim. Cosmochim. Acta **61**, 4769 (1997)

J.M. Hayes, J.M. Kaplan, K.M. Wedeking, in *Earth's Earliest Biosphere: Its Origin and Evolution*, ed. by J.W. Schopf (Princeton University Press, Princeton, 1983), p. 93

J.M. Hayes, Marine Geol. **113**, 111 (1993)

J.M. Hayes, H. Strauss, A.J. Kaufman, Chem. Geol. **161**, 103 (1999)

J.M. Hayes, in *Stable Isotope Geochemistry*, ed. by J.W. Valley, D.R. Cole (Mineralogical Society of America, Washington, 2001), p. 225

Y. Hebting, P. Schaeffer, A. Behrens, P. Adam, G. Schmitt, P. Schneckenburger et al., Science **312**, 1627 (2006)

J.I. Hedges, R.G. Keil, Marine Chem. **49**, 81 (1995)

J.I. Hedges, G. Eglinton, P.G. Hatcher, D.L. Kirchman, C. Arnosti, S. Derenne et al., Org. Geochem. **31**, 945 (2000)

J.I. Hedges, J.A. Baldock, Y. Gelinas, C. Lee, M. Peterson, S.G. Wakeham, Nature **409**, 801 (2001)

N. Hertkorn, R. Benner, M. Frommberger, P. Schmitt-Kopplin, M. Witt, K. Kaiser et al., Geochim. Cosmochim. Acta **70**, 2990 (2006)

T. Hill, P. Lewicki, Statistics: Methods and Applications, A comprehensive reference for science, industry, and data mining, StatSoft, Inc., Tulsa, Oklahoma, 2006, p. 854

H. Holland, Phil. Trans. R. Soc. B **361**, 903 (2006)

H.D. Holland, *The Chemical Evolution of the Atmosphere and Oceans* (Princeton University Press, Princeton, 1984), p. 598

K. Horikoshi, W.D. Grant (eds.), *Extremophiles: Microbial Life in Extreme Environments* (Wiley, New York, 1998), p. 322

B. Horsfield, H.J. Schenk, K. Zink, R. Ondrak, V. Dieckmann, J. Kallmeyer et al., Earth Planet. Sci. Lett. **246**, 55 (2006)

W.B. Hughes, A.G. Holba, L.I.P. Dzou, Geochim. Cosmochim. Acta **59**, 3581 (1995)

J.M. Hunt, *Petroleum Geochemistry and Geology* (Freeman, New York, 1996), p. 743

H.E. Innes, A.N. Bishop, P.A. Fox, I.M. Head, P. Farrimond, Org. Geochem. **29**, 1285 (1998)

C.G. Johnston, J.R. Vestal, Appl. Environ. Microbiol. **57**, 2308 (1991)

T. Kaneda, Microbiol. Rev. **55**, 288 (1991)

D.M. Karl, D.F. Bird, K. Bjorkman, T. Houlihan, R. Shackelford, L. Tupas, Science **286**, 2144 (1999)

O.E. Kawka, B.R.T. Simoneit, Org. Geochem. **22**, 947 (1994)

D.S. Kelley, J.A. Karson, G.L. Früh-Green, D.R. Yoerger, T.M. Shank, D.A. Butterfield et al., Science **307**, 1428 (2005)

M.J. Kennedy, D.R. Pevear, R.J. Hill, Science **295**, 657 (2002)

W. Konings, S.-V. Albers, S. Koning, A. Driessen, Antonie Van Leeuwenhoek **81**, 61 (2002)

J. Koster, H.M.E. VanKaamPeters, M.P. Koopmans, J.W. deLeeuw, J.S. Sinninghe Damsté, Geochim. Cosmochim. Acta **61**, 2431 (1997)

N.I. Krinsky, Free Radic. Biol. Med. **7**, 617 (1989)

N.I. Krinsky, Pure Appl. Chem. **66**, 1003 (1994)

N. Ladygina, E.G. Dedyukhina, M.B. Vainshtein, Proc. Biochem. **41**, 1001 (2006)

S. Larter, H. Huan, J. Adams, B. Bennett, O. Jokanola, T. Oldenburg et al., AAPG Bull. **90**, 921 (2006)

A. Lazcano, Space Sci. Rev. (2007, this volume)

M.P. Lesser, Annu. Rev. Physiol. **68**, 253 (2006)

G.A. Logan, M.C. Hinman, M.R. Walter, R.E. Summons, Geochim. Cosmochim. Acta **65**, 2317 (2001)

D.R. Lovley, S.J. Giovannoni, D.C. White, J.E. Champine, E.J.P. Phillips, Y.A. Gorby, S. Goodwin, Arch. Microbiol. **159**, 336 (1993)

R.G. Luthy, G.R. Aiken, M.L. Brusseau, S.D. Cunningham, P.M. Gschwend, J.J. Pignatello et al., Environ. Sci. Technol. **31**, 3341 (1997)

A.S. Mackenzie, T.M. Quigley, AAPG Bull. **72**, 399 (1988)

P.R. Mahaffy, Space Sci. Rev. (2007, this volume), doi: 10.1007/s11214-007-9223-1

T.B. Mahajan, J.E. Elsila, D.W. Deamer, R.N. Zare, Orig. Life Evol. Biospheres (Formerly Orig. Life Evol. Biosphere) **33**, 17 (2003)

G.M. Marion, C.H. Fritsen, H. Eicken, M.C. Payne, Astrobiology **3**, 785 (2003)

L.M. Mayer, Geochim. Cosmochim. Acta **58**, 1271 (1994)

L.M. Mayer, Marine Chem. **92**, 135 (2004)

T.M. McCollom, J.S. Seewald, Geochim. Cosmochim. Acta **65**, 3769 (2001)

T.M. McCollom, Geochim. Cosmochim. Acta **67**, 311 (2003)

G.D. McDonald, M.C. Storrie-Lombardi, Astrobiology **6**, 17 (2006)

K.D. McKeegan, L.A. Leshin, in *Stable Isotope Geochemistry*, ed. by J.W. Valley, D.R. Cole (Mineralogical Society of America, 2001), p. 279

D.M. McKirdy, R.E. Cox, J.K. Volkman, V.J. Howell, Nature **320**, 57 (1986)

D.L. Melchior, Curr. Top. Membr. Transp. **17**, 263 (1982)

V.A. Melezhik, A.E. Fallick, M.M. Filippov, O. Larsen, Earth Sci. Rev. **47**, 1 (1999)

M.R. Mello, N. Telnaes, P.C. Gaglianone, M.I. Chicarelli, S.C. Brassell, J.R. Maxwell, Org. Geochem. **13**, 31 (1988)

P. Metzger, C. Largeau, Appl. Microbiol. Biotechnol. **66**, 486 (2005)

V.I. Miteva, P.P. Sheridan, J.E. Brenchley, Appl. Environ. Microbiol. **70**, 202 (2004)

J.M. Moldowan, P. Sundararaman, M. Schoell, Org. Geochem. **10**, 915 (1986)

J.M. Moldowan, F.J. Fago, R.M.K. Carlson, D.C. Young, G. an Duvne, J. Clardy et al., Geochim. Cosmochim. Acta **55**, 3333 (1991)

J.M. Moldowan, S.R. Jacobson, Int. Geol. Rev. **42**, 805 (2000)

J.P. Montoya, Geochim. Cosmochim. Acta **67**, A303 (2003)

R.M. Morgan-Kiss, J.C. Priscu, T. Pocock, L. Gudynaite-Savitch, N.P.A. Huner, Microbiol. Mol. Biol. Rev. **70**, 222 (2006)

P.L. Morrill, J.L. Eigenbrode, O.J. Johnson, R.M. Erdil, B. Sherwood Lollar, K.H. Nealson et al., Astrobiology **6**, 238 (2006)

B. Mycke, W. Michaelis, E.T. Degens, Org. Geochem. **13**, 619 (1988)

H. Naraoka, A. Shimoyama, K. Harada, Earth Planet. Sci. Lett. **184**, 1 (2000)

R. Navarro-Gonzalez, K.F. Navarro, J. de la Rosa, E. Iñiguez, P. Molina, L.D. Miranda et al., PNAS **103**, 16089 (2006)

S. Neunlist, M. Rohmer, Biochem. J. **231**, 635 (1985)

R. Noble, R. Alexander, R.I. Kagi, Org. Geochem. **8**, 171 (1985)

G. Ourisson, M. Rohmer, K. Poralla, Annu. Rev. Microbiol. **41**, 301 (1987a)

G. Ourisson, M. Rohmer, K. Poralla, J. Gen. Microbiol. **41**, 301 (1987b)

R.D. Pancost, J.S. Sinninghe Damsté, Chem. Geol. **195**, 29 (2003)

R.J. Parkes, B.A. Cragg, S.J. Bale, J.M. Getliff, K. Goodman, P.A. Rochelle et al., Nature **371**, 410 (1994)

J. Parnell, G.R. Osinski, P. Lee, P.F. Green, M.J. Baron, Geology **33**, 373 (2005)

J.D. Pasteris, J.J. Freeman, B. Wopenka, K. Qi, Q.G. Ma, K.L. Wooley, Astrobiology **6**, 625 (2006)

A.A. Pavlov, J.F. Kasting, Astrobiology **2**, 27 (2002)

A. Pearson, M. Budin, J.J. Brocks, PNAS **100**, 15352 (2003)

J. Peretó, P. López-Garcia, D. Moreira, Trends Biochem. Sci. **29**, 469 (2004)

K.E. Peters, J.M. Moldowan, Org. Geochem. **17**, 47 (1991)

K.E. Peters, C.C. Walters, J.M. Moldowan, *The Biomarker Guide* (Cambridge Univ. Press, Cambridge, 2005), p. 471

S.T. Petsch, T.I. Eglinton, K.J. Edwards, Science **292**, 1127 (2001)

S.T. Petsch, K.J. Edwards, T.I. Eglinton, Org. Geochem. **34**, 731 (2003)

L.C. Price, Geochim. Cosmochim. Acta **57**, 3261 (1993)

P.B. Price, Proc. Natl. Acad. Sci. USA **97**, 1247 (2000)

M. Radke, H. Willsch, D. Leythaeuser, M. Teichmüller, Geochim. Cosmochim. Acta **46**, 1831 (1982)

B. Ransom, R.H. Bennett, R. Baerwald, K. Shea, Mar. Geol. **138**, 1 (1997)

B. Rasmussen, Nature **405**, 676 (2000)

B. Rasmussen, Geology **33**, 497 (2005)

A.G. Requejo, J. Allan, S. Creaney, N.R. Gray, K.S. Cole, Org. Geochem. **19**, 245 (1992)

A.-L. Reysenbach, S.L. Cady, Trends Microbiol. **9**, 79 (2001)

A. Riboulleau, S. Derenne, C. Largeau, F. Baudin, Org. Geochem. **32**, 647 (2001)

M. Rohmer, J.L. Bouvier, G. Ourisson, Proc. Natl. Acad. Sci. USA **76**, 847 (1979)

M. Rohmer, P. Bouvier-Nave, G. Ourisson, J. Gen. Microbiol. **130**, 1137 (1984)

S.J. Rowland, Org. Geochem. **15**, 9 (1990)
P.E. Sauer, T.I. Eglinton, J.M. Hayes, A. Schimmelmann, A.L. Sessions, Geochim. Cosmochim. Acta **65**, 213 (2001)
M. Schoell, M.A. McCaffrey, F.J. Fago, J.M. Moldowan, Geochim. Cosmochim. Acta **56**, 1391 (1992)
J.W. Schopf, M.R. Walter (eds.), *Earth's Earliest Biosphere: Its Origin and Evolution* (Princeton University Press, Princeton, 1983), p. 543
S. Schouten, M.J.L. Hoefs, M.P. Koopmans, H.-J. Bosch, J.S. Sinninghe Damsté, Org. Geochem. **29**, 1305 (1998)
L.K. Schulz, A. Wilhelms, E. Rein, A.S. Steen, Org. Geochem. **32**, 365 (2001)
J. Seckbach (ed.), *Enigmatic Microorganisms and Life in Extreme Environments (Cellular Origin and Life in Extreme Habitats)* (Kluwer, New York, 1999), p. 716
W.K. Seifert, J.M. Moldowan, Geochim. Cosmochim. Acta **42**, 77 (1978)
M.A. Sephton, Nat. Prod. Rep. **3**, 292 (2002)
A.L. Sessions, T.W. Burgoyne, A. Schimmelmann, J.M. Hayes, Org. Geochem. **30**, 1193 (1999)
Y. Shen, R. Buick, D.E. Canfield, Nature **410**, 77 (2001)
P.P. Sheridan, V.I. Miteva, J.E. Brenchley, Appl. Environ. Microbiol. **69**, 2153 (2003)
B. Sherwood Lollar, T.D. Westgate, J.A. Ward, G.F. Slater, G. Lacrampe-Couloume, Nature **416**, 522 (2002)
B. Sherwood Lollar, G. Lacrampe-Couloume, G.F. Slater, J.A. Ward, D.P. Moser, T.M. Gihring et al., Chem. Geol. **226**, 328 (2005)
B.R.T. Simoneit, Appl. Geochem. **17**, 129 (2002)
B.R.T. Simoneit, A.Y. Lein, V.I. Peresypkin, G.A. Osipov, Geochim. Cosmochim. Acta **68**, 2275 (2004)
B.R.T. Simoneit, Mass Spectrom. Rev. **24**, 719 (2005)
J.S. Sinninghe Damsté, W.I.C. Rijpstra, A.C. Kock-van Dalen, J.W. de Leeuw, P.A. Schenck, Geochim. Cosmochim. Acta **53**, 1443 (1989)
J.S. Sinninghe Damsté, A.C.T. Van Duin, D. Hollander, M.E.L. Kohnen, J.W. deLeeuw, Geochim. Cosmochim. Acta **59**, 5141 (1995)
J.S. Sinninghe Damsté, M.P. Koopmans, Pure Appl. Chem. **69**, 2067 (1997)
M.L. Sogin, H.G. Morrison, J.A. Huber, D.M. Welch, S.M. Huse, P.R. Neal et al., Proc. Natl. Acad. Sci. USA **103**, 12115 (2006)
V. Souza-Egipsy, J. Ormo, B.B. Bowen, M.A. Chan, G. Komatsu, Astrobiology **6**, 527 (2006)
R.E. Summons, T.G. Powell, Nature **319**, 763 (1986)
R.E. Summons, Org. Geochem. **11**, 281 (1987)
R.E. Summons, T.G. Powell, Geochim. Cosmochim. Acta **51**, 557 (1987)
R.E. Summons, S.C. Brassell, G. Eglinton, E. Evans, R.J. Horodyski, N. Robinson et al., Geochim. Cosmochim. Acta **52**, 2625 (1988)
R.E. Summons, L.L. Jahnke, in *Biomarkers in Sediments and Petroleum*, ed. by J.M. Moldowan, P. Albrecht, R.P. Philip (Prentice Hall, Englewood Cliffs, 1992), p. 182
R.E. Summons, L.L. Jahnke, J.M. Hope, G.A. Logan, Nature **400**, 554 (1999)
R.E. Summons, A.S. Bradley, L.L. Jahnke, J.R. Waldbauer, Phil. Trans. R. Soc. B **361**, 951 (2006)
R.E. Summons, P. Albrecht, G.D. McDonald, J.M. Moldowan, Space Sci. Rev. (2007, this volume), doi: 10.1007/s11214-007-9256-5
D.Y. Sumner, J. Geophys. Res. – Planets **109** (2004)
K. Takai, M.R. Mormile, J.P. McKinley, F.J. Brockman, W.E. Holben, W.P. Kovacik et al., Environ. Microbiol. **5**, 309 (2003)
H.M. Talbot, D.F. Watson, E.J. Pearson, P. Farrimond, Org. Geochem. **34**, 1353 (2003)
E.W. Tegelaar, J.W. de Leeuw, S. Derenne, C. Largeau, Geochim. Cosmochim. Acta **57**, 3103 (1989)
H.L. ten Haven, J.W. de Leeuw, J. Rullkötter, J.J. Sinninghe, Nature **330**, 641 (1987)
M.M. Tice, D.R. Lowe, Earth – Sci. Rev. **76**, 259 (2006)
B.P. Tissot, D.H. Welte, *Petroleum Formation and Occurrence* (Springer, New York, 1984), p. 699
T.G. Tornabene, R.E. Lloyd, G. Holzer, J. Oro, in *COSPAR, Life Sciences and Space Research*, ed. by R. Holmquist (Pergamon, Oxford, 1980), p. 109
J.L.C.M. van de Vossenberg, A.J.M. Driessen, W.N. Konings, Extremophiles **2**, 163 (1998)
J.L.C.M. van de Vossenberg, A.J.M. Driessen, W.N. Konings, in *Cell and Molecular Responses to Stress: Environmental Stressors and Gene Response*, ed. by K.B. Storey, J.M. Storey (Elsevier, Amsterdam, 2000), p. 71
M.T.J. van der Meer, S. Schouten, J.W. de Leeuw, D.M. Ward, Environ. Microbiol. **2**, 428 (2000)
A. Vershinin, Biofactors **10**, 99 (1999)
G.J.M. Versteegh, P. Blokker, G.D. Wood, M.E. Collinson, J.S. Sinninghe Damsté, J.W. de Leeuw, Org. Geochem. **35**, 1129 (2004)
J.K. Volkman, Org. Geochem. **36**, 139 (2005)
L.I. Vorob'eva, Appl. Biochem. Microbiol. **40**, 217 (2004)

W.B. Whitman, D.C. Coleman, W.J. Wiebe, Proc. Natl. Acad. Sci. USA **95**, 6578 (1998)
A.L. Zerkle, C.H. House, S.L. Brantley, Am. J. Sci. **305**, 467 (2005)
M.Y. Zolotov, E.L. Shock, Geophys. Res. Lett. **32** (2005)
M. Zundel, M. Rohmer, FEMS Microbiol. Lett. **28**, 61 (1985a)
M. Zundel, M. Rohmer, Eur. J. Biochem. **150**, 23 (1985b)

# Chirality and Life

**Laurence D. Barron**

Originally published in the journal Space Science Reviews, Volume 135, Nos 1–4.
DOI: 10.1007/s11214-007-9254-7 © Springer Science+Business Media B.V. 2007

**Abstract** Chirality, meaning handedness, pervades much of modern science, from the physics of elementary particles to the chemistry of life. The amino acids and sugars from which the central molecules of life—proteins and nucleic acids—are constructed exhibit homochirality, which is expected to be a key biosignature in astrobiology. This article provides a brief review of molecular chirality and its significance for the detection of extant or extinct life on other worlds. Fundamental symmetry aspects are emphasized since these bring intrinsic physical properties of the universe to bear on the problem of the origin and role of homochirality in the living world.

**Keywords** Homochirality · Origin of life · Absolute enantioselection · Mirror symmetry breaking

## 1 Introduction

Life as we know it on Earth is a chemical system capable of self-reproduction and evolution. Its main ingredients are liquid water and organic molecules, most of which are of sufficiently low symmetry that they can exist in two distinguishable mirror-image forms like L- and D-alanine displayed in Fig. 1a. Such molecules are said to be *chiral*, meaning left- or right-handed (from the Greek word *cheir* for hand). The word 'chirality' itself was first introduced into science by Lord Kelvin (1904), Professor of Natural Philosophy at the University of Glasgow. A finite cylindrical helix is the archetype for all figures exhibiting chirality: a left-handed helix and its right-handed mirror image cannot be superposed since reflection reverses the screw sense (Fig. 1b). The two nonsuperposable mirror-image forms are called *enantiomers* for molecules and *enantiomorphs* for macroscopic objects such as crystals. As well as its significance in organic stereochemistry and in the structure and behaviour of the molecules of life, chirality pervades many other areas of modern science (Barron 2004). It is especially important in the physics of elementary particles in the context

L.D. Barron (✉)
WestCHEM, Department of Chemistry, University of Glasgow, Glasgow G12 8QQ, UK
e-mail: laurence@chem.gla.ac.uk

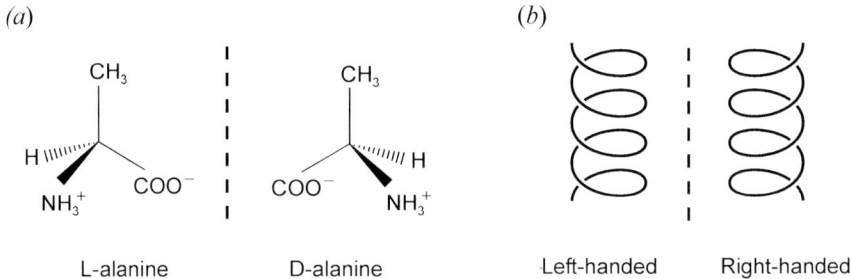

**Fig. 1** (**a**) The two mirror-image enantiomers of alanine (zwitterionic form). (**b**) A left-handed helix and its right-handed mirror image

of parity violation, which may even be linked to the chirality of biomolecules. The problem of chirality and life is a large and often controversial subject, a comprehensive review of which is not possible in the limited space available here. This article provides a selective view that reflects the author's background in physical chemistry.

The realization that some molecules can be chiral emerged from the phenomenon of natural optical activity, first observed by the French scientists Arago and Biot in the early years of the nineteenth century in the form of optical rotation of the plane of polarization of linearly polarized light in samples such as quartz crystals and certain organic liquids. Shortly after its discovery, Fresnel developed a theory of optical rotation based on the differential refraction of right- and left-circularly polarized light, which led him to suggest that optical activity may result from "... a helicoidal arrangement of the molecules of the medium, which would present inverse properties according as these helices were dextrogyrate or laevogyrate." Pasteur extended the concept of chirality (which he called *dissymmetry*) from the realm of optically active crystals to that of the individual molecules which provide optically active fluids or solutions through his epoch-making separation in 1848 of crystals of sodium ammonium paratartrate, an optically inactive form of sodium ammonium tartrate, into two sets which, when dissolved in water, gave optical rotations of equal magnitude but opposite sign. This demonstrated that paratartaric acid was a mixture, now known as a *racemic* mixture, of equal numbers of enantiomeric molecules.

## 2 Homochirality and Life

A hallmark of life's chemistry, and hence a key biosignature for astrobiology, is its *homochirality* (Mason 1988; Bonner 1988; Keszthelyi 1995; Feringa and van Delden 1999; MacDermott 2002; Compton and Pagni 2002), which is well-illustrated by the central molecules of life, proteins and nucleic acids. Proteins consist of polypeptide chains made from combinations of 20 different amino acids (primary structure), all exclusively the L-enantiomers. This homochirality in the monomeric amino acid building blocks of proteins leads to homochirality in higher-order structures such as the right-handed $\alpha$-helix (secondary structure), and the fold (tertiary structure) that is unique to each different protein in its native state (Fig. 2). Nucleic acids consist of chains of deoxyribonucleosides (for DNA) or ribonucleosides (for RNA), connected by phosphodiester links, all based exclusively on the D-deoxyribose or D-ribose sugar ring, respectively (Fig. 3). This homochirality in the monomeric sugar building blocks of nucleic acids leads to homochirality in their secondary structures such as the right-handed B-type DNA double helix.

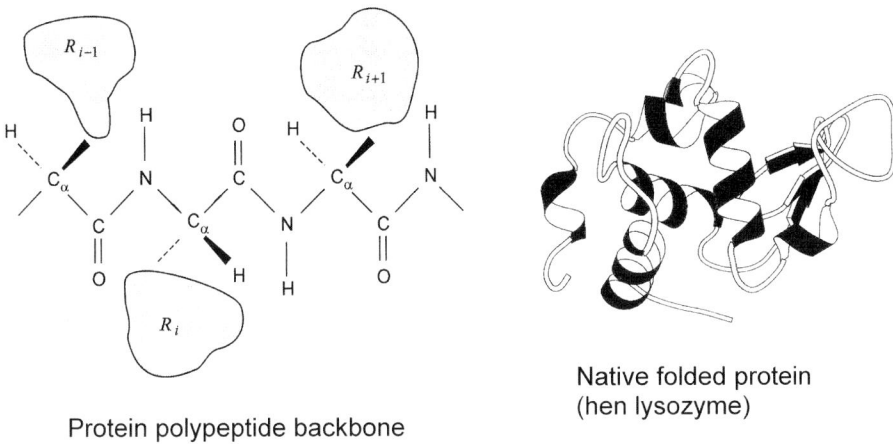

**Native folded protein (hen lysozyme)**

**Protein polypeptide backbone**

**Fig. 2** The polypeptide backbones of proteins are made exclusively from homochiral amino acids (all L). $R_i$ represents side chains such as $CH_3$ for alanine. This generates homochiral secondary structures, such as the right-handed $\alpha$-helix, within the tertiary structures of native folded proteins like hen lysozyme

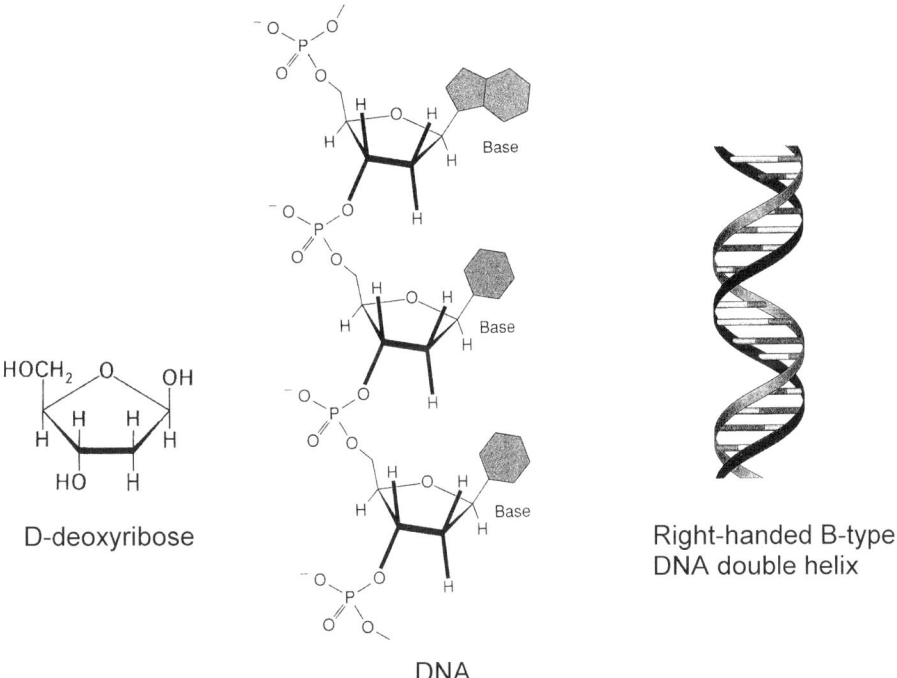

**D-deoxyribose**

**DNA**

**Right-handed B-type DNA double helix**

**Fig. 3** Nucleic acids are made exclusively from homochiral sugars (all D) such as D-deoxyribose for DNA. This generates homochiral secondary structures such as the right-handed B-type DNA double helix

Homochirality is essential for an efficient biochemistry, rather like the universal adoption of right-handed screws in engineering. One example is Fischer's 'lock and key' principle (Behr 1994), which provides a mechanism for stereochemical selection in nature, as in enzyme catalysis. Small amounts of 'non-natural' enantiomers such as the D-forms of some amino acids are in fact found in living organisms where they have specific roles (Fujii 2002; Konno et al. 2007), but they have not been found in functional biopolymers. Since molecules sufficiently large and complex to support life are almost certain to exist in two mirror-image chiral forms, homochirality appears to be a *sine qua non* for any molecule-based life. Furthermore, since no element other than carbon forms such a huge variety of compounds, many of them chiral, the chemistry is expected to be organic. Last but not least, the liquid water that is essential for life on Earth is more than simply a medium: it acts as a 'lubricant' of key biomolecular processes such as macromolecular folding, unfolding and interaction (Westhof 1993). No other liquid solvent has the same balance of vital physico-chemical properties. Hence *homochirality associated with a complex organic chemistry in an aqueous environment* would appear to be as essential for life on other worlds as it is on Earth, but the possibility of completely different scenarios should be kept in mind (Ball 2005).

But is homochirality in the monomeric building-blocks of biopolymers a condition for the initiation of a *prebiotic* chemistry? Homochiral nucleic acid polymers, for example, do not form efficiently in a racemic solution of the monomers (Joyce et al. 1984). Theoretical analysis suggested that addition of a nucleotide of the wrong handedness halts the polymerization (Avetisov et al. 1991), a process called enantiomeric cross-inhibition. However, chiral synthetic supramolecular chemistry (Cornelissen et al. 2001) suggests that homochirality in the chiral monomers may not in fact be a prerequisite for the initiation of a prebiotic homochiral polymer chemistry (Cintas 2002; Weissbuch et al. 2005). Although polyisocyanates, for example, constructed from achiral monomers form helical polymers with equal numbers of right- and left-handed forms, the introduction of a chiral bias in the form of a small amount of a chiral version of a monomer can induce a high enantiomeric excess (ee) of one helical sense over the other (Green et al. 1995). Furthermore, a polyisocyanate constructed from a random copolymerization of chiral monomers containing just a few percent excess of one enantiomer over the other shows a large excess of the helical form generated from homopolymerization of the corresponding enantiopure monomer. Another example of the dramatic influence a small chiral bias may exert was discovered recently in solid–liquid phase equilibria of amino acids: a few percent ee of one enantiomer can lead to very high solution ees, including a virtually enantiopure solution for serine (Klussman et al. 2006). An important feature of this system is that it is based on an equilibrium mechanism, as distinct from far-from-equilibrium mechanisms as previously invoked in kinetically induced amplification via autocatalytic reactions or crystallizations (Kondepudi and Asakura 2001). Closely related is another recent discovery that sublimation of a near-racemic sample of solid serine yields a sublimate that is highly enriched in the major enantiomer (Perry et al. 2007). Another possibility (Cintas 2002) is that the faces of chiral crystals such as quartz, or the chiral faces of nonchiral crystals such as glycine (Weissbuch et al. 2005) or minerals such as calcite (Hazen et al. 2001; Orme et al. 2001) could act as templates for the formation of homochiral polymers. Related to this is the recent demonstration that the chiral rims of racemic $\beta$-sheets can operate as templates for the generation of long homochiral oligopeptides (Rubinstein et al. 2007). Spontaneous resolutions in crystallization have also been mooted as a possible source of prebiotic ees (Avalos et al. 2004a).

Although detection of complete homochirality would provide strong evidence for molecule-based life on other worlds, extant or extinct, partial ees could be equivocal. An important example was the discovery of small but significant ees of the L-forms of $\alpha$-amino acids, which are not found in terrestrial proteins (thereby avoiding the contamination problem), in the Murchison meteorite (Cronin and Pizzarello 1997), with ees for L-isovaline up to 15% (Pizzarello et al. 2003); but this is thought to have arisen from the action of some physical chiral influence rather than being a signature of life (MacDermott 2002; Pizzarello 2006). More generally, the possibility that a large fraction of prebiotic organic molecules were delivered to the early Earth by meteorites and cometary dust grains, and that some of this organic material could be chiral and nonracemic due to processing in the interstellar medium under some physical chiral influence, is an important topic in astrobiology (Brack 1999).

## 3  Chirality and Absolute Enantioselection

### 3.1  Pasteur's Conjecture

From the previous section it would appear that, although complete homochirality in the prebiotic chiral monomers may not be necessary for the initiation of a prebiotic homochiral polymer chemistry, a small initial ee may be required (at least for mechanisms that do not depend on a chiral template such as a mineral surface). This small ee could be produced by some physical chiral influence. Pasteur himself conjectured that molecular chirality in the living world is the product of some universal chiral force or influence in nature. Accordingly, he attempted to extend the concept of chirality to other aspects of the physical world (Pasteur 1884). For example, he thought that the combination of a translational with a rotational motion generated chirality; likewise a magnetic field. Curie (1894) suggested that collinear electric and magnetic fields are chiral. However, as explained in the following, of these only a translating–rotating system, as in a circularly polarized light beam, for example, exhibits 'true chirality'. Pasteur's incorrect belief that a static magnetic field alone is also a source of chirality has been shared by many other scientists. This misconception is based on the fact that a static magnetic field can induce optical rotation (the Faraday effect) in achiral materials; but as Lord Kelvin (1904) emphasized: "the magnetic rotation has no chirality". In a new twist to the story (Barron 2000), a magnetic field was recently used in a more subtle fashion than that conceived by Pasteur by exploiting the novel phenomenon of magnetochiral dichroism (*vide infra*) to induce a small ee.

In view of the great interest it attracts due to its profound implications, the topic of parity violation in the fabric of the universe is given some prominence in this article. If it were ever proved that parity violation was linked in some way to the origin and role of homochirality in the living world, this would provide the ultimate source of a universal chiral force sensed by Pasteur. However, at the time of writing, there is no firm evidence to support the idea. The author has no particular commitment to Pasteur's conjecture, either in the form of parity violation or a more mundane chiral physical influence such as circularly polarized light, rather than to, say, a local strongly chiral environment such as a mineral surface, as the original source of homochirality in life on Earth. On a cosmic scale, enantioselective mechanisms depending on parity violation are the only ones which could predetermine a particular handedness, such as the L-amino acids and D-sugars found in terrestrial life; in all other mechanisms the ultimate choice would arise purely by chance.

### 3.2  Symmetry and Chirality

Chirality is an excellent subject for the application of symmetry principles (Barron 2002, 2004). As well as conventional point group symmetry, the fundamental symmetries of space inversion, time reversal and even charge conjugation have something to say about chirality at all levels: the experiments that show up optical activity observables, the objects generating these observables and the nature of the quantum states that these objects must be able to support. Even the symmetry violations observed in elementary particle physics can infiltrate into the world of chiral molecules, with intriguing implications. These fundamental symmetry aspects, summarized briefly in the following, are highly relevant to considerations of molecular chirality in the context of astrobiology since they bring intrinsic physical properties of the universe to bear on the problem of the origin of homochirality and its role in the origin and special physico-chemical characteristics of life, with implications for the possibility (or otherwise) that life is ubiquitous throughout the cosmos.

### 3.2.1 Spatial Symmetry

Chiral figures are not necessarily *asymmetric*, that is devoid of all symmetry elements, since they may possess one or more proper rotation axes: for example, a finite cylindrical helix (Fig. 1b) has a twofold rotation axis $C_2$ through the mid-point of the coil, perpendicular to the long axis. However, chirality excludes improper symmetry elements, namely centres of inversion, reflection planes and rotation–reflection axes. Hence chirality is supported by the point groups comprising only proper rotations, namely $C_n$, $D_n$, $O$, $T$ and $I$.

### 3.2.2 Inversion Symmetry: Parity, Time Reversal and Charge Conjugation

More fundamental than spatial (point group) symmetries are the symmetries in the laws of physics, and these in turn depend on certain uniformities that we perceive in the world around us. In quantum mechanics, the invariance of physical laws under an associated transformation often generates a conservation law or selection rule that follows from the invariance of the Hamiltonian under the transformation. Three symmetry operations corresponding to distinct 'inversions' are especially fundamental, namely parity, time reversal and charge conjugation (Berestetskii et al. 1982).

Parity, represented by the operator $P$, inverts the coordinates of all the particles in a system through the coordinate origin. This is equivalent to a reflection of the physical system in any plane containing the coordinate origin followed by a rotation through 180° about an axis perpendicular to the reflection plane. If replacing the space coordinates $(x, y, z)$ everywhere in equations describing physical laws (e.g., Newton's equations for mechanics or Maxwell's equations for electromagnetism) leaves the equations unchanged, all processes determined by such laws are said to conserve parity.

Time reversal, represented by the operator $T$, reverses the motions of all the particles in a system. If replacing the time coordinate $t$ by $-t$ everywhere leaves equations describing physical laws unchanged, then all processes determined by such laws are said to conserve time reversal invariance, or to have reversality. A process will have reversality as long as the process with all the motions reversed is in principle a possible process, however improbable (from the laws of statistics) it may be. Time reversal is therefore best thought of as motion reversal. It does not mean going backward in time!

Charge conjugation, represented by the operator $C$, interconverts particles and antiparticles. This operation from relativistic quantum field theory has conceptual value in studies of molecular chirality. It appears in the $CPT$ theorem which states that, even if one or more of $C$, $P$, or $T$ is violated, invariance under the combined operation $CPT$ always holds. The $CPT$ theorem has three important consequences for a particle and its antiparticle: their rest masses and lifetimes are identical, and their electromagnetic properties such as charge and magnetic moment are equal in magnitude but opposite in sign.

A *scalar* physical quantity such as energy has magnitude but no directional properties; a *vector* quantity such as linear momentum **p** has magnitude and an associated direction; and a *tensor* quantity such as electric polarizability has magnitudes associated with two or more directions. Scalars, vectors and tensors are classified according to their behaviour under $P$ and $T$. A vector whose sign is reversed by $P$ is called a *polar* or *true* vector; for example a position vector **r**. A vector whose sign is not changed by $P$ is called an *axial* or *pseudo* vector; for example, the angular momentum $\mathbf{L} = \mathbf{r} \times \mathbf{p}$ (since the polar vectors **r** and **p** change sign under $P$, their vector product **L** does not). A vector such as **r** whose sign is not changed by $T$ is called *time even*; a vector such as **p** or **L** whose sign is reversed is called *time odd*.

*Pseudoscalar* quantities have magnitude with no directional properties, but they change sign under space inversion $P$. An example is the natural optical rotation angle.

### 3.2.3 True and False Chirality

There is no disagreement when the term 'chiral' is applied to a static object displaying distinguishable enantiomers under space inversion $P$ (or mirror reflection), like alanine in Fig. 1a. But when the term is applied to less tangible enantiomorphous systems in which motion is an essential ingredient, time reversal arguments are required to clarify the concept. The hallmark of a chiral system is that it can exhibit time-even pseudoscalar observables, which are supported only by quantum states with mixed parity but which are invariant under time reversal. This leads to the following definition (Barron 1986, 2004):

> True chirality is exhibited by systems existing in two distinct enantiomeric states that are interconverted by space inversion, but not by time reversal combined with any proper spatial rotation.

The spatial enantiomorphism shown by truly chiral systems is therefore time invariant, meaning that time reversal cannot undo the effect of space inversion. Spatial enantiomorphism that is time noninvariant has different characteristics, called 'false chirality' to emphasize the distinction.

Consider an electron, which has a spin quantum number $s = \frac{1}{2}$, with $m_s = \pm\frac{1}{2}$ corresponding to the two opposite projections of the spin angular momentum onto a space-fixed axis. A stationary spinning electron is not a chiral object because space inversion $P$ does not generate a distinguishable $P$-enantiomer (Fig. 4a). However, an electron translating with its spin projection parallel or antiparallel to the direction of propagation has true chirality because $P$ interconverts distinguishable left (L) and right (R) spin-polarized versions by reversing the propagation direction but not the spin sense, whereas time reversal $T$ does not because it reverses both (Fig. 4b). Similar considerations apply to a circularly polarized photon except that photons, being massless, have no rest frame and so always move at the velocity of light.

Neither a static uniform electric field **E** (a time-even polar vector) nor a static uniform magnetic field **B** (a time-odd axial vector), or any combination, constitutes a truly chiral system. However, collinear electric and magnetic fields generate spatial enantiomorphism (Curie 1894): parallel and antiparallel arrangements are interconverted by $P$ and are not superposable. But since they are also interconverted by $T$, the enantiomorphism corresponds to false chirality. Another example of false chirality is a nontranslating cone spinning about its symmetry axis, which becomes truly chiral if translating along the axis of spin (Barron 1986, 2004).

In fact the basic requirement for two collinear vectorial influences to generate true chirality is that one transforms as a polar vector and the other as an axial vector, with both either

**Fig. 4** The effect of parity $P$ and time reversal $T$ on the motions of (**a**) a stationary spinning particle and (**b**) a translating spinning particle

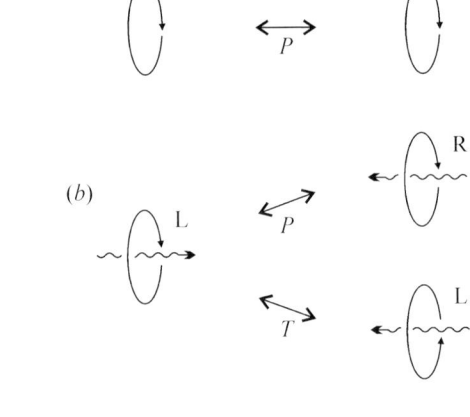

**Fig. 5** The magnetochiral dichroism experiment. The absorption index $n'$ of a medium composed of chiral molecules is slightly different for *unpolarized* light when a static magnetic field is applied parallel (↑↑) and antiparallel (↑↓) to the direction of propagation of the beam

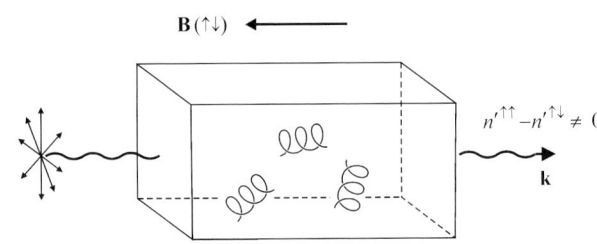

time even or time odd. The second case is exemplified by magnetochiral phenomena (Wagnière and Meir 1982; Barron 2004) where a birefringence and a dichroism may be induced in an isotropic chiral sample by a uniform magnetic field **B** collinear with the propagation vector **k** of a light beam of arbitrary polarization, including unpolarized. The magnetochiral dichroism experiment is illustrated in Fig. 5. Here the parallel and antiparallel arrangements of **B** and **k**, which are interconverted by $P$, are true chiral enantiomers because they cannot be interconverted by $T$ since **B** and **k** are both time odd.

### 3.2.4 Symmetry Violation

Prior to the discovery of parity violation by Lee and Yang in 1956 (Gottfried and Weisskopf 1984), it seemed self-evident that handedness is not built into the laws of nature. If two objects exist as nonsuperposable mirror images, such as the two enantiomers of a chiral molecule, it did not seem reasonable that nature should prefer one over the other. Any difference was thought to be confined to the sign of pseudoscalar observables: the mirror image of any complete experiment involving one enantiomer should be realizable, with any pseudoscalar observable (such as the natural optical rotation angle) changing sign but retaining exactly the same magnitude. Observations of asymmetries in phenomena such as radioactive $\beta$-decay demonstrated that this was not the case for processes involving the weak interactions. It was subsequently realized that symmetry could be recovered by invoking invariance under the combined CP operation in which charge conjugation and space inversion are applied together.

The unification of the theory of the weak and electromagnetic interactions into a single electroweak interaction theory by Weinberg, Salam and Glashow in the 1960s (Gottfried and

Weisskopf 1984), revealed that the absolute parity violation associated with the weak interactions could infiltrate to a tiny extent into all electromagnetic phenomena and hence into
the world of atoms and molecules. This is brought about by a 'weak neutral current' which
generates, inter alia, the following parity-violating electron–nucleus contact interaction term
(in atomic units) in the Hamiltonian of the atom or molecule (Hegstrom et al. 1980):

$$V_{eN}^{PV} = \frac{G\alpha}{4\sqrt{2}} Q_W \{ \boldsymbol{\sigma}_e \cdot \mathbf{p}_e, \rho_N(\mathbf{r}_e) \},$$

where {} denotes an anticommutator, $G$ is the Fermi weak coupling constant, $\alpha$ is the fine
structure constant, $\boldsymbol{\sigma}_e$ and $\mathbf{p}_e$ are the Pauli spin operator and linear momentum operator of
the electron, $\rho_N(\mathbf{r}_e)$ is a normalized nuclear density function and $Q_W$ is an effective weak
charge. Since $\boldsymbol{\sigma}_e$ and $\mathbf{p}_e$ are axial and polar vectors, respectively, and both are time odd,
their scalar product $\boldsymbol{\sigma}_e \cdot \mathbf{p}_e$ and hence $V_{eN}^{PV}$ are time-even pseudoscalars. One manifestation
of parity violation in atomic physics is a tiny natural optical rotation in vapours of free atoms
(Bouchiat and Bouchiat 1997). *CP* invariance means that the equal and opposite sense of
optical rotation would be shown by the corresponding atoms composed of antiparticles.

Chiral molecules support a unique manifestation of parity violation in the form of a lift-
ing of the exact degeneracy of the energy levels of mirror-image enantiomers, known as
the parity-violating energy difference (PVED). Although not yet observed experimentally
using, for example, ultrahigh resolution spectroscopy, this PVED may be calculated (Mac-
Dermott 2002; Quack 2002; Wesendrup et al. 2003). Since, on account of the PVED, the
*P*-enantiomers of a truly chiral object are not exactly degenerate (isoenergetic), they are
not strict enantiomers (because the concept of enantiomers implies the exact opposites). So
where is the strict enantiomer of a chiral object to be found? In the antiworld, of course:
strict enantiomers are interconverted by *CP*! In other words, the molecule with the opposite
absolute configuration but composed of antiparticles should have exactly the same energy
as the original (Barron 2002, 2004), which means that a chiral molecule is associated with
two distinct pairs of strict enantiomers (Fig. 6).

Violation of time reversal was first observed by Christenson et al. in 1964 in decay modes
of the neutral $K$-meson, the $K^0$ (Gottfried and Weisskopf 1984). The effects are very small;
nothing like the parity-violating effects in weak processes, which can sometimes be absolute.
In fact $T$ violation itself was not observed directly: rather, the observations showed *CP*
violation from which $T$ violation was implied from the *CPT* theorem. Direct $T$ violation
was observed in 1998 in the form of slightly different rates, and hence a breakdown in
microscopic reversibility, for the particle to antiparticle process $K^0 \rightarrow K^{0*}$ and the inverse
$K^{0*} \rightarrow K^0$. Since a particle and its antiparticle have the same rest mass if *CPT* invariance
holds, *only the kinetics, but not the thermodynamics, are affected in CP- or T-violating*
*process.* *CPT* invariance may also be used to show that the *CP*-enantiomers of a chiral
molecule which appear in Fig. 6 remain strictly degenerate even in the presence of *CP*
violation (Barron 1994a). Whether or not *CP* violation could have any direct manifestations
in molecular physics is still the subject of debate (Barron 1994a).

The concept that a spinning particle translating along the axis of spin possesses true
chirality exposes a link between chirality and special relativity. Consider a particle with a
right-handed chirality moving away from an observer. If the observer accelerates to a sufficiently high velocity that she starts to catch up with the particle, it will appear to be moving
towards her and so takes on a left-handed chirality. The chirality of the particle vanishes
in its rest frame. Only for massless particles such as photons and neutrinos is the chirality
conserved since they always move at the velocity of light in any reference frame. This relativistic aspect of chirality is a central feature of elementary particle theory, especially in the

**Fig. 6** The two pairs of strict enantiomers (exactly degenerate) of a chiral molecule that are interconverted by *CP*. The structures with atoms marked by *asterisks* are antimolecules built from the antiparticle versions of the constituents of the original molecules

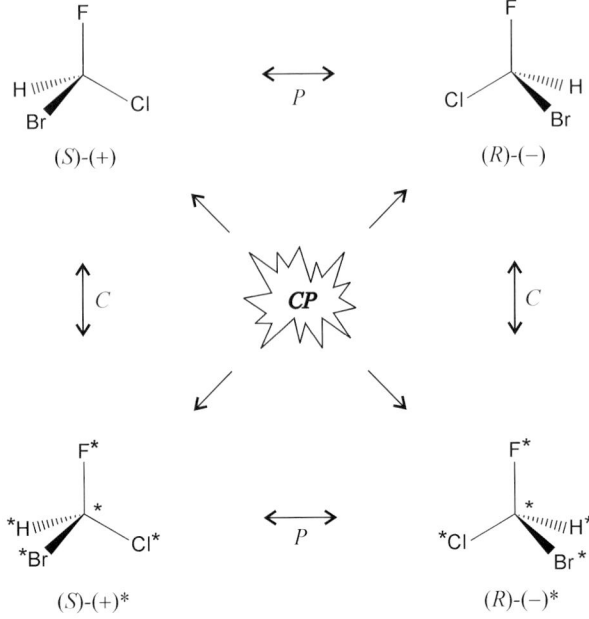

weak interactions where the parity-violating aspects are velocity-dependent (Gottfried and Weisskopf 1984).

### 3.2.5 Symmetry Breaking

The appearance of parity-violating phenomena is interpreted in quantum mechanics by saying that, contrary to what had been previously supposed, the Hamiltonian lacks inversion symmetry due to the presence of pseudoscalar terms such as the weak neutral current interaction. Such symmetry *violation*, sometimes called symmetry nonconservation, must be distinguished from symmetry *breaking* which applies when a system displays a lower symmetry than that of its Hamiltonian (Barron 2004). Natural optical activity, for example, is a phenomenon arising from parity (or mirror symmetry) breaking because a resolved chiral molecule displays a lower symmetry than its associated Hamiltonian: it lacks inversion symmetry (equivalent to mirror symmetry), whereas all the terms in the molecular Hamiltonian (ignoring tiny parity-violating terms) have inversion symmetry. It has been pointed out (Avalos et al. 2004b) that the terms 'chiral symmetry' and 'chiral symmetry breaking', which are widely used to describe the appearance of chirality out of achiral precursors, are inappropriate because chirality is not a symmetry at all in molecular science. Rather, chirality is a special type of reduced spatial symmetry that enables an object to exist in two nonsuperposable mirror-image forms. 'Mirror symmetry breaking' is more correct. In view of its homochirality, life is an example of mirror symmetry breaking on a grand scale!

### 3.3 Physical Chiral Influences

#### 3.3.1 Truly Chiral Influences

The use of an external physical influence to produce an ee in what would otherwise be a racemic product in a chemical reaction is known as an absolute asymmetric synthesis. The production of an ee in more general situations is often referred to as absolute enantioselection. The subject still attracts much interest and controversy (Bonner 1988; Keszthelyi 1995; Avalos et al. 1998; Feringa and van Delden 1999; Compton and Pagni 2002). The considerations of Sect. 3.2 provide a sound foundation for the critical assessment of physical influences that may have provided a small chiral bias during prebiotic chemical evolution.

If an influence is classified as truly chiral it has the correct symmetry characteristics to induce absolute asymmetric synthesis, or some related process such as preferential asymmetric decomposition, in any conceivable situation, although of course the influence might be too weak to produce an observable effect. In this respect it is important to remember the dictum (Jaeger 1930): "The necessary conditions will be that the externally applied forces are a *conditio sine qua non* for the initiation of the reaction which would be impossible without them."

The ability of a truly chiral influence to induce absolute asymmetric synthesis in a reaction process at equilibrium may be illustrated by a simple symmetry argument applied to the following unimolecular process

$$M \quad \underset{k_b}{\overset{k_f}{\rightleftarrows}} \quad R \quad \underset{k_f^*}{\overset{k_b^*}{\rightleftarrows}} \quad M^*$$

in which an achiral molecule R generates a chiral molecule M or its enantiomer $M^*$ and the $k$s are appropriate rate constants. In the absence of a chiral influence, M and $M^*$ have the same energy, so no ee can exist if the reaction reaches thermodynamic equilibrium. Consider a collection of single enantiomers M in the presence of a right-handed chiral influence $(Ch)_R$, say. Under parity $P$, the collection of enantiomers M becomes an equivalent collection of mirror-image enantiomers $M^*$ and the right-handed chiral influence $(Ch)_R$ becomes the equivalent left-handed chiral influence $(Ch)_L$. Assuming parity is conserved, this indicates that the energy of M in the presence of $(Ch)_R$ is equal to that of $M^*$ in the presence of $(Ch)_L$. But because parity (or any other symmetry operation) does not provide a relation between the energy of M and $M^*$ in the presence of the same influence, be it $(Ch)_R$ or $(Ch)_L$, they will in general have different energies. Hence an ee can now exist at equilibrium (due to different Boltzmann populations of M and $M^*$). There will also be kinetic effects because the enantiomeric transition states will also have different energies.

Circularly polarized photons, or longitudinal spin-polarized electrons associated with radioactive $\beta$-decay, are obvious examples of truly chiral influences, and their ability to induce absolute enantioselection has been demonstrated in a number of cases (Bonner 1988; Keszthelyi 1995; Avalos et al. 1998; Feringa and van Delden 1999; Compton and Pagni 2002). Photochemistry with circularly polarized light is especially favourable because it conforms to Jaeger's dictum above. This photochemistry can occur by photoequilibration of a racemic mixture of molecules without loss of reactant, or by selective destruction of one enantiomer over the other. A recent important example for astrobiology was the use of intense circularly polarized synchrotron radiation in the vacuum ultraviolet to induce a significant ee in racemic amino acids in the solid state via enantioselective photodecomposition, which models a realistic situation relevant to organic molecules in interstellar or circumstellar dust grains (Meierhenrich et al. 2005).

Although a magnetic field alone has no chirality and so cannot induce absolute enantioselection, a static magnetic field collinear with a light beam of arbitrary polarization (Fig. 5) is a truly chiral system (Sect. 3.2.3). The ability of a static magnetic field collinear with an unpolarized light beam to induce absolute enantioselection was first demonstrated experimentally by Rikken and Raupach (2000), who generated a small but significant ee at photochemical equilibrium in an initially racemic mixture of an organometallic complex in aqueous solution. This magnetochiral influence can induce absolute enantioselection in all circumstances. Although much weaker than circularly polarized light in this respect, magnetic fields and unpolarized light are more ubiquitous in the cosmos than circularly polarized light. Both of these two truly chiral influences are expected to show an excess of the right- or left-handed variants in local nebular regions, but not when averaged over the entire cosmos (neglecting parity violation which could, for example, generate a tiny cosmic excess of right- or left-circularly polarized light from initially unpolarized light via circular dichroism in atoms and achiral molecules). Which is potentially the more important has yet to be evaluated. The two are not entirely independent, however, in that one source of circular polarization depends on light scattering from interstellar dust grains partially aligned in galactic magnetic fields (Whittet 1992), thought to be responsible for the observation of strong infrared circular polarization in the Orion OMC-1 star-formation region (Bailey et al. 1998).

Being a time-even pseudoscalar, the weak neutral current interaction $V_{eN}^{PV}$ responsible for the PVED is the quintessential truly chiral influence in atomic and molecular physics. It lifts only the degeneracy of the space-inverted ($P$-) enantiomers of a truly chiral system; the $P$-enantiomers of a falsely chiral system such as a non-translating rotating cone remain strictly degenerate. It has attracted considerable discussion as a possible source of biological homochirality (Mason 1988; MacDermott 2002; Quack 2002; Compton and Pagni 2002; Wesendrup et al. 2003). However, it is still not clear whether or not the PVED preferentially stabilizes the naturally occurring L-amino acids and D-sugars. Measurable differences reported in the physical properties of crystals of D- and L-amino acids, and claimed to be due to parity violation, have been shown to arise from traces of different impurities in the enantiomorphous crystals (Lahav et al. 2006). Furthermore, there is no evidence that the PVED has any influence on the distribution of $(+)$ and $(-)$ crystals in spontaneous resolutions (Pagni and Compton 2002).

### 3.3.2 Falsely Chiral Influences

It is important to appreciate that, unlike the case of a truly chiral influence, enantiomers M and M* remain strictly isoenergetic in the presence of a falsely chiral influence such as collinear electric and magnetic fields. Again this can be seen from a simple symmetry argument applied to the earlier unimolecular reaction. Under $P$, the collection of enantiomers M becomes the collection M* and the parallel arrangement, say, of **E** and **B** becomes antiparallel. The antiparallel arrangement of **E** and **B**, however, becomes parallel again under $T$; but these last two operations will have no effect on an isotropic collection of chiral molecules, even if paramagnetic. Hence the energy of the collection M is the same as that of the collection M* in parallel (or antiparallel) electric and magnetic fields.

When considering the possibility or otherwise of absolute enantioselection being induced by a falsely chiral influence, a distinction must be made between reactions that have been left to reach thermodynamic equilibrium (*thermodynamic control*) and reactions that have not attained equilibrium (*kinetic control*). The case of thermodynamic control is quite clear: because M and M* remain strictly isoenergetic in the presence of a falsely chiral influence, such an influence cannot induce absolute asymmetric synthesis in a reaction which has been allowed to reach thermodynamic equilibrium. The case of kinetic control is more subtle. It has been suggested that processes involving chiral molecules in the presence of a falsely chiral influence may exhibit a breakdown of conventional microscopic reversibility, but preserve a new and deeper principal of *enantiomeric* microscopic reversibility (Barron 1987).

Since only the kinetics, but not the thermodynamics, of the process are affected, this suggests an analogy with the breakdown in microscopic reversibility associated with *CP*- or *T*-violation in particle–antiparticle processes (Barron 1987, 1994a, 1994b, 2002).

Since one effect of **E** in a falsely chiral influence such as collinear **E** and **B** is to partially align dipolar molecules (Barron 1987), it is not required if the molecules are already aligned. Hence a magnetic field alone might induce absolute enantioselection if the molecules are pre-aligned, as in a crystal or on a surface, and the process is far from equilibrium (Barron 1994b). This may be significant for the generation of prebiotic ees. However, to date there has been no unequivocal demonstration of absolute enantioselection induced by this or any other falsely chiral influence.

## 4  Detection of Extraterrestrial Chirality

Chiroptical methods such as UV-visible optical rotation and circular dichroism could be used to detect ees of chiral molecules in situ on planets, comets, etc. using miniaturized instruments on spacecraft (MacDermott 2002). However, any signals from, for example, a soil sample on a planetary surface will be tiny and susceptible to artifacts. More attractive are chromatographic methods such as chiral microfabricated capillary electrophoresis which can identify, separate into enantiomers, and estimate ees, of chiral samples such as amino acids (Skelley and Mathies 2003). One complication is that chiral biomolecules such as amino acids are susceptible to racemization. However, racemization would be greatly retarded under the dry frozen conditions on the surface of Mars, for instance, compared with wet conditions (Bada and McDonald 1995). Remote sensing by detection of circular polarization in light reflected from planetary surfaces using instruments on orbiting spacecraft or through telescopes could be indicative of vegetation (Wolstencroft et al. 2002), but any signals will again be very small. A recent study using Earth-based imaging circular polarimetry of the surface of Mars (Sparks et al. 2005) did not find any regions of circular polarization, but because significant chirality of biological origin is not expected on the Martian surface this result is encouraging in that it demonstrated high sensitivity with low artifact levels, and indicates that the method holds promise for other solar system targets and extrasolar planets.

Vibrational Raman spectroscopy is promising for in situ detection and characterization of bio-organic molecules (Popp and Schmitt 2004). Although not currently suitable for in situ studies, a novel chiroptical version called Raman optical activity (ROA) involving measurements with circularly polarized light (Barron 2004; Barron et al. 2007) may be useful for future detection and characterization of chiral molecules of extraterrestrial origin in meteorites, returned samples, etc. ROA can provide the complete three-dimensional structure, including absolute configuration, of small chiral molecules, and can also measure small ees. For large biomolecules, ROA provides information about protein folds, protein and nucleic acid structure of intact viruses, etc. Surface-enhanced or UV resonance versions currently under development may facilitate the application of ROA to trace amounts of material.

## 5  Key Remaining Questions

Apart from the general consensus that homochirality is probably essential for the chemistry of life, the subject of chirality and life in the context of astrobiology is pervaded more by questions than by firm conclusions. In particular: Is a homochiral organic chemistry in an aqueous environment as essential for any molecule-based life on other worlds as it is on

Earth? If so, will it involve analogues of proteins and nucleic acids built from amino acids and sugars, respectively; and will the homochirality be L- for amino acids and D- for sugars as on Earth, suggesting a unique common origin of life (neglecting parity violation)? Is homochirality in the monomeric building blocks necessary for the initiation of a prebiotic homochiral polymer chemistry, or is a small initial ee, possibly generated by a physical chiral influence, sufficient? Might these small initial ees originate in processing of organic material in the interstellar medium? Alternatively, is a chiral template such as a mineral surface always required? Is there any connection between parity violation and biomolecular homochirality?

**Acknowledgements**    I thank Drs. A.J. MacDermott and G.E. Tranter for helpful discussions.

## References

M. Avalos, R. Babiano, P. Cintas, J.L. Jiménez, J.C. Palacios, L.D. Barron, Chem. Rev. **98**, 2391 (1998)

M. Avalos, R. Babiano, P. Cintas, J.L. Jiménez, J.C. Palacios, Orig. Life **34**, 391 (2004a)

M. Avalos, R. Babiano, P. Cintas, J.L. Jiménez, J.C. Palacios, Tet. Asym. **15**, 3171 (2004b)

V.A. Avetisov, V.I. Goldanskii, V.V. Kuz'min, Phys. Today, July, 33 (1991)

J.L. Bada, G.D. McDonald, Icarus **114**, 139 (1995)

J. Bailey, A. Chryosostomou, J.H. Hough, T.M. Gledhill, A. McCall, S. Clark, F. Ménard, M. Tamura, Science **281**, 672 (1998)

P. Ball, Nature **436**, 1084 (2005)

L.D. Barron, J. Am. Chem. Soc. **108**, 5539 (1986)

L.D. Barron, Chem. Phys. Lett. **135**, 1 (1987)

L.D. Barron, Chem. Phys. Lett. **221**, 311 (1994a)

L.D. Barron, Science **266**, 1491 (1994b)

L.D. Barron, Nature **405**, 895 (2000)

L.D. Barron, in *Chirality in Natural and Applied Science*, ed. by W.J. Lough, I.W. Wainer (Blackwell, Oxford, 2002), p. 53

L.D. Barron, Molecular Light Scattering and Optical Activity, 2nd edn (Cambridge University Press, Cambridge, 2004)

L.D. Barron, F. Zhu, L. Hecht, G.E. Tranter, N.W. Isaacs, J. Mol. Struct. **834**, 7 (2007)

J.-P. Behr (ed.) *The Lock and Key Principle* (Wiley, New York, 1994)

V.B. Berestetskii, E.M. Lifshitz, L.P. Pitaevskii, *Quantum Electrodynamics* (Pergamon, Oxford, 1982)

W.A. Bonner, Top. Stereochem. **18**, 1 (1988)

M.A. Bouchiat, C. Bouchiat, Rep. Prog. Phys. **60**, 1351 (1997)

A. Brack, Adv. Space Res. **24**, 417 (1999)

P. Cintas, Ang. Chem. Int. Ed. **41**, 1139 (2002)

R.N. Compton, R.M. Pagni, Adv. At. Mol. Opt. Phys. **48**, 219 (2002)

J.J.L.M. Cornelissen, A.E. Rowan, R.J.M. Nolte, N.A.J.M. Sommerdijk, Chem. Rev. **101**, 4039 (2001)

J.R. Cronin, S. Pizzarello, Science **275**, 951 (1997)

P. Curie, J. Phys. (Paris) **3**, 393 (1894)

B.L. Feringa, R.A. van Delden, Angew. Chem. Int. Ed. **38**, 3418 (1999)

N. Fujii, Orig. Life **32**, 103 (2002)

K. Gottfried, V.F. Weisskopf, *Concepts of Particle Physics*, vol. 1 (Clarendon Press, Oxford, 1984)

M.M. Green, N.C. Peterson, T. Sato, A. Teramoto, R. Cook, S. Lifson, Science **268**, 1860 (1995)

R.M. Hazen, T.R. Filley, G.A. Goodfriend, Proc. Natl. Acad. Sci. USA **98**, 5487 (2001)

R.A. Hegstrom, D.W. Rein, P.G.H. Sandars, J. Chem. Phys. **73**, 2329 (1980)

F.M. Jaeger, *Optical Activity and High-Temperature Measurements* (McGraw-Hill, New York, 1930)

G.F. Joyce, G.M. Visser, C.A.A. van Boeckel, J.H. van Boom, L.E. Orgel, J. van Westresen, Nature **310**, 602 (1984)

Lord Kelvin, *Baltimore Lectures* (C.J. Clay & Sons, London, 1904)

L. Keszthelyi, Quart. Rev. Biophys. **28**, 473 (1995)

M. Klussman, H. Iwamura, S.P. Mathew, D.H. Wells Jr., U. Pandya, A. Armstrong, D.G. Blackmond, Nature **441**, 621 (2006)

D.K. Kondepudi, K. Asakura, Accs. Chem. Res. **34**, 946 (2001)

R. Konno, H. Brückner, A. D'Aniello, G.H. Fisher, N. Fujii, H. Homma (eds.), *D-Amino Acids: A New Frontier in Amino Acid and Protein Research – Practical Methods and Protocols* (Nova Science Publishers, New York, 2007)

M. Lahav, I. Weissbuch, E. Shavit, C. Reiner, G.J. Nicholson, V. Schurig, Orig. Life **36**, 151 (2006)

A.J. MacDermott, in *Chirality in Natural and Applied Science*, ed. by W.J. Lough, I.W. Wainer (Blackwell, Oxford, 2002), p. 23

S.F. Mason, Chem. Soc. Rev. **17**, 347 (1988)

U.J. Meierhenrich, L. Nahon, C. Alcarez, J.H. Bredehöft, S.V. Hoffman, B. Barbier, A. Brack, Ang. Chem. Int. Ed. **44**, 5630 (2005)

C.A. Orme, A. Noy, A. Wierzbicki, M.T. McBride, M. Grantham, H.H. Teng, P.M. Dove, J.J. DeYoreo, Nature **411**, 775 (2001)

R.M. Pagni, R.N. Compton, Cryst. Growth Des. **2**, 249 (2002)

L. Pasteur, Bull. Soc. Chim. France **41**, 219 (1884)

R.H. Perry, C. Wu, M. Nefliu, R.G. Cooks, *Chem. Commun.*, 2007, p. 1071

S. Pizzarello, Accs. Chem. Res. **39**, 231 (2006)

S. Pizzarello, M. Zolensky, K.A. Turk, Geochim. Cosmochim. Acta **67**, 1589 (2003)

J. Popp, M. Schmitt, J. Raman Spectrosc. **35**, 429 (2004)

M. Quack, Ang. Chem. Int. Ed. **41**, 4618 (2002)

G.L.J.A. Rikken, E. Raupach, Nature **405**, 932 (2000)

I. Rubinstein, R. Eliash, G. Bolbach, I. Weissbuch, M. Lahav, Ang. Chem. Int. Ed. **46**, 1 (2007)

A.M. Skelley, R.A. Mathies, J. Chromatogr. **A1021**, 191 (2003)

W.B. Sparks, J.H. Hough, L.E. Bergeron, Astrobiology **5**, 737 (2005)

G.H. Wagnière, A. Meir, Chem. Phys. Lett. **93**, 78 (1982)

R. Wesendrup, J.K. Laerdahl, R.N. Compton, P. Schwerdtfeger, J. Phys. Chem. A **107**, 6668 (2003)

I. Weissbuch, L. Leiserowitz, M. Lahav, Top. Curr. Chem. **259**, 123 (2005)

E. Westhof (ed.), *Water and Biological Macromolecules* (CRC, Boca Raton, 1993)

D.C.B. Whittet, *Dust in the Galactic Environment* (Institute of Physics Publishing, Bristol, 1992)

R.D. Wolstencroft, G.E. Tranter, D.D. Le Pevelen, IAU Symp. **213**, 149 (2002)

# Biosignatures:
## Isotopic Biosignatures

# Multiple-Sulphur Isotope Biosignatures

**Shuhei Ono**

Originally published in the journal Space Science Reviews, Volume 135, Nos 1–4.
DOI: 10.1007/s11214-007-9267-2 © Springer Science+Business Media B.V. 2007

**Abstract** Variations in sulphur isotope ratios have been used as biosignatures in early rock records and Martian meteorites because some microbial sulphur metabolisms are known to produce large magnitude mass-dependent sulphur isotope fractionation. In order to establish the sulphur isotope biosignature, however, it becomes critically important to evaluate abiogenic processes that fractionate sulphur isotope ratios. A brief review is given here for the fundamental systematics and characteristics of multiple-sulfur isotope effects associated with (1) biological, (2) hydrothermal, and (3) photochemical processes. High-precision analysis of all four isotope abundance of sulphur may provide a unique constraint to establish biosignatures in space exploration.

**Keywords** S-33 · S-36 · Multiple-sulphur isotope · Sulfur isotope · Sulphur isotope · Mass-independent fractionation · Archean · Biosignature · Meteorite · Sulfate reduction · Hydrothermal · Photochemical

## 1 Introduction

Sulphur has four stable isotopes, $^{32}S$, $^{33}S$, $^{34}S$ and $^{36}S$ with fractional abundances of approximately 95.04, 0.75, 4.20 and 0.015%, respectively (Ding et al. 2001). Early studies of sulphur isotope geochemistry recognized that microbial sulphur metabolisms, sulphate reduction in particular, produce some of the largest fractionations in sulphur isotope ratios (Harrison and Thode 1958; Kaplan and Rittenberg 1964). Thus, variations in the sulphur isotope ratios have been sought as biosignatures in early rock records (Monster et al. 1979; Ohmoto et al. 1993; Shen et al. 2001) and in Martian meteorites (Greenwood et al. 2000; Shearer et al. 1996). It has also been recognized, however, that hydrothermal (e.g., Ohmoto and Goldhaber 1997) and photochemical (e.g., Zmolek et al. 1999; Farquhar et al. 2001) processes also fractionate sulphur isotope ratios. Therefore, understanding the chemistry

S. Ono (✉)
Department of Earth, Atmospheric and Planetary Sciences, Massachusetts Institute of Technology, 77 Massachusetts Avenue, Cambridge, MA 02139, USA
e-mail: sono@mit.edu

and isotope effects associated with these *abiogenic* processes becomes equally (if not more) critical to establishing sulphur isotope biosignatures. A brief review is given here for the fundamental systematics and characteristics of multiple-sulphur isotope effects associated with (1) biological, (2) hydrothermal, and (3) photochemical processes. Studies for the early Earth rocks and the Martian meteorites are also reviewed as examples of distinguishing biogenic from abiogenic isotope signatures.

## 2 Notation and Background

### 2.1 Notations

Sulphur isotope ratios are typically reported in the delta notation in (1), as deviations with respect to a reference material, which traditionally is the troilite inclusion from the Cañon Diablo meteorite (CDT).

$$\delta^x S = \left( \frac{(^x S/^{32}S)_{\text{sample}}}{(^x S/^{32}S)_{\text{reference}}} - 1 \right) \times 1{,}000 \ (\permil), \tag{1}$$

where $x = 33$, 34 or 36. One can measure and report up to three delta values, $\delta^{33}S$, $\delta^{34}S$ and $\delta^{36}S$. However, the three isotope ratios are correlated for most terrestrial materials (Fig. 1-A) such that

$$\delta^{33}S \approx 0.515 \times \delta^{34}S, \tag{2}$$

and

$$\delta^{36}S \approx 1.90 \times \delta^{34}S. \tag{3}$$

These relationships are so-called mass-dependent isotope fractionation because the relative magnitudes of isotope fractionations are roughly proportional to mass differences (i.e., $^{33}S/^{32}S$ fractionates about half as much $^{34}S/^{32}S$). These relationships are expected from quantum mechanical theory of equilibrium isotope fractionation (Urey 1947; Bigeleisen and Mayer 1947) and will be reviewed in the next section. Because of these mass-dependent relationships, researchers once thought measurements of $\delta^{33}S$ and $\delta^{36}S$ values for terrestrial samples provide no information in addition to $\delta^{34}S$.

Farquhar et al. (2000a), however, reported that many sedimentary sulphide and sulphate minerals older than 2.4 Ga (Giga annum before present) do not follow mass-dependent isotope fractionation law (Fig. 1-B). This discovery surprised many geochemists and fueled interests in measurements of $^{33}S$ and $^{36}S$. The capital delta notations were introduced as measures of deviations from above mass-dependent relationships (i.e., (2) and (3)) (Gao and Thiemens 1991)

$$\Delta^{33}S = \delta^{33}S - 0.515 \times \delta^{34}S, \tag{4}$$

and

$$\Delta^{36}S = \delta^{36}S - 1.90 \times \delta^{34}S. \tag{5}$$

Recent studies use alternative definitions. For $\Delta^{33}S$ values:

$$\Delta^{33}S^* = \delta^{33}S - 1{,}000 \times (\delta^{34}S/1{,}000 + 1)^{0.515} \tag{6}$$

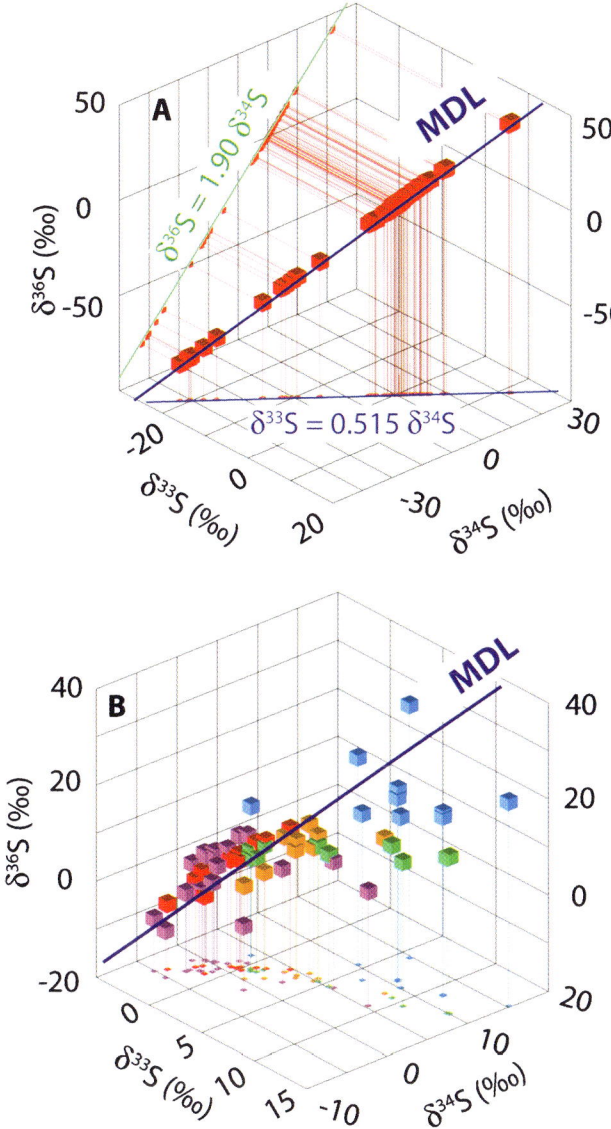

**Fig. 1** Multiple sulphur isotope ratios for post-2.0 Ga terrestrial sulphide and sulphate (**A**) and Archean sulphide (**B**). Data are from Ono et al. (2006b, 2003) and some unpublished data. MDL is the mass-dependent line defined by $\delta^{33}S = 0.515\delta^{34}S$ and $\delta^{36}S = 1.90\delta^{34}S$

(Farquhar et al. 2000a, 2000b, 2000c, 2001), or

$$\Delta^{33}S^* = [\ln(\delta^{33}S/1{,}000 + 1) - 0.515 \times \ln(\delta^{34}S/1{,}000 + 1)] \times 1{,}000 \qquad (7)$$

(Young et al. 2002; Ono et al. 2003, 2006b). Definitions (6) and (7) are often preferred because of logarithmic relationships between mass-dependent isotope fractionations (Hulston and Thode 1965; Young et al. 2002). Definitions (6) and (7) yield practically identical

$\Delta^{33}S*$ values for reasonable ranges of $\delta^{34}S$ ($\sim \pm < 100\%_o$), and are noted here as $\Delta^{33}S*$ to distinguish them from the conventional linear notation in (4).

## 2.2 Origin of the Semi-Classical Mass-Dependent Isotope Effect

The theory of isotope effects was developed nearly six decades ago (Urey 1947; Bigeleisen and Mayer 1947; see also Van Hook 1969; Richet et al. 1977). Mass-dependent law(s) can be derived from this semi-classical isotope fractionation theory (e.g., Hulston and Thode 1965; Matsuhisa et al. 1978; Young et al. 2002; Schauble 2004). The purpose of this section is to briefly review how tightly equilibrium mass-dependent fractionation is constrained among different sulphur molecules and different temperatures within the framework of semi-classical isotope fractionation theory. Mass-dependent laws derived from other types of isotope effects (e.g., diffusion) are reviewed elsewhere (e.g., Young et al. 2002).

Consider an isotope exchange reaction between two diatomic molecules with one sulphur atom,

$$^{32}SO + {}^xSH \leftrightarrow {}^xSO + {}^{32}SH. \tag{8}$$

Equilibrium constant ($K$) can be written as

$$K = \frac{[^xSO][^{32}SH]}{[^{32}SO][^xSH]}. \tag{9}$$

For this example, the isotope fractionation factor ($\alpha$) is

$$^x\alpha_{SO-SH} = \frac{[^xSO]/[^{32}SO]}{[^xSH]/[^{32}SH]}. \tag{10}$$

The isotope fractionation factor at equilibrium can be derived from the ratios of the partition function ($Q$),

$$^x\alpha_{SO-SH} = \frac{{}^xQ_{SO}}{Q_{SO}} \frac{Q_{SH}}{{}^xQ_{SH}}. \tag{11}$$

The partition function ratio is derived as

$$\frac{{}^xQ}{Q} = \frac{\sigma}{{}^x\sigma} \frac{{}^xu}{u} \frac{e^{-{}^xu/2}}{e^{-u/2}} \frac{1 - e^{-u}}{1 - e^{-{}^xu}}, \tag{12}$$

where $\sigma$ is the symmetry number (=1 for this example), $u = hcv/k_BT$, and $h$, $c$ and $k_B$ are Plank's constant, speed of light, and Boltzmann constant, respectively, and $T$ is temperature and $v$ is vibrational frequency of the molecule (Urey 1947; Bigeleisen and Mayer 1947).

When $u$ is small (e.g., high temperature) and $u - {}^xu$ is small (for isotopes except D/H), then (12) can be approximated to (Urey 1947; Bigeleisen and Mayer 1947; Van Hook 1969, equation III-14)

$$\ln \frac{{}^xQ}{Q} \approx -\frac{1}{24}\left({}^xu^2 - u^2\right). \tag{13}$$

If one applies the harmonic oscillator approximation,

$$^xv^2 = \frac{K}{4\pi^2}\left(\frac{1}{{}^xm} - \frac{1}{M}\right), \tag{14}$$

**Table 1** Calculated partition function ratios (in $1{,}000 \ln(^{x}Q/^{32}Q)$) for simple sulphur containing diatomic gas and $H_2S$ and $SO_4^{2-}$

| | $^{33}Q/^{32}Q$ | $^{34}Q/^{32}Q$ | $^{36}Q/^{32}Q$ | $^{33}\theta$ | $^{36}\theta$ |
|---|---|---|---|---|---|
| **0°C** | | | | | |
| HS | 2.80 | 5.44 | 10.28 | 0.5158 | 1.8911 |
| CS | 9.86 | 19.13 | 36.25 | 0.5153 | 1.8946 |
| OS | 10.28 | 19.95 | 37.80 | 0.5152 | 1.8951 |
| $H_2S$ | 6.78 | 13.15 | 24.88 | 0.5158 | 1.8913 |
| $SO_4^{2-}$ | 47.10 | 91.53 | 173.80 | 0.5146 | 1.8989 |
| **1,000°C** | | | | | |
| HS | 0.30 | 0.59 | 1.11 | 0.5159 | 1.8906 |
| CS | 0.68 | 1.33 | 2.51 | 0.5158 | 1.8909 |
| OS | 0.67 | 1.30 | 2.46 | 0.5158 | 1.8909 |
| $H_2S$ | 0.71 | 1.39 | 2.62 | 0.5159 | 1.8907 |
| $SO_4^{2-}$ | 3.00 | 5.82 | 11.02 | 0.5156 | 1.8922 |

where $^{x}m$ and $M$ are atomic weights for $^{x}S$ and a pairing atom, respectively, and $K$ is the force constant that is assumed to be independent of isotope substitutions.

From (13) and (14),

$$^{33}\theta = \frac{\ln(^{33}Q/^{32}Q)}{\ln(^{34}Q/^{32}Q)} = \frac{1/^{33}m - 1/^{32}m}{1/^{34}m - 1/^{32}m}. \tag{15}$$

Here, (15) shows that, within approximation by (13), the relationship between intraelemental partition functions is independent of temperature and vibrational frequency. The $^{33}\theta$ values calculated by (15) is 0.5159 (and $^{36}\theta = 1.890$), when the value is calculated with accurate atomic weights. The approximation is the most accurate for high temperatures (i.e., small $u$), but it deviates at low temperatures and particularly at near crossover temperature where direction of isotope effect switches (Matsuhisa et al. 1978; Deines 2003). Equation (15) leads to the well-known mass-dependent relationship that describes mass-dependent law as a power function (Hulston and Thode 1965; Young et al. 2002),

$$^{33}\alpha = {}^{34}\alpha^{^{33}\theta}. \tag{16}$$

The more elaborated estimation can be applied without using the approximation of (13) and corrected for anharmonicity. The $^{33}\theta$ values of diatomic sulphur molecules (HS, CS and OS) calculated including anharmonicity correction following by Richet et al. (1977) are tabulated in Table 1. Also shown are the partition function ratios for $H_2S$ and $SO_4^{2-}$ calculated by modified Urey–Bradley force field, following a scheme described by Schauble (2004). Farquhar et al. (2003), Farquhar and Wing (2003) and Johnston et al. (2007) also listed $^{33}\theta$ values for a variety of sulphur containing molecules at low temperature. These studies and Table 1 show relatively constant $^{33}\theta$ (from 0.5149 to 0.5159) and $^{36}\theta$ (1.8907 to 1.8989) values for a variety of molecules at 0 and 1,000°C.

## 3  Sulphur Isotope Variations in Terrestrial and Extraterrestrial Materials

### 3.1  Sulphur Isotope Variations in the Solar System

In contrast to other light stable isotopes systems, $^{1,2}$H, $^{12,13}$C, $^{14,15}$N and $^{16,17,18}$O, sulphur isotope ratios of all classes of meteorites exhibit a relatively narrow range compared to terrestrial materials (Figs. 2 and 3). Sulphur in troilite inclusions in iron meteorites, ordinary and enstatite chondrites is isotopically homogeneous with $\delta^{34}$S values of $0 \pm 1.2\permil$ (most of them yield $\pm 0.5\permil$) with respect to CDT (Gao and Thiemens 1991, 1993b; Figs. 2 and 3). Relatively large $\delta^{34}$S values are reported for a series of achondrites ($\pm 1.8\permil$; Rai et al. 2005), and carbonaceous chondrites (from $-2.3$ to $6.1\permil$; Gao and Thiemens 1993a) (Fig. 3).

Materials derived from the Earth's mantle, such as primary sulphide in mid-oceanic and ocean island basalts, yield $\delta^{34}$S values typically within $\pm 1\permil$ with respect to CDT (Sakai et al. 1982, 1984; Kusakabe et al. 1990). Lunar basalts are also $\pm 1\permil$ in $\delta^{34}$S (Thode and Rees 1971; Farquhar and Wing 2005). These homogeneous isotope compositions for meteorites, the Earth and moon rocks suggest sulphur isotope distribution in the solar system is homogeneous, providing a baseline for sulphur isotope ratios of bulk planets in the solar system.

Isotope variations are measured in minor phases in some meteorites. Anomalous $^{33}$S and $^{36}$S compositions are reported from minor sulphur phases of some meteorites. Trace sulphur in the metallic phase of iron meteorites exhibits excess $^{33}$S and $^{36}$S derived from spallation reaction of Fe (Hulston and Thode 1965; Gao and Thiemens 1991). Martian meteorites (SNCs and ALH84001) yield relatively larger variations in $\delta^{34}$S and $\Delta^{33}$S as a result of photochemical and/or hydrothermal processes (Figs. 2 and 3; Farquhar et al. 2000b; Greenwood et al. 2000; Shearer et al. 1996).

### 3.2  Post-Archean Terrestrial Sulphide and Sulphate

Sulphur on present-day Earth shows a variation of over $100\permil$ in $\delta^{34}$S, due to the relatively large isotope effects associated with biological metabolisms using sulphur redox chemistry (sulphate reduction, thiosulphate disproportionation etc., see reviews by Canfield 2001). Sulphate reduction, followed by precipitation of sedimentary sulphide, preferentially removes $^{32}$S from seawater sulphate. This leads to $\delta^{34}$S value of seawater sulphate to be $+21\permil$ with

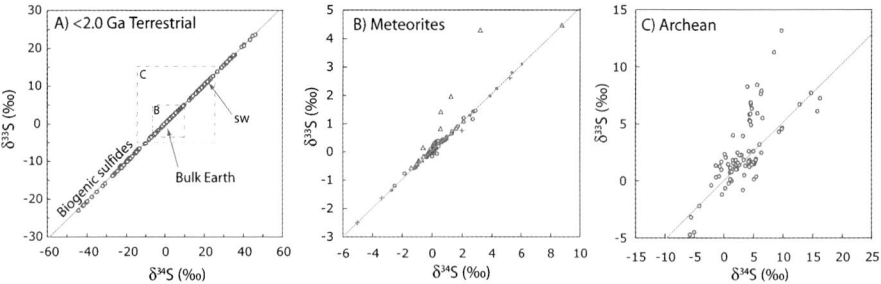

**Fig. 2**  Plots of $\delta^{33}$S and $\delta^{34}$S for (**A**) post-2.0 Ga sulphate and sulphide, (**B**) meteorite sulphur, and (**C**) Archean sulphate and sulphide minerals. *Dashed boxes* in (**A**) show the scales of figures (**B**) and (**C**). SW in (**A**) is the isotope composition of present day seawater sulphate. *Open triangles* and crosses in (**B**) are sulphur extracted from the metallic phase of iron meteorites (Gao and Thiemens 1991) and Martian meteorites (Farquhar et al. 2000a, 2000b, 2000c), respectively. Data sources: (**A**), Johnston et al. (2005a, 2005b) and Ono et al. (2006b); (**B**) Gao and Thiemens (1991, 1993a, 1993b), Farquhar et al. (2000b, 2000c) and Rai et al. (2005); (**C**) Farquhar et al. (2000a) and Ono et al. (2003)

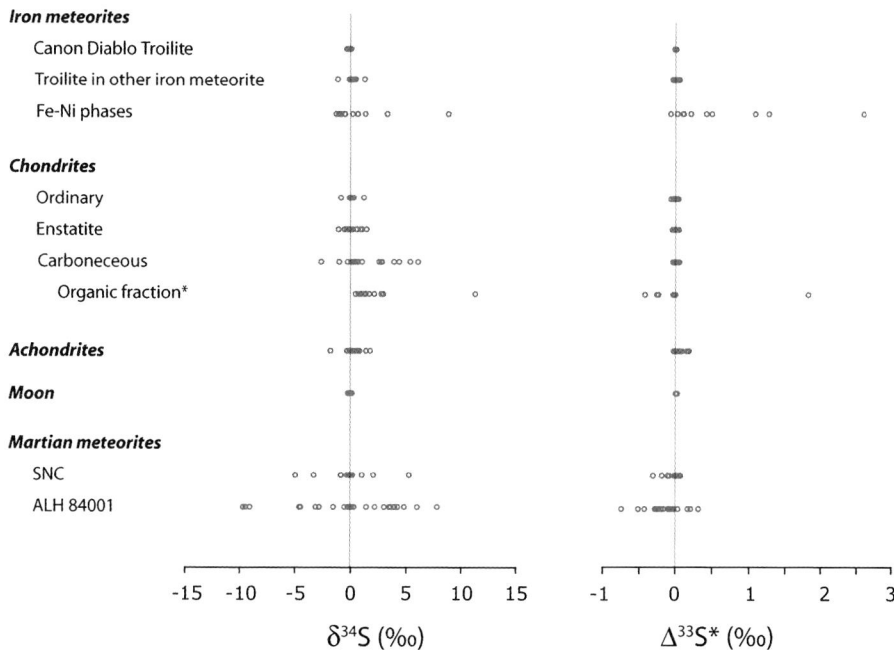

**Fig. 3** The $\delta^{34}$S and $\Delta^{33}$S values for meteorite materials. Data sources: iron meteorite (Gao and Thiemens 1991); Achondrite (Acapulcoite-Lodranite, Ureilite, Aubrite, HED, Bencubbin; Farquhar et al. 2000b; Rai et al. 2005); Chondrite (Gao and Thiemens 1993a, 1993b); Moon (Thode and Rees 1971; Farquhar and Wing 2005); SNC meteorite (Farquhar et al. 2000c); ALH84001 (Shearer et al. 1996; Greenwood et al. 2000), *organic fractions of carbonaceous chondrite (Cooper et al. 1997; Gao and Thiemens 1993a)

respect to the bulk Earth (Holser and Kaplan 1966; Garrels and Lerman 1981; Claypool et al. 1980). Because microbial sulphur isotope fractionation is mass-dependent, the multiple-sulphur isotope composition of terrestrial sulphur follows a linear mass-dependent line (MDL, Figs. 1-A and 2). Recent high-precision studies using $SF_6$ as the mass-spectrometric analyte, however, demonstrate that systematic deviations from MDL do exist for post-2.0 Ga terrestrial sulphur (Johnston et al. 2005a, 2005b; Ono et al. 2006b) (Fig. 4). These small (a few tenths of a permil) $\Delta^{33}$S values can be explained by mass-dependent processes (Farquhar et al. 2003; Johnston et al. 2005a, 2005b; Ono et al. 2006b), and hold new biogeochemical information that cannot be attained by conventional $\delta^{34}$S analyses.

## 3.3 Archean Sulphur MIF and the Early Earth's Atmosphere

An emerging focus of interest on the Archean sulphur isotope record is the occurrence of mass-independent compositions, indicated by nonzero $\Delta^{33}$S values, ranging from $-2.5$ to $+8.2‰$ (Farquhar et al. 2000a, 2000b, 2000c; Mojzsis et al. 2003; Ono et al. 2003; Whitehouse et al. 2005; Ono et al. 2006a). This large-magnitude mass-independent signal disappears by the early Proterozoic, and the specific characteristics of this signal have been thought to carry important implications for the evolution of Earth's atmosphere and sulphur cycle during the Archean and early Proterozoic (Fig. 5).

The sulphur mass-independent fractionation (MIF) signatures in Archean rocks most likely originated from photochemical reactions triggered by solar UV radiation in an atmosphere devoid of $O_2$ and the ozone shield. Under the early Earth's anoxic atmosphere,

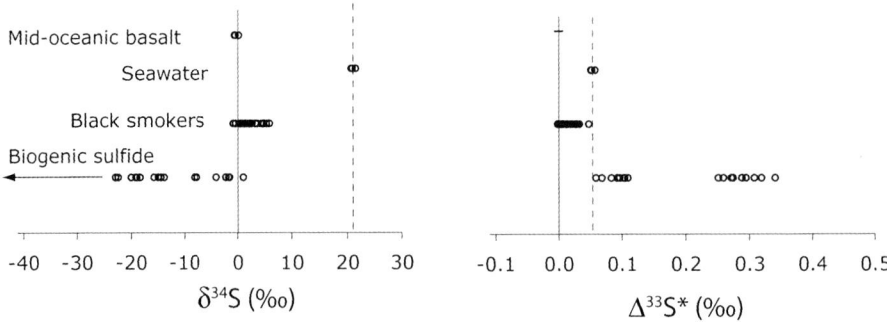

**Fig. 4** The $\delta^{34}$S and $\Delta^{33}$S* values for post-2.0 Ga terrestrial materials. Data sources: Mid-oceanic basalt (Sakai et al. 1982); seawater and biogenic sulphide (Ono et al. 2006b); black smokers (Ono et al. 2007)

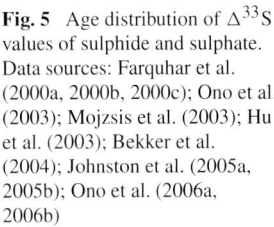

**Fig. 5** Age distribution of $\Delta^{33}$S values of sulphide and sulphate. Data sources: Farquhar et al. (2000a, 2000b, 2000c); Ono et al. (2003); Mojzsis et al. (2003); Hu et al. (2003); Bekker et al. (2004); Johnston et al. (2005a, 2005b); Ono et al. (2006a, 2006b)

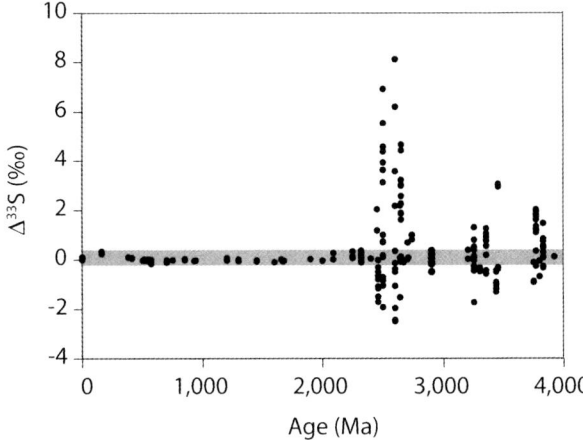

aerosols of $S_8$ and $H_2SO_4$ could have been deposited on the ground, each carrying atmospheric signatures with distinct $\Delta^{33}$S values. This contrasts with today's oxic atmosphere, where atmospheric sulphur exits almost exclusively as $H_2SO_4$ aerosols. The Archean $pO_2$ should have been below 1 ppm to satisfy this atmospheric condition (Pavlov and Kasting 2002). The Archean oceans were also mostly anoxic with aqueous sulphur present as a variety of species (e.g., $SO_4^{2-}$, $S_2O_3^{2-}$, $S_n^{2-}$, $H_2S$). This also contrasts to the present-day oxidized oceans, where sulphur exists mostly as dissolved sulphate (Farquhar et al. 2000a; Ono et al. 2003). The chemistry of sulphur in the atmosphere as well as oceans is critical for the preservation of sulphur MIF signatures from the atmosphere, and oceans, to the sediments.

## 4 Three Domains of Sulphur Isotope Fractionation

### 4.1 Hydrothermal Processes

In high-temperature fluids ($>\sim 300°C$), rapid exchange of sulphur isotopes between sulphate and sulphide lead to equilibrium isotope fractionation,

$$^x SO_4^{2-} + H_2{}^{32}S \leftrightarrow {}^{32}SO_4^{2-} + H_2{}^x S, \tag{17}$$

**Fig. 6** Sulphur isotope fractionation factors between $SO_4$ and $H_2S$ as a function of temperature (bold line). Right $y$ axis shows time required for 90% isotope exchange for fluid with pH $\sim 5$ and $\sum S = 10$ mmol/L based upon the rate law by Ohmoto and Lasaga (1982). The dashed line with arrow represents a model hydrothermal path for the cooling rate $10^{-2}$°C/hour from Ohmoto and Lasaga (1982)

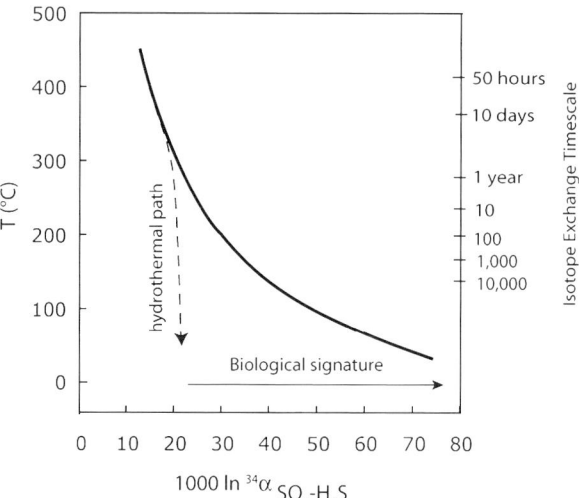

with $SO_4^{2-}$ enriched in $^{34}S$ by 15 and 21‰ at 400 and 300°C, respectively (Fig. 6; Ohmoto and Lasaga 1982). Isotope fractionations among sulphide species (e.g., $H_2S$ and metal sulphides) are relatively minor (typically $<$ a few ‰ at these temperatures) (Ohmoto and Goldhaber 1997). Major controls on sulphur isotope variations in hydrothermal systems, therefore, are (1) mass-balance (i.e., sulphate/sulphide ratio) and (2) the isotope fractionation factor between sulphate and sulphide; (1) is determined by the redox and (2) by the temperature of the fluid.

The sulphate–sulphide pair occurs, for example, at seafloor hydrothermal systems where seawater, carrying sulphate, circulates through high-temperature, rock-buffered (reducing) environments (Shanks 2001). Sulphate is quantitatively reduced when fayalite (or more precisely, the fayalite component of olivine) is present at a temperature above 250°C (Shanks et al. 1981)

$$6Fe_2SiO_4 + SO_4^{2-} + 2H^+ \rightarrow 4Fe_3O_4 + H_2S + 6SiO_2. \tag{18}$$

Typical isotope compositions of vent fluid $H_2S$ and sulphide minerals (2 to 8‰) reflect mixing of seawater-derived sulphur ($+21‰$) and mantle-derived sulphur ($\sim 0‰$) that is leached out from rocks (Shanks 2001; Ono et al. 2007).

Another geologic environment that contains the sulphate–sulphide pair is oxidized magma (e.g., associated with arc volcanism and I-type magma), where disproportionation reactions of magmatic $SO_2$, such as

$$4SO_2 + 4H_2O \rightarrow 3SO_4^{2-} + H_2S + 6H^+, \tag{19}$$

produce the sulphate–sulphide pair. Isotope fractionations (in $\delta^{34}S$) of 20‰ (or more) are common for the sulphate–sulphide pair in oxidized magmatic systems, including epithermal ore deposits (Rye 1993), and in hydrothermal springs and crater lakes (Kusakabe et al. 2000).

It is important to learn how large the sulphur isotope fractionation could be for hydrothermal systems. This, in turn, defines how large the isotope fractionation has to be to become a biosignature. Ohmoto and Lasaga (1982) showed that the kinetics of isotope exchange for reaction (17) and the cooling rate of the fluid determine at what temperature the isotope exchange reaction is *quenched* and thus the magnitude of isotope fractionation that is preserved. At a fluid cooling rate of $10^{-2}$ °C/hour, for example, the quenching temperature is $\sim300$°C, with a fractionation factor of 21‰ (Fig. 6). This temperature depends on the fluid chemistry (pH and total sulphur concentration). Slower cooling rates would also result in lower quenching temperatures and larger fractionation factors (Ohmoto and Goldhaber 1997).

## 4.2 Biological Enzymatic Processes and Kinetic Complexity

Early experimental studies show bacterial sulphate reduction by living cells and cell-extracts produce an isotope fractionation up to 46‰ in $\delta^{34}S$ between substrate sulphate ($SO_4^{2-}$) and product sulphide ($H_2S$) (Kaplan and Rittenberg 1964). Oxidation of $H_2S$ to $S^0$ and $SO_4^{2-}$ as well as disproportionation of intermediate sulphur species ($SO_3^{2-}$, $S^0$ and $S_2O_3^{2-}$), involves isotope effects as large as 20 to 45‰ in $^{34}S/^{32}S$ (Kaplan and Rittenberg 1964; Fry et al. 1985, 1988; Habicht et al. 1998; Canfield 2001). In natural environments, these sulphur metabolisms are important players in anaerobic respiration of organics, leading to precipitation of iron sulphide minerals. Isotope variations in sedimentary sulphide reflect multiple cycles of sulphate reduction, partial-oxidation, and disproportionation, resulting in isotope fractionation of up to 70‰ between seawater sulphate and sedimentary sulphide (Figs. 2 and 4).

The biochemical pathway of microbial sulphate reduction follows several intermediate species and can be described as (Rees 1973; Cypionka 1995; Farquhar et al. 2003; Canfield 2001; Canfield et al. 2006):

$$SO_{4(out)} \overset{\varphi_1,\alpha_1}{\underset{\varphi_2}{\rightleftarrows}} \left[ SO_{4(in)} \overset{\varphi_6}{\underset{\varphi_7}{\rightleftarrows}} APS \overset{\varphi_4,\alpha_4}{\underset{\varphi_5}{\rightleftarrows}} SO_3 \right] \overset{\varphi_3,\alpha_3}{\rightarrow} H_2S, \tag{20}$$

where brackets represent a cell wall, and $SO_{4(out)}$, $SO_{4(in)}$, APS and $SO_3$ are sulphate ions inside and outside of the cell, adenylylsulphate and sulphite ions, respectively. Mass flows among each reservoir are shown as $\varphi$ (in, e.g., moles/sec) and $\alpha$ is the isotope fractionation factor associated with each reaction step where it applies. The first step is transport of $SO_4^{2-}$ inside the cell, which is thought to involve a small isotope effect (Rees 1973). In the second step, $SO_4^{2-}$ is activated by ATP (adenosine 5′-triphosphate) to form APS and PPi (pyrophosphate),

$$SO_4^{2-} + 2H^+ + ATP \leftrightarrow APS + PPi, \tag{21}$$

and APS is reduced to sulphite and AMP (adenosine 5′-monophosphate),

$$APS + H_2 \leftrightarrow HSO_3^- + AMP + H^+ \tag{22}$$

The reactions (21) and (22) are reversible (Cypionka 1995), and step (22) involves isotope fractionation (Rees 1973). The reduction of $SO_3$ to $H_2S$ may proceed as one step,

$$HSO_3^- + 3H_2 + H^+ \rightarrow HSO_3^- + H_2O, \tag{23}$$

or through trithionate ($S_3O_6^{2-}$) and thiosulphate ($S_2O_3^{2-}$) as reaction intermediates (Drake and Akagi 1978). The trithionate and thiosulphate were observed by in vitro experiments (Drake and Akagi 1978) but it is still unresolved whether these are reaction intermediates or reaction by-products (Chambers and Trudinger 1975). Implications of the trithionate pathway for Rees' isotope model were discussed by Brunner and Bernasconi (2005) and Johnston et al. (2007).

**Fig. 7** Multiple-sulphur isotope solution of Rees-Farquhar model of a microbial sulphate reduction flow network (reproduced from Farquhar et al. 2003 and Johnston et al. 2007). The overall isotope fractionation between substrate sulphate and product sulphide is a function of material flow (ratios of forward vs. backward reactions) around reaction branching points, represented as $f_3$ and $f_5$. The model assumes $^{33}\alpha = {}^{34}\alpha^{0.5145}$ and solved for $^{34}\alpha_1$, $^{34}\alpha_3$, and $^{34}\alpha_4$ for 1.003, 0.975 and 0.975, respectively

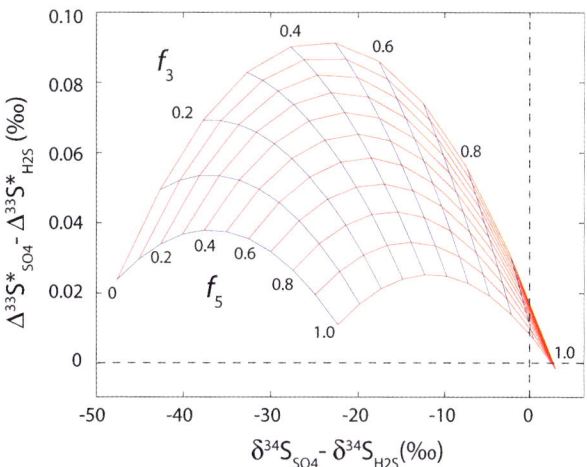

The magnitude of isotope fractionation by microbial sulphate reduction depends upon temperature (Kaplan and Rittenberg 1964; Canfield et al. 2006), the sulphate reduction rate (Ohmoto 1992), sulphate concentration (Harrison and Thode 1958; Habicht et al. 2002), the nature of the carbon source (Kleikemper et al. 2004) and species of microbes (Detmers et al. 2001). One reason for this variability is related to the multistep sulphate reduction process described in scheme (20). In this enzymatic reaction scheme, the overall fractionation factor between product (sulphide) and reactant (sulphate) depends not only the magnitudes of the isotope effects that are *intrinsic* to each enzymatic reaction (i.e., $\alpha_1$, $\alpha_3$, $\alpha_4$) but also the relative rates of mass flow at each branching point (i.e., $\varphi_1$, $\varphi_2$, $\varphi_3$, ...). This behavior of biological isotope effects is called *kinetic complexity* of enzymatic reactions (Northrop 1975; Cleland 2006) or the biological *flow-network* (Hayes 2001). The steady-state solution, assuming constant concentrations for all intermediates, for Rees's sulphate reduction scheme (20) can be written as

$$\frac{H_2{}^x S / H_2{}^{32}S}{{}^x SO_4 / {}^{32}SO_4} = \alpha_\Sigma = \frac{\alpha_1 \alpha_3 \alpha_4}{\alpha_3 \alpha_4 f_3 + (1 - f_3)(\alpha_3 f_5 + 1 - f_5)}, \tag{24}$$

where, $f_3 = \varphi_3 / \varphi_1$, and $f_5 = \varphi_3 / \varphi_4$. As a general rule, intrinsic isotope effects are fully expressed ($\alpha_\Sigma \approx \alpha_1 \alpha_3 \alpha_4$) when sulphide reduction is rate limiting (i.e., $f_3 \approx f_5 \approx 0$). In contrast, only $\alpha_1$ is expressed when quantitative reduction occurs with minimum back flow (i.e., $f_3 \approx 1$).

Equation (24) can be used to trace internal mass flows of sulphur by assigning appropriate intrinsic fractionation factors (Rees 1973; Brunner and Bernasconi 2005; Canfield et al. 2006). Canfield et al. (2006) carried out a series of batch culture experiments to follow $f_3$ and $f_5$ as a function of temperature. Another approach is to use the multiple isotope system in order to reduce one variable, initially suggested for $^{1,2,3}$H system by Northrop (1975). Farquhar et al. (2003) and Johnston et al. (2005a) used $^{33}S/^{32}S$ and $^{34}S/^{32}S$ system simultaneously to solve $f_3$ and $f_5$ as a function of $\delta^{34}S$ and $\Delta^{33}S$ (Fig. 7).

One characteristic feature of Rees-Farquhar's flow network (Farquhar et al. 2003) is that the product $H_2S$ is expected to yield a slightly positive $\Delta^{33}S^*$ value relative to the starting sulphate (i.e., $\Delta^{33}S^*_{H_2S} - \Delta^{33}S^*_{SO_4} > 0$). This is a fundamental property of reaction branching as discussed by Ono et al. (2006b), and is characteristic to any kinetic (irreversible) processes that involve reaction intermediates. This multiple-isotope approach is suggested to be a new tool that can be used to decouple hydrothermal (equilibrium) versus biological (kinetic) processes (Ono et al. 2007; Sect. 6.2).

### 4.3 Atmospheric Processes and Mass-Independent Isotope Effects

Traditional mass-dependent isotope fractionation originates from changes in vibrational energy levels upon isotope substitution (Urey 1947; Bigeleisen and Mayer 1947). Isotope fractionation that does not follow mass-dependent fractionation laws is called mass-independent fractionation.

The best studied mass-independent isotope effect is the ozone recombination reaction

$$O + O_2 + M \leftrightarrow O_3^{\ddagger} + M \rightarrow O_3 + M, \tag{25}$$

where M is a third molecule that absorbs excess kinetic energy. Thiemens and Heidenreich (1983) reported anomalous isotope effects for ozone formed from spark discharge of $O_2$, and Mauersberger et al. (1999), in a series of innovative experiments, obtained isotopomer-specific rate constants for ozone forming reactions. The experiments by Mauersberger et al. (1999) fueled theoretical works applying the RRKM theory (Gao and Marcus 2001) and high level quantum mechanical computations (Babikov et al. 2003a, 2003b).

Although theories are still under development, two factors appear to contribute anomalous isotope effects in ozone formation. One is the zero-point energy effect, where the rate coefficients correlate with the zero-point energy change of the oxygen molecules participating in the isotope exchange reactions (Janssen et al. 2001). This effect contributes an anomalously large (tens of percent) isotope effect in some ozone forming reactions. The origin of the effect is explained at the quantum level by a series of papers by Babikov et al. (2003a, 2003b). In a strict sense, zero-point energy shift of oxygen itself is mass-dependent (twice as energy shift between $^{18}O^{16}O$ vs. $^{16}O^{16}O$ relative to $^{17}O^{16}O$ vs. $^{16}O^{16}O$). The origin of the isotope effects, however, is energy level differences between two entrance channels of ozone forming reactions (Babikov et al. 2003a, 2003b), and is different from conventional frequency shift.

Another effect, the "$\eta$ factor" (Gao and Marcus 2001), originates from the symmetry breaking by the ozone isotope substitution (Thiemens and Heidenreich 1983). When the oxygen atom located at the end position of the ozone molecule is substituted by $^{18}O$ or $^{17}O$, the resulting ozone becomes asymmetric. It is thought that an increase in the energy level density of the metastable state of ozone ($O_3^{\ddagger}$) accompanies end-position isotope substitution may stabilize asymmetric ozone in preference to symmetric ozone (Thiemens and Heidenreich 1983; Gao and Marcus 2001). The resulting isotope effect is expected to be mass-independent because the isotope effect is not due to the difference in mass but is due to difference in the geometry of the molecule.

Much less is known about the physical origin of sulphur mass-independent fractionation. Signatures of sulphur mass-independent isotope fractionations are characteristics for Archean materials (Fig. 5) but small-magnitude sulphur MIF has been measured for sulphate aerosols trapped in polar ice following gigantic volcanic eruptions (Savarino et al. 2003; Baroni et al. 2007) and aerosols collected directly from the atmosphere (Romero and Thiemens 2003). These modern sulphur-MIF originated in the stratosphere and are preserved only when the signatures are not diluted by background tropospheric sulphur. Photolysis of $SO_2$ is thought to be the source reaction for these observed MIF signatures because laboratory $SO_2$ photolysis experiments produce mass-independent isotope signatures in end products ($H_2SO_4$, $S^0$) and left over $SO_2$ (Farquhar et al. 2001). Farquhar et al. (2001) suggested that photolysis itself ($SO_2 \rightarrow SO + O$) is the source of MIF. Other candidates for MIF source reaction include $S_3$ recombination reactions $(S + S_2 \rightarrow S_3)$ that may cause ozone-like symmetry-dependent isotope effects (Francisco et al. 2005). In any case, high-level potential energy calculations (Babikov et al. 2003a, 2003b) together with sophisticated experimental studies (e.g., femtosecond pump-probe spectroscopy) (Knappenberger and Castleman 2004, 2005) will be required to understand the origin of the mass-independent sulphur isotope effect at the quantum level, and to link multiple-sulphur isotope systematics with a specific sulphur photochemical reaction.

## 5  Strategies to Decouple Photochemical Versus Biological Isotope Signatures

### 5.1  The Late Archean (2.7 to 2.5 Ga) Sulphur Isotope Records

Systematic analyses of multiple-sulphur isotope ratios of late Archean basins in Australia and South Africa revealed global (intercontinental) trends reflecting a photochemical signature ($\delta^{33}S \approx 1.6\delta^{34}S$) that is overprinted with local (intrabasin) sulphur cycling by microbial sulphur metabolisms (Ono et al. 2003; Figs. 8 and 9). Late Archean microbial sulphur cycles can be traced in multiple-sulphur isotope ratios as they (1) dilute the atmospheric mass-independent isotope signatures by mixing sulphur reservoirs and (2) produce mass-dependent isotope fractionation. Both multiple-sulphur and carbon isotope systematics correlate with depositional environments (e.g., shallow- to deep-water sedimentary facies) and this can be used to reconstruct Archean microbial ecosystems in their geological context (Beukes et al. 1990; Ono et al. 2003).

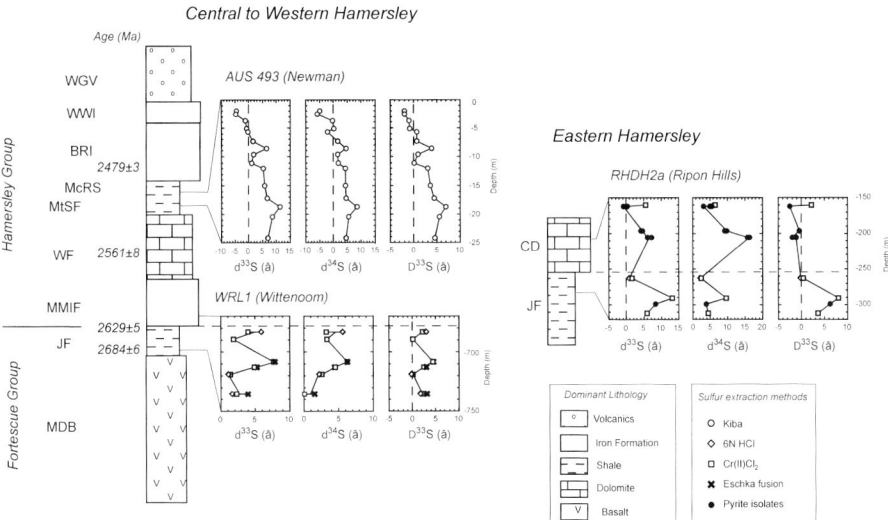

**Fig. 8**  Stratigraphic records of multiple sulphur isotope ratios in shale and carbonate rocks in the late Archean Hamersley Basin, Western Australia (from Ono et al. 2003). WGV = Woongarra Volcanics; WWI = Weeli Wolli Iron Formation; BRI = Brockman Iron Formation; McRS = Mt. McRae Shale; MtSF = Mt Sylvia Formation; WF = Wittenoom Formation; JF = Jeerinah Formation; MDB = Maddina Basalt; and CD = Carawine Dolomite. Reprinted from *Earth and Planetary Science Letters*, 213, Ono et al., New insights into Archean sulfur cycle from mass-independent sulfur isotope records from the Hamersley Basin, Australia. 15–30, with permission from Elsevier

Fig. 9 Multiple sulphur isotope systematics for the late Archean sediments in the Hamersley Basin. (**A**) Atmospheric MIF and mass-dependent fractionation by microbial sulphate reduction (SRB) is shown in *dashed* and *solid arrows*, respectively. *Gray fields* represent isotopic composition of volcanic $SO_2$, $S_8$ and $H_2SO_4$ aerosols, and seawater sulphate (SS). (**B**) Similar to (**A**) but isotope data are plotted (from Ono et al. 2003). Reprinted from *Earth and Planetary Science Letters*, 213, Ono et al., New insights into Archean sulfur cycle from mass-independent sulfur isotope records from the Hamersley Basin, Australia. 15–30, with permission from Elsevier

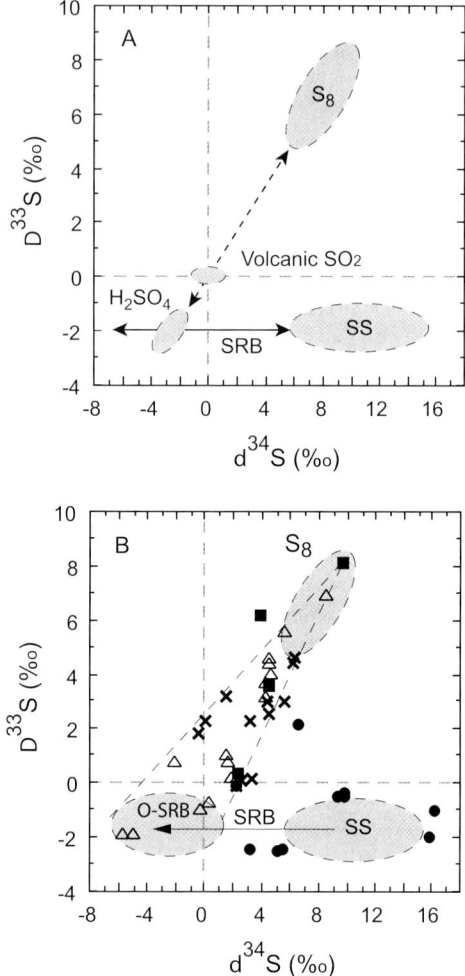

# 6  Strategies to Decouple Hydrothermal Versus Biological Isotope Signatures

## 6.1  Microbial Sulphate Reduction at 3.47 Ga?

The oldest evidence of microbial sulphate reduction was given by Shen et al. (2001) who reported sulphur isotope fractionation as large as 21‰ between pyrite ($FeS_2$) and barite ($BaSO_4$) from the 3.47 Ga-old barite beds in the Pilbara district, Western Australia. These barite–pyrite pairs were later found to be mass-dependently fractionated (i.e., no signatures of photochemical origin of isotope fractionation) and thus, the measured isotope fractionation is likely hydrothermal or biological in origin (Runnegar et al. 2002). A biological origin for the measured isotope fractionation is supported by geologic and mineralogical evidence that suggest the barite was originally gypsum precipitated in an evaporitic environment (Buick and Dunlop 1990), although some question this interpretation (Runnegar et al. 2002). This is one example where geologic context is of critical importance in establishing isotope biosignatures. In particular, the thermal history of the system provides critical infor-

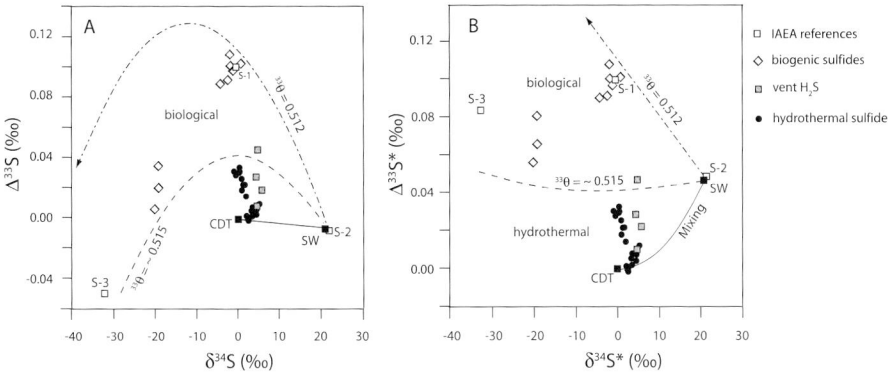

**Fig. 10** Comparison of hydrothermal sulphide and biological sulphide in $\delta^{34}$S vs. $\Delta^{33}$S coordinates (**A**) and in $\delta^{34}$S* vs. $\Delta^{33}$S* coordinates (**B**). *Solid lines* represent mixing between bulk Earth (CDT) and seawater sulphate (SW). *Dashed lines* show isotope compositions for sulphide with $^{33}\theta$ of 0.515 and 0.512. These are representative for hydrothermal equilibrium fractionation and biological fractionation, respectively. Hydrothermal sulphides plot between the mixing line and fractionation line with $^{33}\theta = 0.515$, whereas biogenic pyrites plot between the two fractionation lines ($^{33}\theta = 0.515$ and $^{33}\theta = 0.512$). *Filled circles* = metal sulphide from hydrothermal system; *gray square* = vent H$_2$S from EPR 9–10°N; *open triangle* = biogenic pyrite measured by Ono et al. (2006b); *open square* = international reference materials (IAEA S-1, S-2, S-3 and CDT). From Ono et al. (2007). Reprinted from *Geochimica et Cosmochimica Acta*, 71, Ono et al., S-33 constraints on the seawater sulfate contribution in modern seafloor hydrothermal vent sulfides. 1170–1182, with permission from Elsevier

mation to assess possible hydrothermal overprinting of an earlier low temperature isotope fractionation.

## 6.2 Triple sulphur ($^{32}$S, $^{33}$S, $^{34}$S) Isotope Approach

Microbial sulphate reduction by thermophilic Archea (*Archeaeoglobus fulgidus*) and sulphate reducing bacteria (*Desulfospira jorgensenii* and *Desulfobacterium autotrophicum*) are characterized by $^{33}\theta$ values between 0.510 and 0.513 (Farquhar et al. 2003; Johnston et al. 2005a, 2007) as a result of the kinetic complexity discussed in Sect. 4.2. This slope is analyticaly resolved from that expected for equilibrium (hydrothermal) fractionation (i.e., $\delta^{33}$S $= 0.515\delta^{34}$S) (Farquhar et al. 2003; Johnston et al. 2007).

The $\Delta^{33}$S* values of biogenic sulphide reported by Ono et al. (2006b) are higher than seawater sulphate, and plot between fractionation lines defined by $^{33}\theta = 0.515$ and $^{33}\theta = 0.512$ (Fig. 10). This is consistent with what is expected from the Rees-Farquhar model of biologic sulphate reduction (Fig. 6). The $\Delta^{33}$S values of seafloor hydrothermal sulphides, in contrast, plot below seawater sulphate because they are formed by mixing of seawater-derived and mantle-drived sulphur (Ono et al. 2007) (Fig. 10). This suggests that the multiple-sulphur isotope system can be used as a powerful tool to detect signatures of microbial sulphate reduction even when their $\delta^{34}$S values are inconclusive.

## 7 Application to Search for Life on Mars

Signatures of mass-independent fractionations are observed in Martian meteorites for oxygen isotopes ($^{16,17,18}$O) (Karlsson et al. 1992; Romanek et al. 1998; Farquhar et al. 1998) and sulphur isotopes (Farquhar et al. 2000b; Greenwood et al. 2000). The multiple-sulphur isotope systematics of the late Archean sulphides (Ono et al. 2003) would provide an analogue for the sulphide–sulphate minerals on Mars, with atmospheric photochemistry producing mass-independent fractionation and biology acting to remove the mass-independent signatures but also superimpose mass-dependent fractionation (Farquhar et al. 2001). The same systematics would apply for the multiple-oxygen isotope ratios of sulphate minerals (Jarosite), where biologically mediated process in an aqueous phase dilute oxygen mass-independent signatures by exchanging oxygen between sulphate and water. Therefore, multiple-sulphur isotope ratio, together with oxygen isotope measurements and geologic context, will provide unique and fundamental information about the Martian sulphur cycle.

**Acknowledgements**   This manuscript was improved by comments from James Farquhar and Edwin Schauble. I also thank Penny Morrill, Boz Wing, and Doug Rumble for discussion, and Agouron Institute, Carnegie Institution of Washington, NASA Astrobiology Institute, NSF-OCE (#0622983) for funding support.

# References

D. Babikov, B.K. Kendrick, R.B. Walker, R.T. Pack, P. Fleurat-Lesard, R. Schinke, J. Chem. Phys. **119**, 2577–2589 (2003a)

D. Babikov, B.K. Kendrick, R.B. Walker, R. Schinke, R.T. Pack, Chem. Phys. Lett. **372**, 686–691 (2003b)

M. Baroni, M.H. Thiemens, R.J. Delmas, J. Savarino, Science **315**, 84–87 (2007)

A. Bekker, H.D. Holland, P.L. Wang, D. Rumble, H.J. Stein, J.L. Hannah, L.L. Coetzee, N.J. Beukes, Nature **427**, 117–120 (2004)

N.J. Beukes, C. Klein, A.J. Kaufman, J.M. Hayes, Econ. Geol. Bull. Soc. Econ. Geol. **85**, 633–690 (1990)

J. Bigeleisen, M.G. Mayer, J. Chem. Phys. **15**, 261–267 (1947)

B. Brunner, S.M. Bernasconi, Geochimica Cosmochimica Acta **69**, 4759–4771 (2005)

R. Buick, J.S.R. Dunlop, Sedimentology **37**, 247–277 (1990)

D.E. Canfield, Rev. Mineral. Geochem. **43**, 609–636 (2001)

D.E. Canfield, C.A. Olesen, R.P. Cox, Geochimica Cosmochimica Acta **70**, 548–561 (2006)

L.A. Chambers, P.A. Trudinger, J. Bacteriol. **123**, 36–40 (1975)

G.E. Claypool, W.T. Holser, I.R. Kaplan, H. Sakai, I. Zak, Chem. Geol. **28**, 199–260 (1980)

W.W. Cleland, in *Isotope Effects in Chemistry and Biology*, ed. by A. Kohen, H.-H. Limbach (CRC, Boca Raton, 2006), pp. 915–930

G.W. Cooper, M.H. Thiemens, T.L. Jackson, S. Chang, Science **277**, 1072–1074 (1997)

H. Cypionka, in *Sulfate-Reducing Bacteria*, ed. by L.L. Barton, vol. 8 (Plenum, New York, 1995), pp. 151–184

P. Deines, Chem. Geol. **199**, 179–182 (2003)

J. Detmers, V. Bruchert, K.S. Habicht, J. Kuever, Appl. Environ. Microbiol. **67**, 888–894 (2001)

T. Ding, S. Valkiers, H. Kipphardt, P. de Bievre, P.D.P. Taylor, R. Gonfiantini, R. Krouse, Geochimica Cosmochimica Acta **65**, 2433–2437 (2001)

H.L. Drake, J.M. Akagi, J. Bacteriol. **136**, 916–923 (1978)

J. Farquhar, H. Bao, M. Thiemens, Science **289**, 756–759 (2000a)

J. Farquhar, T.L. Jackson, M.H. Thiemens, Geochimica Cosmochimica Acta **64**, 1819–1825 (2000b)

J. Farquhar, D.T. Johnston, B.A. Wing, K.S. Habicht, D.E. Canfield, S. Airieau, M.H. Thiemens, Geobiology **1**, 27–36 (2003)

J. Farquhar, J. Savarino, S. Airieau, M.H. Thiemens, J. Geophys. Res. **106**, 1–11 (2001)

J. Farquhar, J. Savarino, T.L. Jackson, M.H. Thiemens, Nature **404**, 50–52 (2000c)

J. Farquhar, M.H. Thiemens, T. Jackson, Science **280**, 1580–1582 (1998)

J. Farquhar, B. Wing, Earth Planet. Sci. Lett. **213**, 1–13 (2003)

J. Farquhar, B.A. Wing, Proc. Lunar Planet. Sci. **XXXVI**, 2380 (2005)

J.S. Francisco, J.R. Lyons, I.H. Williams, J. Chem. Phys. **123**, 54302 (2005)

B. Fry, H. Gest, J.M. Hayes, FEMS Microbiol. Lett. **27**, 227–232 (1985)

B. Fry, H. Gest, J.M. Hayes, Appl. Environ. Microbiol. **54**, 250–256 (1988)

X. Gao, M.H. Thiemens, Geochimica Cosmochimica Acta **55**, 2671–2679 (1991)

X. Gao, M.H. Thiemens, Geochimica Cosmochimica Acta **57**, 3159–3169 (1993a)

X. Gao, M.H. Thiemens, Geochimica Cosmochimica Acta **57**, 3171–3176 (1993b)
Y.Q. Gao, R.A. Marcus, Science **293**, 259–263 (2001)
R.M. Garrels, A. Lerman, Proc. Nat. Acad. Sci. USA **78**, 4652–4656 (1981)
J.P. Greenwood, S.J. Mojzsis, C.D. Coath, Earth Planet. Sci. Lett. **184**, 23–35 (2000)
K.S. Habicht, D.E. Canfield, J. Rethmeier, Geochimica Cosmochimica Acta **62**, 2585–2595 (1998)
K.S. Habicht, M. Gade, B. Thamdrup, P. Berg, D.E. Canfield, Science **298**, 2372–2374 (2002)
T.M. Harrison, H.G. Thode, Trans. Faraday Soc. **53**, 84–92 (1958)
J.M. Hayes, Rev. Mineral. Geochem. **43**, 225–277 (2001)
W.T. Holser, I.R. Kaplan, Chem. Geol. **1**, 93–135 (1966)
G.X. Hu, D. Rumble, P.L. Wang, Geochimica Cosmochimica Acta **67**, 3101–3118 (2003)
J.R. Hulston, H.G. Thode, J. Geophys. Res. **70**, 3475–3484 (1965)
C. Janssen, J. Guenther, K. Mauersberger, D. Krankowsky, Phys. Chem. Chem. Phys. **3**, 4718–4721 (2001)
D.T. Johnston, J. Farquhar, B.A. Wing, A.J. Kaufman, D.E. Canfield, K.S. Habicht, Am. J. Sci. **305**, 645–660 (2005a)
D.T. Johnston, B.A. Wing, J. Farquhar, A.J. Kaufman, H. Strauss, T.W. Lyons, L.C. Kah, D.E. Canfield, Science **310**, 1477–1479 (2005b)
D.T. Johnston, J. Farquhar, D.E. Canfield, Geochimica Cosmochimica Acta **71**, 3929–3947 (2007)
I.R. Kaplan, S.C. Rittenberg, J. Gen. Microbiol. **34**, 195–212 (1964)
H.R. Karlsson, R.N. Clayton, E.K. Gibson Jr., T.K. Mayeda, Science **255**, 1409–1411 (1992)
J. Kleikemper, M.H. Schroth, S.M. Bernasconi, B. Brunner, J. Zeyer, Geochimica Cosmochimica Acta **68**, 4891–4904 (2004)
K.L. Knappenberger, A.W. Castleman, J. Chem. Phys. **122**, 154306 (2005)
K.L. Knappenberger Jr., A.W. Castleman Jr., J. Chem. Phys. **121**, 3540–3549 (2004)
M. Kusakabe, S. Mayeda, E. Nakamura, Anonymous, Earth Planet. Sci. Lett. **100**, 275–282 (1990)
M. Kusakabe, Y. Komoda, B. Takano, T. Abiko, J. Volcanol. Geotherm. Res. **97**, 287–307 (2000)
K. Mauersberger, B. Erbacher, D. Krankowsky, J. Gunther, R. Nickel, Science **283**, 370–372 (1999)
Y. Matsuhisa, J.R. Goldscmith, R.N. Clayton, Geochimica Cosmochimica Acta **43**, 1131–1140 (1978)
S.J. Mojzsis, C.D. Coath, J.P. Greenwood, K.D. McKeegan, T.M. Harrison, Geochimica Cosmochimica Acta **67**, 1635–1658 (2003)
J. Monster, P.W.U. Appel, H.G. Thode, M. Schidlowski, C.M. Carmichael, D. Bridgwater, Geochimica Cosmochimica Acta **43**, 405–413 (1979)
D.B. Northrop, Biochemistry **14**, 2644–2651 (1975)
H. Ohmoto, in *Early Organic Evolution: Implications for Mineral and Energy Resources*, ed. by M. Schidlowski (Springer, 1992), pp. 378–397
H. Ohmoto, M.B. Goldhaber, in *Geochemistry of Hydrothermal Ore Deposits*, ed. by H.L. Barnes (Wiley, 1997)
H. Ohmoto, T. Kakegawa, D.R. Lowe, Science **262**, 555–557 (1993)
H. Ohmoto, A.C. Lasaga, Geochimica Cosmochimica Acta **46**, 1727–1745 (1982)
S. Ono, N.J. Beukes, D. Rumble, M.L. Fogel, South Afr. J. Geol. **109**, 97–108 (2006a)
S. Ono, J.L. Eigenbrode, A.A. Pavlov, P. Kharecha, D. Rumble, J.F. Kasting, K.H. Freeman, Earth Planet. Sci. Lett. **213**, 15–30 (2003)
S. Ono, W.C. Shanks, O. Rouxel, D. Rumble, Geochimica Cosmochimica Acta **71**, 1170–1182 (2007)
S. Ono, B.A. Wing, D. Johnston, J. Farquhar, D. Rumble, Geochimica Cosmochimica Acta **70**, 2238–2252 (2006b)
A.A. Pavlov, J.F. Kasting, Astrobiology **2**, 27–41 (2002)
V.K. Rai, T.L. Jackson, M.H. Thiemens, Science **309**, 1062–1065 (2005)
C.E. Rees, Geochimica Cosmochimica Acta **37**, 1141–1162 (1973)
P. Richet, Y. Bottinga, M. Javoy, Annu. Rev. Earth Planet. Sci. **5**, 65–110 (1977)
C.S. Romanek, U.C. Perry, A.H. Treiman, R.A. Socki, J.H. Jones, E.K. Gibson, Meteorit. Planet. Sci. **33**, 775–784 (1998)
A.B. Romero, M.H. Thiemens, J. Geophys. Res. Atmospheres **108**, 4524–4524 (2003)
B. Runnegar, C.D. Coath, J.R. Lyons, K.D. McKeegan, Anonymous, *12th Goldschmidt Conference*, A656, 2002
R.O. Rye, Econ. Geol. Bull. Soc. Econ. Geol. **88**, 733–752 (1993)
H. Sakai, T.J. Casadevall, J.G. Moore, Geochimica Cosmochimica Acta **46**, 729–738 (1982)
H. Sakai, D.J. Des Marais, A. Ueda, J.G. Moore, Geochimica Cosmochimica Acta **48**, 2433–2441 (1984)
J. Savarino, A. Romero, J. Cole-Dai, S. Bekki, M.H. Thiemens, Geophys. Res. Lett. **30**, 2131–2131 (2003)
E.A. Schauble, Rev. Mineral. Geochem. **55**, 65–111 (2004)
W.C. Shanks III, J.L. Bischoff, R.J. Rosenbauer, Geochimica Cosmochimica Acta **45**, 1977–1995 (1981)
W.C. Shanks, Rev. Mineral. Geochem. **43**, 469–525 (2001)
C.K. Shearer, G.D. Layne, J.J. Papike, M.N. Spilde, Geochimica Cosmochimica Acta **60**, 2921–2926 (1996)

Y. Shen, R. Buick, D.E. Canfield, Nature **410**, 77–81 (2001)
M. Thiemens, J.E. Heidenreich, Science **219**, 1073–1075 (1983)
H.G. Thode, C.E. Rees, Earth Planet. Sci. Lett. **12**, 434–438 (1971)
H.C. Urey, J. Chem. Soc., 562–581 (1947)
A.W. Van Hook, in *Isotope Effects in Chemical Reactions*, ed. by C.J. Collins, N.S. Bowman. American Chemical Society Monograph (New York, 1969)
M.J. Whitehouse, B.S. Kamber, C.M. Fedo, A. Lepland, Chem. Geol. **222**, 112–131 (2005)
E.D. Young, A. Galy, H. Nagahara, Geochimica Cosmochimica Acta **66**, 1095–1104 (2002)
P. Zmolek, X. Xu, T.L. Jackson, M. Thiemens, W.C. Trogler, J. Phys. Chem. **103**, 2747–2480 (1999)

# Stable Isotope Ratios as a Biomarker on Mars

**Mark van Zuilen**

Originally published in the journal Space Science Reviews, Volume 135, Nos 1–4.
DOI: 10.1007/s11214-007-9268-1 © Springer Science+Business Media B.V. 2007

**Abstract** As both Earth and Mars have had similar environmental conditions at least for some extended time early in their history (Jakosky and Phillips in Nature 412:237–244, 2001), the intriguing question arises whether life originated and evolved on Mars as it did on Earth (McKay and Stoker in Rev. Geophys. 27:189–214, 1989). Conceivably, early autotrophic life on Mars, like early life on Earth, used irreversible enzymatically enhanced metabolic processes that would have fractionated stable isotopes of the elements C, N, S, and Fe. Several important assumptions are made when such isotope fractionations are used as a biomarker. The purpose of this article is two-fold: (1) to discuss these assumptions for the case of carbon and to summarize new insights in abiologic reactions, and (2) to discuss the use of other stable isotope systems as a potential biomarker. It is concluded that isotopic biomarker studies on Mars will encounter several important obstacles. In the case of carbon isotopes, the most important obstacle is the absence of a contemporary abiologic carbon reservoir (such as carbonate deposits on Earth) to act as isotopic standard. The presence of a contemporary abiologic sulfate reservoir (evaporite deposits) suggests that sulfur isotopes can be used as a potential biomarker for sulfate-reducing bacteria. The best approach for tracing ancient life on Mars will be to combine several biomarker approaches; to search for complexity, and to combine small-scale isotopic variations with chemical, mineralogical, and morphological observations. An example of such a study can be a layer-specific correlation between $\delta^{13}C$ and $\delta^{34}S$ within an ancient Martian evaporite, which morphologically resembles the typical setting of a shallow marine microbial mat.

**Keywords** Carbon isotopes · Sulfur isotopes · Biomarker · Early life · Evolution · Mars

## 1 Introduction

The possibility of early life on Mars is an important focus of current research in astrobiology. Several lines of evidence suggest that there was a warmer climate and liquid water on

M. van Zuilen (✉)
Géobiosphère Actuelle et Primitive, Institut de Physique du Globe de Paris, Paris, France
e-mail: vanzuilen@ipgp.jussieu.fr

the surface during the earliest part of Martian history (the Noachian representing the time interval from 4.5 to 3.5 Ga ago). Indirect evidence for this consists of dendritic networks of valleys and strong erosion features on ancient craters (Golombek and Bridges 2000). Direct evidence consists of ancient evaporitic deposits, that were found during the latest Mars missions (Squyres et al. 2004). The warmer climate necessary to explain these features was likely maintained by a substantial atmosphere of greenhouse gases (Pollack et al. 1987). Near the end of the Noachian, however, the climate became colder and dryer. This follows from morphological evidence; e.g., the initiation of U-shaped valley forms at the downstream end of V-shaped valleys. Such features suggest an evolution from a water-related to ice-related erosional process. This abrupt climate change is the result of a combination of processes, including the decline of volcanic outgassing after the formation of Tharsis, impact-induced ejection of atmospheric gases, and solar-wind stripping of the upper atmosphere (Jakosky and Phillips 2001). In summary, it is likely that the warm climate on early Mars at least persisted to about 3.5 Ga ago, and early forms of life could have evolved in shallow water bodies during this time. It is now reasonably well established that life had originated on Earth by 3.5 Ga ago (Allwood et al. 2006; Rosing 1999; Shen et al. 2001). The very poor and sporadic preservation of the early Archean (ca. 4.0– 3.5 Ga) rock record on Earth obscures any further studies of morphological, chemical, and isotopic evidence of early life forms. The search for life on the surface of Mars is therefore of particular importance for our general understanding of the origin and early evolution of life.

But how do we recognize a potential trace of early life on the surface of Mars? In order to use established biomarkers and tracers, the general assumption needs to be made that biological processes on early Mars would have developed in approximately the same way as they did on early Earth. The most ubiquitous morphological evidence for early life on Earth comes in the form of stromatolite structures (Allwood et al. 2006, and references therein). Stromatolites are laminated, accretionary structures, which are commonly regarded to have formed by the sediment-binding or direct carbonate precipitating activities of microbial mats or biofilms that consist mainly of photosynthesizing bacteria. It has been suggested before that the future search for life on Mars should focus on such morphologic features of stromatolites (Cady et al. 2003; McKay and Stoker 1989). However, it should be emphasized that in many cases such structures are controversial even in the rock record on Earth. Several abiologic processes can mimic the observed morphological features of stromatolites, including evaporitic precipitation, soft-sediment deformation, or silicious sinter formation around hot springs (Lowe 1994). Apparently, simple morphological observation is not enough, and inferences about stromatolite-based life should be supported by careful description of geological context (Allwood et al. 2006) and additional tracers such as microfossils (Schopf 1983; Schopf and Klein 1992), molecular biomarkers (Brocks et al. 1999), and stable isotope ratios (Schidlowski 2001; Shen et al. 2001).

In this paper I will focus specifically on the general interpretation of stable isotope ratios and their potential use as an independent tracer for ancient life. There have been previous assessments of the use of carbon isotope fractionation as a biomarker for the search for life on early Mars (Rothschild and DesMarais 1989; Schidlowski 1992). The purpose of this paper is two-fold: (1) to summarize new insights in abiologic reactions and to expand the discussion on carbon isotopes, and (2) to draw some general conclusions on the use of stable isotopes as a biomarker and briefly discuss the potential for using the sulfur stable isotope ratio as a Martian biomarker.

## 2 Stable Isotope Ratios as a Biomarker

Conceivably early autotrophic life on Mars, like early life on Earth, used irreversible enzymatically enhanced metabolic processes to convert atmospheric $CO_2$ and/or $CH_4$ to organic compounds (e.g., photosynthesis, methanotrophy) or to gain energy from redox reactions (e.g., sulfate-reduction, nitrification–denitrification). Many such metabolic processes cause fractionation of stable isotopes, leading to isotopic "signatures". Stable isotope analysis has been used extensively to trace the history of life on Earth. The carbon isotope ratio of sedimentary organic remains (kerogen) has been used as a tracer for $CO_2$-fixing autotrophic organisms such as photosynthesizers and methanogens (Hayes 2001; Zerkle et al. 2005). The nitrogen isotope ratio of kerogen has been used as a tracer for, e.g., chemoautotrophic bacteria in deep sea hydrothermal vent environments (Pinti et al. 2001). The sulfur isotope ratio of sulfide deposits has been used as a tracer for sulfate-reducing organisms (Canfield and Raiswell 1999; Shen et al. 2001), and recently attempts have been made to use iron isotopes of Fe-oxides and Fe-carbonates to identify iron-oxidizing anoxygenic photosynthesizers and dissimilatory iron-reducing organisms (Johnson et al. 2004). Such isotopic studies are based on the following general assumptions: (1) biologic processes generate a significant and well-defined degree of isotopic fractionation, relative to an abiologic standard; (2) this degree of isotopic fractionation cannot be achieved by a geologically plausible abiologic process; (3) the surface reservoirs (geosphere, hydrosphere, atmosphere, biosphere) of the element in question have been characterized isotopically and have remained constant over geologic time; (4) secondary geologic processes did not blur or destroy the original degree of isotopic fractionation; and (5) the isotopic signature is syngenetic with and indigenous to the rock record. Unfortunately, not all of these assumptions are met when studying the early rock record on Earth. Due to this, many controversies still exist regarding the traces of early life. It remains to be seen to what extent stable isotope systematics can be used to trace possible life on other planets.

## 3 Carbon Isotope Systematics

### 3.1 Assumption 1: Biologic Processes Cause Isotopic Fractionation

Carbon isotope composition is presented as a per mil (‰) enrichment of $^{13}C$ over $^{12}C$ in a sample (delta notation expressed as $\delta^{13}C = ([(^{13}C/^{12}C)_{sample}/(^{13}C/^{12}C)_{ST}] - 1) \times 1,000$) relative to a standard that has the isotopic composition of average marine carbonate (ST = Vienna Pee Dee Belemnite, VPDB). Nier and Gulbransen (1939) were the first to show that plants prefer to concentrate the lighter of the two stable carbon isotopes, leading to biomass with a negative $\delta^{13}C$. Since then many autotrophic pathways (the most important being photosynthesis and methanogenesis) have been recognized that discriminate against $^{13}C$. This discrimination is a result of a kinetic isotope effect inherent to irreversible enzymatic $CO_2$-fixation. The metabolic products of certain $CO_2$-fixing organisms (methanogens, acetogens) are also isotopically fractionated, and can be consumed and further fractionated by other groups of organisms (such as methanotrophs). Detailed discussions of different biological pathways involved in carbon isotope fractionation were given by Hayes (2001) and Zerkle et al. (2005). Figure 1 shows a compilation of the carbon isotope composition of different groups of organisms. Most geologic processes cause much smaller fractionation of carbon isotopes, since the mechanism of fractionation is based on isotope equilibrium. For instance, atmospheric $CO_2$ with an average mantle value of $\delta^{13}C = -5‰$ (Des Marais

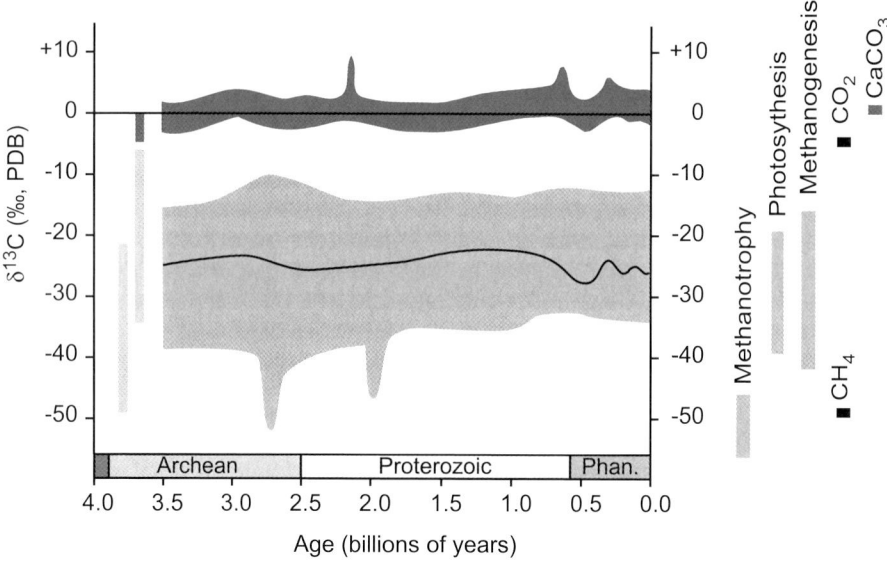

**Fig. 1** Overview of $\delta^{13}$C over geologic time. An average offset of 25‰ exists between the sedimentary organic carbon reservoir and the carbonate carbon reservoir, which can be traced to 3.5 Ga ago. The *light-grey area* represents the total observed scatter in $\delta^{13}$C$_{org}$ over time, the *dark-grey area* represents the scatter in $\delta^{13}$C$_{carb}$ over time. Figure redrawn from Schidlowski (2001)

and Morre 1984) will dissolve in seawater and partition in carbonic acid, bicarbonate, and carbonate. Carbonate precipitating from seawater at 20°C will be slightly enriched in $^{13}$C, due to the isotope equilibrium $\Delta_{carb-CO_2} = +5‰$ at that temperature, leading to an average carbonate $\delta^{13}$C = 0‰. The two main sinks of carbon that are preserved in the sedimentary record are therefore the remains of biologic material (carbonaceous matter) with an average $\delta^{13}$C value of $-25‰$, and carbonate deposits with an average $\delta^{13}$C of 0‰ (relative to carbonate standard VPDB).

### 3.2 Assumption 2: Abiologic Processes Do not Produce Significant Isotope Fractionation

Several geologically plausible processes exist that generate carbonaceous matter (hydrocarbons, polycyclic aromatic hydrocarbons, kerogen, graphite) in an entirely abiologic way; e.g., thermal disproportionation of siderite (McCollom 2003; van Zuilen et al. 2002), reactions between $CO_2$ and $CH_4$ or direct reduction of $CO_2$ in metamorphic fluids (Luque et al. 1998; Naraoka et al. 1996), and especially Fischer-Tropsch (FT-) synthesis during hydrothermal alteration of ultramafic rocks (McCollom and Seewald 2006). This latter process is of particular interest because hydrocarbon production has been observed in modern mid-ocean ridge hydrothermal systems (Holm and Charlou 2001), where many forms of primitive chemoautotrophic life occur. Such environmental settings were particularly abundant on early Earth and form one of the likely sites for the origin and evolution of early life (Nisbet and Sleep 2001). It is therefore crucial to obtain an understanding of abiologic processes that occur in such settings, and especially to determine whether such processes generate isotope fractionation that is of the same order as that of biologic processes. An important controversy is discussed in the following on the possible role of FT-synthesis of isotopically light hydrocarbons in Archean seafloor hydrothermal feeder dikes. Some of the oldest traces of life on

**Box: Abiologic formation of organic compounds:**

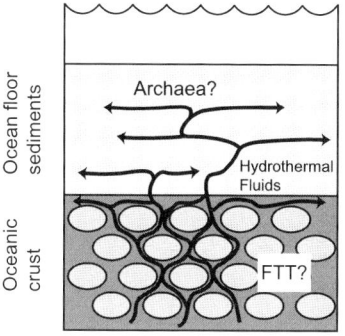

Hydrothermal alteration of ultramafic rocks (e.g. olivine; $Fo_{88}$= 88 Forsterite, 12 Fayalite) leads to serpentinization and production of $H_2$. Under these conditions dissolved $CO_2$ can be reduced to $CH_4$ and hydrocarbons by a Fischer-Tropsch type (FTT) synthesis.

Olivine ($Fo_{88}$) + $H_2O$ = Serpentine + Brucite + Magnetite + $H_2$

$12\ CO_2 + 37\ H_2 = C_{12}H_{26} + 24\ H_2O$

**Fig. 2** (**a**) A typical Archean seafloor hydrothermal setting. (**b**) Reaction schematic of Fischer-Tropsch synthesis occurring in a hydrothermal setting

Earth have been found in 3.5 Ga old chert horizons (predominantly the Dresser Formation at North Pole Dome, the Apex chert, and the Strelley Pool chert) occurring in the lower Warrawoona Group, eastern Pilbara Craton, Western Australia. Carbonaceous microstructures resembling fossilized bacteria were reported from the 3.5 Ga old Apex chert (Schopf 1993; Schopf et al. 2002). At the time this chert was thought to represent a shallow marine depositional setting in which photosynthetic bacteria could thrive. The low $\delta^{13}C$ of these kerogenous structures further suggested that autotrophic life, and in particular photosynthesis was already present early in Earth's history. However, it was subsequently shown by Brasier et al. (2002) that the rock samples studied by Schopf (1993) actually represent a hydrothermal feeder dike that terminates in a bedded chert horizon. It is highly unlikely that photosynthetic bacteria occurred in such a sub-seafloor hydrothermal setting (Fig. 2a). Instead, Brasier et al. (2002) suggested that serpentinization by circulating $CO_2$-rich fluids at depth in the ocean floor basaltic crust could have prompted hydrocarbon formation by Fischer-Tropsch (FT-) type reactions (Fig. 2b). FT-reactions indeed seem to produce a range of hydrocarbons (Foustoukos and Seyfried 2004) with a $\delta^{13}C$ range that is similar to biologic material (Lancet and Anders 1970; McCollom and Seewald 2006) and in the presence of $NH_3$ can produce complex molecules that resemble building blocks of life (Chang et al. 1983; Kung et al. 1979). Lindsay et al. (2005) subsequently studied chert feeder dikes that terminate in the Strelley Pool Chert, and observed low $\delta^{13}C$ carbonaceous clumps and wisps only within a specific depth range below the ocean floor horizon. They suggested that this depth range corresponds to the optimal conditions for FT- reactions ($P$ ca. 500 kbar, $T$ ca. 300°C). Ueno et al. (2004) observed low $\delta^{13}C$ carbonaceous structures in the chert feeder dike swarms of the Dresser Formation and did not rule out FT-reactions as a possible source. However, these authors noted a relatively low hydrothermal temperature (ca. 100–200°C) and the absence of an effective catalyst mineral phase. They therefore suggested that chemolithoautotrophic organisms may actually have been present in these feeder dikes (Fig. 2a).

The notion that hydrothermal processes generated carbonaceous material in cherts of the Warrawoona Group has led to some important reinterpretations of previously recognized microfossil life. Many aspects of hydrothermal systems and associated conditions for chemoautotrophic life remain to be studied. This is also true for FT-synthesis under hydrothermal conditions; it remains to be determined whether significant quantities of kerogen can be formed from aqueous processing of FT-produced hydrocarbons (Marshall et al. 2007). Nevertheless, as Lindsay et al. (2005) suggested, the abiotic organic output of hydrothermal systems may overwhelm the isotopic and morphologic signatures of primitive life that are present, and therefore make it the most difficult environments in which to recognize a record of the early biosphere.

### 3.3 Assumption 3: Carbon Reservoirs Remained Constant Over Time

The carbon isotope ratio of sedimentary organic matter depends on the relative size of reservoirs and fluxes of the exogenic part of the carbon cycle. The details of the exogenic carbon cycle, including the effects of erosion and outgassing, transport, chemical transformation, and sedimentation were discussed in detail by Des Marais (2001). A brief summary is given here of the major implication of relative shifts in the main reservoirs of carbon. The exogenic carbon cycle in its most simplified form consists of an atmospheric $CO_2$ source (primarily volcanic outgassing), and two major sinks; abiologic carbon in the form of carbonate deposits (chemical precipitates) and biologic carbon in the form of organic-rich sediments. The carbon isotope ratios of these reservoirs are depicted here as $\delta^{13}C_{in}$, $\delta^{13}C_{carb}$, and $\delta^{13}C_{org}$, respectively. If all volcanic carbon entering the atmosphere is ultimately removed as carbonate-deposits or organic-rich sediments, the exogenic carbon cycle remains in a steady-state. In this case an isotope mass balance is established between the carbonate- and organic carbon-reservoirs:

$$\delta^{13}C_{in} = F_{carb} \times \delta^{13}C_{carb} + F_{org} \times \delta^{13}C_{org},$$

where $F_{carb}$ and $F_{org}$ are the relative fluxes of carbonate deposition and organic sedimentation. The amount of organic sedimentation represents the combination of biologic productivity, organic decomposition, and burial. It is assumed that the $\delta^{13}C_{in}$ does not change over time and reflects the carbon isotope ratio of the Earth's mantle $CO_2$ ($\delta^{13}C = -5\permil$). The isotopic difference ($\varepsilon^{13}C_{carb-org} = \delta^{13}_{carb} - \delta^{13}C_{org}$) between the carbonate reservoir and sedimentary organic reservoir depends on three factors: (1) the isotopic equilibration between inorganic carbon species $CO_2$, $HCO_3^-$, $CO_3^{2-}$; (2) the metabolic pathways of carbon fixation as described in the previous paragraph; and (3) the effects of transformation of organic carbon by bacterial reworking and diagenesis. It is assumed that these effects remain constant over time, and that the isotopic difference between the two reservoirs is primarily determined by biologic carbon fixation. It can then be shown that the absolute value of $\delta^{13}C_{org}$ depends on the relative amounts of carbonate precipitation ($F_{carb}$) and sedimentation/burial of organic matter ($F_{org}$). This relationship is shown in Fig. 3. The carbonate carbon reservoir on Earth has been shown to be isotopically constant over the large part of the rock record (see Fig. 1, some interesting exceptions exist in the Proterozoic). Compilations of observed organic $\delta^{13}C$ ratios of different ages have shown a consistent average offset from $\delta^{13}C_{carb}$ of approximately $-25\permil$. This isotopic difference $\varepsilon_{carb-org} = 25\permil$ over geologic time suggests that the relative amounts of $F_{carb}$ and $F_{org}$ have not changed much. This leads to the conclusion that, besides the fossil record, the geological carbon isotope record is important uninterrupted evidence for the presence of a substantial biosphere over geological time down to the early Archean (Schidlowski 2001). Any stable isotope study that depends on absolute isotopic ratios thus inherently assumes that fluxes between reservoirs did not change over time. This is an important consideration for future application of stable isotope systematics on Mars.

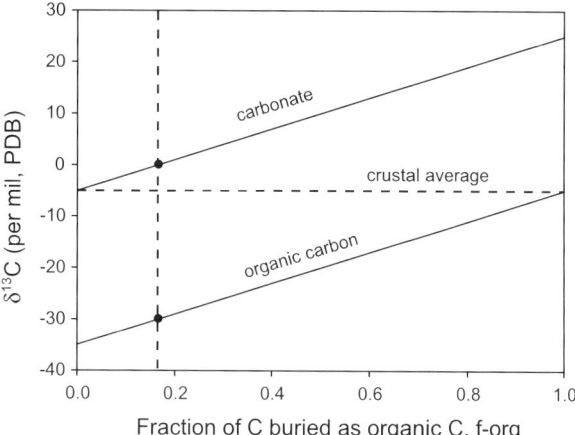

**Fig. 3** Schematic representation of the relative fluxes of organic carbon and carbonate to the sedimentary reservoir. The absolute $\delta^{13}C$ of sedimentary organic carbon depends on the relative fluxes. The isotopic difference between the organic carbon reservoir and the carbonate reservoir, however, remain the same. Schematic after Des Marais (2001)

## 3.4 Assumption 4: Secondary Processes Did not Alter the Original Isotope Ratio

Thermal alteration during diagenesis and prograde metamorphism will change the $\delta^{13}C$ of sedimentary carbonaceous matter. Preferential breaking of the $^{12}C-^{12}C$ bond over the $^{12}C-^{13}C$ bond during thermal cracking of organic matter leads to loss of isotopically light carbon compounds and a $^{13}C$-enriched residue. As carbonaceous matter is altered, it will also lose heteroatoms such as H, N, O, and S. Several studies have shown the inverse correlation between $^{13}C$-enrichment and H/C-ratio (Strauss et al. 1992; Watanabe et al. 1997). In general it can be assumed that Archean carbonaceous material that has experienced lower greenschist metamorphism (ca. 300°C) has obtained a $\delta^{13}C$ value that is enriched relative to the original value by about 3‰. Such material will also have lost most of its hydrogen, and other heteroatoms such as nitrogen. At higher temperatures, above greenschist facies, isotope exchange between carbonates and organic matter occurs, leading to a loss of original isotope signature (Kitchen and Valley 1995). In addition to this simple progressive metamorphism, carbon isotope ratios can be strongly influenced by isotope exchange with fluids during hydrothermal alteration. It has been shown (Robert 1988) that the $\delta^{13}C$ of carbonaceous matter inversely correlates with $\delta^{18}O$ of the surrounding quartz matrix in hydrothermally altered and metamorphosed Archean cherts; the lowest $\delta^{13}C$ values being found in cherts that were least altered by water–rock interaction. This correlation can be attributed to isotopic exchange between carbonaceous matter and dissolved $CO_2$ in hydrothermal fluids. The processes listed here shift the $\delta^{13}C$ of sedimentary biological material to higher values, greatly complicating the recognition of an original biologic signature.

## 3.5 Assumption 5: The Isotopic Signature Is Syngenetic and Indigenous

Evidence for early life can be convincing only if it can be demonstrated that a biomarker actually formed part of the original rock, and was not introduced at a later stage. This is a serious problem for Archean sediments and is particularly important for carbon isotope studies. Many early studies relied on whole-rock isotope analysis. If only a very small amount of reduced carbon is present in the rock, it cannot be determined from the isotope ratio alone whether this represents ancient life, or whether it represents the isotopic signature of extant organisms that live within cracks and pore-space within the sample. For instance, van Zuilen et al. (2002) showed in a stepped-combustion experiment that low $\delta^{13}C$ values in highly metamorphosed banded iron formations from the 3.8 Ga old Isua Supracrustal Belt, southern West Greenland are due to post-metamorphic addition of organic material. Indeed, it was shown by Westall and Folk (2003) that endolithic coccoids occur within cracks of BIFs in the Isua Supracrustal Belt. An important improvement in biomarker and stable isotope studies of early life is the use of drill core samples. Such samples are relatively fresh; they have been protected from weathering processes and are less prone to biologic surface contamination. However, as was noted by Brocks et al. (1999), drill core samples have other potential contamination problems in the form of groundwater that contains biolipids, and petroleum products from drilling activities. Many recent drilling projects for early life studies aim specifically at the characterization and exclusion of such nonindigenous contamination issues.

## 4 Implications for Mars

Despite the problems outlined here, the carbon isotope record throughout Earth history is one of the most important pieces of evidence for the early evolution of life on the planet. The question, as outlined in the introduction, is whether such isotope systematics can be applied to Mars. There are several important differences between Earth and Mars with respect to carbon isotope systematics. The transformation of organic compounds due to diagenesis and metamorphism are much less of a problem on Mars. Geologic activity ended about 3.5 Ga ago, and Mars never had an active cycle of subduction and recycling such as the one on Earth. Therefore it can be assumed that the $\delta^{13}C$ of organic matter would not have changed much since its formation. However, the strong oxidizing conditions on the surface of Mars would have destroyed any organic compounds. Preserved organic matter would therefore be restricted to deep subsurface sedimentary sequences. The serious problems of contamination as encountered on Earth (anthropogenic source as well as extant deep biosphere), are not as important on the surface of Mars. Contamination of organic compounds can only be introduced by the Mars rovers and analytical equipment. Proper sterilization and blank test runs will enable a good control of Earth-based contamination. From a practical point of view, subsurface stable isotope analysis by rover-based analytical tools is an excellent means for the search for ancient life on Mars.

There are two major problems that will prevent straightforward interpretation of $\delta^{13}C$ from organic compounds on Mars. The first important factor is the addition of organic compounds to the Martian surface by meteorite impacts early in its history (first 600–800 Ma). Carbonaceous chondrites contain organic material that is formed by abiologic processes, such as FT-synthesis (Lancet and Anders 1970). The $\delta^{13}C$ of this organic carbon source is in the order of $-15$ to $-20‰$, relative to the carbonate VPDB standard (Sephton et al. 2003). On Mars, where geologic recycling did not occur, such a meteorite-derived reservoir of organic debris would have been incorporated in sediments and be preserved. It will be very difficult to determine whether organics that are found on Mars are the remains of biology or the debris of carbonaceous chondrites.

The second important factor is the exogenic carbon cycle on early Mars. As was pointed out in Sect. 3.3, the absolute carbon isotope ratio of sedimentary organic material depends on the sources, sinks, and relative fluxes of carbon species on the planet. In case of Earth this can be simplified to one source (mantle $CO_2$ that is transferred to the atmosphere), two sinks (sedimentary organic carbon and carbonate deposits), and relatively stable fluxes over geologic time. In the case of Mars, any stable carbon isotope analysis of carbonaceous matter will have to be interpreted with the same assumption; sources and sinks of carbon have

**Fig. 4** Schematic showing the major processes that shaped the Martian atmosphere. Conditions for early life would have been particularly favorable during the middle- to late-Noachian. During this time impact ejection, solar wind stripping (Jakosky and Phillips 2001), and adsorption on Martian dust (Rahn and Eiler 2001) would all have affected the early Martian atmosphere. Some of these processes had a strong influence on the carbon isotope ratio of atmospheric $CO_2$, leading to enrichment in $^{13}C$ over time

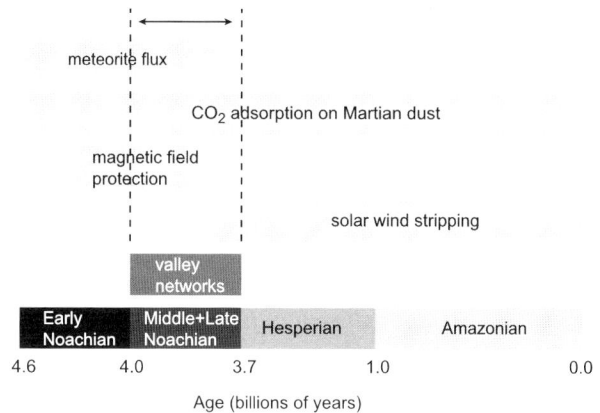

stayed the same throughout geologic history. This is a weak assumption, since it is known that the initial Martian $CO_2$-rich atmosphere underwent several important changes (Fig. 4). As was mentioned in the introduction, the abrupt climate change early in Martian history (first 600–800 Ma) is the result of a decline in volcanic outgassing after the formation of Tharsis, impact-induced ejection of atmospheric gases, and solar-wind stripping of the upper atmosphere (Jakosky and Phillips 2001). Impact-induced ejection (or impact-erosion) did not affect the isotope ratio of the remaining atmospheric $CO_2$, whereas solar-wind induced sputtering of the upper atmosphere caused a strong effect, leading to isotopic enrichment of the remaining atmospheric $CO_2$ (Krasnopolsky et al. 1996). Another process that could have caused isotopic enrichment of atmospheric $CO_2$, is the adsorption of $CO_2$ to Martian dust (Rahn and Eiler 2001). The poorly defined decline in volcanic outgassing and the possible deposition of carbonates during the late- and middle-Noachian further complicate the interpretation of the Martian carbon cycle and the associated isotope fractionations. With this variety of processes it becomes very difficult to determine the $\delta^{13}C$ of atmospheric $CO_2$ during the Noachian, unless there is an uninterrupted representative $\delta^{13}C$ record preserved in carbonate deposits.

In summary, during the Noachian—the time interval in which life may have originated and evolved on Mars—only contemporaneous carbonate deposits can provide a solid base to which $\delta^{13}C$ of organic material should be compared. Although carbonate has been found in SNC-meteorites, at this point no significant carbonate deposits have been found on the surface of Mars.

## 5 The Possible Use of Other Stable Isotopes as a Biomarker on Mars

The crucial aspect of carbon isotope systematics on Mars at this point is the presence or absence of carbonate deposits or any other inorganic carbon phase that can be used as a standard against which possible biologic matter can be compared. This is also the problem for other stable isotope systems. The $\delta^{15}N$ of atmospheric nitrogen-species and the $\delta^{34}S$ of atmospheric sulfur-species on Mars have likely been changed by the same processes that influenced the $\delta^{13}C$ of atmospheric $CO_2$. The use of stable isotope systematics as a tracer for early life therefore depends on abiologic standards that are contemporaneous (formed when the biologic matter or metabolic product formed), and that are representative for the major reservoirs (atmosphere, hydrosphere) from which organisms derived their nutrients.

For sulfur isotopes there is at least one important abiologic sulfur reservoir present on the Martian surface that is accessible for isotope analysis, and that can act as an abiologic reference point for studies of biologic sulfur cycling. Abundant sulfate-salts have been found by the Mars Exploration Rover Opportunity in sedimentary rocks of Meridiani Planum (Squyres et al. 2004). Such sulfate-salts would be representative of the sulfur isotope ratio ($\delta^{34}$S) of dissolved sulfate in ancient standing bodies of water. Dissimilatory sulfate-reducing bacteria preferentially process light over heavy sulfur isotopes, leading to strongly isotopically depleted (negative $\delta^{34}$S) sulfide products such as pyrite (Canfield and Raiswell 1999). The isotopic difference between pyrite and associated sulfate-salts ($\delta^{34}S_{pyr}-\delta^{34}S_{salt}$) could thus act as a trace of ancient life, if certain abiologic processes that fractionate sulfur isotopes (e.g., hydrothermal reduction of sulfate to sulfide) can be ruled out. In some cases such abiologic processes can be excluded on geological grounds, allowing direct interpretation of $\delta^{34}$S values and the recognition of ancient sulfate-metabolism (Canfield and Raiswell 1999; Shen et al. 2001). Recently Philippot et al. (2007) used the isotope ratios of both $^{34}$S/$^{32}$S and $^{33}$S/$^{32}$S in micropyrites from a 3.5 Ga old rock to further discriminate between sulfate-reducing organisms and sulfur-disproportionating organisms. The analysis of multiple isotopes of sulfur ($^{32}$S, $^{33}$S, $^{34}$S, and $^{36}$S) enables the identification of small deviations of $^{33}$S and $^{36}$S from ideal mass-dependent relationships, which further serves to discriminate between biologic and abiologic processes (Ono et al. 2006).

For nitrogen isotopes there is a possible abiologic reservoir present on the Martian surface that can act as a contemporaneous standard. Phyllosilicates of Noachian age have been found on the Martian surface (Chevrier et al. 2007). In certain minerals of this type ammonia ($NH_4^+$) can substitute for potassium ($K^+$) (Papineau et al. 2005). The nitrogen isotope ratio of organic compounds can therefore be compared to an abiologic standard (structurally bound ammonia) in ancient Martian phyllosilicate deposits, and potentially act as a biomarker for certain nitrogen-fixing organisms. The major problem for nitrogen isotopes, however, is the question of whether biologic processes produce an isotopic fractionation that is sufficiently strong and distinguishable from common abiologic processes; e.g., FT-synthesis (Chang et al. 1983; Kung et al. 1979).

For iron isotopes the situation is slightly different from that of C, N, and S, since the geochemical cycling of this element does not involve the Martian atmosphere. The major source of iron would have been igneous Martian crust, and the major sinks iron oxides (predominantly hematite), sulfates (e.g. jarosite), and possibly carbonates (e.g. siderite). It should thus be simple to find a contemporaneous abiologic standard for iron isotope studies. However, as is the case for nitrogen, the application of iron isotopes as a biomarker is hampered by the fact that biologic processes [e.g., oxidation of dissolved Fe(II) by anoxygenic photosynthesizing bacteria, (Croal et al. 2004)] cannot easily be distinguished from abiologic processes [e.g., direct oxidation of dissolved Fe(II) in standing bodies of water (Bullen et al. 2001)].

## 6 Conclusions

This paper provided a short discussion of the assumptions that form the basis for carbon stable isotope studies of early life. For the ancient rock record on Earth, metamorphism and contamination form the crucial obstacles. On Mars it can be predicted that the absence of specific abiologic reservoirs (e.g., carbonate deposits) forms the biggest obstacle for future studies. In this respect it will be worthwhile to look into the use of other stable isotope systems for which there are confirmed ancient abiologic reservoirs. The most obvious candidate is the sulfur isotope ratio of sulfide deposits in sulfate-rich evaporites. The general problem in stable isotope studies is the ever-present possibility that abiologic processes cause an isotopic fractionation that is similar to that of biologic processes. The best approach for tracing ancient life on Mars is therefore to search for complexity, and to combine small-scale isotopic variations with chemical, mineralogical, and morphological observations. An example of such a study can be a layer-specific correlation between $\delta^{13}C$ and $\delta^{34}S$ within an ancient Martian evaporite, which morphologically resembles the typical setting of a shallow marine microbial mat.

# References

A.C. Allwood, M.R. Walter, B.S. Kamber, C.P. Marshall, I.W. Burch, Nature **441**, 714–718 (2006)
M. Brasier, O.R. Green, A.P. Jephcoat, A. Kleppe, M.J. Van Kranendonk, J.F. Lindsay, A. Steele, N.V. Grassineau, Nature **416**, 76–81 (2002)
J.J. Brocks, G.A. Logan, R. Buick, R.E. Summons, Science **285**, 1033–1036 (1999)
T.D. Bullen, A.F. White, C.W. Childs, D.V. Vivit, M.S. Schulz, Geology **29**, 699–702 (2001)
S.L. Cady, J.D. Farmer, J.P. Grotzinger, J.W. Schopf, A. Steele, Astrobiology **3**, 351–368 (2003)
D.E. Canfield, R. Raiswell, Am. J. Sci. **299**, 697–723 (1999)
S. Chang, D.J. Des Marais, R. Mack, S.L. Miller, G.E. Stratbearn, in *Earth's Earliest Biosphere*, ed. by J.W. Schopf (Princeton University Press, Princeton, 1983)
V. Chevrier, F. Poulet, J.-P. Bibring, Nature **448**, 60–63 (2007)
L.R. Croal, C.M. Johnson, B.L. Beard, D.K. Newman, Geochimica Cosmochimica Acta **68**, 1227–1242 (2004)
D.J. Des Marais, in *Stable Isotope Geochemistry*, ed. by J.W. Valley, D.R. Cole (Mineralogical Society of America, Washington, 2001)
D.J. Des Marais, J.G. Morre, Earth Planet. Sci. Lett. **69**, 43–47 (1984)
D.I. Foustoukos, W.E.J. Seyfried, Science **304**, 1002–1005 (2004)
M.P. Golombek, N.T. Bridges, J. Geophys. Res. **105**, 1841–1853 (2000)
J.M. Hayes, in *Stable Isotope Geochemistry*, ed. by J.W. Valley, D.R. Cole (Mineralogical Society of America, Washington, 2001)
N.G. Holm, J.L. Charlou, Earth Planet. Sci. Lett. **191**, 1–8 (2001)
B.M. Jakosky, R.J. Phillips, Nature **412**, 237–244 (2001)
C.M. Johnson, B.L. Beard, E.E. Roden, D.K. Newman, K.H. Nealson, *Geochemistry of Non-traditional Stable Isotopes*. Mineralogical Society of America, 2004
N.E. Kitchen, J.W. Valley, J. Metamorph. Geol. **13**, 577–594 (1995)
V.A. Krasnopolsky, M.J. Mumma, G.L. Bjoraker, D.E. Jennings, Icarus **124**, 553–568 (1996)
C. Kung, R. Hayatsu, M.H. Studier, R.N. Clayton, Earth Planet. Sci. Lett. **46**, 141–146 (1979)
M.S. Lancet, E. Anders, Science **170**, 980–982 (1970)
J.F. Lindsay, M.D. Brasier, N. McLoughlin, O.R. Green, M. Fogel, A. Steele, S.A. Mertzman, Precambr. Res. **143**, 1–22 (2005)
D.R. Lowe, Geology **22**, 387–390 (1994)
F.J. Luque, J.D. Pasteris, B. Wopenka, M. Rodas, J.F. Barrenechea, Am. J. Sci. **298**, 471–498 (1998)
C.P. Marshall, G.D. Love, C.E. Snape, A.C. Hill, A.C. Allwood, M.R. Walter, M.J. Van Kranendonk, S.A. Bowden, S.P. Sylva, R.E. Summons, Precambr. Res. **155**, 1–23 (2007)
T.M. McCollom, Geochimica Cosmochimica Acta **67**, 311–317 (2003)
T.M. McCollom, J.S. Seewald, Earth Planet. Sci. Lett. **243**, 74–84 (2006)
C.P. McKay, C.R. Stoker, Rev. Geophys. **27**, 189–214 (1989)
H. Naraoka, M. Ohtake, S. Maruyama, H. Ohmoto, Chem. Geol. **133**, 251–260 (1996)
A.O. Nier, E.A. Gulbransen, J. Am. Chem. Soc. **61**, 697–698 (1939)
E.G. Nisbet, N.H. Sleep, Nature **409**, 1083–1091 (2001)
S. Ono, B. Wing, D. Johnston, J. Farquhar, D. Rumble, Geochimica Cosmochimica Acta **70**, 2238–2252 (2006)
D. Papineau, S.J. Mojzsis, J.A. Karhu, B. Marty, Chem. Geol. **216**, 37–58 (2005)
P. Philippot, M. van Zuilen, K. Lepot, C. Thomazo, J. Farquhar, M.J. van Krandenok, Science **317**, 1534–1537 (2007)
D.L. Pinti, K. Hashizume, J. Matsuda, Geochimica Cosmochimica Acta **65**, 2301–2315 (2001)
J.B. Pollack, J.F. Kasting, S.M. Richardson, K. Poliakoff, Icarus **71**, 203–224 (1987)

T. Rahn, J.M. Eiler, Geochimica Cosmochimica Acta **65**, 839–846 (2001)

F. Robert, Geochimica Cosmochimica Acta **52**, 1473–1478 (1988)

M.T. Rosing, Science **283**, 674–676 (1999)

L.J. Rothschild, D. DesMarais, Adv. Space Res. **9**, 159–165 (1989)

M. Schidlowski, Adv. Space Res. **12**, 101–110 (1992)

M. Schidlowski, Precambr. Res. **106**, 117–134 (2001)

J.W. Schopf, *Earth's Earliest Biosphere, Its Origin and Evolution* (Princeton University Press, Princeton, 1983)

J.W. Schopf, Science **260**, 640–646 (1993)

J.W. Schopf, C. Klein, *The Proterozoic Biosphere: A Multidisciplinary Study* (Cambridge University Press, Cambridge, 1992)

J.W. Schopf, A. Kudryavtsev, D.G. Agresti, T.J. Wdowiak, A.D. Czaja, Nature **416**, 73–76 (2002)

M.A. Sephton, A.B. Verchovsky, P.A. Bland, I. Gilmour, M.M. Grady, I.P. Wright, Geochimica Cosmochimica Acta **67**, 2093–2108 (2003)

Y. Shen, R. Buick, D.E. Canfield, Nature **410**, 77–81 (2001)

S.W. Squyres, J.P. Grotzinger, R.E. Arvidson, J.F. Bell, 3rd, W. Calvin, P.R. Christensen, B.C. Clark, J.A. Crisp, W.H. Farrand, K.E. Herkenhoff, J.R. Johnson, G. Klingelhofer, A.H. Knoll, H.Y. McSween Jr., R.V. Morris, J.W. Rice Jr., R. Rieder, L.A. Soderblom, Science **306**, 1709–1714 (2004)

H. Strauss, D.J. Des Marais, R.E. Summons, J.M. Hayes, in *The Proterozoic Biosphere*, ed. by J.W. Schopf, C. Klein (Cambridge University Press, New York, 1992)

Y. Ueno, H. Yoshioka, S. Maruyama, Y. Isozaki, Geochimica Cosmochimica Acta **68**, 573–589 (2004)

M.A. van Zuilen, A. Lepland, G. Arrhenius, Nature **418**, 627–630 (2002)

Y. Watanabe, H. Naraoka, D.J. Wronkiewicz, K.C. Condie, H. Ohmoto, Geochimica Cosmochimica Acta **61**, 3441–3459 (1997)

F. Westall, R.L. Folk, Precambr. Res. **126**, 313–330 (2003)

A.L. Zerkle, C.H. House, S.L. Brantley, Amer. J. Sci. **305**, 467–502 (2005)

# In-Situ Instrumentation

# Viking Biology Experiments: Lessons Learned and the Role of Ecology in Future Mars Life-Detection Experiments

**Andrew C. Schuerger · Benton C. Clark**

Originally published in the journal Space Science Reviews, Volume 135, Nos 1–4.
DOI: 10.1007/s11214-007-9194-2 © Springer Science+Business Media B.V. 2007

**Keywords** LR · GeX · PR · GCMS · Soil chemistry · Microbial ecology

## 1 Introduction

The Viking missions to Mars landed in two areas of the northern plains, at Chryse Planitia (22.5° N, 48° W) and Utopia Planitia (48° N, 226° W). Onboard the twin landers were Biology Instruments, containing three separate experiments: the Gas-Exchange (GeX), Pyrolytic Release (PR) and Labeled Release (LR) experiments. In addition, there was a soil analyzer based on X-ray Fluorescence Spectrometry (XRFS) to detect the concentration of elements in samples and, most importantly, a Gas Chromatograph / Mass Spectrometer (GCMS) specifically designed to measure organic compounds in the soil. Together, these instruments were used to assay soils on Mars for biological activity, the presence of organic compounds, and the bulk elemental composition of soils.

## 2 Gas-Exchange (GeX) Experiment

The GeX experiment measured the compositional changes of the atmosphere in the headspace over a humidified and then moistened regolith sample. By measuring dynamic changes of evolved gases, the objective of the GeX was to detect the production of gases that were derived from microbial metabolic activity from hydrated soils. The GeX instrument could measure $H_2$, $N_2$, $O_2$, $CO$, $NO$, $CH_4$, $CO_2$, $N_2O$, and $H_2S$; plus the inert gases

A.C. Schuerger (✉)
Space Life Sciences Lab, Department of Plant Pathology, University of Florida, Kennedy Space Center, FL 32899, USA
e-mail: acschuerger@ifas.ufl.edu

B.C. Clark
Space Exploration Systems, Lockheed Martin, Denver, CO 80201, USA
e-mail: benton.c.clark@lmco.com

Ne, Ar, and Kr (Brown et al. 1978; Oyama and Berdahl 1977). Furthermore, the GeX procedure was conducted in both a "humid" mode and a separate "wet" mode in which water vapors were first allowed to interact with regolith without directly contacting the martian samples, followed by additional injections of nutrients that then wetted the samples. The GeX experiments were conducted at 200 mbar total pressure (composed of 7 mbar Mars atmosphere plus added levels of $CO_2$, He, and Kr) at 8–15°C (Klein et al. 1976; Klein 1977, 1978).

Immediately after humidification of the 1 cc sample of regolith, a large amount of $O_2$, and to a lesser extent $CO_2$, were released; the evolution of these gases stabilized very quickly and reached their maximum concentrations at 2.5 hrs (Oyama and Berdahl 1977). After the samples were fully hydrated with the additions of nutrient media, slow decreases of $CO_2$ and $O_2$ were observed over several sols during the continued wet mode. No further release of $O_2$ or other potentially metabolically derived gases were observed after the soils were wetted with the aqueous nutrients. The release of $O_2$ upon humidification had never been observed before in prelaunch tests with terrestrial or lunar soils (Oyama et al. 1976), and was a surprise for the Viking team (Klein 1978). One unique response was also noted when fresh nutrients were periodically added to the GeX reaction vessel. After each subsequent addition of fresh nutrients, the rate of $CO_2$ uptake always slowed down and became more sluggish (Klein 1978). Oyama et al. (1976) observed this response in prelaunch tests with sterile terrestrial samples and interpreted this as an indication of a dissipative chemical reaction.

The arguments against the biological interpretation of the GeX results are: (a) the release of $O_2$ after humidification was extremely rapid, (b) the addition of aqueous nutrients to the samples during the wet mode resulted in no further liberation of $O_2$, (c) $O_2$ also was released from a sample after being "sterilized" at 145°C, and (d) the reabsorption of $CO_2$ over time following fresh injections of media matched prelaunch tests with sterile terrestrial soils (Klein 1978). The most likely chemical interpretation of the GeX results suggest that the presence of a peroxide or superoxide released $O_2$ upon humidification and that metal oxides, hydroxides, or carbonates in the samples, or created by interaction with water, resulted in a moderately or strongly basic solution which subsequently reabsorbed the $CO_2$ (Klein 1978; Klein et al. 1976).

## 3 Pyrolytic Release (PR) Experiment

The PR, or carbon assimilation, experiment tested the possibility that putative martian microorganisms could take up labeled $^{14}CO_2$ and $^{14}CO$ gases during either light or dark reactions. Martian regolith (0.25 g sample enclosed within a 4 cc incubation chamber) was exposed to the labeled gases, with or without added water vapor, and illuminated by a simulated martian spectrum (335–1000 nm) at a flux of 20% of that predicted for the Viking sites (Horowitz et al. 1977; Klein et al. 1976; Klein 1978). The reactions were allowed to continue for 120 hrs at temperatures between 8 and 26°C and at pressures of approximately 10 mbar. Both light and dark-incubated tests were conducted. At the end of the incubation phases, the chambers were vented at 120°C to separate the residual unreacted $^{14}CO_2$ and $^{14}CO$ gases from potentially fixed organics formed during the light or dark reactions. The pyrolyzed organics volatilized from the soils at 625°C were passed through a gas chromatography (GC) column maintained at 120°C which retained organic fragments larger than $CH_4$, but allowed the unreacted $^{14}CO_2$ and $^{14}CO$ gases to pass. The columns contained Chromosorb P (75%) and CuO (25%). Once the pyrolyzed organics were adhered to the GC column, it was subsequently heated to 650°C in order to release the organic compounds while simultaneously oxidizing them to $CO_2$ by reaction with the CuO.

Results indicated that labeled $^{14}CO_2$ and/or $^{14}CO$ gases were fixed in the martian regolith at very low, but significant, amounts equivalent to approximately 7 pmole of CO or 26 pmole of $CO_2$. The reactions that fixed either $^{14}CO_2$ or $^{14}CO$ were not inhibited after heating the samples to 90°C, but were reduced by 90% when the samples were first "sterilized" at 175°C. In addition, the reactions were not affected by the addition of water vapor to the reaction vessel, but were enhanced in the light. Although these results could be consistent with a biological interpretation, in general there are several arguments (Horowitz et al. 1977; Klein 1978) against such reasoning: (1) pre-heating samples to 90°C had no inhibitory effects on the reaction, and pre-heating to 175°C did not completely abolish it, (2) the presence of $H_2O$ vapor, which would be expected to promote biological reactions, appeared to be either inhibitory to the reactions, or had no effect (Horowitz et al. 1977), and (3) the failure of the GCMS to detect soil organics (Biemann et al. 1976) suggests that organics do not buildup in the regolith, and, thus, seem incompatible with the albeit small PR fixation rates of $^{14}CO_2$ or $^{14}CO$ gases by soil microorganisms.

## 4 Labeled Release (LR) Experiment

The LR procedures for the Viking landers (Levin and Straat 1977, 1979a, 1979b, 1981) measured evolved $^{14}CO_2$ gas given off by $^{14}C$-labeled carbohydrates reacting with constituents of the martian regolith. The widespread importance of the Krebs cycle in aerobic metabolism and the Embden–Meyerhof pathway in anaerobic metabolism makes the use of $^{14}C$-labeled metabolites highly efficient in detecting microbial metabolism because both pathways produce carbon dioxide (Levin et al. 1964). Metabolic activity from a wide range of diverse microorganisms (26 species) can be detected with the LR assay including: *Bacillus subtilis*, *Micrococcus* spp., *Pseudomonas* spp., *Staphylococcus epidermidis*, and *Streptomyces* spp. (Levin 1963). Early results with the LR procedures indicated that 1–2 μCi/ml of $^{14}C$-labeled nutrient solution was ideal for detecting as low as 10–12 viable bacteria per ml within 1–3 hrs (Levin 1963; Levin et al. 1959, 1964). In addition, the evolved $^{14}CO_2$ as measured by the LR assay was linearly correlated to cell density, such that each doubling in cell number per sample resulted in an approximate doubling in the counts-per-minute (cpm) from evolved $^{14}CO_2$ gas (Levin et al. 1959, 1964).

In brief, approximately 0.5 g of Mars regolith was placed within a reaction chamber of the Viking LR experiments, and the system closed, thus, capturing the ambient Martian atmosphere of 95% $CO_2$ and 5% of trace gases at approximately 7 mbar. The Viking LR reaction chamber was equilibrated to 10°C, and a series of helium and nutrient solution injections (115 μl each) were conducted to yield a final total pressure after the first injection of 92 mbar, and after the 2nd injection of 116 mbar (Levin and Straat, 1979a, 1979b). The evolved $^{14}CO_2$ gas was recorded over several sols following each injection of nutrients.

The responses given in Fig. 1 are from VL-1 (Levin and Straat 1977) and show results from two active cycles compared to a heat-sterilization cycle which heated the sample to 160°C for 3 hrs prior to initiating the LR assay. The results gave "classic" biological responses for the two active cycles, and a negative response for the heat-sterilized cycle (cycle #2). These results were accepted as strong evidence for biological activity in martian regolith by Levin and Straat (1977, 1979a, 1979b, 1981). The controversy on this interpretation is based on two key points: (i) following a 2nd injection of the $^{14}C$-labeled nutrient solution, the radioactivity decreased instead of increased as would be expected if the $^{14}CO_2$ gas was derived exclusively from biological activity (Levin and Straat 1977), and (ii) no organics were found in the Mars regolith with the Viking GCMS experiment (Biemann et al. 1976;

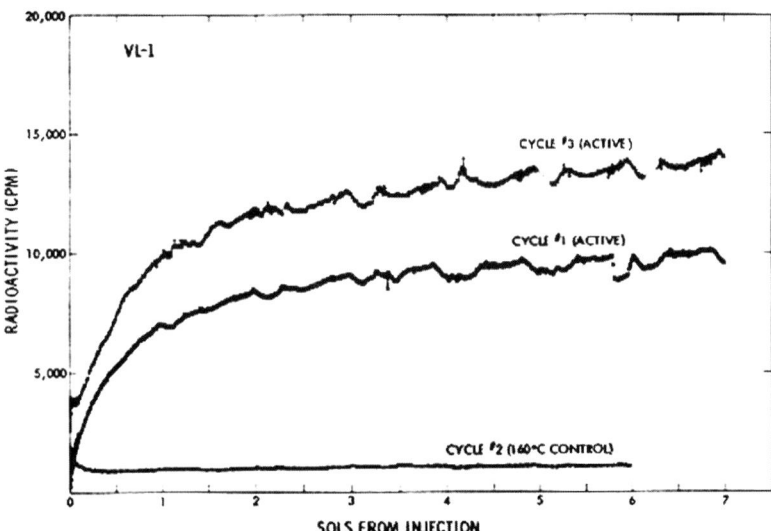

**Fig. 1** Viking Lander 1 (VL-1) LR data. Soil in cycle #2 was heat-sterilized for 3 hrs at 160°C before initiating the LR assay (from Levin and Straat 1977)

Biemann and Lavoie 1979). Although Levin and Straat have responded to these and other criticisms (Levin and Straat 1979a, 1979b, 1981), the biological interpretation of the Viking LR response is considered unverified by most of the exobiological community.

## 5 Gas Chromatography / Mass Spectrometer (GCMS) Experiment

Martian regolith samples were heated in a series of steps up to 500°C in the Viking GCMS experiment to investigate the evolution of volatile compounds derived from in situ organics present in the martian regolith. It was presumed that organics would be present in the martian regolith due to extraterrestrial sources of carbon in cosmic dust, carbonaceous meteorites, and comets (Biemann et al. 1976; and more recently Flynn and McKay 1990; Bland and Smith 2000). However, results from the Viking GCMS experiments indicated that only $H_2O$ and $CO_2$ were detected (Biemann et al. 1976; Biemann and Lavoie 1979), which implied that no organic molecules were present in the martian regolith at the parts per billion level (Biemann et al. 1976). Samples were obtained from the GCMS instruments on Viking down to a depth of no more than 10 cm. These results contributed to the strong arguments made against the biological interpretations of the Viking biology experiments; namely, if organics are not present in the martian surficial soils, then organic-based life must not exist at these test sites. In retrospect, it has been shown by Benner et al. (2000) that the Viking GCMS would not have detected certain refractile compounds such as nonvolatile salts of benzenecarboxylic acids, which might have formed from oxidation of biological materials. It has also been pointed out that the shunt valving used to dump relatively large amounts of $H_2O$ and $CO_2$ released from the soil would also have bypassed low molecular weight organics around the detection systems.

## 6 Lessons Learned from Viking Life-Detection Experiments

The presence of oxidants in the martian soils created an unexpected and unwelcome complication which was not part of the original design of the life-detection investigations. Not only have oxidants on Mars apparently scrubbed out most, or possibly all, of the exogenously delivered organic compounds, but they provided reactants for the organic materials supplied to the Viking GeX and LR experimental modules.

Ubiquitous sulfates in martian soils and the subsequently known presence of $H_2$ gas in the martian atmosphere (Krasnopolsky and Feldman 2001) begs the question of whether sulfate reducing microorganisms might somewhere be present on Mars. Unfortunately, this style of metabolism was not strongly anticipated and none of the Viking life-detection experiments were optimized to detect it. In particular, the incubation chambers were too small to capture sufficient quantities of trace gases, such as $H_2$, to assay for their efficacies as metabolites (Clark 1979). Ironically, the Viking landers had high-pressure $H_2$ gas onboard, but this was used only for the GCMS experiment and was not plumbed to the life-detection modules. Thus, not all possible physiological pathways known from terrestrial microorganisms were included in the Viking experiments. However, such a comprehensive approach to a life-detection assay is a quite daunting task for a power-, volume-, and mass-limited payload like the Viking Biology Instrument. As our knowledge of both Mars and of terrestrial microbiology progresses, we should be able to design more sophisticated payloads to test a greater diversity of microbial physiologies than were tested by Viking, but it will always remain difficult (and perhaps impossible) to test all potential scenarios on any given mission.

Table 1 lists 18 published papers that propose alternative non-biological explanations for the Viking results. Reviewing these papers one might be left with an acute feeling that life-detection experiments on other planetary bodies are fraught with alternative interpretations,

**Table 1** Published non-biological explanations for the Viking results

| | |
|---|---|
| GeX Release of $O_2$ upon humidification: | |
| $KO_2$; $ZnO_2$ | Ponnamperuma et al. 1977 |
| $CaO_2$ | Ballou et al. 1978 |
| $MnO_2$ | Blackburn et al. 1979 |
| $O_2$ trapped in micropores | Nussinov et al. 1978; Plumb et al. 1989 |
| Chemisorbed $H_2O_2$ | Huguenin et al. 1979 |
| O plasma | Ballou et al. 1978 |
| Activated halides | Zent and McKay 1994 |
| LR decomposition of added nutrients: | |
| $H_2O_2$ | Hunten 1979; Oro and Holzer 1979 |
| | Ponnamperuma et al. 1977 |
| | Bullock et al. 1994 |
| Peroxonitrite ($NOO_2^-$) | Plumb et al. 1989 |
| Smectite clays | Banin and Margulis 1983 |
| $O_2^-$ | Yen et al. 2000 |
| Lack of organics in martian soils: | |
| $UV + TiO_2$ | Chun et al. 1978; Pang et al. 1982 |
| Glow discharge from dusts | Mills 1977 |
| Feroxyhyte ($\delta$-FeOOH) | Burns 1980 |
| UV alone | Stoker and Bullock 1997 |

making it nearly impossible to reach a consensus on the results. However, the Viking life-detection controversy between a biological interpretation (see papers by Levin and Straat) and a non-biological interpretation (see papers listed in Table 1) would not have been as difficult to resolve if all four Viking life-detection experiments (GeX, PR, LR, and GCMS) had delivered positive, supportive, and complementary results.

Thus, perhaps the greatest lessons learned from the Viking life-detection experiments are: (1) be prepared for unusual data in which the expectations derived from terrestrial pre-flight experiments might not always be achieved due to unknown factors present at the planetary test site, (2) a positive conclusion that life has been detected on another planetary body is likely to require a wide diversity of positive results from different experimental procedures before alternative conclusions can be ruled out, and (3) evidence for a positive detection of life on another planet must be strong enough to rule out any abiological processes. If abiological processes are as likely to explain the results as biological processes, the scientific community will more likely use the concept of "Occam's Razor" to accept the simpler abiological explanation of the results. Thus, the burden of proof is always much greater for the positive assertion of finding life elsewhere in the Solar System than it is for a null result.

Another example of similar lessons learned for a "life-detection claim" is the debate on how to interpret the proposed biological signs found in the SNC meteorite ALH84001 (Thomas-Keprta et al. 2001; McKay et al. 1996). Initially, five lines of evidence were presented by McKay et al. (1996) that supported a biological interpretation of the observed objects. These were: (a) the chemistry and mineralogy of carbonate globules observed in cracks within previously shocked rock, (b) presence of polycyclic aromatic hydrocarbons (PAHs) associated with the carbonate globules, (c) coexistence of magnetite and iron sulfides within partially dissolved carbonates that exhibited structures similar to inclusions found in magnetotactic terrestrial bacteria, (d) formation age of the carbonate globules younger than the age of the igneous rocks, and (e) SEM and TEM images of carbonate globules with features resembling terrestrial microorganisms, biogenic carbonate structures, or microfossils. But since the publication of the original paper in 1996, a number of other researchers have presented evidence that purely abiological processes could explain the same results (Bradley et al. 1996; Golden et al. 2000; Zolotov and Shock 2000). Although it is still plausible that the putative evidence for life in ALH84001, as described by McKay et al. (1996) were created by biological processes on Mars, it is likely that the general scientific community will remain skeptical until more unequivocal evidence can be found in support of the biological interpretation of features in ALH84001. In both the Viking and ALH84001 controversies, the general scientific community has opted for "abiological mechanisms" as the explanations of results because of a built-in bias towards (a) simpler explanations (Occam's Razor), and (b) the belief that the burden of proof for a positive conclusion of new life must be, by necessity, very high to avoid false positives.

The lead of the Life Detection Team on Viking, Dr. Harold Klein, had expressed the opinion prior to the mission that a step-by-step approach to Mars, including an initial reconnaissance of its surface state, would have been preferable to attempting detection of life on the very first mission (Klein et al. 1976). This turned out to be more true than anyone first imagined, and points to the importance of a thorough understanding of the chemistry, mineralogy, and oxidizing state of the martian surface materials as part of any future life-detection measurements. Much has been learned, but much also remains unknown. Future missions to the surface, such as the Phoenix, Mars Science Laboratory (MSL), and ExoMars missions will make the first-ever direct measurements of pH, quantitative $H_2O$ content, soluble anions and cations, and clay content of soils. A Mars surface sample return (MSSR) mission is currently on a far-distant funding horizon, and a previous finalist in the first round of Mars Scout competitions, SCIM (Sample Collection for Investigation of Mars), was a mission which proposed to sample airborne dust and gas by skimming the middle atmosphere of Mars and returning the samples to Earth for detailed laboratory analysis. Even if all of these missions are successfully flown, the experimental payloads may still not perform all the baseline measurements that will be necessary for future in situ analyses of abiotic and putative biotic activities in martian soils.

## 7 Soil Chemistry

The surprising compositional uniformity of the soils at two Viking landing sites (Clark et al. 1982) separated by thousands of kilometers, has been extended to all five sites now visited by landed missions to the surface of Mars (Yen et al. 2005). This apparently universal unit of globally-distributed dust and soil particles has been produced by a combination of igneous and sedimentary processes. In spite of the extremely fine particle size of grains in the atmosphere, with mean diameters of only a few micrometers, the soils appear to be less chemically weathered than might be expected, with considerable components of primary igneous minerals such as olivine and pyroxene, as well as a lack of ferric oxyhydroxides (e.g., goethite) (Morris et al. 2006). Alteration products would be expected if aqueous processing were significant. In contradistinction to these discoveries is the high concentrations of salts in martian soils (Clark and VanHart 1981; Yen et al. 2005; Ming et al. 2006). However, it has been difficult to identify precise chemical forms of the sulfates, chlorides and trace bromides that have been inferred from element enrichments of S, Cl and Br. Detection of $MgSO_4$ mobilizations in duricrust at Viking sites (Clark et al. 1982), in trenches into Gusev plains soils (Wang et al. 2006), and in Meridiani outcrops (Clark et al. 2005) are indicative that at least some of the sulfates in soils are in the form of highly soluble $MgSO_4$. Calcium and ferric sulfates have been found in certain other occurrences on Mars. The situation with chlorides is also problematical, and although NaCl is the more stable form, a preponderance of $Mg^{++}$ over $Na^+$ cations could implicate $MgCl_2$.

The oxidants in martian soils (Oyama and Berdahl 1977; Hunten 1979) constituted one of the greatest surprises of the Viking missions, but in retrospect can be explained as a small but very important component of peroxides or possibly grain-surface oxide enrichments which result from interaction with photochemically-produced oxidizing species in the thin martian atmosphere which allows short wavelength solar UV to penetrate down to the surface (Hunten 1979). For this reason, it is possible that the Cl and Br may be in the form of chlorates and bromates, respectively, rather than the chlorides and bromides common on Earth (Clark et al. 2005).

This brief summary of the soil chemistries highlights the need for soil chemical analyses carried out concomitantly with future life-detection experiments on Mars in particular, and on all other planetary bodies in general. As discussed above, the Viking biology experiments were fraught with difficult interpretations due in part to unexpected or inconsistent results from the GeX, PR, LR, and GCMS experiments and, in part, due to only a rudimentary understanding of the chemistries of the soils being tested. Even today, many very fundamental soil chemistry measurements such as pH, electrical conductivity, redox potential, bioavailability of ions, and concentrations of biotoxic elements are lacking for Mars (Schuerger et al. 2002). Most of these parameters have dramatic effects on terrestrial organisms, and are likely to be equally important to a putative Mars microbiota. Thus, soil chemistry analyses must be preformed in concert with life-detection protocols to provide the best possible chance to accurately interpret results from the biological investigations.

## 8 Ecological Considerations

Several additional considerations specific to the detailed methods used for the Viking biology experiments are warranted. First, the GeX and LR experiments were run at significantly higher pressures (200 and $\approx$100 mbar, respectively) than are found on the surface of Mars. Recent evidence suggests many common terrestrial microorganisms recovered from spacecraft might be unable to grow below 25 mbar under simulated martian conditions (Schuerger et al. 2006; Schuerger and Nicholson 2006), and by logical extension, it might also be possible that there exists an upper pressure threshold for a putative extant martian surface microbiota; a threshold that might have been exceeded by the GeX and LR assays. Although it is speculative to ascribe pressure thresholds to martian life that has not yet been observed, we emphasize the issue of pressure to draw attention to the ecological conditions under which a putative surface microbiota will have evolved. Thus, we suggest that future life-detection experiments with surface samples consider that pressure may be an important ecological factor that must be controllable in experimental hardware. Samples should be run, at least in part, at conditions that mimic those found on the surface of present-day Mars. Furthermore, pressure has likely changed during recent obliquity cycles, rising to perhaps as much as 25 mbar at high obliquity (Fanale et al. 1982), and, thus, pressure ranges should be tested from at least 7–25 mbar.

The temperature ranges used in the PR, GeX, and LR experiments were all well above 0°C (+8 to +15°C), and may not have reflected the temperature maximums at the two Viking sites. Surface temperatures at both Viking sites failed to rise above 0°C during the more than two martian years in which temperatures were measured (Kieffer 1976). Similar to pressure, future life-detection experiments should recreate actual temperature conditions present at the site being sampled within the incubation chambers. Both pressure and temperature are briefly emphasized here, but other environmental conditions are also important including soil pH, soil redox potential, bioavailability of ions, soil moisture, soil electrolytes, soluble salts, and biotoxic compounds or elements in solution (Schuerger et al. 2002). In fact, there is substantial logic behind running comprehensive soil chemistry tests at a site prior to initiating life-detection assays in order to (a) recreate appropriate incubation conditions within the life-detection payloads and (b) to enhance the interpretation of data derived from the life-detection experiments.

For example, martian soils may be acidic, neutral or alkaline. All values of pH are possible due to the fact that most common geologic minerals are intrinsically alkaline when reacting with $H_2O$ (from release of $Mg^{++}$ and $Ca^{++}$ cations), yet sulfur in the form of sulfates is very common on Mars. Ferric sulfates are acidic, and the formation processes for other sulfates, from either sulfides in soils or S-containing gases released to the atmosphere, also result in strong acidity. Thus, future experiments should not only monitor pH but also carry sufficient reactants to modify pH to investigate optimum growth conditions for microorganisms.

Likewise, the presence of soluble salts in soils will result in different ionic strengths depending on the relative proportions of soil and water in an incubation chamber. By monitoring electrical conductivity and/or ionic content, conditions can be modified by varying the amounts of $H_2O$ and chelators added. On Earth, various organisms are adapted for optimum growth within a relatively restricted range of ionic strength, some for fresh water but others flourishing at salt concentrations near saturation (such as peripheral biotic zones at the Dead Sea or hypersaline regimes in salt ponds). Thus, there would be an advantage in a life-detection payload to possess the capacity to first determine the soluble salts at a site followed by the capability to recreate those salts and concentrations within a life-detection incubation chamber.

Some specific nutrients may be limiting. For example, it has been hypothesized that the ultimate limitation on the vigor and extent of a martian ecology may be sources of nitrogen (Stoker et al. 1993). Although the presence of phosphorus has been well established by in situ measurements, no techniques capable of detecting either nitrates or reduced nitrogen compounds in the soil have been available. These and other expected nutrients could be added. Likewise, supplying active gases such as hydrogen ($H_2$), methane ($CH_4$), and/or ammonia ($NH_3$), none of which are currently at significant concentrations in the contemporaneous martian atmosphere, may stimulate metabolic responses from dormant organisms.

Finally, detailed analysis adds further doubt to the possibility of stable liquid water at the surface of Mars (Hecht 2002; Beaty et al. 2006). Thus, the previous assays might have targeted the wrong niche or may have been overly biased with respect to liquid water. Results derived from the Viking GeX and LR experiments in particular were conducted under conditions that deviated widely from the extant conditions at the martian surface. This point was emphasized by Klein et al. (1976) when describing the PR experiments on Viking: "The (PR) experiment is carried out under actual martian conditions, insofar as these can be attained within the Viking spacecraft, the premise being that, if there is life on Mars, it is adapted to martian conditions and is probably maladapted to extreme departures from those conditions." Furthermore, it is intriguing that the only Viking Biology experiment that exhibited a weak, but significant, positive response for biology, was the PR experiment (Biemann et al. 1976; Klein et al. 1976). Could it be possible that biology was present on Mars (i.e., the weak PR positive result), but the enigmatic results from the GeX and LR experiments run at higher temperatures and much higher pressures than are found at the Viking sites obscured the biological response?

## 9 Conclusions

Based on what was known at the time, the Viking biology experiments were a robust set of logical assays that failed to convince the wider scientific community that life existed at the sampled sites. Since Viking, significant progress has been made in our understanding of the physical, geological, and hydrological conditions on the surface of Mars; and on the understanding of the extreme conditions in which terrestrial microbial life can survive and grow. Although it is likely that incubation conditions for life-detection experiments on Mars will extend into pressure, temperature, and moisture conditions not generally found on the surface, it is not known whether these conditions will inhibit the activity of a putative Mars microbiota. Thus, martian ecological considerations must be included in the design of future life-detection payloads. In addition, we advocate that all life-detection experiments be accompanied by robust soil chemistry experiments in order to gain a concurrent understanding of geochemical conditions in hydrated soils.

It is recognized that the list of potential experimental conditions briefly discussed above are unlikely to be all included within a single life-detection package. However, one can easily envision several life-detection payloads on a given mission asking a range of questions searching for an extant microbiota on Mars (e.g., there were three life-detection experiments on Viking: PR, LR, GeX, plus organics analysis by GCMS). Thus, the primary purpose of this brief review has been to emphasize that ecological considerations must be embraced such that life-detection experiments are conducted, in part, under contemporaneous environmental conditions found at the sampled site. Deviations from these norms should be "reasonable" at first and then more extreme if life is not detected. Thus, the capacity for maintaining a wide diversity of environmental conditions within incubation chambers in future life-detection payloads should be emphasized.

## References

E.V. Ballou, P.C. Wood, T. Wydeven, M.E. Lehwalt, R.E. Mack, Nature **271**, 645–646 (1978)

A. Banin, L. Margulis, Nature **305**, 523–525 (1983)

D. Beaty et al., Astrobiology **6**(5), 677–732 (2006)

S.A. Benner, K.G. Devine, L.N. Matveeva, D.H. Powell, Proc. Natl. Acad. Sci. **97**, 2425–2430 (2000)

K. Biemann, J.M. Lavoie, J. Geophys. Res. **84**, 8385–8390 (1979)

K. Biemann, J. Oro, L.E. Orgel, A.O. Nier, D.M. Anderson, P.G. Simmonds, D. Flory, A.V. Diaz, D.R.
        Rushneck, J.A. Biller, Science **194**, 72–76 (1976)

T.R. Blackburn, H.D. Holland, G.P. Ceasar, J. Geophys. Res. **84**, 8391–8394 (1979)

P.A. Bland, T.B. Smith, Icarus **144**, 21–26 (2000)

J.P. Bradley, R.P. Harvey, H.Y. McSween Jr., Geochimica et Cosmochimica Acta **60**, 5149–5155 (1996)

F.S. Brown, H.E. Adelson, M.C. Chapman, O.W. Clausen, A.J. Cole, J.T. Cragin, R.J. Day, C.H. Debenham,
        R.E. Fortney, R.I. Gilje, D.W. Harvey, J.L. Kropp, S.J. Loer, J.L. Logan, W.D. Potter, G.T. Rosiak, Rev.
        Sci. Instrum. **49**, 139–182 (1978)

M.A. Bullock, C.R. Stoker, C.P. McKay, A.P. Zent, Icarus **107**, 142–154 (1994)

R.G. Burns, Nature **285**, 647 (1980)

S.F.S. Chun, K.D. Pang, J.A. Cutts, J.M. Ajello, Nature **274**, 875–876 (1978)

B.C. Clark, Orig. Life **9**, 241–249 (1979)

B.C. Clark, D. VanHart, Icarus **45**, 370–378 (1981)

B.C. Clark, A.K. Baird, R.J. Weldon, D.M. Tsusaki, L. Schnabel, M.P. Candelaria, J. Geophys. Res. **87**,
        10059–10067 (1982)

B.C. Clark et al., Earth Planet. Sci. Lett. **240**, 73–94 (2005)

F.P. Fanale, J.R. Salvail, W.B. Banerdt, R.S. Saunders, Icarus **50**, 381–407 (1982)

G.J. Flynn, D.S. McKay, J. Geophys. Res. **95**(B9), 14 497–14 509 (1990)

D.C. Golden, D.W. Ming, C.S. Schwandt, R.V. Morris, S.V. Yang, G.E. Lofgren, Meteorit. Planet. Sci. **35**,
        457–465 (2000)

M.H. Hecht, Icarus **156**, 373–386 (2002)

N.H. Horowitz, G.L. Hobby, J.S. Hubbard, J. Geophys. Res. **82**, 4659–4662 (1977)

R.L. Huguenin, K.J. Miller, W.S. Harwood, J. Mol. Evol. **14**, 103–132 (1979)

D.M. Hunten, J. Mol. Evol. **14**, 71–78 (1979)

H.H. Kieffer, Science **194**, 1344–1346 (1976)

H.P. Klein, J. Geophys. Res. **82**, 4677–4680 (1977)

H.P. Klein, Icarus **34**, 666–674 (1978)

H.P. Klein, N.H. Horowitz, G.V. Levin, V.I. Oyama, J. Lederberg, A. Rich, J.S. Hubbard, G.L. Hobby, P.A.
        Straat, B.J. Berdahl, G.C. Carle, F.S. Brown, R.D. Johnson, Science **194**, 99–105 (1976)

V.A. Krasnopolsky, P.D. Feldman, Science **294**, 1914–1917 (2001)

G.V. Levin, Adv. Appl. Microbiol. **5**, 95–133 (1963)

G.V. Levin, P.A. Straat, J. Geophys. Res. **82**, 4663–4667 (1977)

G.V. Levin, P.A. Straat, J. Mol. Evol. **14**, 167–183 (1979a)

G.V. Levin, P.A. Straat, J. Mol. Evol. **14**, 185–197 (1979b)

G.V. Levin, P.A. Straat, Icarus **45**, 494–516 (1981)

G.V. Levin, V.R. Harrison, W.C. Hess, A.H. Heim, V.L. Stauss, J. Am. Water Works Assoc. **51**, 101–104
        (1959)

G.V. Levin, A.H. Heim, M.F. Thompson, D.R. Beem, Life Sci. Space Res. **2**, 124–132 (1964)

D.S. McKay, E.K. Gibson Jr., K.L. Thomas-Keprta, H. Vali, C.S. Romanek, S.J. Clemett, D.F.X. Chillier,
        C.R. Maechling, R.N. Zare, Science **273**, 924–930 (1996)

A.A. Mills, Nature **268**, 614 (1977)

D.W. Ming et al. J. Geophys. Res. **111**(E02S12) (2006). doi:10.1029/2005JE002560

R.V. Morris et al. J. Geophys. Res. (2006). doi:10.1029/2005JE002584

M.D. Nussinov, Y.B. Chernyak, J.L. Ettinger, Nature **274**, 859–861 (1978)

J. Oro, G. Holzer, J. Mol. Evol. **14**, 153–160 (1979)

V.I. Oyama, B.J. Berdahl, J. Geophys. Res. **82**, 4669–4676 (1977)

V.I. Oyama, B.J. Berdahl, G.C. Carle, M.E. Lehwalt, H.S. Ginoza, Orig. Life **7**, 313–333 (1976)

K.D. Pang, S.F. Chun, J.M. Ajello, Z. Nansheng, L. Minji, Nature **295**, 43–46 (1982)

R.C. Plumb, R. Tantayanon, M. Libby, W.W. Xu, Nature **338**, 633–635 (1989)

C. Ponnamperuma, A. Shimoyama, M. Yamado, T. Hobo, R. Pal, Science **197**, 455–457 (1977)

A.C. Schuerger, W.L. Nicholson, Icarus **185**, 143–152 (2006)

A.C. Schuerger, D. Ming, H. Newsom, R.J. Ferl, C.P. McKay, Life Support Biosphere Sci. **8**, 137–147 (2002)

A.C. Schuerger, B. Berry, W.L. Nicholson, 37th Lunar and Planetary Science Conference, Houston, TX,
        13–17 March, 2006, Abstract 1397

C.R. Stoker, M.A. Bullock, J. Geophys. Res. **102**(E5), 10881–10888 (1997)

C.R. Stoker, J.L. Gooding, T. Roush, A. Banion, D. Burt, B.C. Clark, G. Flynn, O. Gwynne, in *Resources of Near-Earth Space*, ed. by J. Lewis, M.S. Matthews, M.L. Guerrieri (University of Arizona Press, Tucson, 1993), pp. 659–708.

K.L. Thomas-Keprta, S.J. Clemett, D.A. Bazylinski, J.L. Kirschvink, D.S. McKay, S.J. Wentworth, H. Vali, E.K. Gibson Jr., M.F. McKay, C.S. Romanek, Proc. Natl. Acad. Sci. **98**, 2164–2169 (2001)

A. Wang et al., J. Geophys. Res. (2006). doi:10.1029/2005JE002513

A.S. Yen et al., Nature **436**, 49–69 (2005)

A.S. Yen, S.S. Kim, M.H. Hecht, M.S. Frant, B. Murray, Science **289**, 1909–1912 (2000)

A.P. Zent, C.P. McKay, Icarus **108**, 146–157 (1994)

M.Y. Zolotov, E.L. Shock, Meteorit. Planet. Sci. **35**, 629–638 (2000)

# Morphological Biosignatures from Subsurface Environments: Recognition on Planetary Missions

B.A. Hofmann

Originally published in the journal Space Science Reviews, Volume 135, Nos 1–4.
DOI: 10.1007/s11214-007-9147-9 © Springer Science+Business Media, Inc. 2007

**Abstract** The Earth is inhabited by life not just at its surface, but down to a depth of kms. Like surface life, this deep subsurface life produces a fossil record, traces of which may be found in the pore space of practically all rock types. The (palaeo)subsurface of other planetary bodies is therefore a promising target in the search for another example of life. Subsurface filamentous fabrics (SFFs), i.e. mineral encrustations of a filament-based textural framework, occur in many terrestrial rocks representing present or ancient subsurface settings. SFF are interpreted as mineral encrustations on masses of filaments/pseudofilaments of microbial origin. SFF are a common example of the fossil record of subsurface life. Macroscopic (pseudostalactites, U-shapes) and microscopic (filaments) characteristics make SFF's a biosignature that can be identified with relative ease. SFF in the subsurface are probably about as common and easily recognizable as are stromatolites in surface environments. Close-up imagers ($\sim$50 micron/pixel resolution) and microscopes ($\sim$3 micron/pixel resolution) on upcoming Mars lander missions are crucial instruments that will allow the recognition of biofabrics of surface- and subsurface origin. The resolution available however will not allow the recognition of small ($\sim$1 micron) individual mineralized microbial cells. The microscopy of unprepared rock surfaces would benefit from the use of polarizing filters to reduce surface reflectance and enhance internally reflected light. Tests demonstrate the potential to visualize mineralized filaments using this procedure.

**Keywords** Biosignatures · Subsurface biosphere · Mars

## 1 Introduction

The terrestrial subsurface is a now well-recognized habitat for heterotrophic and chemosynthetic microbial life forms (Amy and Haldeman 1997; Gold 1992; Stevens 1997; Stevens and McKinley 1995). Given the availability of a source of chemical energy, the limit for

B.A. Hofmann (✉)
Natural History Museum Bern, Bernastrasse 15, 3005 Bern, Switzerland
e-mail: beda.hofmann@geo.unibe.ch

life essentially is given by the increasing temperature towards depth, roughly the 120°C isotherm (Kashevi and Lovley 2003). Voids in rocks that may range from microscopic pores to large caves (Boston et al. 2001a) provide space and facilitate access to energy sources for microbes. Present and past subsurface environments on other planetary bodies therefore are promising targets in the search for another example of life just like surface sediments. Subsurface conditions on different planetary bodies, including Earth, likely are less different and more stable than their surface counterparts. It is likely that pores and cavities in the Martian subsurface at depths corresponding to temperatures >0°C are filled with water (Burr et al. 2002; Clifford and Parker 2001). Such environments likely are quite similar to anaerobic deep subsurface environments on Earth where the presence of life is accepted (Aitken et al. 2004; McKinley et al. 2000; Moser et al. 2005; Parkes et al. 2005). While the presence of a subsurface biosphere is accepted, evidence of a fossil record of this subsurface life has been rather scarce (Furnes et al. 1999; Furnes and Staudigel 1999; Hofmann and Farmer 2000; Kretzschmar 1982; Reitner 2004; Schumann et al. 2004; Trewin and Knoll 1999; Westall et al. 2006a, 2006b).

## 2 Subsurface Filamentous Fabrics (SFF): A Fossil Record of Subsurface Biosphere(s)?

Based on large numbers of samples from ~200 localities, a so far largely ignored type of mineral fabric from subsurface environments has been recognized to be widespread (Hofmann and Farmer 2000). The structural building elements of SFF are filament threads, <1 to about 3 micron in diameter, showing multiple encrustations by a variety of minerals. Encrustations of filaments in various large-scale geometric arrangements result in characteristic macroscopic textures, such as layered internal sediments with matted fabrics, vertically arranged threads, streamers, U-loops (Boston et al. 2001b) indicating flexibility of filaments before mineral encrustation occurred, but also in irregular masses without macroscopic expression. Fabrics similar to SFF are known from caves where a microbial origin of "stalactitic snottites", closely resembling mineralized SFF, is supported by gene sequence analysis (Hose et al. 2000), and from oxidizing shipwrecks where "rusticles", similar again to SFF, form in a fully submersed environment (Cullimore and Johnston 2000). In these recent cave- and shipwreck environments sulfur- and iron oxidizing microorganisms appear to be the main organisms responsible for the formation of macroscopic fabrics. SFF may have formed in any subsurface environments colonized by microbial life, independent of the geological nature of the host rock.

Highly variable degrees of cementation of filamentous fabrics result in macroscopic aspects ranging from "hair-like tufts" to massive rocks without recognizable surface expression. Such mineral fabrics may be relatively easy to recognize at resolutions already implemented (Mars Exploration Rovers, Beagle 2) or foreseen for upcoming Mars landing missions (MSL, Phoenix, ExoMars-Pasteur). Terrestrial occurrences of SFF are found in a variety of rocks, mainly cavities in volcanics (57% of occurrences), in the oxidation zone of sulfide ore deposits (28%), and in sedimentary environments (10%). SFF, in particular the so-called moss-agates, have attracted the attention of researchers for a long time (Bowerbank 1842; Daubenton 1782; Razumovsky 1835). Descriptions of single occurrences during the past 25 years favor a biological explanation (Baele 1998; Feldmann et al. 1997; Hofmann 1989; Kretzschmar 1982; Reitner 2004; Schumann et al. 2004; Trewin and Knoll 1999). Morphologically similar filamentous fabrics also are known from numerous hydrothermal vent sites, and generally are biologically interpreted (Fortin et al. 1998;

Juniper and Sarrazin 1995). It is proposed here that SFF represent an expression of sub-surface life that is as widespread in and characteristic of subsurface environments as are stromatolites in Precambrian sediments. SFF represent a type of biosignature with similar potential and limitations. The geological age of SFF ranges from sub-recent to Precambrian, with possible Archean examples (Hofmann et al. 2006).

*Macroscopic characteristics of SFF:*   While some occurrences form massive blocks of chalcedony without distinctive macroscopic surface features, in many cases SFF exhibit characteristic morphologies resulting from the mineralization of filament bundles or fila-ment streamers (Fig. 1a,b): Stromatolite-like matted fabrics resulting in layered, sometimes agate-like banding; pseudostalactites, vertically oriented tubular structures with a innermost core diameter of a few microns, much too small for stalactites which have mm-sized cen-ters; association of numerous filaments to "ropes" or "stalks", and U-loops: gravity-bent filaments attached at two ends, indicating a high initial flexibility (Fig. 1b). Minerals in-volved in the encrustation of filaments (Fig. 2) are all aqueous precipitates and include Fe-hydroxides, hematite, coronadite, Pb-phosphates, Fe-rich clays, pyrite, quartz, opal, calcite, zeolites, and others.

**Fig. 1   a** Example of subsurface filamentous fabric (SFF) from Jebel Irhoud, Morocco. Goethite-encrusted filaments show only thin overgrowth of quartz and calcite, allowing recognition of fabric. Width of sample 18 cm. **b** Subsurface filamentous fabric from Sidi Rahal, Morocco. Sample is shown in original orientation showing gravity-oriented vertical encrusted filaments (single attachment) and U-loops formed from filaments/filament strands attached at 2 points. Field of view is 6 cm, diameter or encrusted filaments 0.15 to 0.5 mm. Photographs by Peter Vollenweider

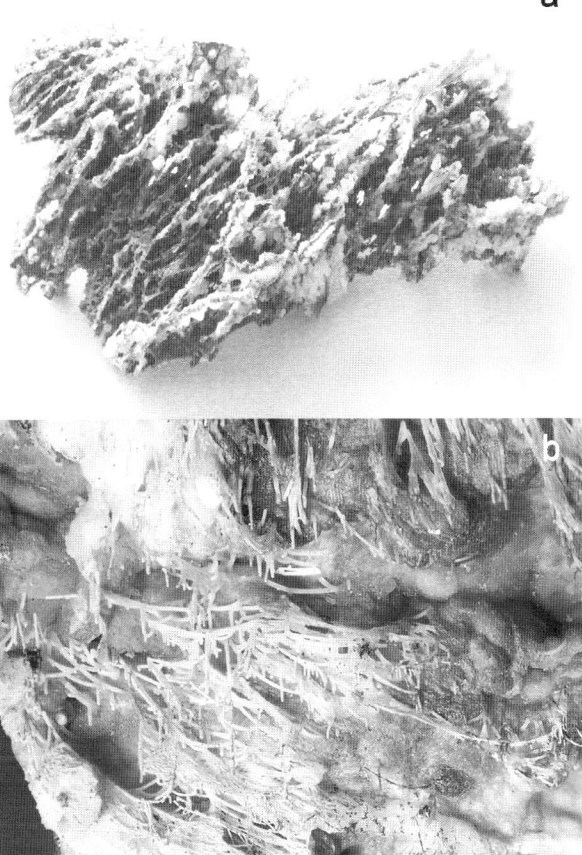

**Fig. 2** Subsample of SFF from
Jebel Irhoud, Morocco (Fig. 1a).
**a** Untreated sample in air;
**b** calcite encrustation dissolved
with HCl, in air; (**c**) as (**b**) but in
water immersion, strongly
reducing surface scattering and
showing filamentous texture.
Many filament strand are strongly
curved indicating initial
flexibility (*arrows*). Field of view
is 23 mm

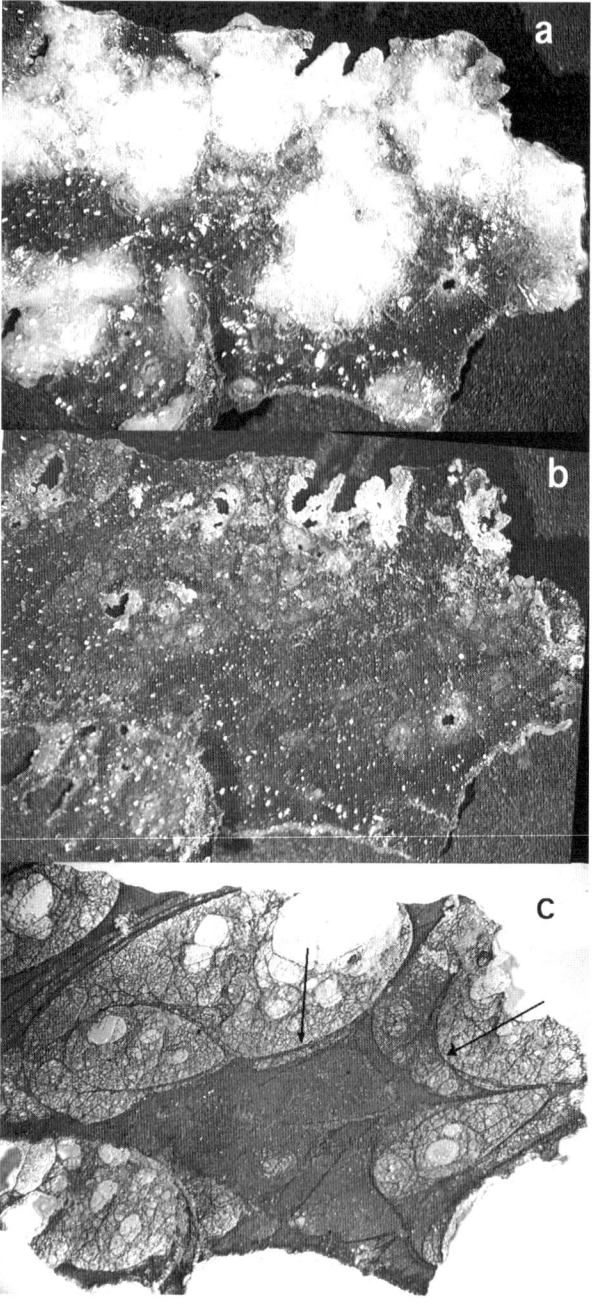

*Microscopic characteristics of SFF:*    Investigations of SFF by optical microscopy in thin
sections and by SEM on natural and etched surfaces shows that filaments with characteristic
diameters near 1 micron are the basis and textural oldest constituent of SFF. All other phases
were formed later as encrustations on the filaments or as massive cavity fill. The shape

of filaments is irregular and highly bent, and differs from mineral fibers that show little bending and whose diameter is much more variable. Besides simple filaments, forms closely resembling the extracellular stalk of *Gallionella ferruginea* have been observed, mainly in one area in Iceland, in association with the more common filaments (Feucht et al. 2006; Hofmann and Farmer 2000).

*Biosignature potential of filaments and SFF:* Morphological biosignatures (as chemical and isotopic signatures) alone will always have to be met with a level of caution (Garcia-Ruiz et al. 2002, 2003; Jones et al. 2004). The recognition of filaments is less ambiguous than that of coccoid microbes, and filaments tend to occur in large numbers in subparallel configuration, leading to macroscopic "streamers" (Hallberg et al. 2006). Biological and non-biological filamentous forms may be discriminated based on evidence of flexibility, characteristics of bending and diameter (absolute values and range) of single filaments. Filamentous microbes also tend to form macroscopically recognizable biofabrics. Filamentous shapes are found among primitive bacteria and archaea and it appears likely that the development of filamentous forms occurred early in the history of life, as filaments have been reported from Archean rocks (Rasmussen 2000; Walsh 1992, 2004; Westall et al. 2006a, 2006b). The presence of filamentous forms can thus be expected among the earliest life forms on Earth, and potentially, Mars. Filaments and filament-based biofabrics thus represent a very robust morphological biosignature. Filament-like structures may also form due to the mineralization of extracellular polymeric substances/microbial mucus (Jones et al. 2004; Westall et al. 2000b). While such filament-like forms are not fossils s.s., they nevertheless represent viable biosignatures.

*Distinction between subsurface biological filaments and non-biological look-alikes:* Biological filaments are thin (typically few microns diameter) threads that may reach a length of hundreds of microns. Non-biological threads also exist and may be mistaken for biological ones. How can they be distinguished? Among the possible non-biological threads fibrous minerals (e.g. asbestos, torodokite, palygorskite) and natural glass fibers (volcanic glass) appear the most similar forms, potentially giving rise to fabrics reminiscent of SFF, but of non-biological origin. A comparison between threads of certainly biological origin (recent filamentous microbes) and non-biological threads (Hofmann and Farmer, in prep.) indicates that biological filaments are less variable in thickness, and show more changes of direction, and a higher degree of bending, than non-biological ones. Also, flexibility indicated by the formation of U-loops and formation of mats is not expected in non-biological filaments. The presence of carbon in filaments and its isotopic composition may yield further hints of biogenicity, but such information may be obliterated due to oxidation or migration of organic matter.

## 3 Detection of morphofossils, including SFF, on planetary missions

Due to the ambiguity inherent to all morphological signatures of microbial life, such signatures will need confirmation using other characteristics such as isotopic or organic geochemistry. However, the potential of easy recognition renders non-microscopic morphological signatures an interesting target in planetary exploration (Brack et al. 1999; Westall et al. 2000a). Unsuccessful (Beagle 2), successful (Mars Exploration Rovers MER) and planned Mars landers (ESA's ExoMars, NASA's Mars Science Laboratory) all were/are equipped with imaging systems able to visualize biofabrics such as stromatolites and SFF.

**Table 1**  Resolutions of close-up and microscopic imaging instruments on Mars missions

| Lander | Instrument | Resolution (micrometer/pixel) |
|---|---|---|
| Beagle 2 | Close-up lens | 50 |
| Beagle 2 | Microscope | 4 |
| Mars Exploration Rovers | Microscopic imager | 30 |
| Mars Science Laboratory | MAHLI | 12.5–75 |
| Phoenix | Robotic arm camera | 23–122 |
| Phoenix | MECA | 4 |
| ExoMars | Close-up imager (CLUPI) | 15 |
| ExoMars | Microscope | ∼3 |
| For comparison | | |
| Standard petrographic microscope[a] | 40× obj. | 0.31 |
| Standard petrographic microscope[a] | 100× obj. | 0.12 |

[a]Based on a 1024 pixel frame analogous to MER/Beagle 2

While the panoramic camera systems typically have resolutions in the millimeter range, close-up imagers such as the close-up lens on Beagle 2, the Microscopic imagers on the MER Rovers, the Close-up imager (CLUPI) on ExoMars and the Mars hand lens imager (MAHLI) on MSL are able to visualize biofabrics with sufficient resolution to allow a detailed comparison with terrestrial counterparts. Typical resolutions for imaging systems on Mars missions are given in Table 1. The main purpose of intermediate resolution imaging (in the order of 50 microns/pixel) is the interpretation of rock and soil textures, but the potential of recognizing biofabrics is an important feature as well. Even higher resolutions can be achieved by using microscopes such as on Beagle 2 (Thomas et al. 2004) and foreseen for the Phoenix lander (MECA) and ESA's ExoMars mission (Table 1).

When using high magnifications in optical microscopy in laboratory situations, thin sections for transmitted light, typically 20 micrometers thick, or highly polished samples for reflected light are used. In the next generation of planetary missions such sample preparation will not be available. When unprepared samples have to be used for microscopy, as foreseen in the ExoMars mission, tricks may be needed to overcome some of the disadvantages brought about by the irregular nature of sample surfaces. Field geologist typically wet the surface of rock samples exactly for this purpose. Sample wetting or immersion in a liquid appears impractical on planetary missions. However, a similar effect can be obtained by using doubly polarized light in a manner identical to the observation of internal reflections in reflected light microscopy (Figs. 2, 3, 4). For the imaging of 3D surfaces of rocks, sample preparation is not needed. Test conducted with unprepared rock surfaces and resolutions amenable to microscopes foreseen for upcoming Mars missions demonstrate that SFF filaments can be made visible under such non-ideal conditions (Fig. 4).

Several cases of fossilization of microbial fabrics in open-space infills, in terms of accessibility to close-up and microscopic imaging, are known from Earth and could similarly exist on other planetary bodies, e.g. Mars:

*3D textures in open space:*  Slightly mineralized filaments in cavities (Figs. 1, 2). Textures of this type can be investigated with close-up and microscopic imagers without sample

**Fig. 3** Surface of Mars meteorite Sayh al Uhaymir 094, natural surface. **a** Automontage of image stack taken in plane polarized light, mainly light scattered from the surface. **b** Automontage of image stack taken in crossed polarized light, showing mainly light reflected from below the surface of the sample. Field of view is 770 micrometer

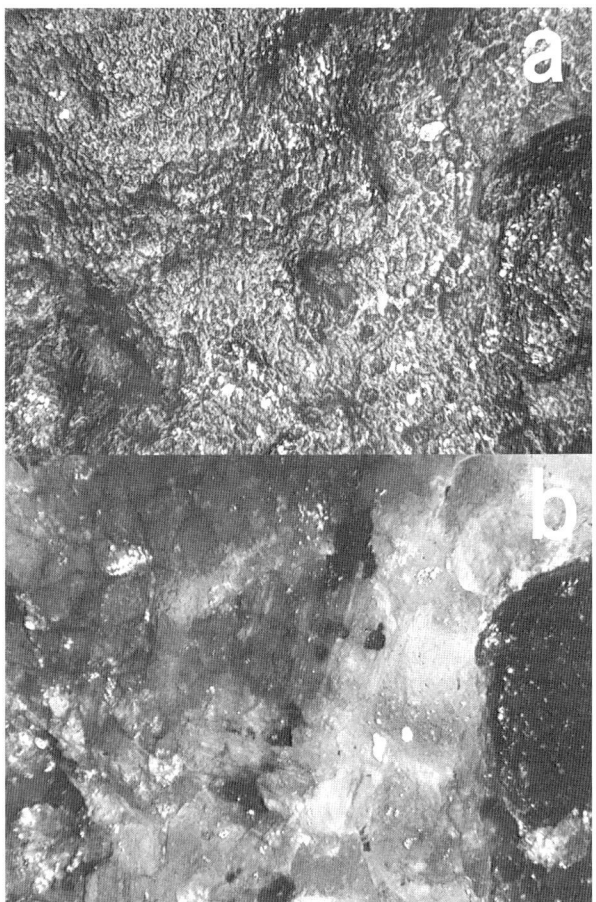

preparation, relying 100% on surface scattered light. Due to significant relief, stacking of images and recombination into a single image is needed.

*Partially cemented 3D textures:* Textures formed in open space but showing strong cementation, e.g. microbial mats of filaments with thick overgrowths of minerals. Surface textures may indicate the presence of microbial textures, even though a detailed imaging based on the surface is not possible. Cross sections are then needed, potentially provided by fractures or drillcores. Surface scattered light only provides preliminary information, while details depend on light from below the mineral surface.

*Completely cemented 3D textures:* Textures formed in open space, but later on completely cemented by minerals (Fig. 4). The surface of such materials may not indicate any biological textures at all, while these may be perfectly preserved as inclusions in the minerals. Surface scattered light only provides information about the enclosing mineral(s), while light from below the surface must be analyzed for detailed information.

In order to get optical access to both the surface morphology and textural evidence inside rocks, two types of imaging situations are therefore required:

**Fig. 4** Surface chert (fine-grained low-T quartz) containing inclusions of goethite-encrusted filaments. SFF from basaltic host rocks, Breiddalur, eastern Iceland, natural surface. **a** Automontage of image stack taken in plane polarized light, mainly light scattered from the surface. **b** Automontage of image stack taken in crossed polarized light, showing mainly light reflected from below the surface of the sample. Filaments are not visible in (**a**), but clearly discernible in (**b**). Field of view is 770 micrometer

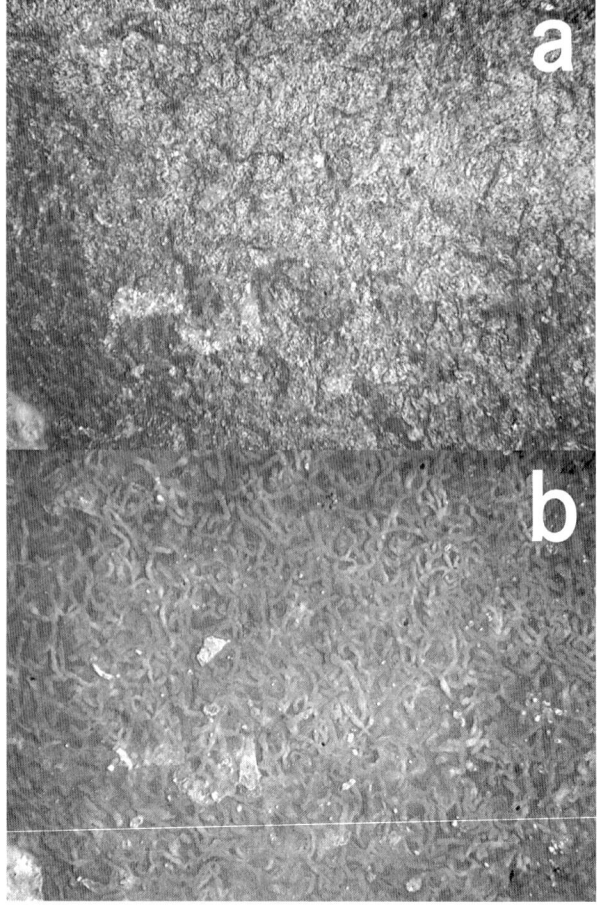

(1) Imaging of the surface scattered light, comparable to standard visual observations of solid rock surfaces with the naked eye, a hand-lens or binocular lenses (Figs. 2a,b, 3a, 4a).

(2) Imaging of light from below the rock/mineral surface, comparable to macroscopic observations of a wet or polished rock surface (avoiding directly reflected light), or microscopic observation of a thin section, or of a polished surface using doubly crossed polarized light in a standard reflected light microscope. This can be achieved without sample preparation or surface wetting by using doubly crossed polarized light (Figs. 3b, 4b). The incident beam is linearly polarized and passes a second polarizing filter after the sample, allowing only the transit of light whose direction of polarization has been changed due to passing through minerals, this effectively eliminated all surface scattered light. This method of observation reduces the intensity of the observed light by a factor of ∼3–20 in a standard petrographic microscope, leading to correspondingly increased exposure times.

The importance of colour: Mineral constituents of Martian rocks display strongly different colours: Iron oxides and hydroxides are brown to red-brown, bluish in case of wind-polished hematite surfaces, pyroxenes and olivines are greenish brown, feldspars and many salt minerals are colourless. While in an image of a Mars rock (Fig. 3b) the different minerals are hard to identify, in a colour image the difference between feldspars and mafic minerals is strongly accentuated. Also, color enhances the contrast between iron-stained filaments in Fig. 4b and the grey matrix.

Apart of biosignatures characteristic of surface-bound life (e.g. stromatolites, wrinkle structures) subsurface biosignatures such as SFF have a good potential to be discovered, if present at landing sites, with the close-up and microscopic imagers foreseen on upcoming Mars missions. Zones containing significant amounts of aqueous minerals, as recognizable from orbit (Bibring et al. 2006) would be preferred areas for prospection for SFF. Colour information is important for the understanding of the areal distribution of minerals in close-up or microscopic images.

## 4 Conclusions

Apart from life in sedimentary surface environments, dominantly relying on photosynthesis, a subsurface biosphere exists on Earth and possibly also exist(ed) on Mars and other planetary bodies. Subsurface filamentous fabrics (SFF), common in low-T subsurface palaeoenvironments from the sub-recent to the Archean, are interpreted as the fossil record of this subsurface biosphere, being a type of fabric formed by the mineralization of filament-like structures of microbial origin (filaments or mucoid pseudofilament). SFF, as well as biofabrics from sedimentary environments (e.g. stromatolites) are amenable to imaging at intermediate resolutions (~50 micron/pixel). Details of microbial textures in SFF, stromatolites or other biofabrics may be visualized using microscopic imagers, even though the resolution of instruments foreseen for upcoming missions, and the lack of sample preparation, will not allow the detection of single fossilized cells. Microscopic imaging may be greatly enhanced by automatic combination of image stacks (partially focused) to a single in-focus image, and the use of polarizing filters to enhance the proportion of light reflected from below the surface of a sample.

## References

C.M. Aitken, D.M. Jones, S.R. Larter, Nature **431**, 291–294 (2004)
P.S. Amy, D.L. Haldeman, *The Microbiology of the Terrestrial Deep Subsurface* (Lewis, Boca Raton, 1997), 356 pp
J.M. Baele, Ann. Soc. Géol. Nord **6**(2), 127–133 (1998)
J.-P. Bibring, Y. Langevin, J.F. Mustard, F. Poulet, R. Arvidson, A. Gendrin, B. Gondet, N. Mangold, P. Pinet, F. Forget, T.O. Team, Science **312**, 400–404 (2006)
P.J. Boston, M.N. Spilde, D.E. Northup, L.A. Melim, Lunar planetary Science Conference XXXII. Paper 2015, 2001a
P.J. Boston, M.N. Spilde, D.E. Northup, L.A. Melim, D.S. Soroka, L.G. Kleina, K.H. Lavoie, L.D. Hose, L.M. Mallory, C.N. Dahm, L.J. Crossey, R.T. Schelble, Astrobiology **1**(1), 25–55 (2001b)
J.S. Bowerbank, Ann. Mag. Nat. Hist. **10**(1842), 9–18, 84–91 (1842)
A. Brack, B. Fitton, F. Raulin, ESA SP-1231, 1999, p. 188
D.M. Burr, J.A. Grier, A.S. Mcewen, L.P. Keszthelyi, Icarus **159**, 53–73 (2002)
S.M. Clifford, T.J. Parker, Icarus **154**, 40–79 (2001)
D.R. Cullimore, L. Johnston, Can. Chem. News **Nov/Dec**, 14–15 (2000)
L.J.M. Daubenton, Mémoires de l'Académie Royale des Sciences, 1782, pp. 667–673
M. Feldmann, J. Neher, W. Jung, F. Graf, Ecologae Geologicae Helvetiae **90**, 541–556 (1997)
C. Feucht, B. Hüsser, B.A. Hofmann, 5th European workshop on Astrobiology, Budapest, Oct. 2005. Int. J. Astrobiol. **5**(86) (2006)
D. Fortin, F.G. Ferris, S.D. Scott, Am. Mineral. **83**, 1399–1408 (1998)
H. Furnes, K. Muehlenbachs, O. Tumyr, T. Torsvik, I.H. Thorseth, Terra Nova **11**(5), 228–233 (1999)
H. Furnes, H. Staudigel, Earth Planet. Sci. Lett. **166**, 97–103 (1999)

J.M. Garcia-Ruiz, A. Carnerup, A.G. Christy, N.J. Welham, S.T. Hyde, Astrobiology **2**(3), 335–351 (2002)

J.M. Garcia-Ruiz, S.T. Hyde, A.M. Carnerup, A.G. Christy, M.J. Van Kranendonk, N.J. Welham, Science **302**, 1194–1197 (2003)

T. Gold, Proc. Natl. Acad. Sci. USA **89**, 6045–6049 (1992)

K.B. Hallberg, K. Coupland, S. Kimura, D.B. Johnson, Appl. Environ. Microbiol. **72**(3), 2022–2030 (2006)

B. Hofmann, Nagra Technical Report 88-30, Baden, Switzerland, 195 pp., 1989

B.A. Hofmann, J.D. Farmer, Planet. Space Sci. **48**, 1077–1086 (2000)

B.A. Hofmann, Y. Krüger, A.E. Fallick, M. Eggimann, M.J. Van Kranendonk, 6th European Workshop on Astrobiology, Lyon, 16–18 October 2006. Int. J. Astrobiol. 2006 (in press).

L.D. Hose, A.N. Palmer, M.V. Palmer, D.E. Northup, P.J. Boston, H.R. Duchene, Chem. Geol. **169**, 399–423 (2000)

B. Jones, K.O. Konhauser, R.W. Renaut, R.S. Wheeler, J. Geol. Soc. Lond. **161**, 983–993 (2004)

S.K. Juniper, J. Sarrazin, in *Seafloor Hydrothermal Systems*, ed. by S.E. Humphris, R.A. Zierenberg, L.S. Mullineaux, R.E. Thomson (American Geophysical Union, Washington, 1995), pp. 178–193

K. Kashevi, D.R. Lovley, Science **301**, 934 (2003)

M. Kretzschmar, Facies **7**, 237–260 (1982)

J.P. McKinley, T.O. Stevens, F. Westall, Geomicrobiol. J. **17**, 43–54 (2000)

D.P. Moser, T.M. Gihring, F.J. Brockman, J.K. Fredrickson, D.L. Balkwill, M.E. Dollhopf, B.S. Lollar, L.M. Pratt, E. Boice, G. Southam, G. Wanger, B.J. Baker, S.M. Pfiffner, L.-H. Lin, T.C. Onstott, App. Environ. Microbiol. **71**(12), 8773–8783 (2005)

R.J. Parkes, G. Webster, B.A. Cragg, A.J. Weightman, C.J. Newberry, T.G. Ferdelman, J. Kallmeyer, B.B. Jørgensen, I.W. Aiello, J. Fry, Nature **436**, 390–394 (2005)

B. Rasmussen, Nature **405**, 676–679 (2000)

G. Razumovsky, Bull. Géol. Soc. Fr. **6**, 165–168 (1835)

J. Reitner, in *Geobiologie. 74. Jahrestagung der Paläonto-logischen Gesellschaft, Göttingen, 02. bis 08. Oktober 2004. Kurzfassungen der Vorträge und Poster*, ed. by J. Reitner, M. Reich, G. Schmidt (Universitätdrucke Göttingen, Göttingen, 2004)

G. Schumann, W. Manz, J. Reitner, M. Lustrino, Geomicrobiol. J. **21**, 241–246 (2004)

T.O. Stevens, in *The Microbiology of the Terrestrial Deep Subsurface*, ed. by A. Penny, D. Haldeman (Lewis, Boca Raton, New York, 1997), pp. 205–223

T.O. Stevens, J.P. McKinley, Science **270**, 450–454 (1995)

N. Thomas, B.S. Lüthi, S.F. Hviid, H.U. Keller, W.J. Markiewicz, T. Blümchen, A.T. Basilevsky, P.H. Smith, R. Tanner, C. Oquest, R. Reynolds, J.-L. Josset, S. Beauvivre, B. Hofmann, P. Rüffer, C.T. Pillinger, Planet. Space Sci. **52**, 853–866 (2004)

N.H. Trewin, A.H. Knoll, Palaios **14**, 288–294 (1999)

J.N. Walsh, Precambr. Res. **54**, 271–293 (1992)

M.M. Walsh, Astrobiology **4**(4), 429–437 (2004)

F. Westall, A. Brack, B. Hofmann, G. Horneck, G. Kurat, J. Maxwell, G.G. Ori, C. Pillinger, F. Raulin, N. Thomas, B. Fitton, P. Clancy, D. Prieur, D. Vassaux, Planet. Space Sci. **48**, 181–202 (2000a)

F. Westall, C.E.J. De Ronde, G. Southam, N. Grassineau, M. Colas, C. Cockell, H. Lammer, Philos. Trans. Roy. Soc. B **361**, 1857–1875 (2006a)

F. Westall, S.T. De Vries, W. Nijman, V. Rouchon, B. Orberger, V.K. Pearson, J. Watson, A. Verchovsky, I. Wright, J.-N. Rouzaud, D. Marchesini, A. Severine, Geol. Soc. Am. Spec. Pap. **405**, 105–131 (2006b)

F. Westall, A. Steele, J. Toporski, M. Walsh, C. Allen, S. Guidry, D. Mckay, E. Gibson, H. Chafetz, J. Geophys. Res. **105**(E10), 24511–24527 (2000b)

# Exploration of the Habitability of Mars: Development of Analytical Protocols for Measurement of Organic Carbon on the 2009 Mars Science Laboratory

**Paul Mahaffy**

Originally published in the journal Space Science Reviews, Volume 135, Nos 1–4.
DOI: 10.1007/s11214-007-9223-1 © Springer Science+Business Media B.V. 2007

**Abstract** The mission goal of the 2009 Mars Science Laboratory is to assess the habitability of a region on Mars. This large rover incorporates an Analytical Laboratory that contributes to this mission objective by means of a detailed characterization of mineralogy and chemistry. The Sample Analysis at Mars instrument suite in the Analytical Laboratory provides the capability to analyze volatiles released from rocks and soils and gases directly sample from the atmosphere. A primary focus of this suite is the detection and identification of organic molecules. The protocols for the extraction and analysis of organics under development for this mission are described as are experiments carried out on Mars analog samples to evaluate these methods.

**Keywords** Mars · Organics · Mars Science Laboratory · Sample Analysis at Mars Investigation · Astrobiology

## 1 Introduction

This paper describes the development and ongoing refinement of analytical protocols for the Mars Science Laboratory (MSL) in situ measurement of organic molecules in gases released from solid samples or gases sampled directly from the atmosphere. The Sample Analysis at Mars (SAM) investigation is the primary MSL analytical tool for the organics analysis. SAM is a suite of instruments that includes a quadrupole mass spectrometer (QMS), a gas chromatograph (GC), and a tunable laser spectrometer (TLS). Section 2 gives an overview of the MSL science objectives in relationship to recent, ongoing, or planned Mars observations and its relationship to the search for past or present microbial life outside our planet. Section 3 provides a summary of the motivation for the specific SAM measurements. Surface in situ measurements by SAM are expected to extend the present state of understanding of Martian organics, currently derived from surface in situ measurements, orbital- or terrestrial-based experiments, or from the study of Martian meteorites. Section 4 gives a summary overview

P. Mahaffy (✉)
Atmospheric Experiment Laboratory, NASA Goddard Space Flight Center, Greenbelt, MD 20771, USA
e-mail: Paul.R.Mahaffy@nasa.gov

O. Botta et al. (eds.), *Strategies of Life Detection*. DOI: 10.1007/978-0-387-77516-6_18

of the planned implementation of the MSL science payload and sample processing tools. Section 5 describes the measurement protocols under development and test by the SAM team for organics measurements, and Sect. 6 gives examples of how the protocols are evaluated using Mars analog materials. A description of the planned SAM investigation of noble gas and non-carbon light element isotopic investigations is outside the scope of this paper, as is a detailed description of the SAM instrumentation that is presently under development for space flight.

## 2 Overview of MSL Science Objectives

A central goal of the Mars exploration program of several nations is to search for evidence of extant or extinct life on Mars and investigate the ability of that planet to sustain life. Recent success of the National Aeronautics and Space Administration (NASA) and European Space Agency (ESA) missions to Mars give an unprecedented total in 2006 of four operating orbiting spacecraft and two operating surface landers at the red planet. Data from these missions are providing a wealth of new information and a new understanding of the present state of Mars and its history that provides an increasingly firm foundation for the study of its past and present habitability and its potential to sustain life. Of particular note is the discovery of vast Polar Regions, rich in near-surface water ice (Boynton et al. 2002), geologically recent fluid flows that formed gullies (Malin and Edgett 2000), possible trace levels of the disequilibrium species methane in the Martian atmosphere (Krasnopolsky et al. 2004), and mineralogical and morphological evidence of aqueous alteration of surface materials (Squyres et al. 2004).

As a significant step toward the search for life on Mars, the top level science goal for the MSL mission is to explore and quantitatively assess a potential habitat on Mars. Three major objectives and specific measurement sets for this mission were stated by NASA after an extended period of mission definition and iteration by groups of scientists and engineers working closely together. The objectives are to

- **Assess the biological potential of at least one target environment (past or present)** by determining the nature and inventory of organic carbon compounds, taking an inventory of the chemical building blocks of life (C, H, N, O, P, S), and identifying features that may record the actions of biologically relevant processes.
- **Characterize the geology of the landing region at all appropriate spatial scales** by investigating the chemical, isotopic, and mineralogical composition of martian surface and near-surface geological materials and interpreting the processes that have formed and modified rocks and regolith.
- **Investigate planetary processes that influence habitability** by assessing long-timescale (i.e., 4 billion year) atmospheric evolution processes and by determining the present state, distribution, and cycling of water and $CO_2$.

These MSL objectives realize a subset of a larger set of priority measurement objectives for the long-term scientific exploration of Mars that have been defined and updated over a period of several years by the Mars Exploration Program Analysis Group (MEPAG).

## 3 The Chemical and Isotopic Composition of Martian Volatiles

### 3.1 Sources and Sinks of Martian Organic Molecules

A primary motivation for the search for organic molecules on Mars is to understand if there is molecular evidence of pre-biotic or biotic activity, perhaps preserved from more than several billion years ago when the martian climate may have been much more Earth-like, with a thicker atmosphere, warmer surface temperatures, and persistent lakes or oceans. On Earth, tectonic recycling largely destroys molecular signatures of early life. In contrast, the more rapid cooling of Mars may have quenched such recycling and enabled preservation of early biotic or pre-biotic chemistry. For example, if preserved amino acids were found in Mars sedimentary deposits, their distribution could suggest a biotic or abiotic source mechanism. On the other hand, oxidants such as superoxide radicals (Yen et al. 1999), reactive surface complexes such as oxidized halides (Zent and McKay 1994), or radiation processing (Kminek and Bada 2006) may have transformed or destroyed martian organic molecules, thus reducing the diversity of such compounds from an earlier era and our ability to describe an early chemical history. Galactic cosmic rays penetrate only meters into the martian surface. Thus, it is possible that organics from an early wet Mars that may have been buried by aqueous, aeolian, impact, or volcanic transport of material and only recently exposed, may provide prime sites for the MSL search for organic molecules. Recent orbital spectroscopic evidence of phyllosilicates formed by aqueous alteration (Bibring et al. 2006; Murchie et al. 2007) has revealed several prime targets for a MSL landing site and for the MSL search for organic molecules.

Early exogenous sources of organics on Mars from carbonaceous asteroids and comets are expected to be similar to those that may have seeded Earth with pre-biotic compounds. These compounds may have been important in the origin of life on Earth. An estimated IDP influx of $10^6$ to $10^7$ kg/year resulting in several to tens of percent of the total mass of the Martian regolith has been predicted (Flynn and McKay 1990) and recent models (Bland and Smith 2000) have predicted that meteorites greater than 10 grams are likely to be more abundant on Mars than any place on Earth, including the Antarctic meteorite-rich blue ice fields. These authors predict hundreds to hundreds of thousands of small meteorites delivered per square kilometer. The Ni enrichment in the bright dust observed by the MER chemical investigations has been described (Yen et al. 2005) as consistent with 1.2% contribution from chondritic (CI) meteoritic material. Although C-chondrites contain most of their carbon in a kerogen-like macromolecular form (Cronin et al. 1998), they also contain a wide range of extractable compounds including amino acids, nucleobases and many other compound types (Botta and Bada 2002). Distributions of compounds or their oxidation products such as carboxylic acids that might plausibly be direct products of chondritic material might suggest that extensive biological production and processing of carbon compounds did not take place.

### 3.2 The Search for Organics in the Atmosphere of Mars

The Mariner 9 infrared spectrometer was able to obtain spectra in the 200 to 2,000 cm$^{-1}$ (5–50 μm) spectral range for nearly a year in 1971 and 1972 and establish substantially reduced upper limits (Macguire 1977) for methane, ethane, ethylene, and acetylene shown in Table 1. More recent reports of methane mixing ratio observed from ground-based or from Mars express are also listed in this table. It should be noted that due to the very low methane abundance, these detections are very near the sensitivity limit of both the orbital and ground-based instruments.

**Table 1** Organic species mixing ratios or upper limits in the martian atmosphere

| Species | Reported mixing ratio or upper limit (UL) | Notes |
|---|---|---|
| $CH_4$ | 10 ppb ($\pm 5$) | (Formisano et al. 2004), Mars Express PFS, average reported, variation between 0 and 30 ppb observed |
| $CH_4$ | UL = 20 ppb | (Macguire 1977), Mariner 9 IR spectrometer |
| $CH_4$ | 11 ppb ($\pm 4$) | (Krasnopolsky et al. 2004), ground based |
| $CH_4$ | UL = 7 ppb | Villanueva et al. (2006) |
| $C_2H_6$ | 400 ppb | (Macguire 1977), Mariner 9 IR spectrometer |
| $C_2H_4$ | 500 ppb | (Macguire 1977), Mariner 9 IR spectrometer |
| $C_2H_2$ | 2 ppb | (Macguire 1977), Mariner 9 IR spectrometer |
| $CH_2O$ | 3 ppb | Krasnopolsky et al. (1997) |
| $CH_2O$ | $< 5 \times 10^{-7}$ | Korablev et al. (1993) (tentative detection / Phobos) |

No atmospheric organics were reported by the Viking entry mass spectrometer at an altitude of approximately 135 km (Nier et al. 1976) or the Viking Gas Chromatograph Mass Spectrometer (Biemann et al. 1976) from the surface of Mars.

### 3.3 The Search for Organics in Solid Phase Martian Materials

The focus of the Viking mission was to determine if there was life on Mars. In addition to the specific life-detection experiments designed to measure microbial metabolism (Klein et al. 1976; Oyama et al. 1977; Levin 1997) that are discussed in this volume (Schuerger and Clark 2007), one of the primary science objectives of the Viking GCMS (Anderson et al. 1972; Biemann 1974) was to search for organic molecules or other volatiles released in pyrolysis of Martian fines to up to 200°C, 350°C, or 500°C that might be associated with microbial life. The sensitivity of the Viking GCMS for those organic molecules that could be transmitted through its GC column and through a palladium hydrogen separator to its 12–200 dalton magnetic sector mass spectrometer was a function of the attenuation of the gas directed into the mass spectrometer. This gas flow was limited during portions of the GC run to prevent saturation of the small vacuum ion pump. However, during the most sensitive period of operation, the Viking GCMS system would have been able to detect molecules at the several parts per billion (mass ratio to solid sample heated). Nevertheless, no organic molecules attributed to a Martian source were identified by either lander from either surface samples or from samples collected several centimeters below the surface by the Viking arm and scoop.

The negative Viking GCMS result for organic molecules can be qualified by the following observations: (1) the sensitivity of the GCMS for light organic molecules was reduced by a factor of $\sim 1{,}000$ for light molecules by the design of the gas-processing system that protected the vacuum ion pumps; (2) several classes of polar organic molecules such as carboxylic acids that are likely oxidation products (Benner et al. 2000) of aliphatic and aromatic hydrocarbons would not have been transmitted through the Viking GC column; (3) the Viking sample-acquisition system could only sample loosely consolidated fines instead of less permeable materials that might have been better protected from atmospheric oxidants. Nevertheless, these results provide motivation to use the extraordinary remote sensing tools presently available from orbital platforms to identify sites that are better candidates than the Viking landing sites for preservation of organic molecules that can be safely be accessed by a mobile landing platform. No surface organic molecules have been identified, to date, from orbit.

Several studies have been directed at identification of organics in meteorites that were likely removed from Mars by impact ejection (McSween 1994). These include reports of polycyclic aromatic hydrocarbons (PAHs) in the Antarctic martian meteorite ALH 84001 (McKay et al. 1996) detected by resonance ionization time-of-flight mass spectrometry and the organic volatile products benzene, toluene, C2 alkylbenzene, and benzonitrile detected by pyrolysis GCMS in Nakhla (Sephton et al. 2002). No organic pyrolysis products were detected by the later authors in their samples of ALH 84001 while EET A79001 was also found to release aromatic organics. The extent of terrestrial contribution to the organic material Martian meteorites is necessarily a primary concern in this type of study and has received considerable attention for PAHs (Becker et al. 1997; Clemmett et al. 1998), amino acids (Bada et al. 1998), and for acid-insoluble organic material (Jull et al. 2000). Approaches to understanding the extent of terrestrial contamination can include analysis of the environment from where the meteorite was collected, a search for molecules expected to be produced abiotically, and precision carbon and hydrogen isotope measurements.

## 3.4  Distinguishing Sources of Martian Organics

One source of organic compounds delivered to Mars is certainly meteoritic and interplanetary dust particle infall. Organic compounds contained in these materials would be expected to be present in the martian regolith in the absence of their chemical oxidation in the martian surface environment. If a sufficiently rich suite of organic molecules are detected on Mars, both the distribution of chemical structures and isotopic composition will be employed to help establish their source. Organic molecules produced abiotically in space and exposed to radiation processing show distinct structural and isotopic characteristics. For example, they exhibit more highly branched carbon chain structures than those organic compounds that are the products of biological processes and exhibit a more uniform variation of abundance with molecular weight. For extraterrestrial organic matter delivered to Earth, differences in the stable carbon and hydrogen isotope ratios are also used as a tool to help distinguish these organics from the often dominating terrestrial organic matter. Although carbon isotopes alone may not provide sufficient discrimination, these measurements—combined with other information such as the D/H ratio and the structural information—can often provide a good indication of their extraterrestrial origin (Sephton and Botta 2005).

The classes of organic compounds delivered to Mars from space may resemble those found in meteorites delivered to Earth, some of which are rich in organic compounds. For example, in the Murchison (CM2) carbonaceous chondrite a wide range of compound classes are found. These include more than 80 amino acids, nucleobases, sugar-related compounds, polycyclic aromatic hydrocarbons, carboxylic acids, as well as alcohols, aldehydes, ketones, and aliphatic and aromatic compounds (Sephton and Botta 2005). A primary objective of the SAM investigation will be to determine if these compound types are preserved in the near-surface materials in the chemical environment of the MSL landing site. Life on Earth imprints specific patterns in molecular structure, such as an enhancement in linear vs. branched carbon chains, specific chirality, and even/odd enhancements produced by enzymatic processing (Summons et al. 2007). On Earth extant life can be distinguished by homochirality of the amino acid building blocks of proteins. One of the six SAM GC columns will have the capability to separate a number of chiral species, such as amines. In addition to structural patterns imprinted on sets of organic molecules during their formation, environmental processing results in evolution of these patterns (Eigenbrode 2007, this volume). If sufficient abundance of organic molecules are present, the patterns in molecular structure will be diagnostic of their source. In this preliminary phase of exploration of the abundance of organics on Mars, the focus of the SAM GCMS experiment is to identify the widest possible range of organic compounds within the constraints of our pyrolysis and our substantially more limited one-step solvent extraction and derivatization processing.

Of particular concern for martian organic analyses is terrestrial contamination. Organic compounds derived from Earth could be introduced to the samples before or during in situ sample acquisition and processing by the MSL and could potentially compromise the analysis of martian organics. Thus, the MSL science and engineering teams are working to mitigate this potential problem through a multi-pronged approach. Organic materials such as lubricants and epoxies used in the construction of the MSL systems are screened and whenever possible analyzed using similar pyrolysis GCMS techniques to those that will be employed on the surface of Mars. Materials that will contact or come in close proximity with the martian samples during acquisition and processing are most carefully selected and analyzed. A plan has been formulated to employ witness plates to collect organic materials emanating from MSL components during all stages of its assembly and qualification. Inside the SAM suite, the cells that will accept samples can be heated to ~1,000°C prior to sample delivery to drive off residual organic materials that might have migrated to its surfaces. Several organic free blanks spiked with an easily identified fully fluorinated molecule are planned to be processed inside the SAM suite and occasionally through the entire sample manipulation system of the rover. This experiment will establish the level of contamination picked up during mechanical processing to produce the fine-grained material used by SAM and the fluorinated molecule will provide an externally delivered standard as a check on both sample delivery integrity and instrument performance. For practical reasons, the budget for terrestrial contamination is set at low parts per billion for several organic compound classes of greatest interest (e.g., benzene or aromatic hydrocarbons, 8 ppb; carbonyl or hydroxyl compounds, 10 ppb; amino acids, 1 ppb; amines or amides, 8 ppb; non-aromatic hydrocarbons, 8 ppb) although the sensitivity of the GCMS for stable organic compounds that evolve during pyrolysis and that transmit through the GC column is sub-parts per billion for a well-baked mass spectrometer analyzer. The target upper limit for total terrestrial reduced carbon in any MSL-processed sample delivered to SAM is 40 ppb. If organic molecules are not abundant at the MSL landing site, the definitive identification as indigenous to Mars of trace species detected will depend on how well the MSL developers are able to meet this contamination requirement.

## 3.5 Methane on Mars

Several of the various Mars atmospheric methane mixing ratios or upper limits that have been reported from both ground-based observations (Villanueva et al. 2006; Krasnopolsky et al. 2004) and from the Planetary Fourier Spectrometer PFS on the Mars Express spacecraft (Formisano et al. 2004) are shown in Table 1. Methane is a potential biomarker because it can be produced from extant or extinct microbial sources, but there are also plausible abiotic methane sources including serpentinization at depths of several kilometers below the martian surface, volcanic emissions, or exogenous delivery from primitive bodies such as cometary sources. While analysis of the likelihood of each of these sources continues to be analyzed (Atreya et al. 2006; Krasnopolsky 2006), it is clear that a considerably improved data base regarding methane source locations and temporal variability is needed to understand the sources and sinks of methane on Mars. Both the PFS and the ground-based data suggest a spatial and temporal variability that may indicate that the source flux is much greater than the fluxes suggested by the average mixing ratio and the predicted photochemical destruction rate.

It has been suggested that atmospheric chemistries (Atreya et al. 2006) that are the consequence of dust storm or dust devil electric field induced reactions (Delory et al. 2006) can produce the oxidant $H_2O_2$ in sufficient abundance to precipitate onto the martian surface. This oxidant could contribute to the destruction of reduced carbon compounds delivered by IDP and meteoritic infall to the martian surface and, in fact, this atmospheric chemistry driven by the dust devils and dust storms might also provide a sink for atmospheric methane. Identifying the sources and sinks of $H_2O_2$ could be critically important for understanding organic preservation in the martian environment.

## 4 Mars Science Laboratory Planned Capabilities

The selected MSL payload incorporates ten selected investigations. The locations of the instruments that acquire MSL scientific data and of various other essential MSL elements are shown in Fig. 1. The ten instruments or instrument suites fall into one of four categories: (1) remote sensing, (2) contact, (3) environmental, or (4) analytical laboratory. The Mars Exploration Rovers (MER) Spirit and Opportunity also contained remote-sensing and contact instruments, but both the environmental and analytical laboratory instruments are new

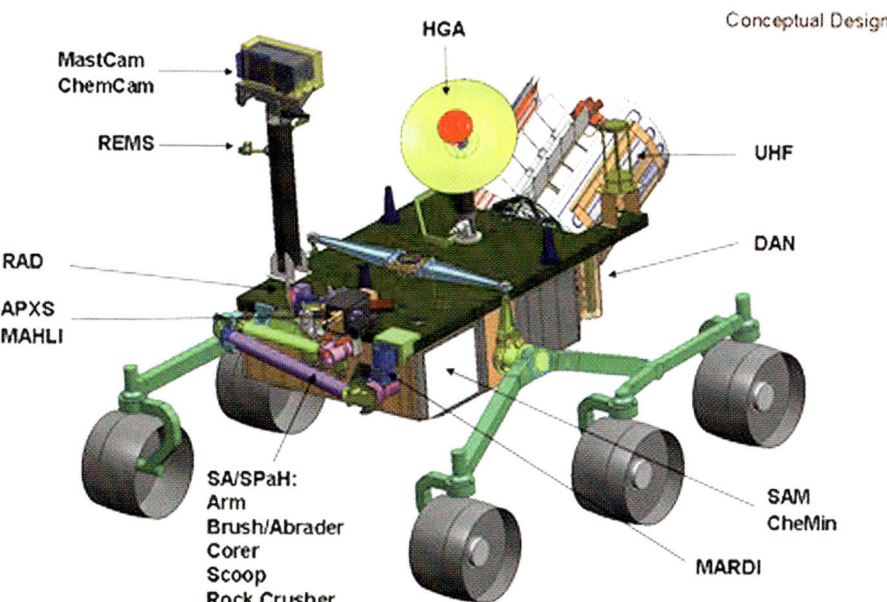

**Fig. 1** The locations of MSL elements are illustrated in this conceptual design. For scale the wheel height is 1/2 meter and the mast instruments are ∼2 meters above the surface. Clockwise from bottom center these are the Sample Acquisition/Sample Handling and Processing (SA/SPaH) system, the Mars Hand Lens Imager (MALI), the alpha particle X-ray spectrometer (APXS), the Radiation Assessment Detector (RAD), the Rover Environmental Monitoring Station (REMS), the Chemistry and Micro-Imaging (ChemCam) experiment, the Mast Camera (MastCam), the High Gain Antenna (HGA), the ultra-high-frequency (UHF) communication element, the Dynamic Albedo of Neutrons (DAN) investigation, the Sample Analysis at Mars (SAM) instrument suite, the Chemistry and Mineralogy (CheMin) instrument, and the Mars Descent Imager (MARDI)

measurement tools for a mobile Mars lander. During Mars surface exploration the high-resolution imaging system (MastCam) and laser induced breakdown spectrometer (Chem-Cam) will provide the first indications of sample morphology and chemistry. A more detailed screening of samples that can be reached by the MSL arm will be obtained by the contact instruments. These are the alpha particle X-ray spectrometer (APXS), a more sensitive version of the MER elemental analysis instruments (Gellert et al. 2005), and MAHLI, the microscopic imager (Edgett et al. 2005). In the meantime, environmental data on atmospheric temperatures, pressure, and humidity as well as the subsurface hydrogen abundance are, respectively, obtained by the meteorology instrument (REMS) and DAN, the neutron detector (Hassler et al. 2006). The radiation experiment (RAD) will provide data that are relevant for both scientific exploration and for future human exploration. The rapid sample screening will enable efficient use of the relatively slow operation of coring a sample from a rock, grinding it to powder suitable for use in the instruments in the analytical laboratory and the mineralogy (CheMin) and organics and isotopes (SAM) measurements. CheMin (Blake et al. 2007) provides definitive identification of minerals with both X-ray diffraction (XRD) and X-ray fluorescence (XRF) capability. In addition to the organic analysis protocols described below SAM carries out a range of noble gas and trace species composition and isotope measurements. These are not described here, other than to note that the SAM TLS is able to measure atmospheric methane to a sub ppb mixing ratio and the $^{13}C/^{12}C$ ratio in methane to an estimated accuracy of 3 per mil for a mixing ratio of 10 ppb or higher.

## 5 SAM Measurement Protocols for Organic Analysis

SAM consists of three instruments: a quadrupole mass spectrometer (QMS), a gas chromatograph (GC), and a tunable laser spectrometer (TLS) supported and integrated by the elements listed in Figs. 2 and 3. These elements include an internal sample manipulation system and a high-capacity pumping station. SAM is designed to measure volatile trace-gas species, including atmospheric gases or organics thermally or chemically extracted from solid-phase materials, such as rocks or fines. Three fundamentally different approaches are employed for the measurement of organics in solids delivered to SAM by the sample acquisition/sample preparation and handling unit (SA/SPaH).

### 5.1 Pyrolysis

The primary method for the detection of organic molecules by SAM is pyrolysis. This approach samples the gas thermally evolved from a small aliquot of sample delivered from the SA/SPaH to one of 60 quartz cups of the sample manipulation system. Each quartz cup can accommodate up to 0.5 cm$^3$ of sample and the incremental volume of sample delivered from the SA/SPaH is approximately 0.05 cm$^3$. This enables a specified volume of sample to be delivered to a cup and possible reuse of cups in an extended mission by deposition of fresh sample on the devolatilized residue in a cup. For direct analysis via the QMS or TLS, the sample in the quartz cup is heated from ambient to ~1,000°C with a programmable temperature ramp. As gases are released, they are swept through the gas manifold by a helium carrier gas for detection by the spectrometers. Typically, water of hydration is released from samples early in the temperature ramp. Moderately volatile organics are released in the 300°C to 600°C region due to thermal desorption and the breakdown of macromolecular organic matter. At higher temperatures, gases evolve due to the further pyrolysis of refractory organics and the breakdown of minerals. For example, carbonates and sulfates

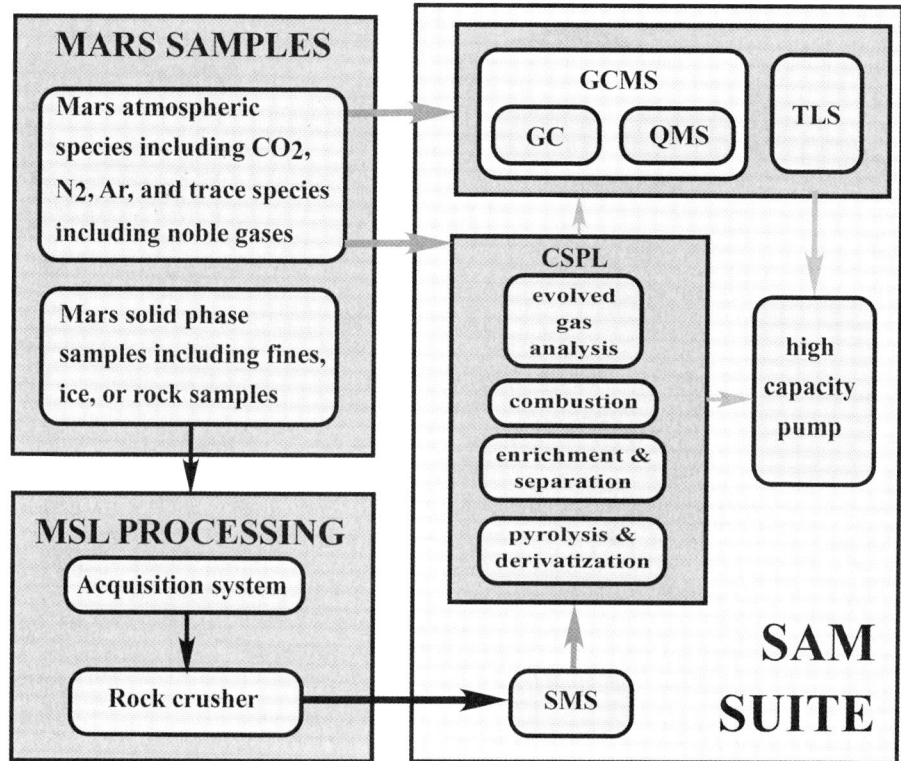

**Fig. 2** The gas and solid sample flow through the SAM suite. The instruments or major components listed are the sample manipulation system (SMS), the wide-range pump, the tunable laser spectrometer (TLS), the quadrupole mass spectrometer (QMS), and the gas chromatograph (GC). The organic extraction is achieved in the chemical separation and processing laboratory (CSPL)

thermally dissociate to $CO_2$ and $SO_2$, respectively, at temperatures greater than $500°C$. The temperature at which minerals degrade is often diagnostic of the mineral type.

In addition to direct analysis of the evolved gases, there is an option in SAM to direct the gas flow over a high-surface-area adsorbent to trap organic molecules, thus separating organic from inorganic volatiles for subsequent analysis by GCMS. Passing the trapped and released organic volatiles through one of six GC columns effectively separates different molecules allowing for individual detection in the QMS. As a consequence of pyrolysis, the complex refractory organic molecules embedded in a mineral matrix or thermally unstable species may break down during thermal processing to produce lower molecular weight or more stable pyrolysis products. Thus, information on the parent organic molecules must be inferred from the patterns of stable products evolved with temperature. For example, pyrolysis of microbial material typically evolves amines, but the more fragile amino acids are destroyed (Glavin et al. 2006). Although the performance of the flight version of the SAM GCMS is not yet established, our tests on the SAM prototype instrument suggests that limit of detection will be $\sim 10–14$ to $10–13$ mole depending on compound and instrument background. The mass range of the QMS is 2–535 dalton to sample a wide range of organic compounds.

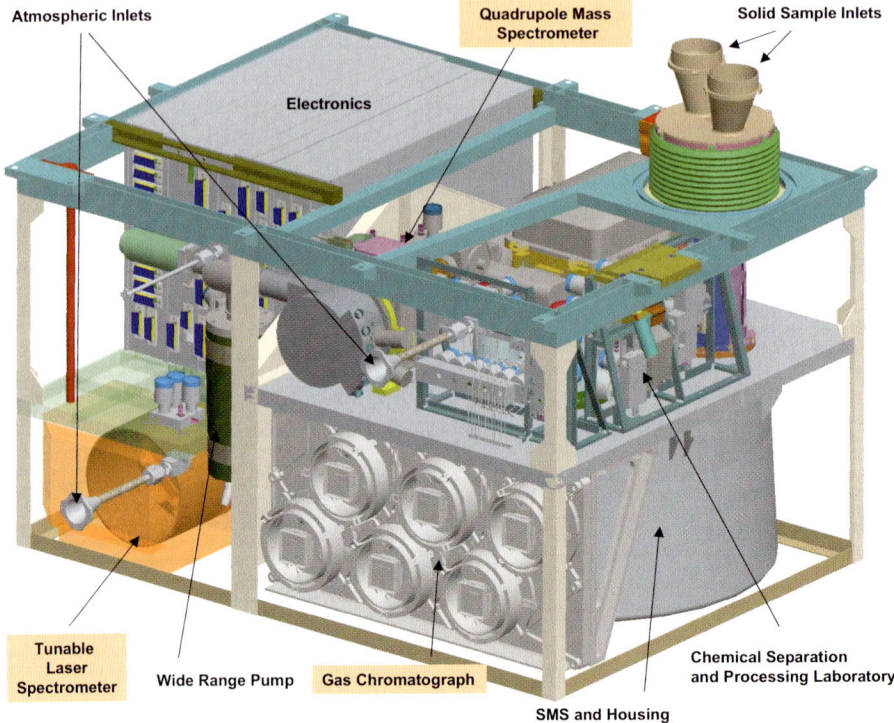

**Fig. 3** The layout of the various SAM components is illustrated. The mass of the suite is greater than 30 kg and the dimensions of the frame in length, width, and height are 53 cm, 42 cm, and 31 cm, respectively

## 5.2 Combustion

The second tool used by SAM to understand the state of carbon in Mars rocks and fines is combustion. Reduced carbon in samples delivered to the SAM sample cups is planned to be oxidized by stepped combustion using isotopically pure $^{16}O_2$ and analyzed in the QMS and the TLS for the $^{13}C/^{12}C$ ratio in the $CO_2$ product. The $^{13}C/^{12}C$ ratio in organic matter is used as a biomarker on Earth because organisms prefer the lighter $^{12}C$ isotope and typically incorporate 2–4% more $^{12}C$ into their cells than is present in the $CO_2$ carbon source. The utility of these measurements will depend on the ability of SAM to reveal the isotopic composition of the most important reservoirs of inorganic carbon for comparison with organic carbon. Two such reservoirs are the atmosphere itself and carbonates that may also be found in the rocks or sedimentary materials that MSL may be able to sample. The combined evolved gas and atmospheric sampling should enable this comparison.

## 5.3 Solvent Extraction and Derivatization

In addition to the dry pyrolysis experiments, a small number of SAM sample cups are dedicated to a simple single-step solvent extraction and chemical derivatization process. Resource constraints of MSL preclude a more ambitious fluid extraction and analysis approach. Depending on the chemistries encountered in Mars rocks and soils, this technique may be effective in enabling an analysis of several classes of molecules that could be of biotic or prebiotic relevance including amino acids, amines, carboxylic acids, and nucleobases. Without derivatization these compounds would not elute from the columns of the gas chromatograph under the protocols applied on SAM. Extraction of the organics from the powdered sample delivered to the SAM cells employs dimethylformamide (DMF) and the derivatization agent is a silylation reaction utilizing N,N-Methyl-tert.-butyl(dimethylsilyl)trifluoroacetamide (MTBSTFA).

The selected volumes of derivatization agent and solvent are mixed together during the SAM integration and then hard-sealed into a metal cup. The top of the cup consists of an electron-beam welded foil that can be punctured on the surface of Mars using the vertical motion of the sample cup into a foil puncture station. The cup is then placed under the SAM sample inlet tube so that these fluids can be mixed with the Mars powdered sample. The cup is next delivered to the pyrolysis station where the desired reaction temperature ($\sim 80°$C) in the sample is set. The hard seal into the pyrolysis chamber ensures that vapor does not escape to space during this reaction time. After several tens of minutes of reaction, much of the solvent is evaporated to space through a microvalve and a heated vent tube and the chemical products produced by the derivatization reaction are flash heated into the injection traps of the gas chromatograph.

The SAM team has exposed the selected solvents and derivatization agents to more radiation than would be expected over the course of the mission to ensure that the radiation-induced chemistry is negligible over the nominal SAM mission. While it is possible that excessive water in the Mars sample or silylation side reactions with salts or clays may substantially compromise the detection of amino acids and carboxylic acids, this "one-pot" extraction/derivatization method has been successfully applied by the SAM team (Buch et al. 2006) to several terrestrial Mars analogs including low-bioload Atacama samples previously studied by pyrolysis and other techniques (Navarro-González et al. 2003). The volume of the SAM wet cells enables a substantial excess of derivatization agent to be supplied that will mitigate the impact of side reactions. An internal standard (a fluorinated amino acid compound in a separately punctured dry chamber) will allow us to evaluate if these undesired side reactions have fully or partially prevented the reaction with the organics of interest. See Buch et al. (2006) for a preliminary report on the development of the protocol used in SAM and its application to Atacama Mars analog samples.

## 6 Mars Analog Studies

A variety of terrestrial or meteoritic samples have been examined either in the laboratory or in field studies using breadboard SAM GCMS components to evaluate the effectiveness of the protocols described earlier. These include dry desert soils from the most lifeless region of the Atacama desert, Hawaii Mars analog soils from Mauna Kea, arctic hot spring fossil samples, and meteoritic samples.

Selected regions of the Atacama desert are among the driest on Earth and near-surface soils have been sampled that contain no culturable heterotrophic bacteria (Navarro-González et al. 2003) using standard culture techniques with only benzene and formic acid detected as organic pyrolysis products. Nevertheless, trace levels of amino acids such as alanine, glycine, valine, and proline could be extracted in our laboratory from these soils using the solvent extraction and derivatization protocol described earlier. Also extracted and derivatized from these Atacama samples were oxalic acid, benzoic acid, and acetic acid.

A range of geologically diverse sites in the protected islands of Svalbard have recently been studied as Mars analog terrestrial environments. These islands contain both Devonian

**Fig. 4** (**a**) The field pyrolysis extraction GCMS spectrum of a dolomite-cemented, hydrocarbon-rich sample shows a variety of organic compounds. Structures for several of the highest probability matches to the National Institute of Standards and Technology (NIST) mass spectrometer data base are illustrated for several of these peaks. This sample was obtained from a sedimentary outcrop in the Ebbadalen Formation in Billefjorden. This area contains a diversity of Carboniferous sandstone, evaporate, and carbonate deposits and was one of several sites examined during the AMASE 06 field campaign. The hydrocarbon chain compounds are the most abundant pyrolysis products. (**b**) Subsequent laboratory derivatization of another sample of the dolomite using the SAM one-cup derivatization protocol. Although we still need to confirm that the spectrum does not include a contribution from handling contamination, the pattern in molecular weight of the fatty acids illustrates a molecular biomarker pattern that SAM is designed to identify on Mars. TBDMS designates the tert.-butyldimethylsilyl derivatized acid for silylation of non-amenable GC compounds

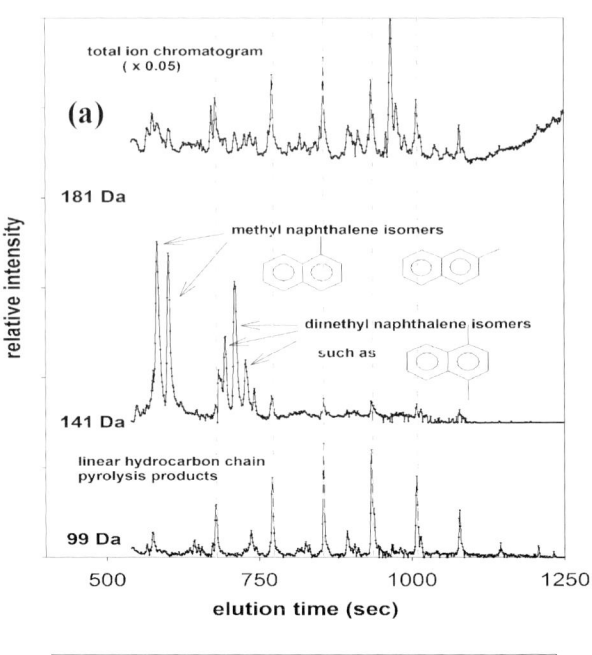

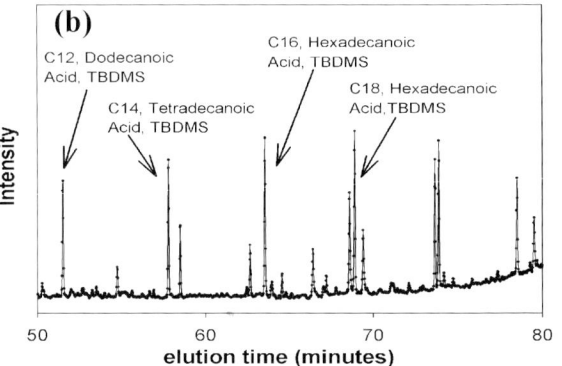

sandstones and sediments and more recent Carboniferous to Tertiary deposits as well as a variety of volcanic and hot spring environments with extensive carbonate deposits. The 2006 Arctic Mars Analogue Svalbard Expedition (AMASE 06) was supported both by NASA and several Norwegian research institutions to field test candidate in situ instruments and approaches to sample screening and selection. Both of SAM's extraction protocols with one-cup extraction/derivatization and pyrolysis were tested in the field during AMASE 06. Figure 4 illustrates field data from a pyrolysis run where several milligrams of a dolomite cemented deposit found in the Ebbadalen Formation in Billefjorden were heated and the evolved gases sent to a GC column that is under evaluation for SAM use on MSL. This sample was found to be very rich in organic compounds with a variety of hydrocarbon types including aromatic compounds and a variety of compounds containing carbon, hydrogen, oxygen, and hydrogen in both linear and heterocyclic configurations. Derivatization analysis of this sample showed a much simpler GCMS spectrum indicating a significant abundance of carboxyl moieties with a distinct biomarker signature in their molecular weight distribution.

# 7 Summary

The MSL sample acquisition and processing system under development combined with the SAM extraction protocols of pyrolysis and derivatization together with TLS measurements of methane and GCMS measurements of heavier organic molecules provide tools for analysis of a wide range of compounds in the martian atmosphere and in a variety of solid near-surface samples. The data acquired from these analyses are expected to provide key insight into the habitability of Mars by potential extinct or extant life.

**Acknowledgements**   Ms. Kirsten Firstad and Dr. Oliver Botta collaborated in the collection and analysis of the Svalbard samples described in Fig. 4 and Dr. Andrew Steele leads the Svalbard Expedition. Both the Mars analog field work and the SAM development are enabled by NASA support.

# References

D.M. Anderson, K. Biemann, L.E. Orgel, J. Oro, T. Owen, G.P. Shulman, P. Toulmin, H.C. Urey, Icarus **16**, 111 (1972)

S.K. Atreya, P.R. Mahaffy, A.S. Wong, Planet. Space Sci. **55**, 358–369 (2006)

J.L. Bada, D.P. Glavin, G.D. McDonald, L. Becker, Science **279**, 362–365 (1998)

L. Becker, D.P. Glavin, J.L. Bada, Geochim. Cosmochim. Acta **61**, 475–481 (1997)

L. Becker, B. Popp, T. Rust, J.L. Bada, Earth Planet. Sci. Lett. **167**, 71–79 (1999)

S.A. Benner, K.G. Devine, L.N. Matveeva, D.H. Powell, PNAS **97**, 2425–2430 (2000)

J.-P. Bibring, Y. Langevin, J.F. Mustard, F. Poulet, R. Arvidson, A. Gendrin, B. Gondet, N. Mangold, P. Pinet, F. Forget, Science **312**, 400 (2006)

K. Biemann, Orig. Life Evol. Biospheres **5**, 457–462 (1974)

K. Biemann, J. Oro, P. Toulmin III, L.E. Orgel, A.O. Nier, D.M. Anderson, D. Flory, A.V. Diaz, D.R. Rushneck, P.G. Simmonds, Science **194**, 72–76 (1976)

P.A. Bland, T.B. Smith, Icarus **144**, 21–26 (2000)

D.F. Blake, P. Sarrazin, D.L. Bish, S.J. Chipera, D.T. Vaniman, S. Collins, S.T. Elliott, A.S. Yen, Lunar Planet. Inst. Conf. Abstr. **38**, 1257 (2007)

O. Botta, J.L. Bada, Surv. Geophys. **23**, 411–467 (2002)

W.V. Boynton, W.C. Feldman, S.W. Squyres, T.H. Prettyman, J. Bruckner, L.G. Evans, R.C. Reedy, R. Starr, J.R. Arnold, D.M. Drake, P.A. Englert, A.E. Metzger, I. Mitrofanov, J.I. Trombka, C. d'Uston, H. Wanke, O. Gasnault, D.K. Hamara, D.M. Janes, R.L. Marcialis, S. Maurice, I. Mikheeva, G.J. Taylor, R. Rokar, C. Shinohara, Science **297**, 81–85 (2002)

A. Buch, D.P. Glavin, R. Sternberg, C. Szopa, C. Rodier, R. Navarro-González, F. Raulin, M. Cabane, P.R. Mahaffy, Planet. Space Sci. **54**, 1592–1599 (2006)

S.J. Clemmett et al., Faraday Discuss. **109**, 417–436 (1998)

J.R. Cronin, S. Pizzarello, D.P. Cruikshank, in *Meteorites and the Early Solar System*, ed. by J.F. Kerridge, M.S. Matthews (University of Arizona Press, Tucson, 1998), pp. 819–857

G. Delory, W. Farrell, S.K. Atreya, N. Renno, A.S. Wong, S.A. Cummer, D.D. Sentman, J.R. Marshall, S. Rafkin, D.C. Catling, Astrobiology **6**, 451–462 (2006)

K.S. Edgett, J.F. Bell III, K.E. Herkenhoff, E. Heydari, L.C. Kah, M.E. Minitti, T.S. Olson, S.K. Rowland, J. Schieber, R.J. Sullivan, R.A. Yingst, M.A. Ravine, M.A. Caplinger, J.N. Maki, 36th Annual Lunar and Planetary Science Conference, vol. 36 (2005), p. 1170

J. Eigenbrode, Space Sci. Rev. (2007), this volume, doi: 10.1007/s11214-007-9213-3

G.J. Flynn, D.S. McKay, J. Geophys. Res. **95**, 14497–14509 (1990)

V. Formisano, S. Atreya, T. Encrenaz, N. Ignatiev, M. Giuranna, Science **306**, 1758–1761 (2004)

D.P. Glavin, H.J. Cleaves, A. Buch, M. Schubert, A. Aubrey, J.L. Bada, P.R. Mahaffy, Planet. Space Sci. **54**, 1584–1591 (2006)

R. Gellert, R. Rieder, J. Brückner, B.C. Clark, G. Dreibus, G. Klingelhöfer, G. Lugmair, D.W. Ming, H. Wänke, A. Yen, J. Zipfel, S.W. Squyres, J. Geophys. Res. (Planets) **111**, 2 (2005)

D.M. Hassler, R.F. Wimmer-Schweingruber, R. Beaujean, S. Bottcher, S. Burmeister, F. Cucinotta, R. Muller-Mellin, A. Posner, S. Rafkin, G. Reitz, The Rad Team 36th COSPAR Scientific Assembly, vol. 36 (2006), p. 2720

A.J.T. Jull, J.W. Beck, G.S. Burr, Geochim. Cosmochim. Acta **64**, 3763–3772 (2000)

H.P. Klein, N.H. Horowitz, G.V. Levin, V.I. Oyama, J. Lederberg, A. Rich, J.S. Hubbard, G.L. Hobby, P.A. Straat, B.J. Berdahl, G.C. Carle, F.S. Brown, R.D. Johnson, Science **194**, 99 (1976)

G. Kminek, J. Bada, Earth Planet. Sci. Lett. **245**, 1–5 (2006)

O. Korablev, M. Ackerman, V.I. Moroz, D. Muller, A.V. Rodin, S.K. Atreya, Planet. Space Sci. **41**, 441–451 (1993)

V.A. Krasnopolsky, J.P. Maillard, T.C. Owen, EGS meeting, Niece, France, 2004

V.A. Krasnopolsky, J.P. Maillard, T.C. Owen, Icarus **172**, 537–547 (2004)

V.A. Krasnopolsky, Icarus **180**, 359–367 (2006)

V.A. Krasnopolsky, G.L. Bjoraker, M.J. Mumma, D. Jennings, J. Geophys. Res. **102**, 6525–6534 (1997)

G. Levin, The Viking Labeled Release Experiment and Life on Mars. Proceedings of SPIE, 29 July–1 August 1997, San Diego, California

G.V. Levin, P.A. Straat, J. Geophys. Res. **82**, 4663 (1977)

W.C. Macguire, Icarus **32**, 85–97 (1977)

M.C. Malin, K.S. Edgett, Science **288**, 2330–2335 (2000)

D.S. McKay, E.K. Gibson, K.L. Thomas-Keprta, H. Vali, C.S. Romanek, S.J. Clemett, X.D.F. Chiller, C.R. Maechling, R.N. Zare, Science **273**, 924–930 (1996)

H.Y.J. McSween, Meteoritics **29**, 757–779 (1994)

S. Murchie, the CRISM Science and Engineering Teams, First Results from the Compact Reconnaissance Imaging Spectrometer for Mars (CRISM), Lunar and Planetary Institute Conference Abstracts, vol. 38 (2007), p. 1472

R. Navarro-González, F.A. Rainey, P. Molina, D.R. Bagaley, B.J. Hollen, J. de la Rosa, A.M. Small, R.C. Quinn, F.J. Grunthaner, L. Cáceres, B. Gomez-Silva, C.P. McKay, Science **302**, 1018–1021 (2003)

A.O. Nier, W.B. Hanson, A. Seiff, M.B. McElroy, N.W. Spencer, R.J. Duckett, T.C.D. Knight, W.S. Cook, Science **193**, 786–788 (1976)

V.I. Oyama, B.J. Berdahl, G.C. Carle, Nature **265**, 100–114 (1977)

A.C. Schuerger, B.C. Clark, Space Sci. Rev. (2007), this volume, doi: 10.1007/s11214-007-9194-2

M.A. Sephton, O. Botta, Int. J. Astrobiol. **4**, 269–276 (2005)

M.A. Sephton, I.P. Wright, I. Gilmour, J.W. de Leeuw, M.M. Grady, C.T. Pillnger, Planet. Space Sci. **50**, 711–716 (2002)

T. Summons et al., Space Sci. Rev. (2007), this volume

S.W. Squyres, and 49 CoAuthors, Science **306**, 1698–1702 (2004)

G.L. Villanueva, M.J. Mumma, R.E. Novak, T. Hewagama, M.A. DiSanti, B.P. Bonev, A.M. Mandell, M.D. Smith, Astrobiology **6**, 242 (2006)

A.S. Yen, B. Murray, G.R. Rossman, F.J. Grunthaner, J. Geopphys. Res. **104**, 27,031–27,041 (1999)

A.S. Yen, R. Gellert, C. Schro, R.V. Morris, J.F. Bell, A.T. Knudson, B.C. Clark, D.W. Ming, J.A. Crisp, R.E. Arvidson, D. Blaney, J. Bruckner, P.R. Christensen, D.J. DesMarais, P.A. de Souza, T.E. Economou, A. Ghosh, B.C. Hahn, K.E. Herkenhoff, L.A. Haskin, J.A. Hurowitz, Bl. Joliff, J.R. Johnson, G. Klingelhofer, M.B. Madsen, S.M. McLennan, H.Y. McSween, L. Richter, R. Rieder, D. Rodionov, L. Soderblom, S.W. Squyres, N.J. Tosca, A. Wang, M. Wyatt, J. Zipfel, Nature **436**, 49–54 (2005)

A.P. Zent, C.P. McKay, Icarus **108**, 146–157 (1994)

# Urey: Mars Organic and Oxidant Detector

**J.L. Bada · P. Ehrenfreund · F. Grunthaner · D. Blaney · M. Coleman · A. Farrington · A. Yen · R. Mathies · R. Amudson · R. Quinn · A. Zent · S. Ride · L. Barron · O. Botta · B. Clark · D. Glavin · B. Hofmann · J.L. Josset · P. Rettberg · F. Robert · M. Sephton**

Originally published in the journal Space Science Reviews, Volume 135, Nos 1–4.
DOI: 10.1007/s11214-007-9213-3 © Springer Science+Business Media B.V. 2007

**Abstract**  One of the fundamental challenges facing the scientific community as we enter this new century of Mars research is to understand, in a rigorous manner, the biotic potential both past and present of this outermost terrestrial-like planet in our solar system. Urey:

---

J.L. Bada (✉)
Scripps Institution of Oceanography, La Jolla, CA 92093-0212, USA
e-mail: jbada@ucsd.edu

P. Ehrenfreund
Leiden Institute of Chemistry, PO Box 9502, 2300 Leiden, The Netherlands

F. Grunthaner · D. Blaney · M. Coleman · A. Farrington · A. Yen
Jet Propulsion Laboratory, Pasadena, CA 91109, USA

R. Mathies · R. Amudson
University of California, Berkeley, CA 94720, USA

R. Quinn
SETI Institute, NASA Ames Research Center, Moffett Field, CA 94035, USA

A. Zent
NASA Ames Research Center, Moffett Field, CA 94035, USA

S. Ride
Imaginary Lines, Inc., 9191 Towne Centre Dr, San Diego, CA 92122-122, USA

L. Barron
Department of Chemistry, University of Glasgow, Glasgow G12 8QQ, UK

O. Botta
International Space Science Institute, 3012 Bern, Switzerland

B. Clark
Space Exploration Systems,  Lockheed Martin, Denver, CO 80201, USA

D. Glavin
NASA/Goddard Space Flight Center, Code 915.0, Greenbelt, MD 20771, USA

Mars Organic and Oxidant Detector has been selected for the Pasteur payload of the European Space Agency's (ESA's) ExoMars rover mission and is considered a fundamental instrument to achieve the mission's scientific objectives. The instrument is named Urey in recognition of Harold Clayton Urey's seminal contributions to cosmochemistry, geochemistry, and the study of the origin of life. The overall goal of Urey is to search for organic compounds directly in the regolith of Mars and to assess their origin. Urey will perform a groundbreaking investigation of the Martian environment that will involve searching for organic compounds indicative of life and prebiotic chemistry at a sensitivity many orders of magnitude greater than Viking or other in situ organic detection systems. Urey will perform the first in situ search for key classes of organic molecules using state-of-the-art analytical methods that provide part-per-trillion sensitivity. It will ascertain whether any of these molecules are abiotic or biotic in origin and will evaluate the survival potential of organic compounds in the environment using state-of-the-art chemoresistor oxidant sensors.

**Keywords** Mars · Life detection instrumentation · Space research

# 1 Introduction

Based on results from the NASA Mars Exploration Rovers and the ESA Mars Express mission, there is compelling and accumulating evidence that liquid water bodies were once present on Mars. Although it is unknown how long these watery environments existed, they could potentially have provided a milieu capable of supporting life or its precursor, prebiotic chemistry. The detection of phyllosilicates and sulfate minerals indicates that Mars has had a complex aqueous history with chemical processing giving rise to sedimentary deposits that could preserve organic material (e.g. Poulet et al. 2005; Bibring et al. 2006; Bishop et al. 2005; Thomas et al. 2005).

All known life is based on organic carbon. The 1976 Viking missions detected no organic compounds above a threshold level of a few parts per billion (ppb) in near-surface samples (Biemann et al. 1976). However, experimental testing has now established that the Viking Gas Chromatography/Mass Spectrometry (GCMS) would not have detected key biomolecules, such as amino acids, even if several million bacterial cells per gram were present (Glavin et al. 2001). In addition, oxidation reactions involving organic compounds on the Martian surface would likely produce nonvolatile products such as mellitic acid salts that would have also precluded detection by the Viking instruments (Benner et al. 2000). Thus,

B. Hofmann
Naturhistorisches Museum der Burgergemeinde Bern, 3005 Bern, Switzerland

J.L. Josset
Space Exploration Institute SPACE-X, Neuchatel, Switzerland

P. Rettberg
Institute of Aerospace Medicine Radiation Biology, Linder Höhe, 51147 Köln, Germany

F. Robert
Muséum National Histoire Naturelle, LEME – NanoAnalyses, 75005 Paris, France

M. Sephton
Department of Earth Science and Engineering, Imperial College, London SW7 2AZ, UK

although Viking clearly demonstrated that the levels of organics at the Viking sites are depleted below the expected levels due to meteoritic input, the results did not conclusively demonstrate the absence of organic compounds on the surface of Mars. An additional consideration is that Viking only sampled the first tens of centimeters of the Martian regolith. It is now thought that ionizing radiation and oxidants formed in the surface and/or near-surface atmosphere have largely destroyed any organic compounds within the upper meter of the Martian surface (Kminek and Bada 2006). Thus, drilling to a depth deeper than this zone of radiolysis is required to obtain samples with the best prospect of containing organic molecules (Dartnell et al. 2007). This is one of the primary goals of the Pasteur rover on ExoMars.

## 2 The Urey Instrument Experiments

The primary scientific objectives of the experiments to be conducted by Urey are to investigate the following questions:

- Are organic compounds with a primary amino group (amino acids, amines, nucleobases, amino sugars) and polycyclic aromatic hydrocarbons (PAHs)—our target compounds—derived from either extinct or extant life and/or abiotic sources, detectable in the regolith of Mars?
- Using compositional and chirality characteristics of detected amino acids, can we determine whether they are of biotic or abiotic origin?
- Are organic compounds degraded in near surface environments on Mars via an array of photolytic and heterogeneous chemical processes, and are these processes a central factor in determining the abundance and type of our target compounds in the Mars regolith?

Figure 1 graphically illustrates the target organic compounds for Urey and the relationships between the amino acids, their chirality, and PAHs. These relationships can be used to determine the origins of any target compounds that are detected. The Urey objectives focus on the search for evidence of extant or extinct life on Mars and derive from the most accepted understanding of the results of the Viking experiments: levels of organic compounds in the near-surface region are extremely low (part per billion or less); there is evidence for at least three different oxidants in the Martian regolith at the Viking sites (Klein 1979); and the surprisingly low levels of organic compounds may be related to the long-term interaction

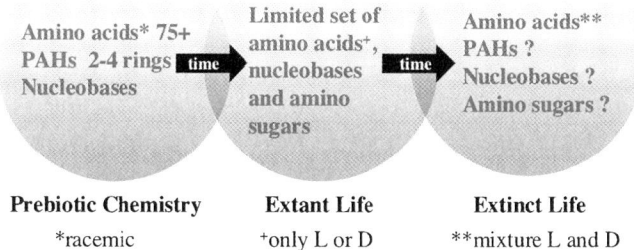

**Fig. 1** The target organic compounds for *Urey* and the relationships between the amino acids, their chirality, and PAHs. Although terrestrial life uses 20 distinct amino acids and 5 nucleobases, life elsewhere could be based on a different ensemble of these compounds, which *Urey* will be able to detect and characterize

with these oxidants or the oxidizing conditions of the Martian surface. Urey will target several key organic molecules at very low concentration levels (1,000 times better than Viking) and use sensors to detect and characterize oxidants present in the same samples. These complementary analyses will examine the hypothesis that oxidants and organic matter on the surface of Mars are inversely correlated.

## 3 Organic Compound Detection and Characterization

In any investigation of organic compounds possibly derived from life, it is important to search for molecules at the core of terrestrial (and presumably extraterrestrial) biochemistry (Pace 2001). Primary examples of such molecules include amino acids and nucleobases. However, abiogenic occurrences of these compounds do exist. More than 70 different amino acids and several amines and nucleobases have been identified in carbonaceous meteorites (Botta and Bada 2002). These compounds can also be readily synthesized in laboratory simulation experiments (Bada 2004). Biological occurrences of these organic compounds exhibit selected characteristic structural forms that facilitate their biochemical roles. Both abiogenic and biogenic organic structures can be detected and differentiated by Urey.

Searching for amino acids and nucleobases is of fundamental importance with respect to assessing the biopotential of Mars. In bacterial cells, amino acids (in the form of proteins) and nucleobases (in the form of nucleic acids) constitute nearly 75% by dry weight of the total organic material. Although it is not certain that an extraterrestrial biology would use the same set of amino acids as on Earth, their presence in certain types of meteorites indicates they were constituents of organic material in the early solar system and thus available for incorporation in living entities elsewhere (Sephton and Botta 2005). In addition, amino acids are robust compounds and would be expected to survive for billions of years in the Martian regolith (Aubrey et al. 2006).

In terrestrial biology, 20 different amino acids are incorporated into proteins and 5 different nucleobases into nucleic acids. Compositional analyses provide important information with respect to whether any of these compounds detected on Mars consist of a limited set of compounds such as found in biology or the more diverse set found in meteorites and prebiotic experiments. One of the most distinctive features of amino acids is their molecular architecture, which gives rise to the property of handedness or chirality. Most amino acids contain asymmetrically substituted carbon atoms that result in left-handed and right-handed versions (enantiomers). Abiotic chemical reactions (in the absence of artificial asymmetry-inducing compounds) always produce a 1 : 1 (racemic) mixture of the two enantiomers. Thus, equal (or nearly equal) amounts of left and right amino acids are found in carbonaceous chondrites that are free of terrestrial contaminants. Likewise, prebiotic amino acid synthesis yields racemic mixtures in absence of an asymmetric source in the system. In contrast, terrestrial biology uses almost totally left-handed (L-enantiomer) amino acids as building blocks for proteins and enzymes. Although some D-amino acids are found in the cellular membranes of some bacteria, L-amino acids are far more abundant in the total cell and in Earth regolith.

On Earth, racemization can slowly convert biological amino acids in geologic samples into a racemic mixture. However, the D/L amino acid ratio in geologic samples has never been found to exceed 1.0. Racemization may be very slow because of the cold, dry environmental conditions and thus homochiral amino acids even in deposits billions of years old would still be preserved today (Bada and McDonald 1995). Urey can "look back in time" at amino acids and use their D/L ratio as a window into when past life may have existed on Mars. Amino acid homochirality is found only in biology and is considered to be an inevitable characteristic of any biochemistry, terrestrial or alien. It is likely a universal biomarker signature of molecule-based life (de Duve 2005). Biochemical reactions require exclusively one chiral form of amino acids, although either form could be used. Thus, life elsewhere could be based on either L- or D-amino acids. The detection of homochiral amino acids, whether all L- or all D-amino acids, would provide evidence for the existence of life on Mars (Bada 2001).

Aromatic macromolecules are the dominant organic material in space (Ehrenfreund et al. 2006). Aromatic material is extremely stable and quite versatile and PAHs are some of the most ubiquitous molecules in the solar system and universe. PAHs and aromatic macro-molecular carbon were among the most abundant materials delivered to the early planets (Ehrenfreund et al. 2002) by meteoritic infall. Thus, PAHs are excellent target molecules in the search for organic material that may have originated elsewhere in the solar system and was subsequently delivered intact to the surface of Mars.

Laser-induced fluorescence is one of the most sensitive organic compound detection techniques. It is far more sensitive than other techniques, such as charged particle detection used in gas chromatography/mass spectrometry (GCMS). Fluorescence methods have the capacity to detect single molecules (Moerner and Orritt 1999), and this is now routinely practiced. One of the most exhaustively characterized fluorescent analytical detection chemistries that has been developed is for the detection of primary amines, in particular those compounds associated with biology as we know it: amino acids (the components of proteins) and nucleobases (the components of RNA and DNA). Reagents such as fluorescamine provide sub-ppb sensitivity for organic compounds with a primary amino group. PAHs are by themselves naturally fluorescent when stimulated with near-UV light and thus can be directly detected at ppb levels.

## 4  Establish Surface and Subsurface Oxidation Mechanisms and Rates

The search for organic compounds on the surface of Mars has proven to be a difficult task. In the Martian atmosphere several trace gases have been detected including hydrogen peroxide (Encrenaz et al. 2004) and methane ($CH_4$, Krasnopolsky et al. 2004; Formisano et al. 2004).

The Viking landers performed an in situ search on the surface of Mars for both life and organic compounds. However, soil measurements revealed that the surface materials at the landing sites were chemically but not biologically active under the conditions of the Viking life-detection experiments. Three of the major experimental results of Viking were: (1) the release of $O_2$ gas when soil samples were exposed to water vapor in the Gas Exchange Experiment (GEx) (Oyama and Berdahl 1977); (2) the ability of the surface material to rapidly decompose aqueous organic material that was intended to culture microbial life in the Labeled Release Experiment (LR) (Levin and Straat 1977); and (3) the apparent absence of organics in samples analyzed by gas chromatography and mass spectroscopy (GCMS) (Biemann et al. 1977). Today, 30 years later, these results remain to be fully explained. The most widely accepted explanation for the GEx and LR results is the presence of oxidants in the Martian soil (Klein, 1978, 1979). Differences in stability of the active agents in the two experiments suggest that the GEx and LR oxidants are different species and that at least three different oxidizing species are needed to explain all of the experimental results (Klein 1979). The combined results of the Viking GEx, LR, and GCMS led to the hypothesis that the GEx and LR oxidants are evidence of the oxidative decomposition of organic compounds in the Martian environment (Klein 1978, 1979).

A key to understanding carbon chemistry on Mars lies not only in identifying soil oxidants, but also in characterizing the dominant reaction mechanisms and kinetics of oxidative processes that are occurring on the planet. These processes may have decomposed or substantially modified any organic material that might have survived from an early biotic period. There are currently a number of hypotheses to explain oxidant formation on Mars and the roles of oxidants in both the Viking biology experiments and the decomposition of organics on the planet's surface. These hypotheses include: UV generation of superoxide radicals (Yen et al. 2000), triboelectric enhancement of $H_2O_2$ production (Atreya et al. 2006), and the deposition of oxidizing acids in the Martian soil (Quinn et al. 2005), to name just a few (for a review, see Zent and McKay 1994). Urey will investigate a broad range of Mars oxidant hypotheses, because it is likely that, to a greater or lesser degree, several of the processes that have been hypothesized are occurring on Mars simultaneously (Bullock et al. 1994). There are undoubtedly a number of complex, photochemically driven oxidative processes on Mars involving interrelated atmospheric, aerosol, dust, soil, and organic chemical interactions.

To a large extent, the role these processes play in the carbon chemistry on Mars is unknown. Urey will establish the correlation between the levels of organic compounds, oxidant concentration, water abundance, and UV flux levels, at the various sampling localities, as well as a function of depth in the subsurface. Urey measurements will also discriminate between different oxidant formation and organic decomposition mechanisms that may be occurring.

## 5 Field Tests

The value of the Urey investigation has been demonstrated in a series of comprehensive field tests conducted in the Panoche Desert Valley (California) and in the Atacama Desert (Chile) (Quinn et al. 2005; Skelley et al. 2005). These sites are among the best Martian regolith analog sites on Earth for studying organic survival and degradation, owing to their uniquely arid and oxidizing environment, the presence of sulfate mineral deposits and low levels of indigenous living organisms and detectable organic compounds (Navarro-Gonzales et al. 2003). The analytical integration of portable versions of Urey's components was demonstrated in analyses at both sites and produced high-quality data now published in the peer-reviewed literature (see Fig. 2). The MOI tests carried out simultaneously at the field sites showed how the characterization of oxidants is critical in the evaluation of the results obtained in the organic compound analyses (Quinn et al. 2005).

## 6 Instrumentation Overview

The overall Urey instrument design is shown in Figs. 3 and 4 and consists of four major subsystems:

- Sub-Critical Water Extractor (SCWE)
- Mars Organic Detector (MOD)
- Micro-Capillary Electrophoresis Unit (µCE)
- Mars Oxidant Instrument (MOI).

Mass properties, power, and energy budgets of subsystems are listed in Tables 1 and 2. The total return data (without Exomars-based compression) is <40 Mbits/EC for a full data set that can be returned over several sols. For a quick turnaround product, returned on the next available pass and appropriate to make a drive-away decision, <1 Mbit total is appropriate.

**Fig. 2** (*Top*) μCE eletropherogram showing amino acids and amines detected in an Atacama sample (from Skelley et al. 2005). Nucleobases were determined to be below the detection limit (<1 ppb) in this sample. (*Bottom*) MOI sensor responses showing the detection of trace levels of oxidizing acids in Atacama dust using the MOI (from Quinn et al. 2005)

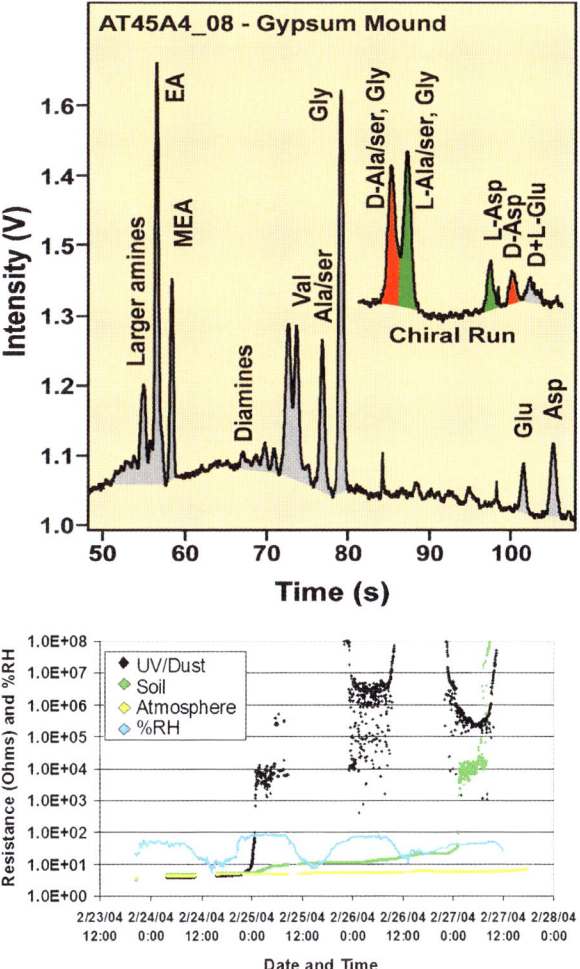

Each of these systems contributes to the overall science output either with analytical processing steps (SCWE/MOD) and/or direct measurements (MOD/μCE/MOI). In addition, integrated electronics and mechanical subsystems tie the analytical components into a cohesive instrument package. The design of Urey is organized around the flow of sample material through the instrument in its various states of analytical processing. Figure 4 illustrates this flow at a conceptual level. The sample to be delivered to the instruments is pulverized prior to delivery. Urey accepts samples of 800 mg. The sample is parsed by the ExoMars sample distribution system into 200 mg for the MOI unit and 600 mg for the SCWE. Solid sample provided to Urey remains inside of the instrument for the duration of the mission.

The SCWE uses subcritical water (20 MPa at 150–325°C) producing a low-dielectric constant solvent that extracts organic compounds from the sample much in the manner of making espresso (Yoshida et al. 1999). Prior to performing the organics extraction, the SCWE flushes salts from the sample with a 30°C "rinse". This flush water is exhausted into a waste collection tank to prevent rover cross-contamination.

**Table 1** Mass properties of the *Urey* instrument

| Component | Mass (g) |
|---|---|
| MOI (including deck unit) | 315 |
| SCWE (wet) | 1972 |
| MOD | 269 |
| µCE | 1423 |
| Sample handling | 443 |
| Total | 4422 |

**Table 2** Power and energy budgets of MOD/µCE and MOI subsystems

| Subsystems | Power | |
|---|---|---|
| MOD/µCE Peak Power: | 36 W | during SCWE heater or sublimation crucible operation |
| MOD/µCE Peak Power: | 6 W | detection readout, standby modes |
| MOD/µCE Energy: | TBD | sequence dependent, not more than 100 W h |
| MOI Peak Power: | 8 W | continuous |
| MOI Energy: | 8 W h | per hour while measuring |

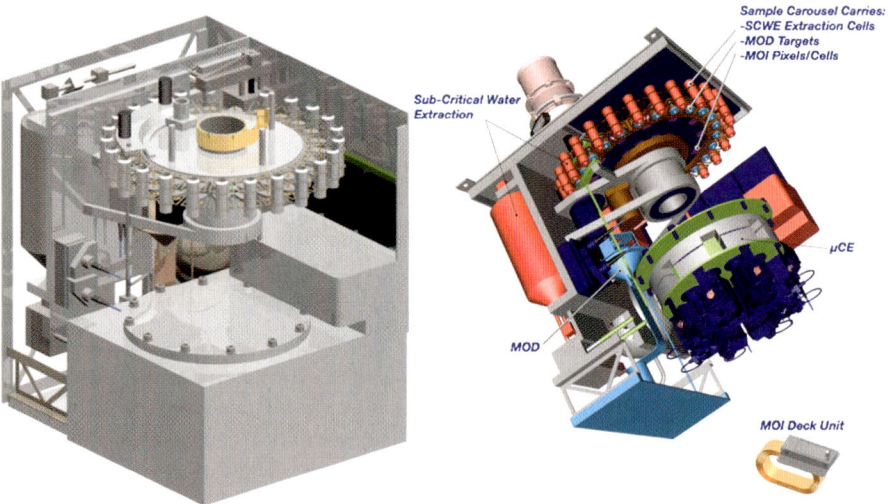

**Fig. 3** Concept drawing of *Urey*: Mars organic and Oxidant Detector

The MOD takes the liquid extract from the SCWE and removes the water by freeze drying and then slowly sublimates the volatile organic compounds to a flourescamine-coated target "puck" held at $-10°C$ by a cold-finger. Using laser-induced fluorescence, it then can excite PAHs or amino acids (bound to the flourescamine) and detect their presence.

The µCE component sends a small amount of liquid to the sublimate on the puck and dissolves some of the sample returning the aliquot to the capillary electrophoresis component

**Fig. 4** Overall sample flow
through *Urey*

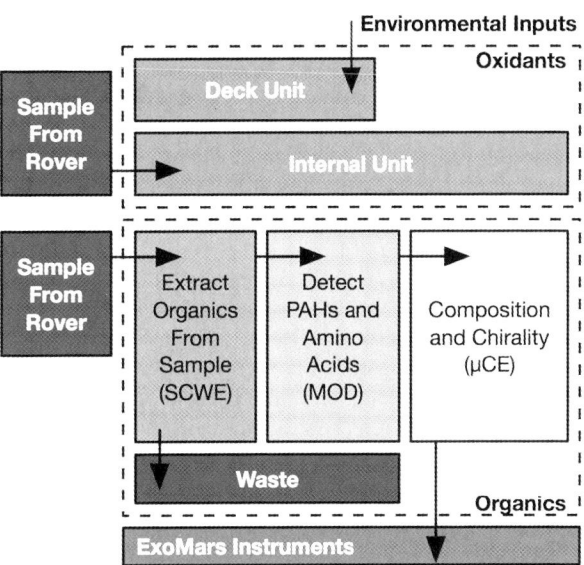

for characterization of composition. Using buffers and other reagents, the μCE component can also detect the chirality of amino acids. The μCE component can also output a small amount of the processed aliquot to other ExoMars instruments for confirming analysis.

MOI is deployed in two separate units. The deck unit uses a filtered sensor configuration to characterize the reactive nature of surface environments and oxidant formation mechanisms by isolating the chemical effects of exposure to dust, UV, and atmospheric gases. This configuration will discriminate between current Mars oxidant hypotheses and measure in situ the effects of UV, dust, and trace gases on the degradation rates of organic compounds. An internal MOI unit is used to bracket the redox potential of the regolith and to characterize oxidative processes that affect the concentration and distribution of organic chemicals in the regolith. The MOI internal unit is configured to examine a split of each regolith sample analyzed by the organics experiment. This configuration will allow the distribution of soil oxidants and oxidative processes to be mapped and correlated with soil organic content. By controlling experimental humidity levels, the MOI internal unit will be able to reproduce (with improved thermal and temporal resolution) the conditions that triggered the high levels of chemical reactivity observed in the Viking experiments.

The fundamental unit of MOI is the "chemical pixel", an $8 \times 2$ array of high-purity gold electrodes patterned on a sapphire substrate. Each electrode is coated with a different film type as specified for each MOI experiment. The electrode gaps and spatial configurations on the array are chosen to maximize the sensitivity of each film type to oxidation. During an experiment, the eight different sensing films are exposed to the environment and eight matched, sealed films serve as controls. The approach is derived from the classical, field-proven "spot test" method of chemical analysis, wherein the identity of unknowns is elucidated through the reaction pattern of the sample with well-characterized test compounds. In the MOI implementation, chemical reactivity levels and oxidation processes are characterized by measuring changes in film electrical resistance as a function of time. A differential measurement approach is used and changes in a sensing film's paired seal reference are used to correct for temperature and other physical effects due to the details of the contact of the thin film to the noble metal of the electrode array.

## 7 Significance of Urey Experiments

The Urey Instrument is an integrated suite which is designed to search the Martian regolith for chiral biomarkers at terrestrial laboratory state-of-the-art detection levels. Urey will search for organic compounds, characterize the biotic composition of discovered organics, and determine chirality. Furthermore, it can explain how oxidants may have affected the original suite of organic compounds and describe how they may have been altered over the geological history of Mars.

A positive result from the Urey investigation, i.e., the detection of our target organic molecules, could be the first convincing demonstration that organic compounds are present on Mars. This would be a major advance in our understanding of the Martian environment and its potential to harbor life. If a convincing signal for the presence of homochiral D-amino acids were found, this would provide the first evidence for the possible presence of unique life that is not related to terrestrial biology. The detection of a homochiral L-amino acid (if different from terrestrial L-amino acids) would provide the same evidence. This result would be of enormous significance to not only NASA and the science community, but to the general public as well. The presence of homochiral L-amino acids would imply that life on Mars and Earth may be related by a common genesis (if these L-amino acids are the terrestrial proteinaceous ones). The alternative possibility, that the amino acids are derived from forward terrestrial contamination, will be quantitatively assessed through controls that determine the degree and composition of residual terrestrial spacecraft contamination. Compositional analyses thus play a pivotal role in the search for life on Mars.

A result indicating no organic target compounds are detectable above the parts-per-trillion detection level will be investigated by the oxidant instrument to establish if the sample environment was capable of preserving organic compounds. This analysis will provide further understanding of the potential for organic compounds to survive the harsh conditions presently found on Mars.

## References

S.K. Atreya et al., Astrobiology **6**, 439–450 (2006)
A. Aubrey, H.J. Cleaves, J.H. Chalmers, A.M. Skelley, R.A. Mathies, F.J. Grunthaner, P. Ehrenfreund, J.L. Bada, Geology **34**, 357–360 (2006)
J.L. Bada, Proc. Natl. Acad. Sci. USA **98**, 797–800 (2001)
J.L. Bada, Earth Planet. Sci. Lett. **226**, 1–15 (2004)
J.L. Bada, G. McDonald, Icarus **114**, 139–143 (1995)
S.A. Benner, K.G. Devine, L.N. Matveeva, D.H. Powell, Proc. Natl. Acad. Sci. **97**, 2425–2430 (2000)
J.-P. Bibring, Omega Team, Science **312**, 400–404 (2006)
K. Biemann, J. Oro, P. Toulmin III, L.E. Orgel, A.O. Nier, D.M. Anderson, P.G. Simmonds, D. Flory, A.V. Diaz, D.R. Rushneck, J.A. Biller, Science **194**, 72–76 (1976)
K. Biemann, J. Oro, P. Toulmin III, L.E. Orgel, A.O. Nier, D.M. Anderson, P.G. Simmonds, D. Flory, A.V. Diaz, D.R. Ruchneck, J.E. Biller, A.L. LaFleur, J. Geophys. Res. **82**, 4641–4658 (1977)
J.L. Bishop, M. Dyar, M. Lane, J. Banfield, Intern. J. Astrobiol. **3**, 275–285 (2005)
O. Botta, J.L. Bada, Surv. Geophys. **23**, 411–467 (2002)
M.A. Bullock, C.R. Stoker, C. McKay, A. Zent, Icarus **107**, 142–152 (1994)
L.R. Dartnell, L. Desorgher, J.M. Ward, A.J. Coates, Geophys. Res. Lett. **34**, LO2207 (2007)
C. de Duve, *Singularities: Landmarks on the Pathways of Life* (Cambridge University Press, NY, 2005)
P. Ehrenfreund, S. Rasmussen, J.H. Cleaves, L. Chen, Astrobiology **6/3**, 490–520 (2006)
P. Ehrenfreund et al., Reports Prog. Phys. **65**, 1427–1487 (2002)
T. Encrenaz et al., Icarus **170**, 424–429 (2004)
V. Formisano, S. Atreya, T. Encrenaz, N. Ignatiev, M. Gluranna, Science **306**, 1758–1761 (2004)
D.P. Glavin, M. Schubert, O. Botta, G. Kminek, J.L. Bada, Earth Planet. Sci. Lett. **185**, 1–5 (2001)
H.P. Klein, Icarus **34**, 666–674 (1978)

H.P. Klein, Rev. Geophys. Space Phys. **17**, 1655–1662 (1979)

G. Kminek, J.L. Bada, Earth Planet. Sci. Lett. **245**, 1–5 (2006)

V. Krasnopolsky, J.P. Maillard, T. Owen, Icarus **172**, 537–547 (2004)

G.V. Levin, P.A. Straat, J. Geophys. Res. **82**, 4663–4668 (1977)

W.E. Moerner, M. Orritt, Science **283**, 1670–1676 (1999)

R. Navarro-Gonzales et al., Science **302**, 1018–1021 (2003)

V.I. Oyama, B.J. Berdahl, J. Geophys. Res. **82**, 4669–4676 (1977)

N.P. Pace, Proc. Natl. Acad. Sci. USA **98**, 805–808 (2001)

F. Poulet, OmegaTeam, Nature **4381**, 623–627 (2005)

R.C. Quinn, A.P. Zent, F.J. Grunthaner, P. Ehrenfreund, C.L. Taylor, J.R.C. Garry, Planet. Space Sci. **53**, 1376–1388 (2005)

M.A. Sephton, O. Botta, Intern. J. Astrobiol. **4**, 269–276 (2005)

A.M. Skelley, J.R. Scherer, A.D. Aubrey, W.H. Grover, R.H.C. Ivester, P. Ehrenfreund, F.J. Grunthaner, J.L. Bada, R.A. Mathies, Proc. Natl. Acad. Sci. USA **102**, 1041–1046 (2005)

M. Thomas, J.D.A. Clarke, C.F. Pain, Aust. J. Earth Sci. **52**, 365–378 (2005)

A.S. Yen, S.S. Kim, M. Hecht, M.S. Frant, B. Murray, Science **289**, 1909–1912 (2000)

H. Yoshida, M. Terashima, Y. Takahashi, Biotechnol. Prog. **15**, 1090–1094 (1999)

A.P. Zent, C.P. McKay, Icarus **108**, 146–157 (1994)

# Raman Spectroscopy—A Powerful Tool for in situ Planetary Science

N. Tarcea · T. Frosch · P. Rösch · M. Hilchenbach ·
T. Stuffler · S. Hofer · H. Thiele · R. Hochleitner ·
J. Popp

Originally published in the journal Space Science Reviews, Volume 135, Nos 1–4.
DOI: 10.1007/s11214-007-9279-y © Springer Science+Business Media B.V. 2007

**Abstract** This paper introduces Raman spectroscopy and discusses various scenarios where it might be applied to in situ planetary missions. We demonstrate the extensive capabilities of Raman spectroscopy for planetary investigations and argue that this technique is essential for future planetary missions.

**Keywords** Raman spectroscopy · Remote Raman spectroscopy · Space-borne Raman spectrometers · In situ planetary science

## 1 Introduction

In the last few years, Raman spectroscopy (Popp and Kiefer 2000) has been recognized as a possible method for in situ planetary analysis (Cochran 1981; McMillan et al. 1996; Sharma et al. 2002; Tarcea et al. 2002; Ellery and Wynn-Williams 2003; Estec et al. 1972; Isreal et al. 1997; Haskin et al. 1997; Korotev et al. 1998; Edwards et al. 1999; Popp et al. 2001; Wang et al. 1999). Two important fields where Raman spectroscopy is used are mineralogical and organic/biological analysis. It has been shown that Raman spectroscopy—and in particular micro-Raman spectroscopy—can contribute to resolving various questions in

N. Tarcea · T. Frosch · P. Rösch · J. Popp (✉)
Institute for Physical Chemistry, University of Jena, Jena, Germany
e-mail: juergen.popp@uni-jena.de

M. Hilchenbach
Max-Plank-Institut für Sonnensystemforschung, Katlenburg-Lindau, Germany

T. Stuffler · S. Hofer · H. Thiele
Kayser-Threde GmbH, Munich, Germany

R. Hochleitner
Mineralogische Staatssammlung, Munich, Germany

J. Popp
Institut für Physikalische Hochtechnologie (IPHT), Jena, Germany

the field of planetary and asteroid investigations because it allows one to address numerous issues, e.g.:

1. Analysis of mineralogical and geochemical materials from a planetary surface in order to understand a planet's evolutionary history.
2. Identification of inorganic, organic or biological compounds, which facilitates the search for past or present life on remote celestial bodies (e.g., Mars). This is the main benefit offered by Raman spectroscopy.
3. Identification of the principal mineral phases (i.e., those making up at least 90% of the material in soils and rocks).
4. Classification of rocks (igneous, sedimentary and metamorphic) and definition of petrogenetic processes.
5. Determination of the oxidation state of planetary elements, e.g., soil, rock surfaces and inside rocks, and the ability to finely differentiate among mineral species.
6. Analysis of the content of volatiles and gaseous inclusions ($H_2O$, $SO_3$, $CO_2$, $NO_2$, $H_2$, $O_2$) in minerals and glasses.
7. Determination of selected minor and trace element contents (e.g., rare earth elements).
8. Determination of reaction kinetics, i.e., oxidation processes on newly exposed surfaces and determination of the reaction products.
9. Morphology of organic inclusions (fossils) and minerals on a micrometer scale obtainable by Raman mapping measurements (Tarcea et al. 2003).
10. Identification of water and ice on, e.g., Mars; identification of secondary minerals, clays, state of carbonaceous matter and hydrated crystals.

Furthermore, Raman spectroscopy in general requires only minimal or no sample preparation. Solid, liquid and gaseous samples can be measured as well as transparent or non-transparent samples. Additionally, samples with different surface textures can be analysed. In short, Raman spectroscopy can be applied to any optically accessible sample (i.e., the sample can be reached by the excitation laser beam and the inelastically scattered photons can be collected). Raman spectroscopy also offers measuring configurations that can accommodate target sizes from 1 $\mu m^2$ (standard laboratory Raman microspectroscopy) up to a few $dm^2$, at ranges from a few mm up to 1 km (Sharma et al. 2003).

Employing Raman spectroscopy for in situ space exploration requires a reliable, automated, sufficiently robust, suitably miniaturized and low-power-consuming instrument capable of addressing the issues enumerated here.

Technical developments have made possible the design of a new generation of small Raman systems which are suitable for robotic deployment on planetary surfaces. Space-borne Raman spectrometers that fulfill these characteristics have been studied, e.g. in the USA for future Mars missions, by Wang and co-workers (1998, 2003). A tiny diode laser (readily available for excitation wavelengths in infrared and visible) serves as the radiation source for this miniaturized Raman spectrometer. For signal detection, a conventional setup was used including spectrometer and CCD. Also, in Germany, a DLR-funded breadboard study called Mineral Investigation by in situ Raman Spectroscopy (MIRAS) was successfully performed under the leadership of the university of Würzburg and Jena in cooperation with industry (Kayser–Threde GmbH) (Popp et al. 2002). For this setup, a diode laser in a Littrow configuration operating at 785 nm was used. An optical tuneable filter (AOTF) was used as the wavelength selecting and deflecting element, and an avalanche photodiode (APD) point detector was used to detect the scattered light.

A prototype of a miniature laser-Raman spectrometer with an 852 nm laser, CCD detector system and confocal microscope was developed by Dickensheets et al. (2000). A remote pulsed laser Raman spectroscopy system for mineral analysis on planetary surfaces was presented by Sharma et al. (2002). UV resonance Raman has been used for easy identification of endolithic organisms and their background mineral matrix; Storrie-Lombardi et al. (1999) discussed this technique for possible use in a future remote planetary mission. A Raman spectrometer was considered as a candidate instrument (Maurice et al. 2004; AURORA/EXOMARS 2005; Popp et al. 2003; Mugnuolo et al. 2000) for the PASTEUR exobiology multiuser facility and planned for the EXOMARS, the first AURORA mission. For the planned 2013 flight to Mars, a combination of Raman and Laser Induced Breakdown Spectrometer (LIBS) in a complex single instrument is planned (Rull and Martinez-Frias 2006).

The progress in the development of appropriate Raman spectroscopic equipment for extraterrestrial research and the advantages of Raman spectroscopy support the idea that this technique is suited for future planetary missions, where the ability to gather information about the mineralogy and the possible presence of organic species will be critical.

## 2 Raman Scattering

Raman spectroscopy is based on the inelastic scattering of laser light by molecules or crystals. When light interacts with matter, most of the incident light is scattered elastically (Rayleigh scattering) with no change in energy. Only a small amount, $10^{-8}$ to $10^{-12}$, of the incident radiation is modulated by the molecular scattering system. Depending on the coupling, the incident photons either gain or lose energy. A sample model of scattering system is shown in Fig. 1. A photon with the energy $h\nu_L$ is incident on the scattering system with the energy level $h\nu_R = E_f - E_i$, where $i$ and $f$ label two quantum states. The Stokes–Raman effect results from the transition from the lower energy level $E_i$ to a higher one ($E_f$).

The anti-Stokes effect transfers energy from the system to the incident light wave, which corresponds to the transition from a higher energy level ($E_f$) to a lower one ($E_i$). Since the anti-Stokes scattering occurs from a thermally excited state ($E_f$) which is, according to Boltzmann statistics, less populated than the ground state ($E_i$), the anti-Stokes is less than the Stokes intensity. In Fig. 2 a typical Stokes and anti-Stokes Raman spectrum from anatase ($TiO_2$) is shown.

Usually the spectral assignment from a Raman spectrum is straightforward. Due to the fingerprint-like information of a vibrational spectrum, and the intrinsic narrow line width of a Raman band, each substance has an easy-to-recognize Raman spectrum. Figure 3 shows

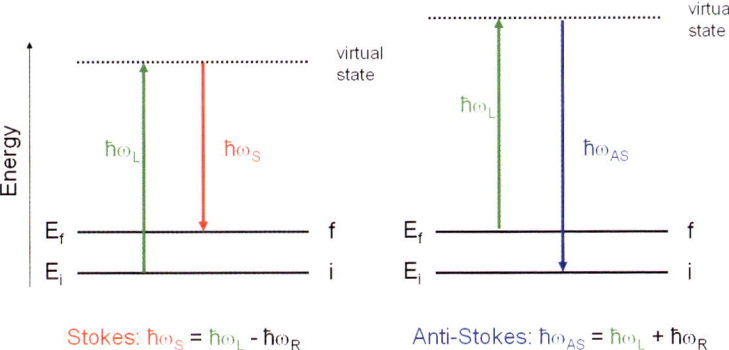

**Fig. 1** Raman scattering process

**Fig. 2** Stokes and anti-Stokes
Raman spectra of anatase (TiO$_2$)

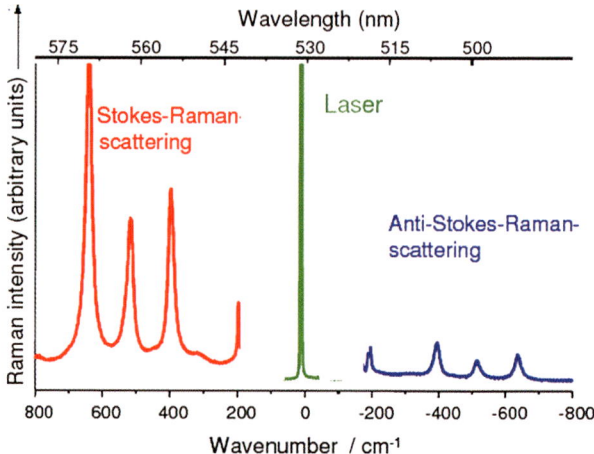

the spectra of some major minerals. Distinctive spectra allow for a quick assignment. Even
for the minerals from the same class, e.g., K feldspar and Na feldspar, there are important
distinct spectral features (e.g. in the region from 300 to 500 cm$^{-1}$).

A schematic diagram of a conventional Raman setup is shown in Fig. 4. A monochro-
matic laser beam is used for excitation, and the resulting wavelength shifts of the scattered
radiation are detected using a dispersion element and a photon detector.

## 3 Instrument

Information about a planetary surface can be gained through orbital and landed Raman spectrometers. Remote Raman spectroscopy from orbiting instruments would be a new tool for
the investigation of planets. Klein et al. (2004) presented the results of various feasibility studies commissioned by the European Space Agency to apply a remote Raman measurement device 10–100 m away from the landing site on a planetary body or even from
the orbiting instruments. Their studies revealed that remote Raman spectroscopy will be
a demanding task. To achieve this goal, the commonly applied laboratory Raman spectroscopy and the well-known Lidar technology need to be combined. Remote-Raman techniques have been evaluated for their potential applications on Mars (Sharma et al. 2002;
Lucey et al. 1998; Sharma et al. 2003; Stoper et al. 2004; Misra et al. 2005). While the application of such a planetary remote Raman device relies heavily on future technical developments, the employment of a miniaturized Raman sensor head embedded directly on a lander
or a Rover is well established (Cochran 1981; McMillan et al. 1996; Sharma et al. 2002;
Tarcea et al. 2002; Ellery and Wynn-Williams 2003; Estec et al. 1972; Isreal et al. 1997;
Haskin et al. 1997; Korotev et al. 1998; Edwards et al. 1999; Popp et al. 2001; Wang et
al. 1999). For the majority of the proposed Raman devices, the basic design is a modular
construction approach. The main components of the final instrument are the laser unit, the
Raman head, the Rayleigh filtering box and the spectral sensor (spectrometer with a match-
ing detector). The modularity offers the possibility of basic components being shared among
different instruments. There is no fixed configuration for the use of such a spectrometer on
a planetary mission. Different configurations are directly linked to the different scenarios of
using such a Raman device:

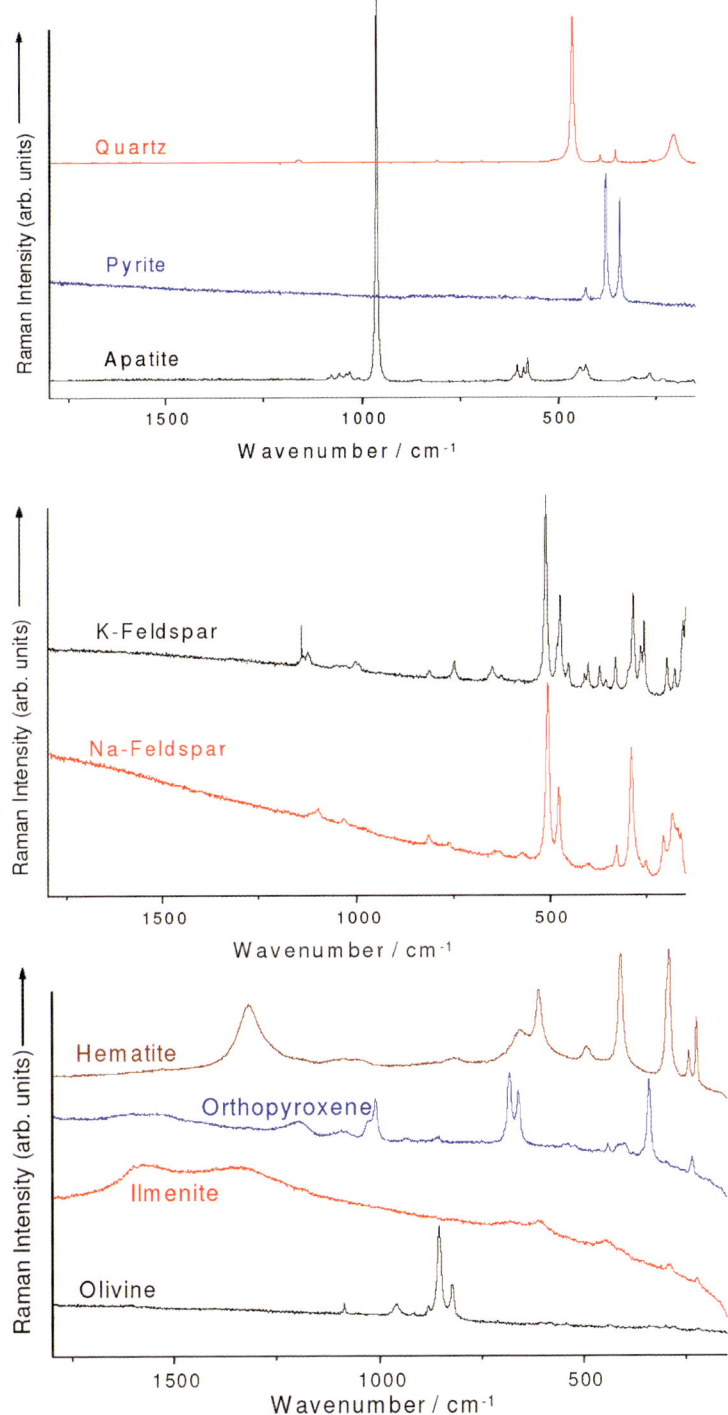

**Fig. 3** Raman spectra of different minerals

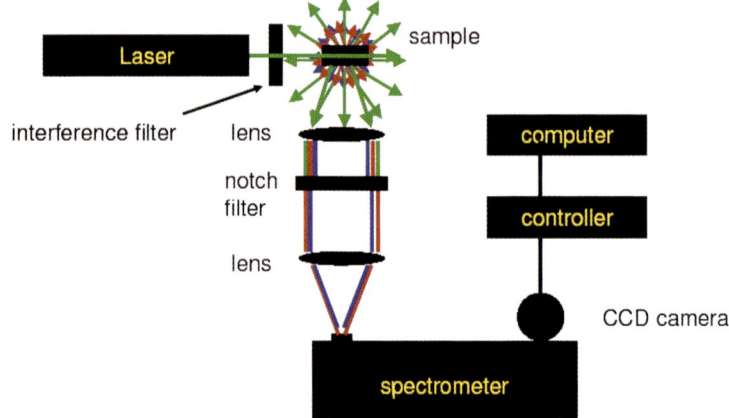

**Fig. 4** Principal schematics of an experimental Raman setup

- The whole instrument can be mounted on the planetary lander having a common system for sample retrieval and sample handling. Samples can be shared with other instruments.
- The sensor head is integrated on a lander robotic arm with no electronics or movable parts in the sensor head. The main box is mounted on the lander platform, including light source and spectrometer. The sensor head and the light source, as well as spectrometer are linked via optical fibre.
- The sensor head is integrated into a MOLE (Mobile Penetrometer) (Richter et al. 2002). No electronics or movable parts are placed into the sensor head, the main box being mounted on the lander platform. Optical fibre make the connections with the sensor head.
- The whole device is integrated on the rover (with an optional robotic arm). No electronics or movable parts would be placed in the sensor head. Depending on the needs of the mission, the Raman device can be adapted easily.

Ellery et al. (2004) argued that, for future Mars lander missions, a specially designed Raman spectrometer will be indispensable. The Raman spectrometer will aid studies of Martian mineralogy and astrobiology and outline astrobiologically relevant features of the Martian environment. Of course, it will also meet the requirements for the detection of biotic residues. Specifically, Ellery et al. (2004) introduced a Raman spectrometer combined with a confocal imager instrument that is extremely compact and low mass; thus, it is ideally suited for onsite planetary applications, in particular astrobiological and mineralogical investigations. In their setup, the main electronics—which may be housed in a lander—are separated from the sensor head, which in turn may be incorporated into a probe such as a "ground-penetrating mole" connected by an optical-fibre-based tether. This strategy allows us to examine environments that would otherwise be inaccessible. The authors suggested that in order to detect biomolecules on the surface of Mars, subsurface penetration using such a mole will be essential; a Raman instrument is ideal for such deployment.

The performance of a flight-ready instrument is not expected to match the performance of a standard laboratory instrument, mainly due to constraints imposed by a very limited mission budget for mass/volume and energy (Popp et al. 2001). To close the gap in performance, the problems which are method-inherent (fluorescence, generally low Raman scattering efficiency, etc.), and those problems which are dealt with by technical artifices in the laboratory (pulsed excitation and gated detection, extremely sensitive detectors, etc.), have to be minimized for a flight-ready instrument without applying costly (in mass/energy terms) technical tricks.

## 3.1  Laser System and Laser Wavelength

To obtain Raman spectra of sufficient quality, the laser beam quality and shape as well as the selected wavelength play a major role. A continuous wave laser illumination during the measurement is preferred to reduce the necessary laser power and also to keep the overall measurement time as short as possible. Diode lasers are efficient (50% conversion rates are normal), compact and inexpensive light sources. They are mass-produced for telecom applications, bar code scanners, CD players or computer optical drives. However, their use for Raman spectroscopy has been limited due to the large spectral laser linewidth and the problematic control of the output frequency. The newly developed laser diodes, e.g., External-Cavity-Diode Lasers (ECDL), Distributed Feed-Back (DFB), Distributed Bragg Reflector (DBR) or even the Laser Diode with External Fibre Bragg Grating partially solve all these problems. Nevertheless. the working characteristics of such a laser vary a great deal with the temperature. Accounting for these laser changes implies the use of an automatic calibration procedure (e.g., a diamond calibration sample) together with very efficient temperature stabilisation for the laser system.

The diode lasers are the best technical option if exciting wavelengths in VIS-NIR are required. Until recently, no compact deep UV diode laser was available on the market which was capable of running at room temperature with an efficiency of more than 15%. A promising solution are the so-called hollow cathode NeCu ion-lasers, which emit @ 248.6 nm; these were recently released by Photon Systems. The laser was developed partially with regard to compact science instruments for NASA technology programs (Storrie-Lombardi et al. 2001).

Raman scattering is only one of multiple physical processes which might take place when light interacts with matter. Some of these processes are competing with the Raman process (e.g., absorption) and/or are interfering with the detection of the weak Raman photons (e.g., fluorescence). For a given sample, the result of the interaction photon-matter is highly dependent on the wavelength of the photons used in interaction. Therefore, the Raman signal yield and the gained information can be maximized by carefully choosing the excitation laser wavelength in a normal Raman experiment.

For a classic Raman measurement, the only way to get rid of one of the main obstacles—florescence excitation—is to tune the laser wavelength to a spectral region where the probability of interference from the florescence signal is minimal. Two approaches are normally used. The first and most widely used is to lower the energy of the incoming photon such that the excitation of the molecule in an electronic state does not take place. Therefore, the wavelength of the laser used for excitation is in the NIR region of the spectrum (from 785 nm up to 1,064 nm). Using this approach for most of the samples (especially the biological samples) avoids fluorescence excitation. On the other hand, avoiding fluorescence in this way for the minerals proves inefficient, since in minerals there is always a certain amount of rare-earth elements and impurities which do have the excited electronic levels at relatively low energies. However, the incoming photon energy cannot go arbitrarily low, since excitation at 830 nm will not allow a full-range Raman spectrum ($4,000–100$ cm$^{-1}$) to be fully recorded with a standard CCD camera based on silicon technology (the cutoff wavelength is $\sim 1,100$ nm).

The second approach used for minimizing the interference of fluorescence with the Raman signal is to shift the excitation wavelength into the deep UV region. At these wavelengths the fluorescence is excited but no fluorescence interference exists when excitation is at wavelengths below about 250 nm. A typical Raman spectral range of 4,000 cm$^{-1}$ occurs at less than 30 nm above the excitation wavelength at 250 nm. Independent of the excitation wavelength, almost no material fluoresces at wavelengths below about 280 nm. This provides complete spectral separation of Raman and fluorescence emission bands resulting in high signal-to-noise measurements.

In addition to having the Raman and fluorescence signals spectrally well separated, if the Raman excitation occurs within an electronic resonance band of a material, the scattering cross-section can be improved by as much as $10^8$. Diamond, nitrites and nitrates, and many other organic and inorganic materials, have strong absorption bands in the deep UV and exhibit resonance enhancement of Raman bands when excited in the deep UV (Chadha et al. 1993).

Comparing the available Raman signal for both cases of NIR and UV excitation one has to observe that the Raman cross-section itself is dependent on the excitation wavelength to the inverse fourth power resulting in higher Raman intensity with shorter wavelength laser excitation. Therefore, an increase of approximately two orders of magnitude in the Raman scattered photons can be obtained by moving from NIR (at 785 nm) to the UV spectral region (248 nm). In addition, the size of the sampling spot for micro-Raman experiments is proportional to the wavelength of the laser beam, and therefore we can achieve a better spatial resolution for Raman mapping experiments when the excitation laser has a shorter wavelength.

However, using the deep UV excitation for a Raman instrument presents several technical difficulties. The off-the-shelf components for a UV laser-based instrument are usually inferior in nominal characteristics when compared with the same components available for VIS/NIR. Overall technical development of UV-suitable components lags behind the developments of components for the NIR spectral region. One of the most important shortcomings in a UV-based Raman system is the poor performance in recording spectral information close to the Rayleigh line, namely the Raman bands up to 500 cm$^{-1}$ from the excitation laser line. This spectral window is of utmost importance since it covers a significant part of the "fingerprint region" in the Raman spectra. For most of the inorganic materials the identification of a component is heavily based on this fingerprint spectral area which usually spans up to 1,200 cm$^{-1}$ (relative wave numbers).

## 3.2 Raman Optical Head and Rayleigh Filters

Figure 5 shows the schematic optical path of a Raman optical head. The laser excitation is transmitted via a lens pair to the sample surface. A clever design of band pass in the excitation path and long pass filters in the reception path allow for minimizing the Raman signal generated within the fibre itself. Unfiltered, these quartz lines overlap with the Raman spectrum of the sample and thus reduce the S/N drastically.

Additionally, plasma lines of the laser will not reach the sample, which would generate second-order spectra. The dichroic mirror or notch filter separates the Raman signal from parasitic or disturbing Raleigh signals and laser plasma lines. The important parameters to be considered in the design of the Raman optical head are:

- High filter efficiency
- High thermal stability and insensitivity to space/Mars environment
- High signal-to-background ratio

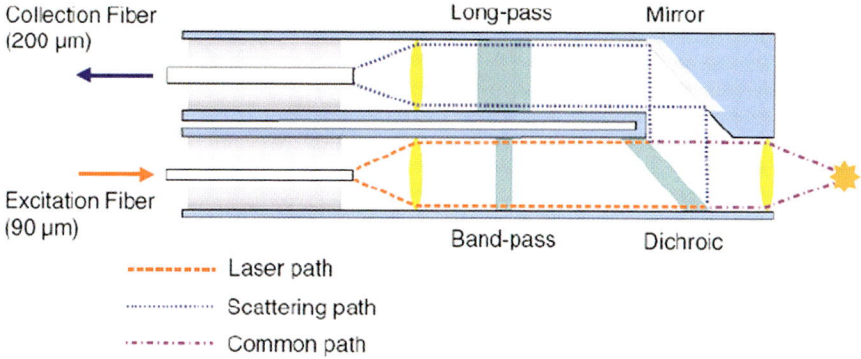

**Fig. 5** Fibre-coupled Raman head, typical commercial fibre probe configuration by InPhotonics

- Maximal transmission for the spectral window of the expected Raman signal
- High numerical aperture to lower the requirements for laser power and spectrometer sensitivity
- Minimal diameter of the lenses used in the probe (which will dictate the overall size and mass of the probe).

For NIR Raman @ 785 nm a number of commercial optical heads are available; e.g., the MR Probe from Kaiser Optical Systems.

Since the Raman instrument needs an imaging lens to focus the spatial resolution, a combination of the Raman instrument with a microscope system (used for white light imaging) can be advantageous.

Similar to the technical challenges facing laser and spectrometer design, the optical design for the beam splitting and notch/edge filters inside the Raman head becomes more challenging with deeper wavelengths. In the VIS and the NIR, a number of commercial, steep-edge notch filters and beam splitters are available. The holographic notch filters at longer wavelengths in particular show supreme performance. A selection with regard to temperature stability and space environment is required. The availability of commercial filters is drastically reduced within the UV wavelength range, especially in the deep UV range; currently, there are no commercial notch filters to meet this need on the market. As compact deep UV lasers are still rare on the market, suitable filters still have to be developed.

An interesting new approach to filtering the laser back reflection between the Raman head and the spectrometer is to introduce a fibre Bragg grating in the reception fibre. Typically these resonant gratings allow extremely sharp reflection peaks if they are used with single-mode fibres. A Bragg grating applied to a multimode fibre is capable of blocking a wavelength band of approx. 1 nm with about 25 dB. This greatly assists the filtering in the optical head (independent from the selected Raman wavelength) and improves the S/N of the signal significantly.

### 3.3 Spectrometer and Detector (Spectral Sensor)

Different types of spectral sensors for Raman spectroscopy have been developed with the aim of meeting the tight requirements imposed by a space mission. Dispersive spectrometers with multichannel detectors (CCDs) (Wang et al. 1999; Dickensheets et al. 2000;

**Fig. 6** Miniaturized
Hadamard–Transform–Spectrometer

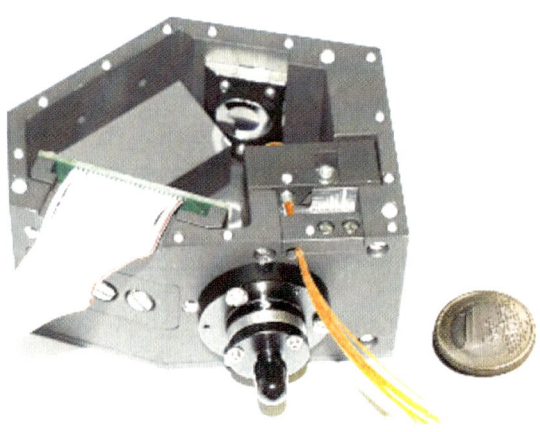

Maurice et al. 2004; AURORA/EXOMARS 2005; Mugnuolo et al. 2000) and also tuneable monochromators with single-channel detectors (APD, PMT) (Tarcea et al. 2002; Popp et al. 2003) have been used as spectral sensors. The vast majority of the developed systems so far are based on dispersive spectrometers, which offer the advantage of multichannel detection when combined with a CCD detector. Scientific detectors with sufficient sensitivity for Raman detection typically have pixel pitches of 13 µm or more. For typical dispersive spectrometers and Echelle spectrometers, where the necessary minimum linear dispersion in the best case is determined by the required spectral resolution and the pixel pitch of the detector, this leads to relatively large linear dispersions and consequently to limitations in compactness.

A novel and promising approach is the Hadamard–Transform–Spectrometer. Its high efficiency is achieved by combining a specially adapted grating with a multislot array, operating as the subsystem aperture. The combination of the multislot array, the adapted grating and the CCD generates so-called "sub-pixels" which enhance the spectral resolution considerably.

To generate subpixel information, the spectrum is measured multiple times with slightly different sampling rasters, shifted by fractions of the physical pixel pitch. The latter is achieved by changing the position of the entry slit with respect to the spectrometer, leading to an equivalent shift of the spectral image on the detector. This can be done by physically shifting the slit and then taking successive measurements (time multiplex) or by arranging slightly shifted slits perpendicular to the direction of dispersion.

The Hadamard–Transform–Spectrometer architecture provides a significantly higher throughput along with improved sensitivity and spectral resolution compared to commercial spectral sensors. The detector consists of a CCD matrix with a length of 25.6 mm. The resolution of 5 cm$^{-1}$ is reached for the whole spectral range with a total length for the Hadamard spectrometer component of about 55 mm (Fig. 6) (Riesenberg et al. 2002). For the detection of very weak signals a cooling of the CCD is necessary. These types of spectral sensors were developed by a team led by Prof. Riesenberg at Institute of Photonic Technology in Jena.

Systems where the spectrometer core is a tuneable monochromator—e.g., an acousto-optic tuneable filter (AOTF) or a liquid-crystal tuneable filter (LCTF), have also been investigated (Tarcea et al. 2002; Popp et al. 2001). An important characteristic of these systems is the ability to transmit images and therefore the use as imaging spectrometers is possible.

# References

AURORA/EXOMARS, Combined Raman/LIBS Spectrometer Elegant Bread Board Technical Proposal Document. TNO Science and Industry, 2005

S. Chadha, R. Manoharan, P. Moënne-Loccoz, W.H. Nelson, W.L. Peticolas, J.F. Sperry, Appl. Spectrosc. **47**, 38–43 (1993)

W.D. Cochran, Adv. Space Res. **1**, 143 (1981)

D.L. Dickensheets, D.D. Wynn-Williams, H.G.M. Edwards, C. Schoen, C. Crowder, E.M. Newton, J. Raman Spectrosc. **31**(7), 633–635 (2000)

H.G.M. Edwards, D.W. Farwell, M.M. Grady, D.D. Wynn-Williams, P. Wright, 1, Planet. Space Sci. **47**, 353–363 (1999)

A. Ellery, D. Wynn-Williams, Astrobiology **3**, 565 (2003)

A. Ellery, D. Wynn-Williams, J. Parnell, H.G.M. Edwards, D. Dickensheets, J. Raman Spectrosc. **35**, 441–457 (2004)

P.A. Estec, J.J. Kovach, P. Waldstein, C. Karr Jr., Proc. Lunar Sci. Conf. **3**, 3047–3067 (1972)

L.A. Haskin, A. Wang, K.M. Rockow, B.L. Jolliff, R.L. Korotev, K.M. Viskupic, J. Geophys. Res. [Planets] **102**(E8), 19293–19306 (1997)

E.J. Isreal, R.E. Arvidson, A. Wang, J.D. Pasteris, B.L. Jolliff, J. Geophys. Res. [Planets] **102**(E12), 28705–28716 (1997)

V. Klein, J. Popp, N. Tarcea, M. Schmitt, W. Kiefer, S. Hofer, T. Stuffler, M. Hilchenbach, D. Doyle, M. Dieckmann, J. Raman Spectrosc. **35**, 433–440 (2004)

R.L. Korotev, A. Wang, L.A. Haskin, B.L. Jolliff, Lunar and Planetary Science XXIX, 29. Lunar and Planetary Science Conference, 1998, pp. 1797–1798

P.G. Lucey et al., *LPSC XXIX* (1998), Abstract #1354

S. Maurice, F. Rull, EXLIBRIS (EXOMARS Laser Induced Breakdown Spectroscopy-Raman Integrated Spectrometers), EGU 1 General Assembly, Abstract EGU04-A-06349, April 2004. See details at http://www.esa.int/specials/aurora/

P.F. McMillan, J. Dubessy, R. Hemley, in *Raman Microscopy*, ed. by G. Turrell, J. Corset (Academic, London, 1996), p. 289

A.K. Misra et al., Spectrochimica Acta A **61**, 2281–2287 (2005)

R. Mugnuolo, F. Angrilli, S. Debei, E. de Marchi, A. Nita, A. Terribili, Presented at the 6th ESA Workshop on Advanced Space Technology for Robotics and Automation (ASTRA 2000), ESA-ESTEC, Noordwijk, The Netherlands, November 2000

J. Popp, W. Kiefer, *Encyclopedia of Analytical Chemistry* (Wiley, 2000), pp. 13104–13142

J. Popp, N. Tarcea, W. Kiefer, M. Hilchenbach, N. Thomas, S. Hofer, T. Stuffler, ESA Publication. ESA SP-496, 2001, pp. 193–196

J. Popp, N. Tarcea, M. Schmitt, W. Kiefer, R. Hochleitner, G. Simon, M. Hilchenbach, S. Hofer, T. Stuffler, in *Proceedings of the Second European Workshop on Exo/Astrobiology*, 2002, p. 339

J. Popp, N. Tarcea, M. Schmitt, W. Kiefer, M. Hilchenbach, J. Whitby, N. Thomas, T. Stuffler, S. Hofer, H. Edwards, F. Rull, Presented at the 37th ESALAB Symposium, ESA-ESTEC, Noordwijk, The Netherlands, 1–3 July 2003

L. Richter, P. Costeb, V.V. Gromovc, H. Kochana, R. Nadalinid, T.C. Nge, S. Pinnaf, H.E. Richtera, K.L. Yungg, Planet. Space Sci. **50**(9), 903–913 (2002)

R. Riesenberg, G. Nitzsche, A. Wuttig, B. Harnisch, *Smaller Satellites: Bigger Business* (Kluwer, 2002), pp. 403–406

F. Rull, J. Martinez-Frias, Spectrosc. Eur. **18**(1), 21 (2006)

S.K. Sharma, S.M. Angel, M. Ghosh, H.W. Hubble, P.G. Lucey, Appl. Spectrosc. **56**, 699 (2002)

S.K. Sharma, P.G. Lucey, M. Ghosh, H.W. Hubble, K.A. Horton, Spectrochimica Acta A **59**, 2391–2407 (2003)

S.K. Sharma et al., Spectrochimica Acta A **59**, 2391–2407 (2003)

Stoper et al., Proc. SPIE **5163**, 99–110 (2004)

M.C. Storrie-Lombardi, A.I. Tsapin, G.D. McDonald, H. Sun, K.H. Nealson, in *Book of Abstracts, 217th ACS National Meeting*, Anaheim, CA, March 21–25, vol. GEOC-069 (American Chemical Society, Washington, 1999)

M.C. Storrie-Lombardi, W.F. Hug, G.D. McDonald, A.I. Tsapin, K.H. Nealson, Rev. Sci. Instrum. **72**(12), 4452–4459 (2001)

N. Tarcea, J. Popp, M. Schmitt, W. Kiefer, R. Hochleitner, G. Simon, M. Hilchenbach, S. Hofer, T. Stuffler, in *Proceedings of the Second European Workshop on Exo/Astrobiology*, 2002, p. 399

N. Tarcea, J. Popp, M. Schmitt, W. Kiefer, T. Stuffler, S. Hofer, G. Simon, R. Hochleitner, B. Hofmann, M. Hilchenbach, *EGS - AGU - EUG Joint Assembly*, Nice, France, April 2003, p. 11790

A. Wang, L.A. Haskin, E. Cortez, Appl. Spectrosc. **52**, 477 (1998)

A. Wang, B.L. Joliiiff, L.A. Haskin, J. Geophys. Res. [Planets] **104**(E4), 8509–8519 (1999)

A. Wang, L.A. Haskin, A.L. Lane, T.J. Wdowiak, S.W. Squyres, R.J. Wilson, L.E. Hovland, K.S. Manatt, N. Raouf, C.D. Smith, J. Geophys. Res. **108**, 5(1) (2003)

# Protein Microarrays-Based Strategies for Life Detection in Astrobiology

Víctor Parro · Luis A. Rivas · Javier Gómez-Elvira

Originally published in the journal Space Science Reviews, Volume 135, Nos 1–4.
DOI: 10.1007/s11214-007-9276-1 © Springer Science+Business Media B.V. 2007

**Abstract** The detection of organic molecules of unambiguous biological origin is fundamental for the confirmation of present or past life. Planetary exploration requires the development of miniaturized apparatus for in situ life detection. Analytical techniques based
on mass spectrometry have been traditionally used in space science. Following the Viking
landers, gas chromatography-mass spectrometry (GC-MS) for organic detection has gained
general acceptance and has been used successfully in the Cassini–Huygens mission to Titan. Microfluidics allows the development of miniaturized capillary electrophoresis devices
for the detection of important molecules for life, like amino acids or nucleobases. Recently,
a new approach is gaining acceptance in the space science community: the application of
the well-known, highly specific, antibody–antigen affinity interaction for the detection and
identification of organics and biochemical compounds. Antibodies can specifically bind a
plethora of structurally different compounds of a broad range of molecular sizes, from amino
acids level to whole cells. Antibody microarray technology allows us to look for the presence of thousands of different compounds in a single assay and in just one square centimeter.
Herein, we discuss several important issues—most of which are common with other instruments dealing with life signature detection in the solar system—that must be addressed in
order to use antibody microarrays for life detection and planetary exploration. These issues
include (1) preservation of biomarkers, (2) the extraction techniques for biomarkers, (3) terrestrial analogues, (4) the antibody stability under space environments, (5) the selection of
unequivocal biomarkers for the antibody production, or (6) the instrument design and implementation.

**Keywords** Planetary exploration · Life-detection instruments · Antibody microarrays ·
Immunosensors

V. Parro (✉) · L.A. Rivas · J. Gómez-Elvira
Centro de Astrobiología (INTA-CSIC), Carretera de Ajalvir km 4, Torrejón de Ardoz, 28850 Madrid,
Spain
e-mail: parrogv@inta.es

# 1 Introduction

The search for unequivocal molecular signatures (biomarkers) resulting from present or past life requires inputs from the scientific and technological communities. Many known organic compounds can be considered as molecular biomarkers because they have been purified and characterized from terrestrial samples of certain biological origin. However, many organics such as amino acids (Pizzarello et al. 1994), some sugars (Cooper et al. 2001), or poly-aromatic hydrocarbons (PAHs) can be produced abiotically, and many of them have been detected in meteorites as well as in space (Oró 1961; Plows et al. 2003). As a consequence, the assignment of an unequivocal biological origin to many organic compounds is a difficult task.

The detection of organic material to contribute to the organic molecule inventory of Mars has become one of the main objectives for astrobiology, and for near future missions like the NASA Mars Science Laboratory (MSL) mission. The Mars Science Laboratory would collect Martian soil samples and rock cores and analyze them for organic compounds and environmental conditions that could have supported microbial life now or in the past. The Sample Analysis at Mars (SAM) is a suite of instruments consisting of a gas chromatograph-mass spectrometer (GC-MS) and a tunable laser spectrometer. It conducts mineral and atmospheric analyses, detects a wide range of organic compounds and performs stable isotope analyses of organics and noble gases (Mahaffy 2007). The SAM GCMS is provided by the technical team that designed and fabricated the Cassini–Huygens GCMS (Niemann et al. 2005; Israël et al. 2005), the first such instrument selected by NASA since the Viking Lander experiments more than three decades ago. Also, the European Space Agency (ESA) has selected several analytical instruments as part of the Pasteur Payload for the Exomars mission (http://esamultimedia.esa.int/docs/Aurora/Pasteur_Newsletter_5.pdf). These include: (1) an infrared microscope (IR), to characterize the structure and composition at grain-size level of the collected samples; (2) a Raman-LIBS to determine the geochemistry; (3) an X-Ray Diffractometer (XRD) to determine the mineralogical composition of the crystalline phases; (4) a Mars Organics and Oxidants Detector (the Urey instrument) to search for amino acids, nucleobases, and PAHs in the collected samples as well as the chemical reactivity of oxidants and free radicals in the soil and atmosphere (Skelley et al. 2005, 2006; Bada et al. 2007); and (4) the Life Marker Chip (LMC), an antibody-based instrument to detect structurally complex organic molecules.

Analytical devices based on immunoassays (Luppa et al. 2001) have been used for the detection of low molecular weight compounds with sensitivities at ppb and ppt levels, from amino acids (Campistron et al. 1986; Silvaieh et al. 2002; Hiasa and Moriyama 2006), insecticides (Mallat et al. 2001a, 2001b), herbicides and pesticides (Rodríguez-Mozaz et al. 2004a, 2004b), PAHs (Alarie et al. 1990), phenols (Parellada et al. 1998), to steroid hormones, antibiotics, or even explosives such as TNT (for a review see Rodríguez-Mozaz et al. 2005). Immunochemical methods for environmental analysis have been applied for many years (Van Emon and Lopez-Avila 1992; Rodríguez-Mozaz et al. 2006), and new immuno-analytical systems are continuously developing (Marquette and Blum 2006). A recent development in the immunosensor field is protein microarray technology (Weller 2005), which allows simultaneous testing of a few to thousands of different analytes. Immunosensors based on antibody microarrays have been and are under development (Weller et al. 1999; Sapsford et al. 2002; Anderson et al. 2006). A number of two-dimensional, array-based multianalyte biosensors using fluoro-immunoassay as their detecting systems have been described for rapid analysis of complex samples (Wadkins et al. 1997; Anderson et al. 2000; Wiese et al. 2001; Ligler et al. 2003; Sapsford et al. 2004), and instrumentation for several applications is being developed (Taitt et al. 2004; Tschmelak et al. 2005a, 2005b; Golden et al. 2005). We have developed an instrument for life detection for astrobiology based on a fluorescent sandwich immunoassay for high molecular weight biomarkers (Parro et al. 2005), and other instrument concepts use similar approaches (Sims et al. 2005). We also recently reported a multiparallel array competitive immunoassay in a microarray format. This immunoassay can be used for the detection of a wide range of molecular weight compounds for astrobiological purposes, from small organics to whole cells, at sensitivities ranging from ppb to even ppt levels (Fernández-Calvo et al. 2006). Because antibodies react only with those molecules containing epitopes with three-dimensional structures equal or highly similar to those used to produce the antibodies, it is possible to get unambiguous results.

## 2 Preservation of Molecular Biomarkers

All living matter, as we know it, consists basically of four types of molecules and their polymers and combinations: carbohydrates, lipids, nucleo-bases and amino acids. Among the factors that contribute to the degradation of the biological matter are: (1) enzymatic and microbial activities; (2) radiation (UV and others); (3) oxidation; (4) metal attacks; (5) Maillard reactions (condensation of the carbonyl group of reducing sugars and primary amino group of amino acids); or (6) high temperatures or extreme pHs. Conditions that avoid or slow such degradation can be: (1) low temperatures, which means lower metabolic rates and catalytic activities; (2) a rapid burial to favor anoxic environments, which confines and protects against UV radiation and oxidation; (3) inclusions in salt crystals and polymerized resins like amber (hypesaline solutions slow the catalytic degradation and favor desiccation); (4) precipitation on the surface of colloid biominerals; (5) anoxic environments rather aerobic ones (the former have lower metabolic rates); (6) rapid dehydration that severely restricts the catalytic activities; (7) mild pH values (extreme pH severely affects molecule preservation favoring depurinization of DNA at low pH or RNA degradation at alkaline one); (8) absence of degradative metal ions (some metal ions or radicals severely affect the biomolecules structure). In general, all types of environmental factors are more important to the overall molecular preservation than the age of the sample (Tuross and Stathoplos 1993). In addition, a minimal concentration of microorganisms in the natural samples (Table 1) is necessary to generate biomarkers (Table 2). This concentration depends on the availability of nutrients and energy sources. Microorganisms in nature are irregularly distributed and often exist in biofilms. Biofilms are surface-attached microbial communities consisting of multiple layers of cells embedded in hydrated matrices (Kierek-Pearson and Karatan 2005), where single or multiple strains, genus, genera or even taxa share the medium and interact with each other. Cells in biofilms are held together by an extracellular matrix composed of a mixture of EPS, proteins, DNA, and water (Zhang et al. 1998). The preservation of these and other molecular biomarkers as well as their structurally modified products by diagenetic processes is critical for unequivocal extant or extinct life detection (see Eigenbrode and Summons 2007).

2.1  Preservation of Biomarker under Martian Conditions

The GC-MS on the 1976 Viking missions to Mars failed to detect organic molecules on the Martian surface. This was surprising due to the relatively high amounts of reduced carbon estimated to be deliberated to Martian surface each year in carbonaceous meteorites (up to $2.4 \times 10^8$ g; Flynn 1996). Many of these compounds are volatile and they should have been

**Table 1** Number of bacteria in natural environments

| Environment | Number of bacteria | |
|---|---|---|
| | (cells/g dw) | (cfu/g dw) |
| Soils | $10^8-10^{10}$ | $10^5-10^8$ |
| Sediments | | |
|     Up to 10 m | $10^7-10^{10}$ | $10^2-10^6$ |
|     Up to 550 m | $<10^6-10^8$ | $<10^2-10^8$ |
| Permafrost sediments | | |
|     Arctic ($>300$ m) | $10^7-10^9$ | $10^2-10^8$ |
|     Antarctic (20 m) | $10^7-10^8$ | $10^1-10^4$ |
| Permafrost buried soils | $10^9$ | $10^4-10^6$ |
| Bottom sediments of lakes | | |
|     Upper layers | $10^8-10^9$ | $10^4-10^5$ |
|     Layers at 3–8 m | $10^8$ | $10^2-10^3$ |
| Central part of the ocean | Cells/ml | Cells/ml |
|     Water | $10^2-10^6$ | 0–10 |
|     Upper layer of sediments | $10^7$ | $10^2-10^6$ |
| Groundwater (up to 2000 m) | $10^3-10^6$ | $10^1-10^5$ |
| Antarctic glaciers (up to 1800 m) | – | 1 cell/l |
| Atacama dessert[a] | – | $10^3-10^6$ |

Dw, dry weight; cfu, colony forming units. From Vorobyova et al. (1997)

[a]From Navarro-González et al. (2003)

detected by the procedure used by the Viking's GC-MS (Sephton et al. 1998). The failure to detect organic compounds in Martian regolith has been interpreted as being a consequence of the presence of a powerful oxidant capable of degrading organics to nonvolatile compounds or even $CO_2$ (Benner et al. 2000). Ultraviolet radiation can cleave water to give H and HO radicals, which in turn can react directly with organic substances, or generate peroxides ($H_2O_2$) or other oxidizing species by combination with other elements in the Martian soil. Quinn et al. (2005) detected a highly oxidative and acidic activity in the dry core of Atacama desert by means of an instrument designed for studying the oxidation state of Martian soils. Their results were consistent with the presence of strong acids in the accumulated dust and emphasized that the abundance of water plays a key role in controlling the oxidizing acid reaction kinetics.

Benner et al. (2000) suggested that organic compounds present on Mars, either from non-biological or hypothetical biological synthesis, could be converted to carboxylic acid derivatives. The expected metastable products should be acetates, oxalate, benzoic acid derivatives or phthalic acid. The salts of organic carboxylic acids are not volatile and, consequently, not easily detected by GC-MS, so even if they had been present, they would not be directly detectable by the Viking GC-MS instrument. Other studies showed that the stabilities of organic macromolecules like tholins (abiotic origin), and humic acids (biological origin) are different in the presence $H_2O_2$ (McDonald et al. 1998). The same work concluded that some organic macromolecules could have been stable on the Martian surface, at least in the polar regions, during the whole history of Mars.

Smaller biomolecules, such as amino acids, purines, and fatty acids, are excellent biomarkers in the search for life on Mars, but they may be much less resistant to oxidative degradation than tholin-like or humic materials. As it has been demonstrated, photodegradation may destroy a significant fraction of organics initially present in the upper layers of the regolith (Oro and Holzer 1979; and Stoker and Bullock 1997). Consequently, a search

**Table 2**  Biomarkers for extant and extinct life detection (see also Parnell et al. 2007)

| Biomarker | Comments |
| --- | --- |
| Extant live biomarkers | |
| **Shape and size**: replication structures (buds, chains of cells, septa, fruiting bodies and spores); some biominerals; macrostructures such as biofilms and stromatolite-like structures. Whole cells | Shape and size are not definitive. Replication structures are definitive indicators of life. Biofilms and stromatolite-like structures could be definitive. |
| **Cellular compartments:** cell walls, membranes organelles. | Structural components |
| **Biochemicals. Biological polymers and metabolites:** EPS, lipopolysaccharydes (LPS), peptidoglycan (PG), teichoic acids, teichuronic acids, proteins, oligo and polysaccharides, nucleic acids, porphyrins, flavins, vitamins, antibiotics, compatible solutes, etc. | Cell wall components or energy and storage compounds. Indicative of actual or recent metabolism. Compatible solutes are small molecular weight metabolites (ectoine, trehalose, betaine, some amino acids, etc.) accumulated intracellularly as a response to high salt or dryness. |
| **Metabolic products**: metahane | They can be inferred through the detection of the enzymes involved in the metabolic process. |
| Extinct live biomarkers | |
| **Aliphatic hydrocarbons**: *n*-Alkanes, algaenans, cycloalkanes, branched alkanes, pristane, phytane, acyclic isoprenoids, phytol, pentamethylicosane highly branched isoprenoids. | From degradation products of aliphatic macromolecules from unicellular green algae (Allard and Templier 2000). Indicative of marine and lacustrine habitats. |
| **Cyclic hydrocarbons:** alkylthiophenes, alkylcyclohexanes and cyclopentanes. | From sulfuration of sedimentary lipids in hypersaline environments. From cyclohexyl fatty acids of thermophilic bacteria |
| **Aromatic carotenoids**: okenane, chlorobactane, etc. Arylisoprenoids derivatives. | Photic zone euxinia. Pigmented (green, brown) chlorobiaceae. Diagenetic and catagenetic products of the above carotenoids |
| **Hopanoids** and other pentacycic triterpanes. | The hopanoids are widely distributed in bacteria and blue green algae, where they are important cell membrane constituents. Found in recent and old sediments. |
| **PAHs** and humic acids | From prebiotic or biological origin. Structural features will be detected. Very complex PAH derived from triglycerides. Soil and marine environments. |
| **Steroids:** sterane, diasterane, norcholestanes | Saturated version of eukaryotic sterols. Diasterane is produced by rearrangement of sterane during diagenesis. |
| **Porphyrins and maleimides**: bacteriochlorophylls, heme group, maleimides. | Photosynthetic bacteria, cytochromes, siderophores. Maleimides are diagenetic product from tetrapyrrole ring of chlorophylls |
| **Amino acids and nucleotides**: All 20 proteinogenic L-aa, modified aa, dNTPs, and other modified nucleotides, purine, pyrimidins, etc. | Thymine dimmers are produced by UV radiation. Some amino acids, purines and pyrimidines could be produced by abiotic chemistry. |

for molecular biomarkers on Mars, should therefore include environments shielded from surface oxidants, such as subsurface samples or the interiors of rocks. Recent work by Kminek and Bada (2006) showed that amino acids can be protected from radiolytic decomposition as long as they are shielded adequately from space radiation. Taking into account an average amino acid radiolysis constant of $0.113$ Mg y$^{-1}$ they estimated that it is necessary to drill to a depth of 1.5 to 2 m to detect the amino acid signature of life that became extinct about three billion years ago. And to be able to detect any remnants in the uppermost half meter of the Martian subsurface, the extinction event would have to be younger than 100–500 million years.

The determination of amino acid enantiomeric excesses is a powerful indicator of their prebiotic or biological origin (see Barron 2007). A microfabricated capillary electrophoresis device for amino acid chirality determination was developed for extraterrestrial exploration (Hutt et al. 1999). This concept has been incorporated into the UREY instrument selected as part of the Pasteur payload for the ESA ExoMars mission (Skelley et al. 2005; see Sect. 1).

## 3 Extraction Methods for Molecular Biomarkers

### 3.1 Extraction of Hydrophobic Compounds (see Summons et al. 2007)

Different extraction methods can be applied according to the type of sample (liquid or solid) and the type of molecules to be extracted. Soxhlet, mainly used with organic solvents like dichlorometane or chloroform (Luque de Castro and Garcia-Ayuso 1998), pressurized liquid extraction (Bjorklund et al. 2000), microwave extraction (Wong et al. 1997) and ultrasonic extraction (Junior et al. 2006) methods give good results for the organic molecules identified as biomarkers in Table 2. Following Soxhlet extraction, Brocks et al. (1999, 2003) identified biomarkers in the fossil record and assigned the environments and their inhabitants on the early Earth.

### 3.2 Extraction of Water Soluble Compounds

Two treatments are commonly applied to samples in the laboratory in order to extract extant life biomarkers: enzymatic or mechanical. Enzymatic methods use hydrolytic enzymes (proteins) with catalytic activity to degrade cell wall polymers of bacteria (e.g., lysozyme) or fungi (e.g., chitinases), etc. The use of enzymatic methods for life-detection experiments is not recommended because, apart from the new elements of complexity added (buffer, stability, activity, etc.), it is a source of contamination with biochemical material. Mechanical methods increase the pressure on the sample to break cells wall and release cellular components into a buffer. Some of these methods are ultrasonication, French press, bead milling or even a coffee grinder. Extant life biomarkers can be extracted simply by sonication in the presence of a biochemical buffer like tris-HCl or Phosphate Buffered Saline (PBS) and some mild detergents, commonly used in molecular biology laboratories. When biomarkers or cells are tightly bound to minerals, more harsh solutions (containing chelators for metals or even denaturing components) have to be employed (Schweitzer et al. 2005a, 2005b).

## 4 Antibody Microarrays for the Detection of Molecular Biomarkers

### 4.1 Antibodies

Antibodies (Ab) are glycoproteins that bind specifically to other molecules (antigens) through a noncovalent, highly specific interaction. Mammals produce several types of anti-

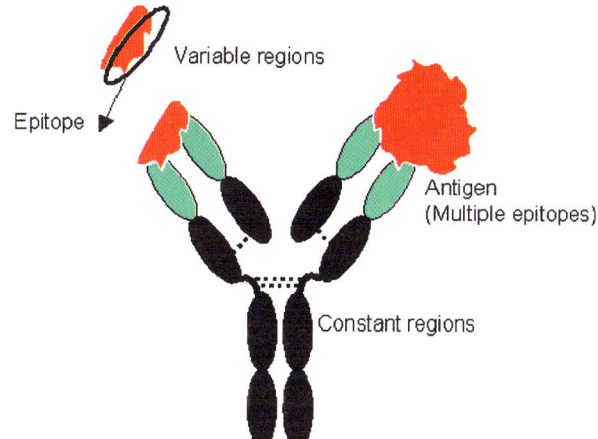

**Fig. 1** Schematic drawing of an IgG (immunoglobulin G) type antibody molecule with the variable regions (*green*) and constant ones (*black*), as well as the antigen (*red*)

bodies, but the most important one in the humoral immune response (that is, the aspect of immunity that is mediated by secreted antibodies produced in the cells of the B lymphocyte lineage) against an antigen (Ag) is the immunoglobulin type G (IgG). IgG antibodies are made up of variable and constant regions (Fig. 1), with the antigen binding activity located in the variable region. The constant region can be modified with substances like fluorophores, other proteins or even other antibodies, without affecting the binding affinity of the Ab to the antigen. There are thousands of commercial antibodies, mainly for biomedical purposes, raised against compounds ranging from small molecules like amino acids, sugars, and lipids, to large polymers and whole cells. Antibodies can show a high level of specificity so that they are able to discriminate between enantiomers of the same molecule (Shabat et al. 1995). They have a broad range of sensitivity, and can exhibit dissociation constants easily at pM (Chen et al. 1999) and even at fM level (Rathanaswami et al. 2005; Boder et al. 2000).

Apart from antibodies, other molecules can be used as specific capturing receptors (Liao et al. 2006):

1. Lectins, which are a group of glycoproteins (generally nonenzymatic) with high binding specificities and affinities for mono and oligosaccharides (Masarova et al. 2004).
2. Affibodies, consisting of in vitro selected binding proteins, share similar properties with antibodies (Hansson et al. 1999).
3. *Aptamers*, DNA or RNA molecules that have been selected from random pools based on their ability to bind other molecules (Collett et al. 2005).
4. The molecularly imprinted polymers (MIPs), applied for many years in chiral separations, are synthetic polymers with recognition sites specific for a target molecule (Ramström and Yan 2004).

All these type of affinity receptors, that can have specificities comparable to those of the antibodies (Renberg et al. 2007; Geiger et al. 1996; Kempe 2000), are under development and seem very promising. However, none of them have the degree of knowledge and applicability as do the immunological techniques, used since 1950s. The technology for antibody production (polyclonal, monoclonal or recombinant) is highly developed and it is supported by a potent industry. In theory, antibodies can be produced against practically all structurally different compounds, from single amino acids to against whole cells.

### 4.1.1 Stability of the Antibodies for Planetary Exploration

Although MIPs are more stable and resistant to harsh conditions, antibodies are also quite stable when they are kept under appropriate conditions. Several parameters like the nature of the solvent, temperature, lyophilization, freezing, salts, light or grafting step can affect the three-dimensional structure of antibodies and consequently the affinity for their antigens. Apart from that, planetary exploration requires functional antibodies under deeply penetrating low-energy radiation, microgravity, temperature cycles and its variations, etc. The pharmaceutical industry has developed methods for antibody stability and it is well known that most antibodies maintain their activity for years when frozen (between $-20°C$ and $-80°C$) in a saline solution or lyophilized and kept at ambient temperature protected from light. Usually stabilizers such as sugars/polyols are added to reduce the degradation of active components during processing and storage (Newman et al. 1993). Lyophilization in the presence of sugars (sucrose or trehalose) is commonly used to provide long-term storage stability to protein pharmaceuticals. It is believed that proteins reach a highly viscous amorphous glassy state with low molecular mobility and low reactivity. Under such a state the stability is highly dependent on storage temperature. Storage at temperatures above the glass transition temperature (usually ca. $50°C$) the glassy material become less viscous, the molecular mobility increases and the stability decreases (Chang et al. 2005). Breen et al. (2001) studied the effect of moisture on the stability of a lyophilized humanized monoclonal antibody formulation by storing vials at temperatures from 5 to $50°C$ for 6 or 12 months. Chemical and physical degradation pathways followed Arrhenius kinetics during storage in the glassy state. High moisture levels did not affect the physical stability when storing below the transition temperature.

Antibodies can also be preserved by drying at ambient temperature and at atmospheric pressure in the presence of trehalose (U.S. Patent 4891319) or other formulations (Spanish Patent ES2180416). We have experimentally determined that printed antibody arrays on microscope slides are stable (keeping more than 90% of their functionality when compared to time zero) for more than six months stored at room temperature, even at $37°C$ in the presence of stabilizing solutions developed by CAB and by the company Biotools S.A. (Parro et al., in preparation). Affibody Co. (Bromma, Sweden) has developed technology to stabilize any protein at high pH conditions (U.S. patent 6831161). Fluorescently labelled and lyophilized antibodies are stable over more than 50 temperature cycles (24 h each) of $-20$ to $+50°C$, and even after an exposure of 150 Gy of high-penetrating, low-energy radiation (Parro et al., in preparation). (Gy = gray. One *gray* is the absorption of one joule of radiation energy by one kilogram of matter.) In agreement with our results, Thompson et al. (2006) reported that no significant alteration in the absorption and emission wavelengths or the quantum yields of two fluorescent dyes was found after a proton and helium ion radiation. Maule et al. (2003) showed that the level of an antibody-antigen binding did not differ between microgravity, Martian gravity and Earth's gravity conditions. It is also known that some antibodies do not stand freezing or lyophilization or can be severely affected by some of the parameters mentioned earlier. Naturally only those showing high performance after multiple tests under space-like conditions will be suitable for planetary exploration.

### 4.2 Terrestrial Analogues for Biomarker and Antigen Selection for Life Detection

On Earth, life is present everywhere liquid water is stable even for only short periods of time. Such temporarily or spatially transient occurrences are typical for extreme environments on Earth, such as hot and cold deserts. These environments are created because of the permanent existence of extreme values of certain physico-chemical parameters like temperature, pressure, pH, redox potential, radiation, water content, salinity, etc.

From a terrestrial point of view, Mars or other extraterrestrial bodies, can be considered as extreme environments because several or all of the mentioned parameters seem to be very far from "normal" values. Several terrestrial environments share physico-chemical characteristics to environments thought to exist or to have existed once on Mars: hydrothermal, acidic iron and sulphur rich, extremely dry deserts like Atacama (Chile), hyperarid deserts or permafrost. The recent discovery of sulphate and iron containing minerals that can be formed only in the presence of water on Mars (Klingelhofer et al. 2004), suggests that iron and sulphur metabolisms could have played a central role both as energy source and for the formation of structural components (metalloproteins, and Fe-S proteins) for an hypothetical Martian biota. The Río Tinto area (southwestern of Spain) has been also considered as a good Martian analogue with regard to iron and sulphur mineralogy (Fernández-Remolar et al. 2004, 2005). Chemolithoautotrophic bacteria like *Leptospirillum ferrooxidans* and *Acidithiobacillus ferrooxidans* (abundant in the Tinto River ecosystem) are very simple in their nutrient requirements (Balashova et al. 1974; Buchanan and Gibbons 1974). Both are able to fix $CO_2$ and $N_2$ (Mackintosh 1978; Norris et al. 1995; Parro and Moreno-Paz 2003), and they obtain energy from iron ($Fe^{2+}$) oxidation. *A. ferrooxidans* is a very versatile bacterium that can chemolithoautotrophically grow not only on $Fe^{2+}/O_2$ or $H_2/O_2$ under aerobic conditions, but also on $H_2/Fe^{3+}$, $H_2/S_0$ or $S_0/Fe^{3+}$ under anaerobic conditions, with $Fe^{3+}$ and $S_0$ being the electron acceptors (Ohmura et al. 2002). Iron and sulfur cycles are taking place in the Río Tinto ecosystem (González-Toril et al. 2003).

Our working hypothesis is that microorganisms living under similar environmental conditions share similar molecular mechanisms to deal with such conditions. These mechanisms will render molecules that could be good biomarker targets to search for life elsewhere. Therefore, in addition to producing antibodies against specific molecules, we have produced polyclonal antibodies against extracts from the acidic, iron- and sulphur-rich Rio Tinto area (Fig. 2). Samples were taken from water, sediments, mineral deposits (sulphate precipitates, jarosite, hematite etc.), and rocks from a pyrite rich subsurface drill core (between 100 to 160

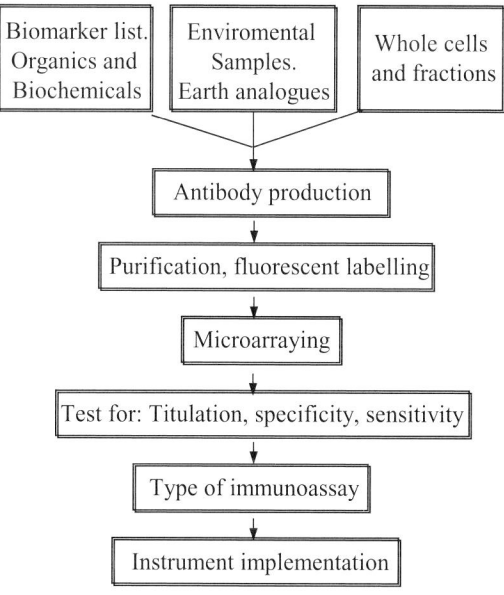

**Fig. 2** General strategy for biomarker selection, antibody production and implementation in instruments for life detection. Biomarkers for antibody production can be selected among those well-characterized organics and biochemicals, from new ones detected in Earth analogues environments, or even from whole cells or subcellular fractions

meters deep, obtained during the field campaign of the NASA-funded Mars Astrobiology Research and Technology Experiment (MARTE) project (http://marte.arc.nasa.gov/; Stoker et al. 2004; Fernández-Remolar et al. 2007). At the same time we are performing biochemical fractionation from pure bacterial cultures, so that we can produce antibodies against different kinds of macromolecules (EPS, anionic polymers, cell wall components, etc.). We have also produced antibodies against pure bacterial cultures (type collections) isolated from cold (Arctic and Antarctic) as well as hydrothermal environments.

### 4.3 Immunoassays with Antibody Microarrays

Protein microarrays allow the covalent binding of thousands of antibodies or other affinity binding molecules to a solid support, such as glass, nitrocellulose, nylon, etc. This large number of binding molecules expands the possibilities for bioaffinity biosensor development (Kusnezow et al. 2003; Chen and Zhu 2006). Microarray immunoassays, as many others immunoassay platforms, can be performed in several formats depending on nature and size of the antigen molecule (target or analyte, Fig. 3). For target molecules containing only one antigenic determinant (xenobiotics, pesticides, steroid hormones, small metabolites, explosives, etc.), a competitive assay has to be done. In a particular type of competitive microarray immunoassay, a limited dilution of fluorescent tracers (antibodies or labelled targets) is incubated with the immobilized capturing molecules (antibodies or a target conjugate) on the microarray. After a washing step, a laser beam excites the fluorochrome, an image is captured by a CCD camera and the signal intensity of every spot is quantified and considered as 100%. Using another method, the tracers and the target containing samples are mixed and

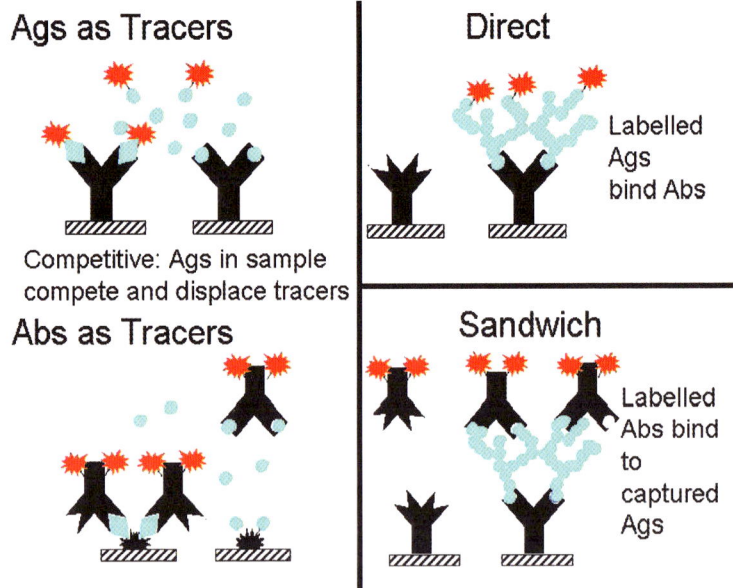

**Fig. 3** Different types of immunoassays. (*Left*) Competitive immunoassay using modified antigens (Ags) (*upper part*), or labelled antibodies (*bottom*) as tracers. In the first case the capturing molecules are antibodies while in the second one are complex molecules (like large proteins conjugated with the antigens). (*Upper right*) Direct immunoassay with fluorescently labelled antigens. (*Bottom right*) Sandwich immunoassay, where label-free antigens are sandwiched by fluorescently labelled antibodies

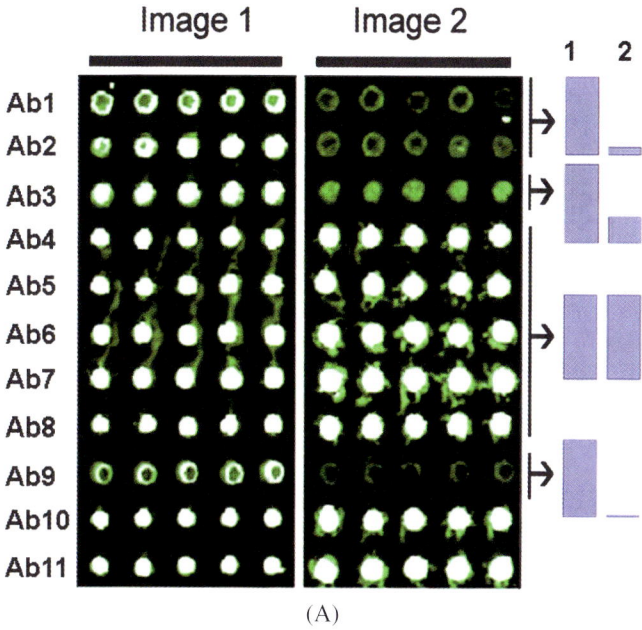

(A)

**Fig. 4** (**A**) A competitive microarray immunoassay for detecting a broad range of molecular size biomarkers. Spotted samples (five replicas each) and antibodies are the same as those published previously (Fernández-Calvo et al. 2006). Here we show two microarray printed on two parallel hybridization chambers of a special device designed to this purpose (**B**). Ab1 to Ab11 corresponded to the next pairs Ag/Ab: Ab1, Terbutryn-HRP/Ab87; Ab2, GlnB peptides/Anti-1492; Ab3, Naphthalene-BSA/Anti-Naph; Ab4, PhenylPhenol-BSA/Anti-PhePhe; Ab5, NifH peptides/Anti-1484; Ab6, NifD peptides/Anti-1485; Ab7, HscA peptides/Anti-1487; Ab8, Fdx peptides/Anti-1490; Ab9, *B. subtilis* spores lysate/Anti-BsuSp; Ab10, Thioredoxin/Anti-Thio; Ab11, Biotinilated-GroEL/Anti-biotin. To obtain image 1, an antibody mixture containing the 11 fluorescently labelled antibodies at appropriate dilution (Fernández-Calvo et al. 2006), was incubated with the microarray at chamber 1 for 10 min at ambient temperature. Simultaneously, an antigenic mixture containing 3.2 pM of terbutryn, 10 ng ml$^{-1}$ of GlnB peptides, 500 ng ml$^{-1}$ of Naphthalene–BSA conjugate, and 10$^7$ *B. subtilis* spores per ml, which was previously incubated with the mixture of the 11 fluorescently labelled antibodies for 10 min, was incubated with the microarray at chamber 2. After a five-minute wash and drying, microarrays were scanned for fluorescence in a GMS-417 array scanner (Affymetrix, Santa Clara, CA). The spot intensities were determined by GenePix 6.0 pro. Software (Axon Instruments, Genomic Solutions). The relative signal intensities expressed as percentage is indicated (*bars to the right*): *1*, from the image 1 (100%) and *2*, from the image 2, showing the loss of signal. (**B**) A multiarray competitive immunoassay (MACIA) device. Multiple parallel chambers act as flow cells closed by a microscope slide containing the same numbers of antibody microarrays. A multichannel pump recirculates the samples

incubated together before added to another microarray in parallel. If a target is present in the sample, it binds and blocks the tracer and consequently there is a loss in signal intensity in the corresponding immobilized capturing molecule (Figs. 3 and 4) with respect to that of the control image. This loss of signal can be quantified and the analyte concentration estimated (Figs. 4 and 5). In the experiments shown in Figs. 4 and 5, antibodies were used as fluorescent tracers, and antigens as immobilized capturing molecules (case at bottom left in Fig. 3). When no antigen or analyte was present in the sample the intensity of the spots was 100% (Fig. 4, image 1). When some antigens were present, the signal intensity of the corresponding spots diminishes proportionally to their concentration (Image 2), that is, the higher the antigen concentration the lower the signal intensity of the spot.

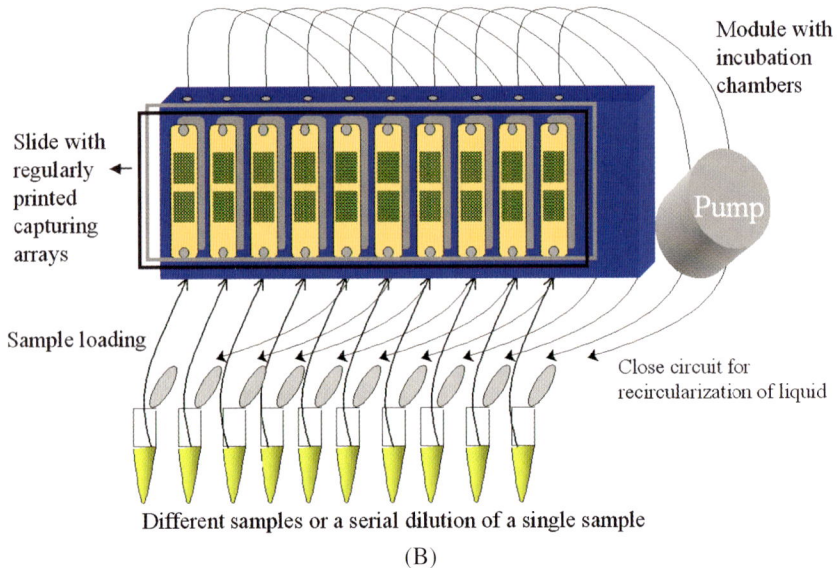

(B)

Fig. 4 (*Continued*)

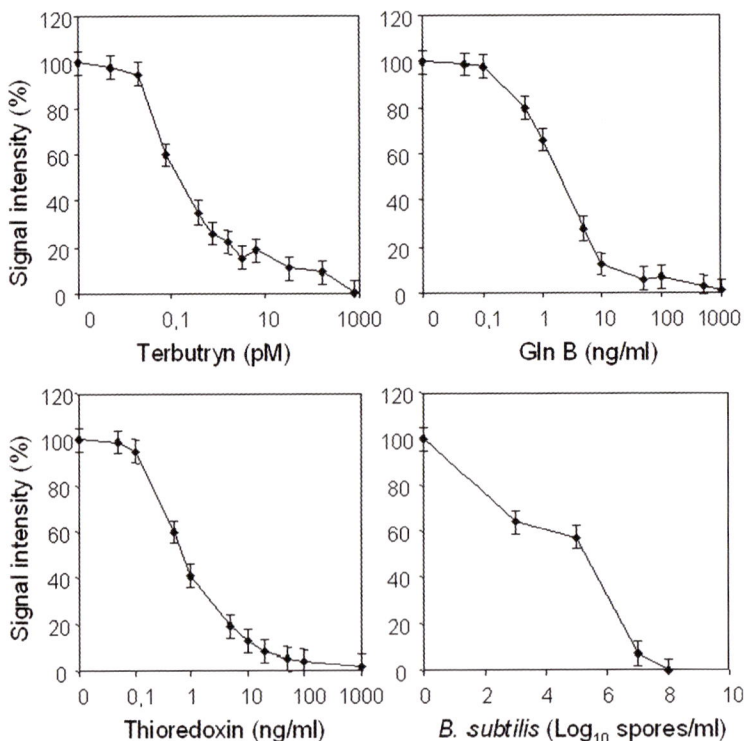

Fig. 5 Calibration curves obtained by multiarray competitive immunoassay (MACIA) for several compounds, for the insecticide terbutyn, a 15 aminoacid peptide (GlnB), a 11.7 kDa protein (thioredoxin) and whole *B. subtilis* spores. Modified and extended experiments from Fernández-Calvo et al. (2006)

A sandwich assay is only valid for those compounds containing at least two epitopes (sites that directly bind to the antibody Fig. 1). A label-free test sample is directly incubated with an antibody microarray, and the printed antibodies (capturing molecules) retain the antigens present in the sample. In a second incubation the same antibody (fluorescent or enzymatically labelled) can bind to other epitopes of the retained antigens to generate a kind of sandwich (Fig. 3). The reaction can be detected after fluorochrome excitation and image capturing. In a direct immunoassay, antigens present in a sample are previously labelled and then incubated with the antibody microarray before a washing step, laser excitation and image capturing (Fig. 3, upper right).

We have demonstrated the feasibility for direct, sandwich, competitive and displacement assays in an antibody microarray format for multiple epitope containing antigens. We also have developed our own antibodies and a multiarray competitive immunoassay (MACIA) for the detection of a wide range of molecular size compounds, from single aromatic ring derivatives or polyaromatic hydrocarbons (PAHs), through small peptides, proteins or whole spores cells (Figs. 4B and 5; Fernández-Calvo et al. 2006).

Antibody arrays are an excellent complementary technique to others like capillary electrophoresis or GC-MS, with some specific strengths:

1. Multiple parallel assays can be run simultaneously.
2. A broad range of molecular sizes can be detected, from aa size to cells.
3. No special external calibration or standards are required for the assay.
4. Many negative and positive controls can be incorporated and analyzed at the same time as samples.
5. Sensitivity can range from ppb to ppt, depending of the affinity constant for each pair antigen–antibody.
6. Results are very easy to analyze.

The presence of putative biomarkers are identified due to the specificity of the probes and their predetermined location. However, when we analyze natural samples, the specificity is not always total because chemical structures of the sample are unknown, may be highly complex and consequently cross-reaction events can be detected. Still antibody microarrays tell us that such a sample is complex and contains compounds highly similar to those used to produce the capturing antibodies. In a sandwich immunoassay, a positive result indicates that the sample contain a molecular structure with at least two antigenic determinant, otherwise sandwich could not be seen.

### 4.3.1 Immunoassays with Organic Solvents

Extinct biomarkers are usually highly hydrophobic and hence soluble only in organic solvents like dichloromethane (DCM), chloroform, methanol, acetonitrile, etc. Gonzalez-Martinez et al. (2006) reported a competitive immunoassay for the detection of the insecticide carbaryl (molecular weight 201.2) at the range of $7.4 \, \mu g \, l^{-1}$ (ppb) in 100% methanol, or at $5.5 \, \mu g \, l^{-1}$ in 20% ethyl acetate–30% methanol–50% water hybrid buffer. A hybrid buffer for water soluble compounds containing up to 15% methanol can work relatively well for the dissolution of both kinds of biomarkers without compromising the antibody properties (Fernández-Calvo et al. 2006). Ideally, the procedure could be based on an extraction buffer that should be able to extract water soluble and organic compounds and leave them accessible to the antibodies. Efforts must be made in this direction, for example, by assaying different solvent combinations at which highly hydrophobic compounds like hopanoids, $n$-alkan, pristine, etc., are soluble enough to be detected and not severely affect the antibody activity.

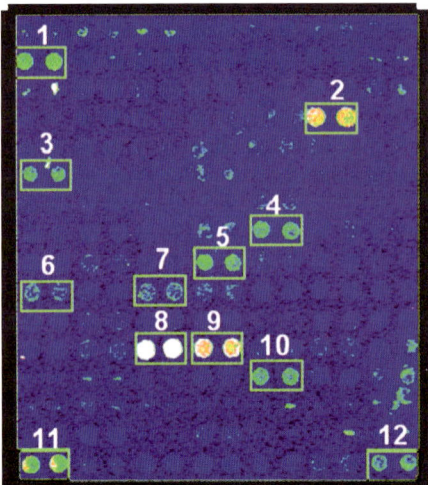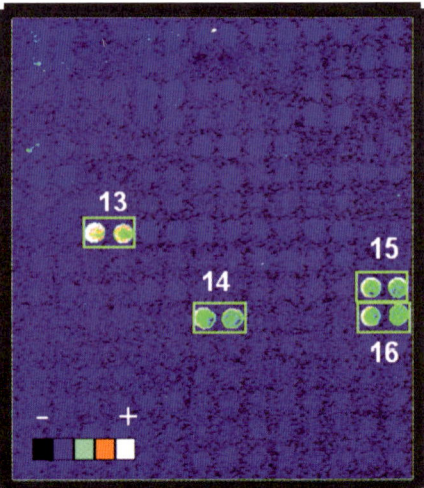

**Fig. 6** Sandwich antibody microarray immunoassay with environmental samples and isolated bacterial strains. The 180 microns in diameter spot microarray (duplicated pattern) contained 101 different antibodies raised against natural extracts, bacteria, proteins or metabolites (Parro et al. 2005; Rivas et al., in preparation). Images show the results obtained after two-hour incubation (15 µl always as a final volume) at ambient temperature with a natural extract from Río Tinto (*left*) or a pure bacterial culture lysate of *Leptospirillum ferriphilum* (equivalent to $10^7$ cells $ml^{-1}$, *right*), and revealed with 5 µg $ml^{-1}$ of their corresponding fluorescent antibody (IgG fraction) for two hours at 4°C. Experimental details are the same as described by Parro et al. (2005). Positive spots (*green–red–white*) over the background (*blue–black*) can be immediately identified and quantified using commercially available software. Compounds present in the natural extract were captured by its own antibody (spot 8), by other raised from a highly related extract (spot 9), by some others raised against *Acidithiobacillus* spp. (spots 4–7, 12), and by others not relevant for this work (spots 1–3). The *L. ferriphillum* extract showed clear positive signal (*right*) with four different antibodies (spots 13–16) raised against different partial fractions of this bacterium (Rivas et al., in preparation). In the inserted scale: − means 0, and + means 65,000, the maximum intensity (saturation)

## 5 Antibody Microarray-Based Instrumentation for in situ Analysis in Astrobiology

Recently, the European Space Agency (ESA) has positively considered an antibody microarray-based instrument called Life Marker Chip (LMC) to become part of the Pasteur payload in the Exomars mission (http://www.aurora.rl.ac.uk/Report_of_Pasteur_9_Sept.pdf). We have demonstrated the feasibility of this approach by using a sandwich immunoassay for the detection of large molecular weight compounds. We incorporated the method into the LMC-like instrument called the Signs Of LIfe Detector (SOLID; Parro et al. 2005). A new field version (SOLID 2, Figs. 7 and 8) has been successfully tested and it was one of the analytical instruments in a simulated robotic drilling in September 2005 (Parro et al. 2007) during the development of the MARTE project (Stoker 2003; Hogan 2005; Whitfield 2004; Stoker et al. 2007). The SOLID instrument consists of a set of operative modules (Figs. 7 and 8) to manipulate, extract, modify (if required) and analyze the samples, while another set of modules supervises the correct performance of the operative ones as well as communication to a remote station. We are currently developing the SOLID 3 prototype following the specifications for the ESA LMC, initially selected as part of Pasteur Payload for the ExoMars mission. SOLID 3 will analyze sample extracts with volumes between 50 and 500 µl obtained with a different device, and will perform both sandwich and competitive fluorescence-based immunoassays.

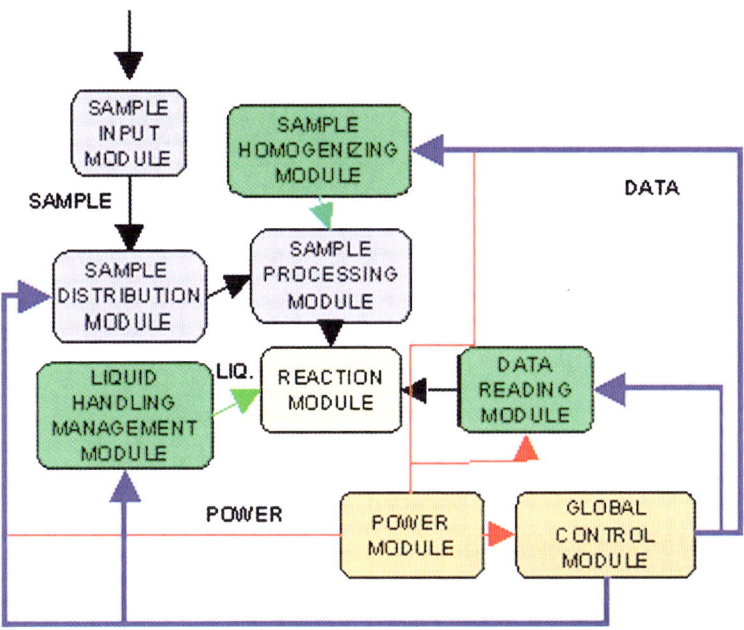

**Fig. 7** Flow chart showing the different functional modules of SOLID instruments (from Parro et al. 2005)

A key issue is the selection of the appropriate set of molecular biomarkers as target for antibody production. Two approaches can be followed for the selection of biomarkers: (1) a direct strategy, in which well-known molecular biomarkers from those listed in Table 2 (see also Parnell et al. 2007) are used to produce antibodies, and (2) a shotgun strategy, consisting of a biochemical extraction and fractionation of astrobiologically relevant environmental samples and then producing antibodies from these extracts (Fig. 2 and part 4.2). Our results with antibodies against natural extracts indicates that they recognized polymeric material composed of polysaccharides and proteins (Fig. 6, Parro et al. 2007; Rivas et al., in preparation). Also, antibodies can be raised against possible terrestrial contaminants, for example, against human antigens, polymers from plastics, or stain used in the instruments, etc., and be used as controls for contamination.

## 6 Conclusion

The high specificity of the ligand-receptor bioaffinity molecular interaction is the fundamental base for the biosensor development. The reaction antigen–antibody has been well known for many years and there is a tremendous technological support for the production of antibodies with different degrees of specificity. The biotechnological, biomedical, environmental industries and especially the biosensor industry take advantage of the bioaffinity reactions, and new and improved assays are continuously appearing. The need for continuous monitoring, portability, robustness, reusability, sensitivity, etc. matches perfectly with the required specifications for the analytical instruments used in planetary exploration. Protein microarrays and, in general, other capturing-molecule microarray can be used for the detection and identification of specific biomarkers for life detection. Instruments capable of doing these type of analysis should be considered for future missions.

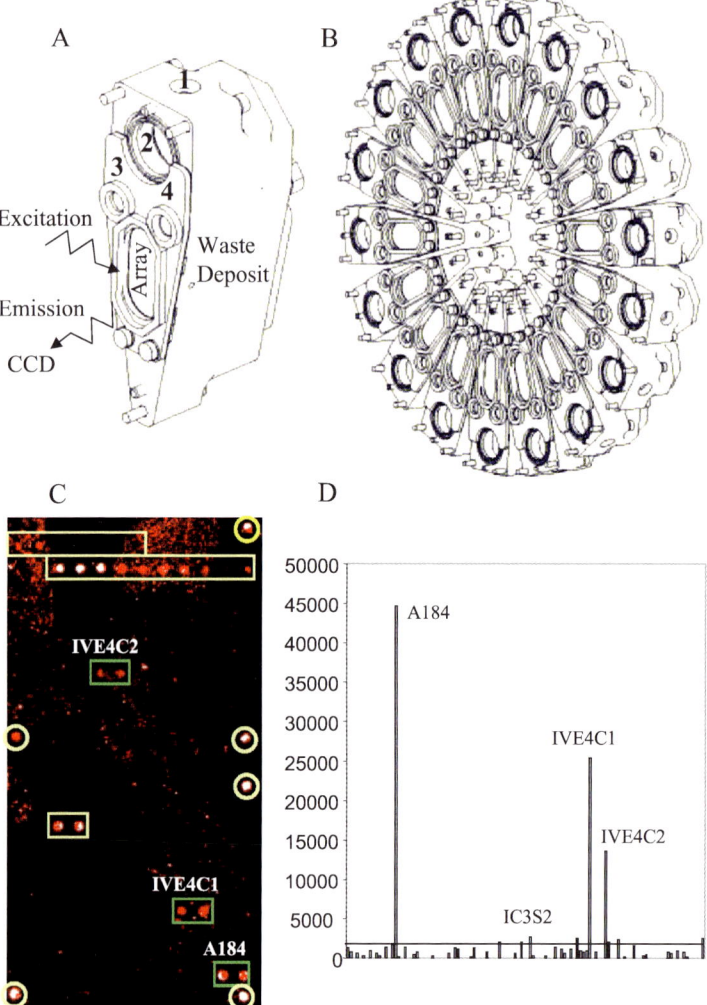

**Fig. 8** Automated sandwich immunoassay with SOLID 2 instrument. (**A**) Scheme of one of the 18 reaction modules shown in (**B**). Each one contains a loading port (*1*), a homogenizing chamber (*2*), additional chambers for reactants and fluorescent antibodies (*3* and *4*), an antibody microarray (*Array*) located in the inner side, and a waste deposit to avoid environmental contamination. After a final wash of the array, a laser excites the fluorochromes on the array and the emitted light is captured by a CCD camera. The captured image is then processed (**C**) and the signal intensities of the spots quantified and analyzed (**D**). In this experiment (**C**) up to $10^6$ cells from the Gram negative bacterium *Acidithiobacillus thiooxidans* were mixed with 0.5 g of a pyrite containing powder from rocks of Río Tinto area. The sample was loaded into SOLID 2 instrument and subjected to a complete sandwich assay: (1) liquid buffer addition (up to 2 ml) and homogenization by sonication, (2) filtering, (3) incubation with a 101 antibody containing microarray, (4) washing, (5) addition of a fluorescently labelled antibody mixture and incubation, (6) washing and (7) image capturing. This microarray contained the same 101 antibodies as the array showed in Fig. 6, apart of internal curve calibration of fluorescent spots (*yellow rectangles*) and additional spots to mark the array frame for easy location (*yellow circles*). Quantification of the spot signals was done with Genepix software (Genomic Solutions). Positive spots corresponded to antibodies A184 (raised against a mixture of whole cells and lysates of *At. thiooxidans*), IVE4C1 (against whole *At. Thiooxidans* cells), and IVE4C2 (Against *At. Thiooxidans* cells after a wash with EDTA). Others, at the limit of detection like IC3S2, corresponded to antibodies raised against natural extracts or some bacterial extracts that might have cross-reacted

**Acknowledgements**    We thank Miriam García Villadangos for excellent technical work, the SOLID team, and professor Juan Pérez-Mercader for continuous and encouraging support. We also acknowledge Carlos Compostizo and Pedro L. Herrero from SENER S.A. This work is supported by Centro de Astrobiología (INTA-CSIC), and the Spanish Ministerio de Educación y Ciencia grant No. ESP2004-05008. V. Parro has a "Ramón y Cajal" contract from the Ministerio de Educación y Ciencia.

# References

J.P. Alarie, J.R. Bowyer, M.J. Sepaniak, A.M. Hoyt, T. Vo-Dinh, Anal. Chimica Acta **236**, 237 (1990)

B. Allard, J. Templier, Phytochemistry **54**, 369 (2000)

G.P. Anderson, S.C. Moreira, P.T. Charles, I.L. Medintz, E.R. Goldman, M. Zeinali et al., Anal. Chem. **78**, 2279 (2006)

G.P. Anderson, K.D. King, K.L. Gaffney, L.H. Johnson, Biosens. Bioelectron. **14**, 771 (2000)

Bada et al., Space Sci. Rev. (2007, this issue). doi:10.1007/s11214-007-9213-3

U.V. Balashova, I. Vedina, G.E. Markosyan, G.A. Zavarzin, Mikrobiologiya **43**, 581 (1974)

Barron, Space Sci. Rev. (2007, this issue). doi:10.1007/s11214-007-9254-7

S.A. Benner, K.G. Devine, L.N. Matveeva, D.H. Powell, Proc. Natl. Acad. Sci. USA **97**, 2425 (2000)

E. Bjorklund, T. Nilsson, S. Bøwadt, Trends Anal. Chem. **19**, 434 (2000)

E.T. Boder, K.S. Midelfort, K.D. Wittrup, Proc. Natl. Acad. Sci. USA **97**, 10701 (2000)

E.D. Breen, J.G. Curley, D.E. Overcashier, C.C. Hsu, S.J. Shire, Pharm. Res. **18**, 1345 (2001)

J.J. Brocks, G.A. Logan, R. Buick, R.E. Summons, Science **285**, 1033 (1999)

J.J. Brocks, R.E. Summons, R. Buick, G.A. Logan, Org. Geochem. **34**, 1161 (2003)

R.E. Buchanan, N.E. Gibbons (eds.), *Bergey's Manual of Determinative Bacteriology*, 8th edn. (The Williams & Wilkens Co., Baltimore, 1974)

G. Campistron, R.M. Buijs, M. Geffard, Brain Res. **376**, 400 (1986)

L.L. Chang, D. Shepherd, J. Sun, D. Ouellette, K.L. Grant, X.C. Tang et al., J. Pharm. Sci. **94**, 1427 (2005)

C.S. Chen, H. Zhu, Biotechniques **40**, 423 (2006)

Y. Chen, C. Wiesmann, G. Fuh, B. Li, H.W. Christinger, P. McKay et al., J. Mol. Biol. **293**, 865 (1999)

J.R. Collett, E.J. Cho, A.D. Ellington, Methods **37**, 4 (2005)

G. Cooper, N. Kimmich, W. Belisle, J. Sarinana, K. Brabham, L. Garrel, Nature **414**, 879 (2001)

Eigenbrode, Summons, Space Sci. Rev. (2007, this volume). doi:10.1007/s11214-007-9252-9

P. Fernández-Calvo, C. Näke, L.A. Rivas, M. García-Villadangos, J. Gómez-Elvira, V. Parro, Planet. Space Sci. **54**, 1612 (2006)

D. Fernández-Remolar, J. Gómez-Elvira, F. Gómez, E. Sebastian, J. Martín, J.A. Manfredi et al., Planet. Space Sci. **52**, 239 (2004)

D.C. Fernández-Remolar, R.V. Morris, J.E. Gruener, R. Amils, A.H. Knoll, Earth Planet. Sci. Lett. **240**, 149 (2005)

D. Fernández-Remolar, O. Prieto-Ballesteros, N. Rodríguez, F. Gómez, R. Amils, J. Gómez-Elvira, C. Stoker, Astrobiology (2007, in press)

G.J. Flynn, Earth, Moon Planets **72**, 469 (1996)

A. Geiger, P. Burgstaller, H. von der Eltz, A. Roeder, M. Famulok, Nucl. Acids Res. **24**, 1029 (1996)

J. Golden, L. Shriver-Lake, K. Sapsford, F. Ligler, Methods **37**, 65 (2005)

M.A. Gonzalez-Martinez, J. Penalva, J.C. Rodriguez-Urbis, E. Brunet, A. Maquieira, R. Puchades, Anal. Bioanal. Chem. **384**, 1540 (2006)

E. González-Toril, E. Llobet-Brossa, E.O. Casamayor, R. Amann, R. Amils, Appl. Environ. Microbiol. **69**, 4853 (2003)

M. Hansson, J. Ringdahl, A. Robert, U. Power, L. Goetsch, T.N. Nguyen et al., Immunotechnology **4**, 237 (1999)

M. Hiasa, Y. Moriyama, Biol. Pharm. Bull. **29**, 1251 (2006)

J. Hogan, Nature **437**, 1080 (2005)

L.D. Hutt, D.P. Glavin, J.L. Bada, R.A. Mathies, Anal. Chem. **71**, 4000 (1999)

G. Israël, C. Szopa, F. Raulin, M. Cabane, H.B. Niemann, S.K. Atreya et al., Nature **438**, 796 (2005)

D.S. Junior, F.J. Krug, M. de Godoi Pereira, M. Korn, Appl. Spectrosc. Rev. **41**, 305 (2006)

M. Kempe, Lett. Pept. Sci. **7** (2000)

K. Kierek-Pearson, E. Karatan, Adv. Appl. Microbiol. **57**, 79 (2005)

G. Klingelhofer, R.V. Morris, B. Bernhardt, C. Schroder, D.S. Rodionov, P.A. de Souza, Jr. et al., Science **1740**, 306 (2004)

G. Kminek, J.L. Bada, Earth Planet. Sci. Lett. **245**, 1 (2006)

W. Kusnezow, A. Jacob, A. Walijew, F. Diehl, J.D. Hoheisel, Proteomics **3**, 254 (2003)

W. Liao, S. Guo, X.S. Zhao, Front. Biosci. **11**, 186 (2006)
F.S. Ligler, C.R. Taitt, L.C. Shriver-Lake, K.E. Sapsford, Y. Shubin, J.P. Golden, Anal. Bioanal. Chem. **377**, 469 (2003)
P.B. Luppa, L.J. Sokoll, D.W. Chan, Clin. Chimica Acta **314**, 1–26 (2001)
M.D. Luque de Castro, L.E. Garcia-Ayuso, Anal. Chimica Acta **369**, 1 (1998)
M.E. Mackintosh, J. Gen. Microbiol. **105**, 215 (1978)
Mahaffy, Space Sci. Rev. (2007, this issue). doi:10.1007/s11214-007-9223-1
E. Mallat, D. Barceló, C. Barzen, G. Gauglitz, R. Abuknesha, Trends Anal. Chem. **20**, 124 (2001a)
E. Mallat, C. Barzen, R. Abuknesha, G. Gauglitz, D. Barceló, Anal. Chimica Acta **426**, 209 (2001b)
C.A. Marquette, L.J. Blum, Biosens. Bioelectron. **21**, 1424 (2006)
J. Masarova, E.S. Dey, B. Danielsson, Pol. J. Microbiol. **53**, 23 (2004)
J. Maule, M. Fogel, A. Steele, N. Wainwright, D.L. Pierson, D.S. McKay, J. Gravit. Physiol. **10**, 47 (2003)
G.D. McDonald, E. De Vanssay, J.R. Buckley, Icarus **132**, 170 (1998)
R. Navarro-González, F.A. Rainey, P. Molina, D.R. Bagaley, B.J. Hollen, J. de la Rosa et al., Science **302**, 1018 (2003)
Y.M. Newman, S.G. Ring, C. Colaco, Biotechnol. Genet. Eng. Rev. **11**, 263 (1993)
H.B. Niemann, S.K. Atreya, S.J. Bauer, G.R. Carignan, J.E. Demick, R.L. Frost et al., Nature **438**, 779–84 (2005)
P.R. Norris, J.C. Murrell, D. Hinson, Arch. Microbiol. **164**, 294 (1995)
N. Ohmura, K. Sasaki, N. Matsumoto, H. Saiki, J. Bacteriol. **184**, 2081 (2002)
J. Oro, G. Holzer, J. Mol. Evol. **14**, 153 (1979)
J. Oró, Nature **190**, 389 (1961)
J. Parellada, A. Narváez, M.A. López, E. Domínguez, J.J. Fernández, V. Pavlov et al., Anal. Chimica Acta **362**, 47 (1998)
J. Parnell, D. Cullen, M.R. Sims, S. Bowden, C.S. Cockell, R. Court et al., Astrobiology **7**, 578 (2007)
V. Parro, M. Moreno-Paz, Proc. Natl. Acad. Sci. USA **100**, 7883 (2003)
V. Parro, J.A. Rodríguez-Manfredi, C. Briones, C. Compostizo, P.L. Herrero, E. Vez et al., Planet. Space Sci. **53**, 729 (2005)
V. Parro, P. Fernández-Calvo, J.A. Rodríguez Manfredi, M. Moreno-Paz, L.A. Rivas, M. García-Villadangos et al., Astrobiology (2007, in press)
S. Pizzarello, X. Feng, S. Epstein, J.R. Cronin, Geochimica Cosmochimica Acta **58**, 5579 (1994)
F.L. Plows, J.E. Elsila, R.N. Zare, P.R. Buseck, Geochimica Cosmochimica Acta **67**, 1429 (2003)
R.C. Quinn, A.P. Zent, F.J. Grunthaner, P. Ehrenfreund, C.L. Taylor, J.R.C. Garry, Planet. Space Sci. **53**, 1376 (2005)
O. Ramström, M. Yan, *Molecularly Imprinted Materials: Science and Technology* (Marcer Dekker/CRC Press, New York, 2004)
P. Rathanaswami, S. Roalstad, L. Roskos, Q.J. Su, S. Lackie, J. Babcook, Biochem. Biophys. Res. Commun. **334**, 1004 (2005)
B. Renberg, J. Nordin, A. Merca, M. Uhlen, J. Feldwisch, P.A. Nygren et al., J. Proteome Res. **6**, 171 (2007)
S. Rodriguez-Mozaz, M.J. Lopez de Alda, D. Barcelo, Anal. Bioanal. Chem. **386**, 1025 (2006)
S. Rodríguez-Mozaz, M.J. López de Alda, M.P. Marco, D. Barceló, Talanta **65**, 291 (2005)
S. Rodríguez-Mozaz, M.P. Marco, M.J. López de Alda, D. Barceló, Anal. Bioanal. Chem. **378**, 588 (2004a)
S. Rodríguez-Mozaz, S. Reder, M.J. López de Alda, G. Gauglitz, D. Barceló, Biosens. Bioelectron. **19**, 633 (2004b)
K.E. Sapsford, P.T. Charles, C.H. Patterson Jr., F.S. Ligler, Anal. Chem. **74**, 1061 (2002)
K.E. Sapsford, Y.S. Shubin, J.B. Delehanty, J.P. Golden, C.R. Taitt, L.C. Shriver-Lake et al., J. Appl. Microbiol. **96**, 47 (2004)
M.H. Schweitzer, J. Wittmeyer, R. Avci, S. Pincus, Astrobiology **5**, 30 (2005a)
M.H. Schweitzer, J.L. Wittmeyer, J.R. Horner, J.K. Toporski, Science **307**, 1952 (2005b)
M.A. Sephton, C.T. Pillinger, I. Gilmour, Geochimica Cosmochimica Acta **62**, 1821 (1998)
D. Shabat, H. Itzhaky, J.L. Reymond, E. Keinan, Nature **374**, 143 (1995)
H. Silvaieh, M.G. Schmid, O. Hofstetter, V. Schurig, G. Gubitz, J. Biochem. Biophys. Methods **53**, 1 (2002)
M.R. Sims, D.C. Cullen, N.P. Bannister, W.D. Grant, O. Henry, R. Jones et al., Planet. Space Sci. **53**, 781 (2005)
A.M. Skelley, H.J. Cleaves, C.N. Jayarajah, J.L. Bada, R.A. Mathies, Astrobiology **6**, 824 (2006)
A.M. Skelley, J.R. Scherer, A.D. Aubrey, W.H. Grover, R.H. Ivester, P. Ehrenfreund et al., Proc. Natl. Acad. Sci. USA **102**, 1041 (2005)
C. Stoker, LPS **XXXIV**, 1076 (2003)
C.R. Stoker, M.A. Bullock, J. Geophys. Res. **102**, 10881 (1997)
C. Stoker, S. Dunagan, T.O. Stevens, R. Amils, J. Gómez-Elvira, D. Fernández et al., LPS **XXXV**, 2025 (2004)

C. Stoker, H. Cannon, S. Dunagan, L.G. Lemke, D. Miller, J. Gómez-Elvira et al., Astrobiology (2007, in press)

Summons et al., Space Sci. Rev. (2007, this issue). doi:10.1007/s11214-007-9256-5

C.R. Taitt, J.P. Golden, Y.S. Shubin, L.C. Shriver-Lake, K.E. Sapsford, A. Rasooly, F.S. Ligler, Microb. Ecol. **47**, 175 (2004)

D.P. Thompson, P.K. Wilson, M.R. Sims, D.C. Cullen, J.M. Holt, D.J. Parker et al., Anal. Chem. **78**, 2738 (2006)

J. Tschmelak, G. Proll, J. Riedt, J. Kaiser, P. Kraemmer, L. Barzaga et al., Biosens. Bioelectron. **20**, 1509 (2005a)

J. Tschmelak, G. Proll, J. Riedt, J. Kaiser, P. Kraemmer, L. Barzaga et al., Biosens. Bioelectron. **20**, 1499 (2005b)

N. Tuross, L. Stathoplos, Meth. Enzymol. **224**, 121 (1993)

J.M. Van Emon, V. Lopez-Avila, Anal. Chem. **64**, 79A (1992)

E. Vorobyova, V. Soina, M. Gorlenko, N. Zalinuva, A. Mamukelashvili, D. Gilichinsky et al., FEMS Microbiol. Rev. **20**, 277 (1997)

R.M. Wadkins, J.P. Golden, F.S. Ligler, J. Biomed. Optics **2**, 74 (1997)

M.G. Weller, A.J. Schuetz, M. Winklmair, R. Niessner, Anal. Chimica Acta **393**, 29 (1999)

M.G. Weller, *Protein Microarrays* (Jones and Bartlett Publishers, Sudbury, 2005) p. 285

J. Whitfield, Nature **430**, 288 (2004)

R. Wiese, Y. Belosludtsev, T. Powdrill, P. Thompson, M. Hogan, Clin. Chem. **47**, 1451 (2001)

M.K. Wong, W. Gu, T.L. Ng, Anal. Sci. **13**, 97 (1997)

X.Q. Zhang, P.L. Bishop, M.J. Kupferle, Water Sci. Technol. **37**, 345 (1998)

# Remote Sensing of (Exo)Planets

# On the "Galactic Habitable Zone"

**Nikos Prantzos**

Originally published in the journal Space Science Reviews, Volume 135, Nos 1–4.
DOI: 10.1007/s11214-007-9236-9 © Springer Science+Business Media B.V. 2007

**Abstract** The concept of Galactic Habitable Zone (GHZ) was introduced a few years ago as an extension of the much older concept of Circumstellar Habitable Zone. However, the physical processes underlying the former concept are hard to identify and even harder to quantify. That difficulty does not allow us, at present, to draw any significant conclusions about the extent of the GHZ: it may well be that the entire Milky Way disk is suitable for complex life.

**Keywords** Bioastronomy · Galactic evolution · Habitable zones

## 1 Introduction

The modern study of the "habitability" of circumstellar environments started almost half a century ago (Huang 1959). The concept of a Circumstellar Habitable Zone (CHZ) is relatively well defined, being tightly related to the requirement of the presence of liquid water as a necessary condition for life-as-we-know-it; the corresponding temperature range is a function of the luminosity of the star and of the distance of the planet from it. An important amount of recent work, drawing on various disciplines (planetary dynamics, atmospheric physics, geology, biology etc.) refined considerably our understanding of various factors that may affect the CHZ; despite that progress, we should still consider the subject to be in its infancy (see, e.g., Chyba and Hand 2005; Gaidos and Selsis 2006; and references therein).

Habitability on a larger scale was considered a few years ago, by Gonzalez et al. (2001), who introduced the concept of Galactic Habitable Zone (GHZ). The underlying idea is that various physical processes, which may favour the development or the destruction of complex life, may depend strongly on the temporal and spatial position in the Milky Way. For instance, the risk of a supernova explosion sufficiently close to represent a threat for life is, in

Strategies for Life Detection, ISSI Bern, April 24–28 2006.

N. Prantzos (✉)
Institut d'Astrophysique de Paris, UMR7095 CNRS, Univ. P. & M. Curie, 98bis, Bd Arago, 75014 Paris, France
e-mail: prantzos@iap.fr

general, larger in the inner Galaxy than in the outer one, and has been larger in the past than at present. Another example is offered by the metallicity (the amount of elements heavier than hydrogen and helium) of the interstellar medium, which varies across the Milky Way disk, and which may be important for the existence of Earth-like planets. Indeed, the host stars of the ~180 extrasolar giant planets detected so far are, on average, more metal-rich than stars with no planets in our cosmic neighbourhood (e.g., Fischer and Valenti 2005). Several other factors, potentially important for the GHZ, were discussed by Gonzalez (2005).

The concept of GHZ is much less well defined than the one of CHZ, since none of the presumably relevant factors can be quantified in a satisfactory way. Indeed, the role of the metallicity in the formation and survival of Earth-like planets is not really understood at present, while the "lethality" of supernovae and other cosmic explosions is hard to assess. The first study attempting to quantitatively account for such effects is made by Lineweaver et al. (2004, hereafter L04), with a detailed model for the chemical evolution of the Milky Way disk. They find that the probability of having an environment favourable to complex life is larger in a "ring" (a few kpc wide) surrounding the Milky Way centre and spreading outwards in the course of the Galaxy's evolution.

We repeat that exercise here in Sect. 2, with a model that reproduces satisfactorily the major observables of the Milky Way disk. In Sect. 3 we discuss the role of metallicity and we quantify it in a different (and, presumably, more realistic) way than L04, in the light of recent simulations of planetary formation. In Sect. 4 we discuss the risk of SN explosions and conclude that it can hardly be quantified at present, in view of our ignorance of how robust life really is. For comparison purposes, though, we adopt the same risk factor as the one defined in L04. In Sect. 5 we present our results, showing that the GHZ may, in fact, extend to the whole Galactic disk today. We conclude that, at the present stage of our knowledge, the GHZ may extend to the entire MW disk.

## 2 The Evolution of the Milky Way Disk

The evolution of the solar neighbourhood is rather well constrained, due to the large body of observational data available. In particular, the metallicity distribution of long-lived stars suggests a slow formation of the local disk, through a prolonged period of gaseous infall (e.g., Goswami and Prantzos 2000; see, however, Haywood 2006 for a different view). For the rest of the disk, available data concern mainly its current status and not its past history; in those conditions, it is impossible to derive a unique evolutionary path. Still, the radial properties of the Milky Way disk (profiles of stars, gas, star formation rate, metallicity, colours etc.), constrain significantly its evolution and point to a scenario of "inside-out" formation (e.g., Prantzos and Aubert 1995). Such scenarios fit "naturally" in the currently favoured paradigm of galaxy formation in a cold dark matter Universe (e.g., Naab and Ostriker 2006).

Some results of a model developed along those lines (Boissier and Prantzos 1999; Hou et al. 2000) are presented in Fig. 1. Note that the evolution of the Galactic bulge (innermost ~2 kpc) is not studied here, since it is much less well constrained than the rest of the disk; the conclusions, however, depend very little on that point. The model assumes a star formation rate proportional to $R^{-1}$ (where $R$ is the galactocentric distance), inspired from the theory of star formation induced by spiral waves in disks. It satisfies all the available constraints for the solar neighbourhood and the rest of the disk; some of those constraints appear in Fig. 1. Among the successes of the model, one should mention the prediction that the abundance gradient should flatten with time, due to the "inside-out" formation scheme (bottom left panel in Fig. 1), a prediction that is quantitatively supported by recent measurements of the metallicity profile in objects of various ages (Maciel et al. 2006).

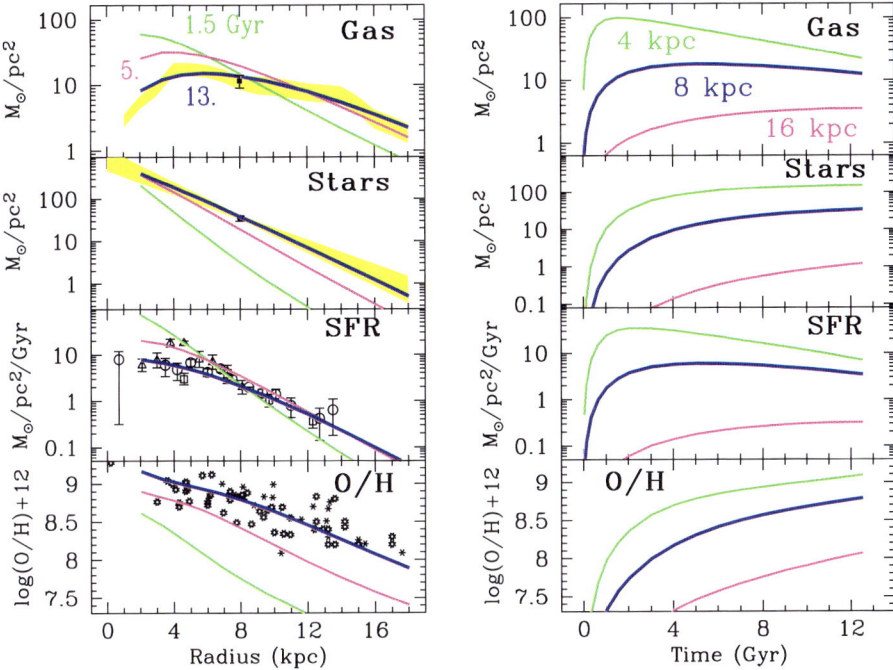

**Fig. 1** The chemical evolution of the Milky Way disk, obtained in the framework of a one-dimensional model with radial symmetry (see Sect. 1 for a description of the model). *Left*, from top to bottom: radial profiles of the surface density of gas, stars, star formation rate (SFR), and of the oxygen abundance. Profiles are displayed at three epochs, namely at 1.5, 5, and 13 Gyr; the latter (thick curve in all figures) is compared to observations concerning the present-day MW disk (yellow shaded areas in the first two figures, data points in the last two). *Right*, from top to bottom: the same quantities are plotted as a function of time for three different disk regions, located at galactocentric distances of 4, 8 and 16 kpc; the second one (thick curves in all panels of the right column) corresponds to the solar neighbourhood

It should be stressed, however, that the output of the model is only as good as the adopted observational constraints. For instance, there is still some uncertainty concerning the level of the oxygen abundance gradient: the "standard" value of $d \log(\mathrm{O/H})/dR = -0.07$ dex/kpc was challenged by evaluations of Daflon and Cunha (2004), who found only half that value. If the latter turns out to be true, some of the model parameters (e.g., the adopted radial dependence of the infall rate and/or the star formation rate) should have to be revised.

In any case, it appears rather well established now that the star formation rate and the metallicity are more important in the inner disk than in the outer one, and that this trend has been even stronger in the past (see left panel in Fig. 1). Some authors have suggested that those two parameters play an important role in the "habitability" of a given region of the Galactic disk (Gonzalez et al. 2001, L04); we discuss briefly that role in the next two sections.

## 3 On the Probability of Having Stars with Earth-Like Planets

Since the first detection of an extrasolar planet around Peg 51 (Mayor and Queloz 1995), more than 230 stars in the solar neighbourhood have been found to host planets (e.g., Butler et al. 2006). The masses of those planets range from 0.015 to 18 Jupiter masses and their distances to the host star lie in the range of 0.03–6 AU. These ranges of mass and distance, however, are at present due to selection effects, resulting from the current limitations of detection techniques. Continuous improvement of those techniques may well reveal the presence of smaller, Earth-like planets at small distances from the host stars (as well as massive ones further away). In fact, the mass distribution of the detected planets is $dN/dM \propto M^{-1.1}$ (Butler et al. 2006) and suggests that Earth-like planets should be quite common (unless a yet-unknown physical effect truncates the distribution from the low mass end).

The unexpected existence of "Hot Jupiters" is usually interpreted in terms of planetary migration (see, e.g., Papaloizou and Terquem 2006 for a review): those gaseous giants can, in principle, be formed at a distance of several AU from their star (where gas is available for accretion onto an already rapidly formed rocky core); their subsequent interaction with the proto-planetary disk leads, in general, to loss of angular momentum and migration of the planets inwards. On their way, those planets destroy the disk and any smaller planets that may have been formed there. Thus, the presence of Hot Jupiters around some stars implies that the probability of life (at least as we know it) in the corresponding stellar system is rather small.

A key feature of the stars hosting planets is their high metallicity (compared to stars with no planets). That feature was first noticed by Gonzalez (1997), who interpreted it in terms of accretion of (metal enriched) planetesimals onto the star. However, subsequent studies found no correlation of the stellar metallicity with the depth of the convective zone of the star, invalidating that idea (see, e.g., Fischer and Valenti 2005, hereafter FV05, and references therein). Thus, it appears that the high metallicity is intrinsic to the star, and it presumably plays an important role to the giant planet formation. FV05 quantified the effect, finding that the fraction of FGK stars with Hot Jupiters increases sharply with metallicity $Z$, as $P_{HJ} = 0.03 \, (Z/Z_\odot)^2$, where $Z$ designates Fe abundance. Note that this function has lower values and is much less steep than the one suggested by Lineweaver (2001, hereafter L01), which reaches a value of $P_{HJ} = 0.5$ at a metallicity of 2 $Z_\odot$; the latter function is used in the calculations of L04.

The impact of metallicity on the formation of Earth-like planets is unknown at present. Metals (in the form of dust) are obviously necessary for the formation of planetesimals, but the required amount depends on the assumed scenario (and its initial conditions, e.g., size of protoplanetary disk). Inspired by the early results on extrasolar planet host stars, several authors argued for an important role of metallicity. Thus, in L01 it is suggested that the probability of forming Earth-like planets should simply have a linear dependence on metallicity $P_{FE} \propto Z$. On the other hand, Zinnecker (2003) argued for a threshold on metallicity, of the order of $1/2 \, Z_\odot$, for the formation of Earth-sized planets.

However, the situation may be more complicated than the one emerging from simple analytical arguments. Recent (and yet unpublished) numerical simulations by the Bern group find that at low metallicities, the decreasing probability of forming giant planets leaves quite a lot of metals to form a significant number of Earth-sized planets (see Fig. 2). Those simulations cover a factor of $\sim 3$ in metallicity (from $\sim 0.6 \, Z_\odot$ to 2 $Z_\odot$) and they roughly reproduce the observations concerning $P_{HJ}(Z)$ (see the earlier discussion); their predictions for $P_{FE}(Z)$ have to be confirmed by further simulations and, ultimately, by observations. They suggest, however, that the formation of Earth-like planets may be quite common, even at low metallicity environments like the outer Galaxy and the early inner Galaxy.

Assuming that $P_{FE}(Z)$ and $P_{HJ}(Z)$ are known, the probability of having stars with Earth-like planets (but not Hot Jupiters, which destroy them) is simply: $P_E(Z) = P_{FE}(Z)(1 - P_{HJ}(Z))$. However, it is clear from the previous paragraphs that neither of the terms of the

**Fig. 2** Fraction of planets around a 1 M$_\odot$ star as a function of their mass (from simulations of Mordasini et al. 2007). Results are displayed for three different values of the metallicity of the star (and of the corresponding proto-planetary disk), as shown on the figure. It is seen that, high metallicities favour larger fractions of massive planets (in rough agreement with observations), while at low metallicities the presence of Earth-like planets is enhanced rather than suppressed (courtesy Y. Alibert)

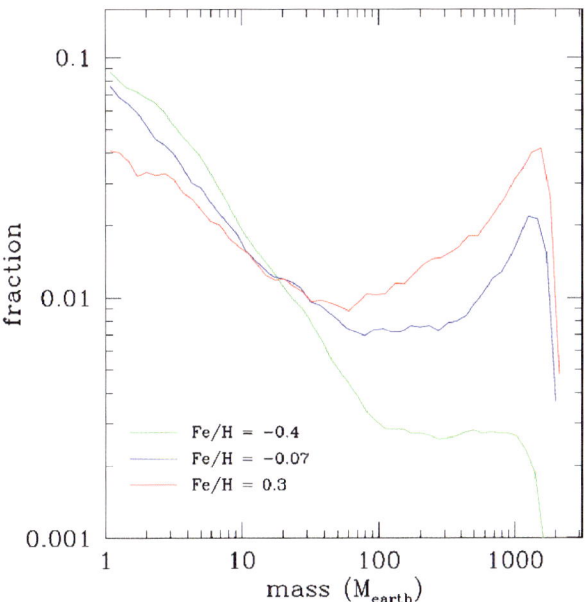

right member of that equation can be accurately evaluated at present.[1] Just for illustration purposes, we adopt in this work a set of metallicity-dependent probabilities different from the one adopted in L04 (who assumed $P_{FE}$ linearly dependent on $Z$, and $P_{HJ}$ strongly increasing with $Z$): we assume that $P_{FE} = \text{const} = 0.4$ (so that the metallicity integrated probability is the same as in L04) for $Z > 0.1\ Z_\odot$ and $P_{HJ}$ from FV05. The two sets appear in Fig. 3, which also displays $P_E$ (bottom panel): with our set, $P_E$ extends to non-zero values at metallicities much lower and higher than in the work of L04. From that factor alone it is anticipated that any GHZ (found through a chemical evolution model) will be much larger with our set of data than with the one of L04.

## 4 On the Probability of Life Surviving Supernova Explosions

The potential threat to complex life on Earth represented by nearby SN explosions was first studied by Ruderman (1974). He pointed out that energetic radiation from such events, in the form of hard X-rays, gamma-rays or cosmic rays, may (partially or totally) destroy the Earth's atmospheric ozone, leaving land life exposed to lethal does of UV fluxes from the Sun. The paper went completely unnoticed, with just one citation for about 20 years. In the last few years, however, a large number of studies were devoted to that topic (for reasons that are not quite clear to the author of this paper). Two factors are, perhaps, at the origin of that interest: the availability of complex models of Earth's atmospheric structure and chemistry; and the discovery of extrasolar giant planets. The latter suggests that Earth-like planets may also be abundant in the Galaxy; however, complex life on them may be a rare

---

[1] In fact, it is not even certain that the migration of Hot Jupiters inwards prohibits the existence of terrestrial planets in the habitable zone: based on recent simulations of planetary system formation, Raymond et al. (2006) find that "...about 34% of giant planetary systems in our sample permit an Earth-like planet of at least 0.3 M$_{Earth}$ to form in the habitable zone".

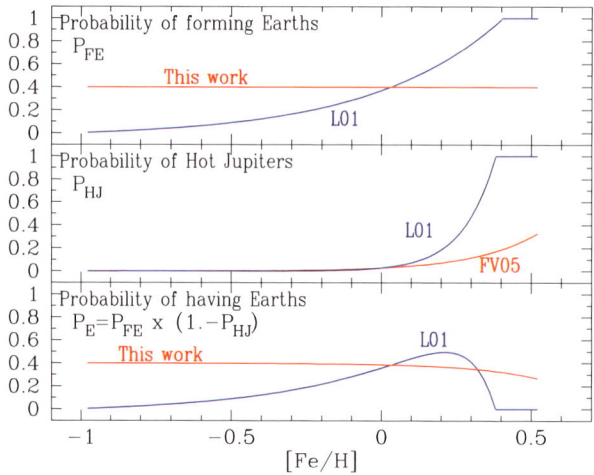

**Fig. 3** Role of metallicity of the proto-stellar nebula in the formation and presence of Earth-like planets around solar-mass stars, according to various estimates. *Top:* The probability of forming Earths has been taken as proportional to metallicity in Lineweaver (2001, L01), while it is assumed to be quasi-independent of metallicity (at least for [Fe/H] > −1) here, after the results displayed in Fig. 2. *Middle:* The probability of forming Hot Jupiters (= destroying Earth-like planets in circumstellar habitable zones) is larger in absolute value and steeper as a function of metallicity in L01 than in Fischer and Valenti (2005, FV05); the latter function is adopted here. *Bottom:* The probability of having Earths, obtained from the two previous ones, in L01 and in this work

phenomenon, because of various cosmic threats like SN explosions. This is the basic idea underlying the concepts of GHZ (Gonzalez et al. 2001) and of the Rare Earth (Brownlee and Ward 2000): complex life in the Universe may be rare, not for intrinsic reasons (i.e., improbability of development of life on a planet), but for extrinsic ones, related to the hostile cosmic environment.

Despite the simplicity of the idea, however, the studies devoted to the topic revealed that it is very difficult to quantify the SN threat to life (or any other cosmic threat). Studies of that kind evaluate the energetic particle flux/irradiation impinging on Earth, which could induce a significant number of gene mutations (based on our understanding of such mutations in various organisms on Earth). Knowing the intrinsic emissivity of energetic particles from a typical SN (which depends essentially on the configuration of the progenitor star and the energetics of the explosion), one may then calculate a lower limit for the distance to the SN, for such fluxes/irradiations to occur. The distribution of the rates of various SN types in the Milky Way (presumably known to within a factor of a few) is then used to evaluate the frequency/probability of such events in our vicinity and elsewhere in the Galaxy.

Numerous studies in the last few years explored several aspects of the scenario, with the use of models of various degrees of complexity for the atmosphere of the Earth (or Mars) and the transfer of high-energy radiation through it (see, e.g., Ejzak et al. 2006 and references therein). Thus, Smith et al. (2004) found that a substantial fraction (∼1%) of the high-energy radiation (X- and $\gamma$-rays) impinging on a thick atmosphere (column density ∼100 g cm$^{-2}$, like the Earth's) may reach the ground and induce biologically important mutations. Ejzak et al. (2006) found that the ozone depletion depends mainly on the total irradiation (for durations in the $10^{-1}$–$10^{8}$ s range) and only slightly on the received flux; this result allows one to deal with both short ($\gamma$-ray bursts)[2] and long (SN explosions) events. They also found that the overall result depends significantly on the shape of the spectrum (with harder spectra being more harmful).

Even if one assumes that such calculations are realistic (which is far from demonstrated), it is hard to draw any quantitative conclusions about the probability of *definitive sterilisation of a habitable planet* (which is the important quantity for the calculation of a GHZ). Even if 100% lethality is assumed for all land animals after a nearby SN explosion, marine life will certainly survive to a large extent (since UV is absorbed from a couple of meters of water). In the case of the Earth, it took just a few hundred million years for marine life to spread on the land and evolve to dinosaurs and, ultimately, to humans; this is less than 4% of the lifetime of a G-type star. Even if land life on a planet is destroyed from a nearby SN explosion, it may well reappear again after a few $10^8$ yrs or so. Life displays unexpected robustness and a cosmic catastrophe might even accelerate evolution towards life forms that are presently unknown.[3] Only an extremely high frequency of such catastrophic events (say, more than one every few $10^7$ Myr) could, perhaps, ensure permanent disappearance of complex life from the surface of a planet.

Gehrels et al. (2003) found that significant biological effects due to ozone depletion— that is, doubling of the UV flux on Earth's surface—may arise for SN explosions closer than $D \sim 8$ pc. Taking into account the estimated SN frequency in the Milky Way (a couple per century), the frequency of the Sun crossing spiral arms ($\sim 10$ Gyr$^{-1}$), and the vertical to the Galactic plane density profile of the supernova progenitor stars (scaleheight $h_{SN} \sim 30$ pc, comparable to the current vertical displacement of the Sun $h_\odot \sim 24$ pc), Gehrels et al. (2003) found that a SN should explode closer than 8 pc to the Sun with a frequency $f \sim 1.5$ Gyr$^{-1}$. Repeating their calculation with a realistic radial density profile for SN resulting from massive stars (see profile of SFR in Fig. 1), we find a number closer to $f \sim 1$ Gyr$^{-1}$. Note, however, the sensitivity of that number to the assumed critical distance ($f \propto D^3$): reducing that distance from 8 pc to 6 pc, would correspondingly reduce the frequency to 0.4 Gyr$^{-1}$. In any case, it appears that no such catastrophic event occurred close to the Earth in the past Gyr or so.

In an attempt to circumvent the various unknowns, L04 quantify the risk for life represented by SN explosions by using the time-integrated rate of SN within 4 Gyr (see Fig. 4). Adopting such a variable avoids considering the spiral arm passage and allows for a rapid implementation of the SN risk factor in a Galactic chemical evolution model; however, the assignment of probabilities is rather arbitrary. For instance, it is assumed that $P < 1$ for TIR$_{SN} = 1$ local units, a rather strange assumption in view of the fact that life on Earth has well survived the local SN rate. Also, it is assumed that $P = 0$ for TIR$_{SN} = 4$ local units, implying that if the local SN time integrated rate were just four times larger than what it has actually been, then our planet would have been permanently unable to host complex life. But such an increase in the SN rate would increase the frequency of "lethal SN" (closer

---

[2] Gamma-ray bursts are more powerful, but also much rarer events than supernovae. Their beamed energy makes them lethal from much larger distances than SN (several kpc in the former case, compared to a few pc in the latter) and several studies are recently devoted to that topic (Ejzak et al. 2006 and references therein). However, they are associated with extragalactic regions of low metallicity (in the few cases with available observations), implying that their frequency in the Milky Way has been probably close to zero in the past several Gyr; the formation of GRB progenitors in low metallicity environments is also favoured on theoretical grounds (e.g., Hirschi et al. 2005).

[3] An illustration is offered by the "Cambrian explosion", with a myriad of complex life forms appearing less than $\sim 40$ Myr after the last "snowball Earth" (which presumably occurred in the Neoproterozoic era, 750 Myr ago).

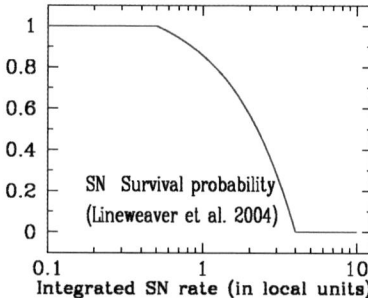

**Fig. 4** The probability that complex life on an Earth-like planet survives a nearby supernova explosion, according to Lineweaver et al. (2004). It is plotted as a function of the time-integrated supernova rate $TIR_{SN} = \int_t^{t+4\ \mathrm{Gyr}} SNR(t')\,dt'$; the latter is expressed in units of the corresponding time-integrated SN rate in the solar neighbourhood and in the last 4 Gyr. The probability is *quite arbitrarily* assumed to be $P = 1$ for integrated SN rates less than 0.5 and $P = 0$ for integrated SN rates larger than 4

than 8 pc, according to Gehrels et al. 2003), from $1\ \mathrm{Gyr}^{-1}$ to $4\ \mathrm{Gyr}^{-1}$, leaving still 250 Myr between lethal events, which is probably more than enough for a "renaissance" of land life. Assuming $P = 0$ for substantially larger TIR values (say, for $TIR_{SN} = 10$ local units) seems a safer bet, but it still constitutes a very imperfect evaluation of the SN threat. For comparison purposes with L04, the "SN risk factor" of Fig. 4 is also adopted here.

## 5  A GHZ as Large as the Whole Galaxy?

Using the chemical evolution model of Sect. 2, the probabilities for forming Earth-like planets that survive the presence of Hot Jupiters of Sect. 3 and the risk factor from SN explosions from Sect. 4, we have calculated the distribution of stars potentially hosting Earth-like planets with complex life in the Milky Way.

The distribution of various probabilities as a function of galactocentric radius appears in Fig. 5, for five different epochs of the Galaxy's evolution, namely 1, 2, 4, 8 and 13 Gyr. Due to the inside-out formation scheme of the Milky Way, all metallicity-dependent probabilities peak early on in the inner disk and progressively increase outwards. The time-integrated SN rate is always higher in the inner than in the outer Galaxy. However, at late times its absolute value in the inner disk is smaller than at early times; as a result, the corresponding probability for surviving SN explosions, which is null in the inner disk at early times, becomes quite substantial at late times.

The bottom line is depicted in the bottom right panel of Fig. 5. It shows that the overall probability of Earths surviving Hot Jupiters and SN has indeed a ring-like shape, quite narrow early-on (as it should, since there is basically no star formation in the outer disk at that time) and progressively "migrating" outwards and becoming more and more extended. Today, that "ring" peaks at $\sim$10 kpc, but it is quite large, since even the inner disk (the molecular ring) the SN risk factor is not much larger (just a factor of a few) than in the solar neighbourhood. Obviously, the GHZ extends to the quasi-totality of the galactic disk today. Figure 6 (left panel) shows the same result as Fig. 5 (bottom right) in a different way (space-time diagram), comparable with Fig. 3 of L04.

In the right panel of Fig. 6, the radial probability of the left panel is multiplied with the corresponding number of stars created up to time $t$. Since there are more stars in the inner disk, this latter probability peaks in the inner Galaxy (in other terms: the left panel displays

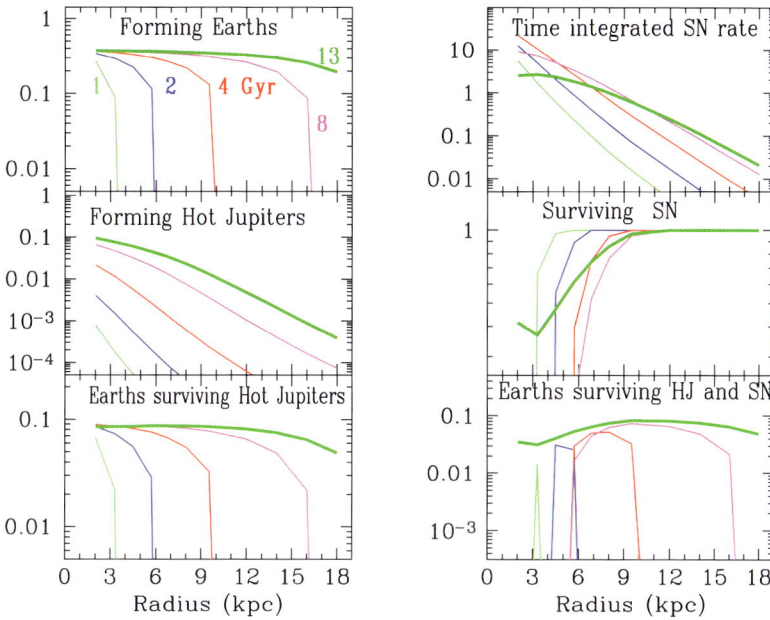

**Fig. 5** Probabilities of various events as a function of galactocentric distance at five different epochs of the Milky Way evolution: 1, 2, 4, 8 and 13 Gyr, respectively (the latter is displayed with a *thick curve* in all figures). The probabilities of forming Earths $P_{FE}$ and Hot Jupiters $P_{HJ}$ and of actually having Earths $P_E = P_{FE}(1 - P_{HJ})$ (*left part* of the figure) depend on the metallicity evolution (see Fig. 3), while the probability of life-bearing planets surviving SN explosions is obtained with the criterion of Fig. 4. Finally, the overall probability for Earth-like planets with life (*bottom right*) defines a ring in the MW disk, progressively migrating outwards; that "probability ring" is quite narrow at early times, but fairly extended today and peaking at about 10 kpc

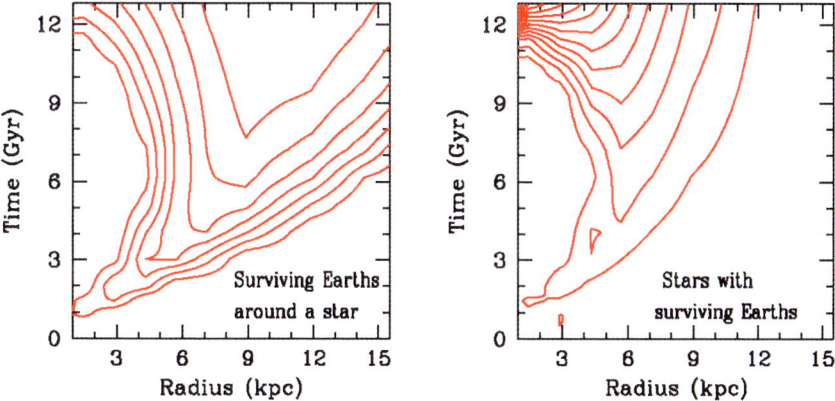

**Fig. 6** *Left:* Probability isocontours in the time versus galactocentric distance plane of having *around one star* an Earth-like planet that survived SN explosions and migration of Hot Jupiters; as in the previous figure, that probability peaks today (∼12 Gyr) at around 10 kpc from the Galactic Centre. *Right:* Multiplying that probability with the corresponding surface density of stars (much larger in the inner disk than in the outer one, because of the exponentially decreasing stellar profile), one finds that it is more likely to find a star with an Earth-like planet that survived SN explosions in the inner Galaxy

the probability of complex life around *one star* at a given position, while the right panel displays the probability of having complex life *per unit volume or surface density* in a given position). Thus, despite the high risk from SN early on in the inner disk, that place becomes later relatively "hospitable". Because of the large density of stars in the inner disk, it is more interesting to seek complex life there than in the outer disk; the solar neighbourhood (at 8 kpc from the centre) is not particularly privileged in that respect.

The results obtained in this section depend heavily on the assumptions made in Sects. 2 and 3, which are far from being well founded at present. It is clear that the contours of a GHZ in the Milky Way cannot be, even approximately, defined either in space or in time; it may well be that most of our Galaxy is (and has been) suitable for life. Thus, the concept of a GHZ may have little or no significance at all. It should be considered, at best, as a broad framework, allowing us to formulate our thoughts/educated guesses/knowledge about a very complex phenomenon such as Life (origin, development and survival) in the Milky Way.

**Acknowledgements**   I am grateful to Franck Selsis for many useful discussions and a careful reading of the manuscript.

## References

S. Boissier, N. Prantzos, Mon. Not. R. Astron. Soc. **307**, 857–876 (1999)
D. Brownlee, P. Ward, *Rare Earth* (Springer, Berlin, 2000)
R.P. Butler, J.T. Wright, G.W. Marcy, Astrophys. J. **646**, 505 (2006), astro-ph/0607493
C. Chyba, K. Hand, ARAA **43**, 31–74 (2005)
S. Daflon, K. Cunha, Astrophys. J. **617**, 1115–1126 (2004)
L. Ejzak, A. Melott, M. Medvedev, B. Thomas (2006), astro-ph/0604556
D. Fischer, J. Valenti, Astrophys. J. **622**, 1102–1117 (2005)
E. Gaidos, F. Selsis (2006), astro-ph/0602008
N. Gehrels et al., Astrophys. J. **585**, 1169–1176 (2003)
G. Gonzalez, Mon. Not. R. Astron. Soc. **285**, 403–412 (1997)
G. Gonzalez, Orig. Life Evol. Biosph. **35**, 555–606 (2005)
G. Gonzalez, D. Brownlee, P. Ward, Icarus **152**, 185–200 (2001)
A. Goswami, N. Prantzos, Astron. Astrophys. **359**, 191–212 (2000)
M. Haywood, Mon. Not. R. Astron. Soc. **371**, 1760–1776 (2006)
R. Hirschi, G. Meynet, A. Maeder, Astron. Astrophys. **443**, 581–591 (2005)
J. Hou, N. Prantzos, S. Boissier, Astron. Astrophys. **362**, 921–936 (2000)
S. Huang, Am. Sci. **47**, 393–402 (1959)
C. Lineweaver, Icarus **151**, 307–313 (2001)
C. Lineweaver, Y. Fenner, B. Gibson, Science **303**, 59–62 (2004)
W. Maciel, L. Lago, R. Costa, Astron. Astrophys. **453**, 587–593 (2006)
M. Mayor, D. Queloz, Nature **378**, 355–359 (1995)
C. Mordasini, Y. Alibert, W. Benz, D. Naef (2007, in preparation)
T. Naab, J. Ostriker, Mon. Not. R. Astron. Soc. **366**, 899–917 (2006)
J. Papaloizou, C. Terquem, Rep. Prog. Phys. **69**, 119–180 (2006)
N. Prantzos, O. Aubert, Astron. Astrophys. **302**, 69–85 (1995)
S. Raymond, A. Mandell, S. Sigursson, Science **313**, 1413–1416 (2006)
M. Ruderman, Science **184**, 1079–1081 (1974)
D.S. Smith, J. Scalo, J.C. Wheeler, Icarus **171**, 229–253 (2004)
H. Zinnecker, in *Proceedings of IAU Symposium, No. 213*, ed. by R. Norris, F. Stootman (Astronomical Society of the Pacific, San Francisco, 2003), p. 45

# Earthshine Observation of Vegetation and Implication for Life Detection on Other Planets

## A Review of 2001–2006 Works

Luc Arnold

Originally published in the journal Space Science Reviews, Volume 135, Nos 1–4.
DOI: 10.1007/s11214-007-9281-4 © Springer Science+Business Media B.V. 2007

**Abstract** The detection of exolife is one of the goals of very ambitious future space missions that aim to take direct images of Earth-like planets. While associations of simple molecules present in the planet's atmosphere ($O_2$, $O_3$, $CO_2$, etc.) have been identified as possible global biomarkers, this paper reviews the detectability of a signature of life from the planet's surface, i.e. the green vegetation. The vegetation reflectance has indeed a specific spectrum, with a sharp edge around 700 nm, known as the "Vegetation Red Edge" (VRE). Moreover, vegetation covers a large surface of emerged lands, from tropical evergreen forest to shrub tundra. Thus, considering vegetation as a potential global biomarker is relevant.

Earthshine allows us to observe the Earth as a distant planet, i.e. without spatial resolution. Since 2001, Earthshine observations have been used by several authors to test and quantify the detectability of the VRE in the Earth spectrum. The vegetation spectral signature is detected as a small "positive shift" of a few percentage points above the continuum, starting at 700 nm. This signature appears in most spectra, and its strength is correlated with the Earth's phase (visible land versus visible ocean). The observations show that detecting the VRE on Earth requires a photometric relative accuracy of 1% or better. Detecting something equivalent on an Earth-like planet will therefore remain challenging, especially considering the possibility of mineral artifacts and the question of "red edge" universality in the Universe.

**Keywords** Earthshine · Earth's spectrum · Biosignature · Vegetation red edge · Global biomarker · Extrasolar planet

## 1 Introduction

*Shall we be able to detect life on an unresolved Earth-like extrasolar planet?* Future space missions, such as Darwin or TPF, will provide us with the first images and low-resolution spectra of these planets, and the question of the presence or not of, ideally, an unambiguous

L. Arnold (✉)
Observatoire de Haute Provence, CNRS, 04870 Saint-Michel-l'Observatoire, France
e-mail: Luc.Arnold@oamp.fr

**Fig. 1** Voyager 1 took this picture of Earth in February 1990 while it was travelling well beyond the orbit of Neptune. Voyager did not take a spectrum of this spatially unresolved view of its mother planet, but this picture illustrates how an Earth-like extrasolar planet might look when imaged in the visible domain by a future space observatory—a pale blue dot. Photo from http://photojournal.jpl.nasa.gov/catalog/PIA02228. Courtesy NASA/JPL-Caltech

signature of life or, more realistically, a set of possible biogenic spectral features in these data will undoubtedly feed an animated debate.

Let us consider a unresolved extrasolar Earth-like planet imaged by a space-based, high-contrast instrument, basically a telescope equipped with a coronagraph that blocks the stellar light by masking the star to allow the observation of the very faint planet near its parent star.[1] The spectrum of the light reflected by the planet, when normalized to the parent star spectrum, gives the planet reflectance spectrum revealing its atmospheric and ground colour, if the latter is visible through a partially transparent atmosphere. Since the planet will remain unresolved (at least with the missions mentioned above), its spectrum will be spatially integrated (i.e., disk-averaged) for the observed orbital phase of the planet.

*What would the spectrum of an unresolved Earth-like planet look like?* A way to answer this question is to consider how the spectrum of our Earth would look if observed from a very large distance, typically several parsecs. This can be done from a space probe traveling deep into the Solar System and looking back at the Earth, as Voyager 1 did in 1990 (Fig. 1) or Mars Express in 2003 (Fig. 2). Note also that an integrated Earth spectrum for a given phase of the planet and also for a given position of an observer far above the Earth can also be done—in principle at least—by integrating spatially resolved spectra from low-orbit satellites.

An alternative method to obtain the Earth-averaged spectrum consists of taking a spectrum of the Moon Earthshine, i.e. Earth light backscattered by the nonsunlit Moon (Fig. 3). A spectrum of the Moon Earthshine directly gives the disk-averaged spectrum of the Earth at the phase seen from the Moon (since the Moon surface roughness "washes out" any spatial information on the Earth's colour).

*What shall we look for in this spectrum?* We shall first look for sets of molecules in the planet atmosphere (like oxygen and ozone) which may be biologic products or by-products. We shall also look for ground colours characteristic of biological complex molecules (like pigments in vegetation). Said in a more general manner, we shall look for missing photons used in a photosynthetic process occurring on the planet. The visible and near-infrared Earthshine spectra published to date clearly show the atmospheric signatures and, at least, tentative signs of ground vegetation which thus appears as an interesting potential global biomarker (Arnold et al. 2002; Woolf et al. 2002; Seager et al. 2005; Montañés-Rodriguez et al. 2005; Montañés-Rodriguez et al. 2006; Hamdani et al. 2006). The question of possible artifacts is, of course, of prime importance (see Sect. 5).

---

[1] An Earth-like planet is $\approx 10^{10}$ fainter at visible wavelengths than its parent star.

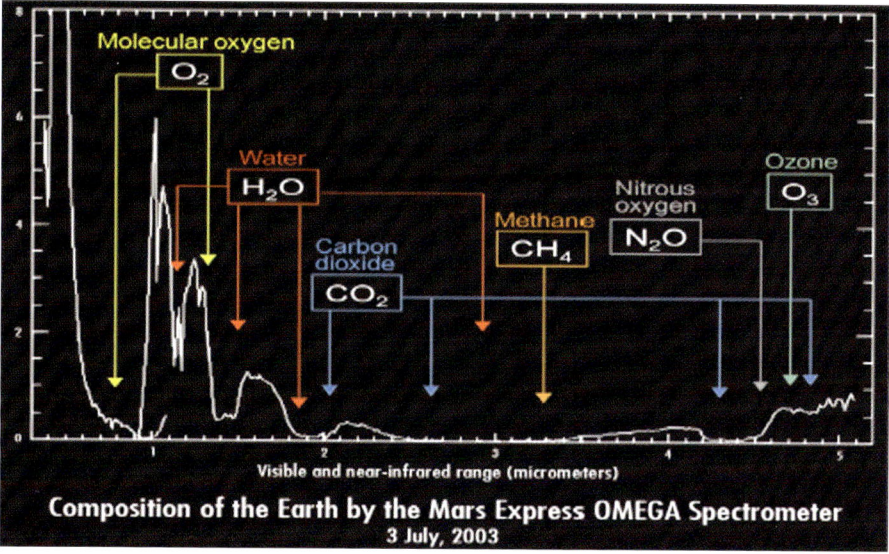

**Fig. 2** Mars Express recorded the Earth spectrum with its OMEGA instrument in July 2003 while it was travelling to Mars. This picture illustrates the possible appearance of an Earth-like extrasolar planet spectrum recorded with a high signal to noise ratio (figure adapted from http://mars.jpl.nasa.gov/express/newsroom/pressreleases/20030717a.html)

**Fig. 3** The Moon, on a morning in September 1999, displaying its crescent, i.e. the sunlit part of the Moon (here overexposed), and also a bright Earthshine over the rest of the Moon disk, i.e. the nonsunlit Moon illuminated by a gibbous Earth. It seems that Leonardo Da Vinci was the first who clearly understood the origin of the phenomenon of Earthshine when studying the geometrical relationship between the Earth, Moon and Sun (Welther 1999). The spectroscopy of the Earthshine directly gives the disk-averaged spectrum of the Earth as seen from the Moon (photo Luc Arnold)

Vegetation indeed has a high reflectivity in the near-IR, higher than in the visible spectrum by a factor of ≈5 (Clark 1999; Fig. 4). This produces a sharp edge around ≈700 nm, the so-called Vegetation Red Edge (VRE). An Earth disk-averaged reflectance spectrum is thus expected to rise by a significant fraction around this wavelength if vegetation is in view from the Moon when the Earthshine is observed.

The following two sections of this paper present the basics about Earthshine spectroscopy and the results of VRE measurements collected between 2001 and 2005. Sections 4 and 5 discuss these results, the perspective and implication for life detection.

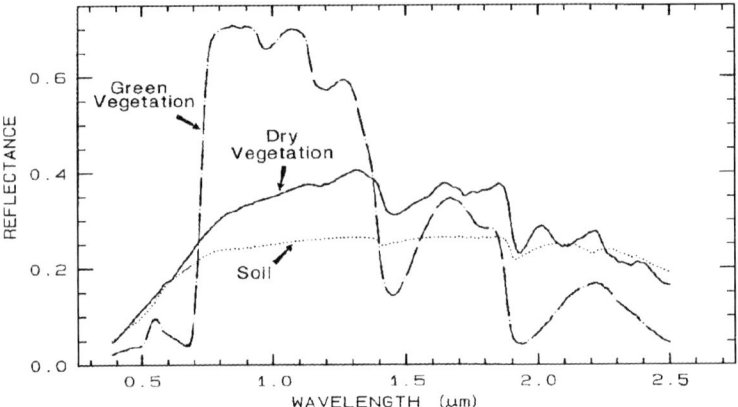

**Fig. 4** Reflectance spectra of photosynthetic (green) vegetation, nonphotosynthetic (dry) and soil (Clark 1999). The so-called vegetation red edge (VRE) is the green vegetation reflectance strong increase from ≈5% at 670 nm to ≈70% at 800 nm

## 2 Basics on Earthshine Spectroscopy and Vegetation Red Edge Signal

It seems that Leonardo Da Vinci was the first who clearly understood the origin of the phenomenon of Earthshine when studying the geometrical relationship between the Earth, Moon and Sun (Welther 1999). The potential of Moon's Earthshine for providing global data on the Earth was identified during the nineteenth century (Flammarion 1877), maybe even earlier. In 1912, Arcichovsky suggested looking for chlorophyll absorption in the Earthshine spectrum, to calibrate chlorophyll in the spectrum of other planets (Arcichovsky 1912). This approach reemerged again within the context of the preparation of the Darwin and TPF missions, when J. Schneider from Paris-Meudon Observatory suggested new observations done at ESO in 1999 and at OHP in 2001 (Arnold et al. 2002). Simultaneous observations were done by Woolf et al. (2002).

*How is Earth spectrum extracted from the Earthshine?* Let us call the Sun spectrum as seen from outside the Earth atmosphere $S(\lambda)$, Earth atmosphere transmittance $AT(\lambda)$, Moonlight $MS(\lambda)$, Earthshine $ES(\lambda)$, Moon reflectance $MR(\lambda)$, and Earth reflectance $ER(\lambda)$. We have

$$MS(\lambda) = S(\lambda) \times MR(\lambda) \times AT(\lambda) \times g_1, \tag{1}$$

$$ES(\lambda) = S(\lambda) \times ER(\lambda) \times MR(\lambda) \times AT(\lambda) \times g_2. \tag{2}$$

The Earth reflectance is simply given by the ratio (2)/(1), i.e.

$$ER(\lambda) = \frac{ES(\lambda) \times g_1}{MS(\lambda) \times g_2}. \tag{3}$$

Simplifying the ratio by $AT(\lambda)$ means that $ES(\lambda)$ and $MS(\lambda)$ should be ideally recorded simultaneously to avoid significant airmass variation and thus an incorrect Rayleigh scattering measurement. The mean of the two $MS$ spectra bracketing $ES(\lambda)$ is thus used to compute $ER(\lambda)$. The $g_i$ terms are geometric factors related to the geometry of the Sun, Earth and Moon triplet. For simplicity, the terms $g_1$ and $g_2$ are set to 1, equivalent to a spectrum normalization. The previous equations assume the sky background has been properly subtracted from the spectra. The data reduction, either for broadband photometry or spectroscopy, was described by Arnold et al. (2002), Qiu et al. (2003), Turnbull et al. (2002), Hamdani et al. (2006). The VRE is extracted from $ER(\lambda)$ and defined by the ratio

$$VRE = \frac{r_I - r_R}{r_R} \tag{4}$$

where $r_I$ and $r_R$ are the near-infrared (NIR) and red reflectance integrated over given spectral domains ($\approx 10$ nm width).

## 3 Review of Results

Table 1 presents the VRE values collected from the literature. Observations roughly confirm what Schneider (2000a, 2000b) and Des Marais et al. (2002) inferred from their previous conjecture, i.e. that vegetation signature is detectable in an integrated (or disk-averaged) Earth spectrum. Observations showed that this signature is weak and variable as suspected, depending, for example, on the ratio between ocean and land in view from the Moon at the time of the observation, or the cloud cover above vegetated area.

Seager et al. (2005) presented two Earthshine spectra recorded on an evening and morning Moon, respectively; thus the two observations had significantly different views of the Earth from the Moon. Although their results remain unfortunately only qualitative, the morning spectrum (South America in view) seems to show a weak signal around 700 nm, while

**Table 1** VRE values from spectroscopy or models. Variations are due to measurements but also to Earth phase (more land or more ocean in view from the Moon at the time of observation; see the text for details)

| VRE (%) | Author | Method |
|---|---|---|
| 5 | Schneider 2000a, 2000b | model |
| $\geq 2$ | Des Marais et al. 2002 | model |
| 4 to 10 | Arnold et al. 2002 | observations |
| 7 to 12 | Arnold et al. 2002 | model |
| 6 to 11 | Arnold et al. 2003 | POLDER data |
| 6 (or 3?) (a) | Woolf et al. 2002 | observations |
| 0 to ? (b) | Seager et al. 2005 | observations |
| 0 | Montañés-Rodriguez et al. 2005 | observations |
| 2 to 3 | Montañés-Rodriguez et al. 2006 | observations |
| 1 to 4 | Hamdani et al. 2006 | observations |
| (c) | Tinetti et al. 2006 | model |
| (d) | Paillet 2006 | model |

(a) Mostly Pacific ocean was in view when this observation was made, therefore the VRE = 6% was maybe overestimated. One of the author (Jucks 2002) said it was probably closer to 3%

(b) Not quantified (weak) uncertain signal when South America in view, but clearly no VRE when mostly ocean in view

(c) NDVI estimator is used rather than VRE. Conclusion is that vegetation remains detectable on a 24 h averaged Earth observed at dichotomy with a realistic cloud cover. High signal-to-noise S/N spectra are considered

(d) Vegetated areas should be 10% from visible cloud-free surface to be detectable in spectra with signal to noise ratio $S/N \geq 10$

the evening spectrum (Pacific ocean in view) remains flat. Hamdani et al. (2006), with spectra recorded from Chile (ESO NTT), also observed a lower VRE of 1% when the view was mainly of the Pacific ocean, and a higher VRE of 3 to 4% when a significant land area is in view. It is worthwhile to note that Europe-based Earthshine observers always have significant land in view from the Moon. Woolf et al. (2002) announced a VRE of 6%, which may be overestimated, considering that a large part of ocean was in view when the observation was done. One of the authors (Jucks 2002) said the VRE may be closer to 3%, although this was apparently never confirmed in a subsequent paper.

Recent results (Hamdani et al. 2006; Montañés-Rodriguez et al. 2006) suggest that the results from our very simple model, described by Arnold et al. (2002), are overestimated too, as well as the VRE measured from POLDER data (Arnold et al. 2003), which clearly are biased by desert, as explained in Sect. 4. It must also be noted that the highest of our measured VRE value, 10%, has an estimated error of $\pm5\%$ (Arnold et al. 2002), suggesting that our most significant values are rather VRE = 4% and 7%, for which the measurements have a better accuracy.

Montañés-Rodriguez et al. (2005) observed the Earthshine on November 19, 2003, and concluded from a first analysis that no sign of vegetation is visible in their spectra. But they later reanalysed their spectra (Montañés-Rodriguez et al. 2006) with just-released global cloud-cover data and concluded that the spectral variations around 700 nm are correlated with cloud-free vegetated area. They obtain VRE values ranging from 2% to 3% while South and North America are in view.

It also worthwhile to mention the other features of the Earth spectrum revealed by the Earthshine observations. In addition to the VRE, the red side (600 : 1,000 nm) of the Earth reflectance spectrum shows the presence of $O_2$ and $H_2O$ absorption bands, while the blue side (320 : 600 nm) clearly shows the Huggins and Chappuis ozone ($O_3$) absorption bands (Hamdani et al. 2006). The higher reflectance in the blue shows that our planet is blue due to Rayleigh scattering in the atmosphere, as nicely demonstrated by Tikhoff (1914) and Very (1915), and confirmed later with accurate Earthshine broadband photometry by Danjon (1936).

Clearly ozone absorption from the wide Chappuis band and Rayleigh scattering strongly impact the global shape of the spectrum and needs to be corrected to access the ground "colour". Once this is done, the resulting spectrum still contain $O_2$ and $H_2O$ bands, but these bands affect only the spectrum locally without compromising the extraction of the VRE (Fig. 5).

## 4  What Did We Learn from ES Observations?

The observations and simulations (Table 1) were all very instructive. If the detection of vegetation in the Earth spectrum is not a surprise, the VRE remains a small spectral feature, in the 0–10% range above the red continuum (depending on Earth phase, clouds, seasons, position of observer, ocean or land in view, etc). Vegetation has a sharp edge at 700 nm but it is easily hidden by clouds (60% typical cloud cover).

Observers know that Earthshine data reduction remains difficult (although possible with great effort!). Data on Earthshine can have low S/N ratio because they are recorded with the Moon often low above the horizon (high air-mass, low Earthshine fluxes with respect to blue sky background), and on the other hand, the detector can easily be saturated when it records the spectrum of the sunlit Moon crescent. Correcting the pollution of Earthshine by light scattered by the bright Moon crescent (Rayleigh scat.) is one of the key points in data

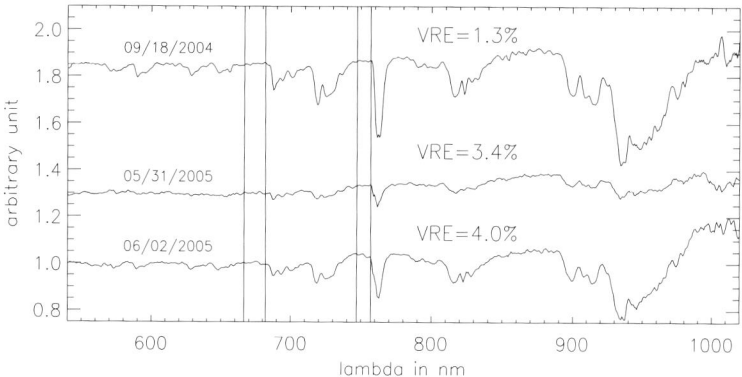

**Fig. 5** Resulting $ER(\lambda)$ corrected for $O_3$ Chappuis absorption, Rayleigh and aerosol scattering, thus ready for the VRE measurement. *Vertical lines* define the two spectral bands used to calculate the VRE. The plots have been shifted vertically for clarity (Hamdani et al. 2006)

reduction. Moon colour also varies with phase, moreover Earthshine and crescent are not observed at the same phase angles! These points were discussed by Hamdani et al. (2006).

*Is broadband photometry, rather than spectroscopy, sufficient to detect vegetation on an unresolved planet?* To quantify the vegetation signature in the spectrum, at least two estimators can be used. The NDVI (Normalized Difference Vegetation Index) (Rouse et al. 1974; Tucker 1979), routinely used for Earth satellite observation, considers the difference, after atmospheric correction, between the reflected fluxes $f$ in broad red and infrared bands, normalized to the sum of the fluxes in these bands,

$$NDVI = \frac{f_I - f_R}{f_R + f_I}. \tag{5}$$

It can also be written in terms of reflectance, as for the VRE above,

$$NDVI = \frac{r_I - r_R}{r_R + r_I}. \tag{6}$$

But it seems that the NDVI and VRE estimators based on two bands are not sufficient to detect vegetation on an unresolved planet. A large Sahara-like desert can indeed produce a signal similar to a smaller but greener patch on the Earth (see Fig. 4). Details were given by Arnold et al. (2003) and Fig. 6 shows the VRE variations over 24 h based on POLDER data (Deschamps et al. 1994) for the Earth and an Earth where lands were all attributed to deserts, with clouds and oceans unchanged: The VRE for the latter still indicates the presence of vegetation! Thus only two photometric bands may not avoid a false positive detection on an unresolved exo-Earth, and it is necessary to have a full spectrum to identify the vegetation red edge around 700 nm, with a spectral resolution $\geq \approx 50$. A spectrum will allow us to distinguish the VRE from a smoother positive slope due to a large desert. The GOME experiment (Burrows et al. 1999) provides spectra well suited to such simulations. It is important to point out that Earth images reconstructed from satellite data are approximation only, although the approximation is probably acceptable for our purpose: Since the data are often recorded for a given solar angle by nadir instruments, it is not possible to take into account effects like shadowing between plants. Nevertheless, at least in principle, a more advanced model can take into account the presence of a hot spot (strong backscattering), or

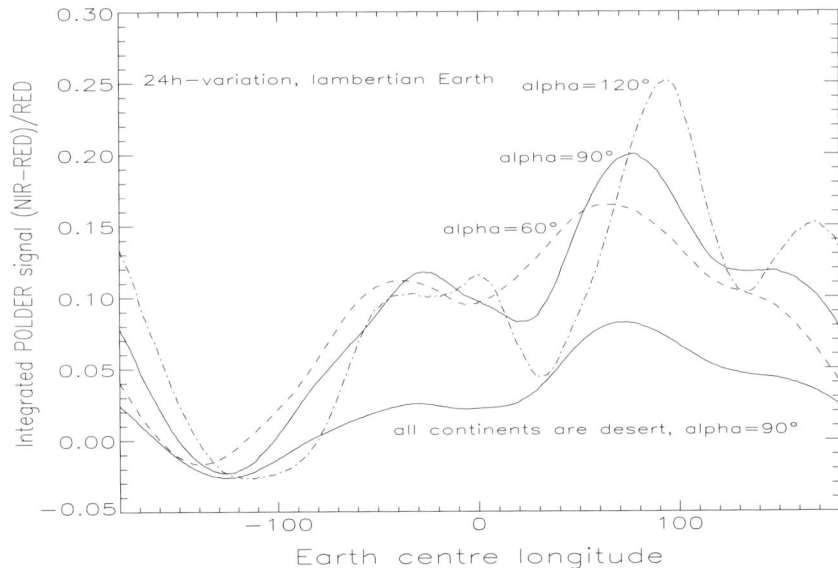

**Fig. 6** VRE variation over a full Earth rotation. The observer is above latitude 0° and measures the VRE while the Earth rotates, for three different Earth phases (60, 90 and 120°; respectively, gibbous, dichotomy and crescent Earth). The *lowest curve* simulates a desert Earth seen at phase 90°: all land pixels have the properties of the Sahara, oceans and clouds are unchanged. The figure shows that the VRE estimator based on two bands of photometry by POLDER is biased: for the Earth seen at a phase angle of 90°, it shows vegetation if it is $\geq\sim 9\%$, while negative values indicate the presence of ocean

more generally, the profile of the Bidirectional Reflectance Distribution Function (BRDF) for each biome and cloud. Otherwise the Earth is considered as a simple Lambertian diffuser.

## 5 Implications for Life Detection on Extrasolar Planets: Perspective and Open Questions

Table 1 shows that the VRE is a small feature (a few % above the continuum) and therefore will require high S/N ratio to be detected. To be detected by a space-based observatory, the exposure times should of the order of 100 hours to reach S/N = 100 with a spectral resolution of 25 for an Earth at 10 pc (Arnold et al. 2002).

*Any chance we might observe a VRE on an Earth-like planet higher than on Earth?* It is quite interesting to note that leaf reflectance of plants increases with leaves thickness (Slaton et al. 2001). Desert plants with fleshy green stems, often without leaf (in the sense of the common conception of a leaf), generally reflect substantially more radiation than do other plants (Gates et al. 1965), up to a factor of 2 at 750 nm. The cloud cover over deserts being smaller than the mean cover, desert plants should in principle contribute significantly to the VRE. A planet without much water (a small ocean of albedo $\approx 0.1$) and few clouds (albedo $\approx 0.6$) would have an albedo dominated by the desert ($\approx 0.3$). The planet albedo would thus be roughly the same as Earth's albedo, i.e. $\approx 0.3$. Therefore, and paradoxically, such a dry planet, where the majority of plants would have evolved toward a wide variety of cold and warm desert plants might display a stronger VRE than the Earth. Quantitatively speaking

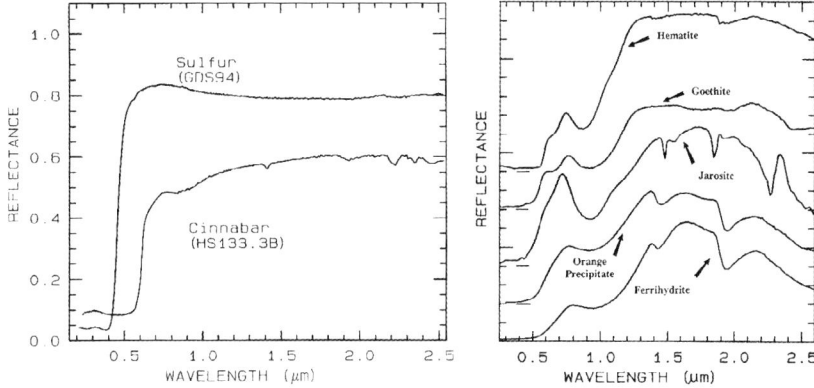

**Fig. 7** Minerals can display "edges" in their reflectance spectrum. The possible confusion of the VRE with mineral spectral features was discussed by Schneider (2004) and Seager et al. (2005), but no exhaustive work on this subject has been done to date. It would indeed be very interesting to know if the reflectance spectrum of the vegetation can be fitted by a relevant combination of spectra of minerals. Spectra are from (Clark 1999)

and based on GOME spectra of Earth biomes, the VRE could reach $\approx$35% for a 50% desert, plant-covered super-continent;[2] it would be $\approx$15% for a tundra-like super-continent. These comfortable numbers—from a detectability point of view—probably are very optimistic and should be regarded as upper limits. Next paragraph may temperate reader's enthusiasm (including mine).

*Is the red edge universal, i.e. inherent to any photosynthetic process in the Universe?* Although on Earth most photosynthetic species show a red edge around 700 nm resulting in a signature visible at a global scale, there are exceptions, like *Rhodopseudomonas* purple bacteria (Blankenship et al. 1995). Thus, strictly speaking, the red edge is not inherent to all photosynthetic species and thus probably not universal. From a life-detection strategy point of view, this observation suggests that, rather than looking for a Earth-like red edge, we should look for a particular ground colour that could not be attributed to a mineral or a combination of minerals. If all mineral artifacts are eliminated, then only a photosynthetic process could be considered to interpret the spectrum (Fig. 7).

Considering that $O_2$ and $O_3$ are produced by photosynthesis on Earth, it seems thus relevant, if $O_2$, $O_3$ and $H_2O$ are detected in the spectrum of an Earth-like planet, to look for the signature of an extrasolar photosynthesis, i.e. a spectral feature—probably weak but, we hope, sharp enough to be detectable and distinguishable from any known mineral— revealing missing photons used in a photosynthetic process. It seems relevant too to look for these missing photons at the wavelength where photons from the mother star are the most abundant, and also where the planetary atmosphere is the most transparent, so these photons can reach the ground. On the Earth, the atmosphere is indeed transparent to visible light and plant pigments involved in photosynthesis strongly absorb in that spectral window. In order to access the ground spectral signature, the atmosphere must be partially clear to allow us to *see* the ground. It will be necessary to remove, at least partially, the atmospheric spectral bands to see that ground, meaning that we will need, at some stage, a model of the planet's atmosphere.

---

[2] I mean no ocean in view.

**Fig. 8** Simulated image of the Earth at 10 light years observed with a 150-km hypertelescope interferometric array made of 150 3-m mirrors working at visible wavelengths (Labeyrie 1999). North and South America are visible. Note that this simulation was done at visible wavelengths, while in the (very) near-IR at 750 nm, vegetated areas would be much brighter and more easily detectable on continents. Spatial resolution at 750 nm would remain the same at visible wavelength with the same hypertelescope flotilla being spread over 225 km instead of 150 km

## 6 Conclusion

Earthshine observations have shown that Earth's vegetation is detectable in the Earth-integrated spectrum. The vegetation signal is only a few percentage points (0 to 5% range) above the continuum. One reason the signal is weak is simply because vegetated areas are often covered by clouds. We speculated about the possible high VRE of a dry planet, i.e. with low cloud cover, but pointed out also that the VRE at 700 nm may not be a universal signature of plants on an extrasolar Earth-like planet.

Clearly *resolved* images of extrasolar planets will help to detect photosynthetic life on these Earth-like planets! But the wonderful instruments that will allow us to see Earth-like planets as small resolved disks are not yet ready to be launched (far from that), although possible designs have been outlined. For example, a 150-km hypertelescope in space—an interferometric sparse array of small telescopes—would provide 40 resolution elements (resels) across an Earth at 10 light years in yellow light (Labeyrie 1999). And a formation of 150 3-m mirrors would collect enough photons in 30 min to freeze the rotation of the planet and produce an image with at least ≈300 resels, and up to thousands depending on array geometry (Fig. 8). At this level of spatial resolution, it will be possible to identify clouds, oceans and continents, either barren or perhaps (we hope) conquered by vegetation.

**Acknowledgements**    The author acknowledges J. Schneider, Paris Observatory, S. Jacquemoud, Paris-7 University, for the stimulating discussions we had during the writing of this paper, and the anonymous reviewer for the remarks that helped to improve the paper.

## References

V.M. Arcichovsky, Auf der Suche nach Chlorophyll auf den Planeten. Annales de l'Institut Polytechnique Don Cesarevitch Alexis a Novotcherkassk **17**(1), 195 (1912)

L. Arnold, S. Gillet, O. Lardière, P. Riaud, J. Schneider, Astron. Astrophys. **352**, 231–237 (2002)

L. Arnold, F.M. Bréon, S. Brewer, J. Guiot, S. Jacquemoud, J. Schneider, In *SF2A-2003: Scientific Highlights 2003*, ed. by F. Combes, D. Barret, T. Contini, L. Pagani (Bordeaux, France, 16–20 June 2003), EDP-Sciences, Paris, Conference Series, pp. 133–136

R.E. Blankenship, M.T. Madigan, C.E. Bauer, *Anoxygenic Photosynthetic Bacteria* (Kluwer, Dordrecht, 1995)

J.P. Burrows, M. Weber, M. Buchwitz, V. Rozanov, A. Ladstter-Weienmayer, A. Richter, R. Debeek, R. Hoogen, K. Bramstedt, K.U. Eichmann, M. Eisinger, D. Perner, J. Atmospheric Sci. **56**(2), 151–175 (1999)

R.N. Clark, Chapter 1: Spectroscopy of rocks and minerals, and principles of spectroscopy, in *Manual of Remote Sensing*, vol. 3, Remote Sensing for the Earth Sciences, ed. by A.N. Rencz (Wiley, New York, 1999), pp. 3–58

A. Danjon, Ann. Obs. Strasbourg **3**, 139–180 (1936)

P.Y. Deschamps, F.M. Bréon, M. Leroy, A. Podaire, A. Bricaud, J.C. Buriez, G. Sèze, IEEE Trans. Geosci. Remote Sens. **32**(3), 598–615 (1994)

D.J. Des Marais, M.O. Harwit, K.W. Jucks, J.F. Kasting, D.N.C. Lin, J.I. Lunine, J. Schneider, S. Seager, W.A. Traub, N.J. Woolf, Astrobiology **2**, 151–180 (2002)

C. Flammarion, *Les terres du ciel* (Librairie Académique Didier & Cie, Paris, 1877), p. 323

D.M. Gates, H.J. Keegan, J.C. Schleter, V.R. Weidner, Appl. Opt. **4**, 12–20 (1965)

S. Hamdani, L. Arnold, C. Foellmi, J. Berthier, M. Billeres, D. Briot, P. François, P. Riaud, J. Schneider, Astron. Astrophys. **460**, 617–624 (2006)

K.W. Jucks, Private communication, 24th April at XXVII European Geophysical Society General Assembly, Nice, France, 21–26 April 2002

A. Labeyrie, Snapshots of alien worlds–The future of interferometry. Science **285**(5435), 1864–1865 (1999)

P. Montañés-Rodriguez, E. Pallé, P.R. Goode, J. Hickey, S.E. Koonin, Astrophys. J. **629**, 1175–1182 (2005)

P. Montañés-Rodriguez, E. Pallé, P.R. Goode, F.J. Martín-Torres, Astrophys. J. **651**, 544–552 (2006)

J. Paillet, Caractérisation spectrale d'exoplanètes telluriques. Ph.D. thesis, Université Paris XI, 2nd oct. 2006 (N° d'ordre 8416)

J. Qiu, P.R. Goode, E. Pallé, V. Yurchyshyn, J. Hicke, P. Montañés Rodriguez, M.-C. Chu, E. Kolbe, C.T. Brown, S.E. Koonin, J. Geophys. Res. (Atmospheres) **108**(D22), 12 (2003)

J.W. Rouse, R.H. Haas, J.A. Schell, D.W. Deering, Final Report, Type III, NASA/GSFC, Greenbelt, 1974

J. Schneider, Exoplanets in *A Encyclopaedia of Astronomy and Astrophysics* (Institute of Physics Publishing, 2000a)

J. Schneider, Private communication, 2000b

J. Schneider, Review of visible versus IR characterization of planets and biosignatures in *Towards other Earths: Darwin/TPF and the Search for Extrasolar Terrestrial Planets*. ESA SP-539 (2004), p. 205

S. Seager, E.L. Turner, J. Schafer, E.B. Ford, Astrobiology **5**, 372–390 (2005)

M.R. Slaton, E.R. Hunt, W.K. Smith, Am. J. Bot. **88**, 278–284 (2001)

G. Tinetti, V. Meadows, D. Crisp, N.Y. Kiang, B.H. Kahn, E. Fishbein, T. Velusamy, M. Turnbull, Astrobiology **6**, 881–900 (2006)

G.A. Tikhoff, Mitteilungen der Nikolai-Hauptsternwarte zu Pulkowo, no 62, Band $V I_2$, 1914, p. 15

C.J. Tucker, Remote Sens. Environ. **8**, 127 (1979)

M.C. Turnbull, W.A. Traub, K.W. Jucks, N.J. Woolf, M.R. Meyer, N. Gorlova, M.F. Skrutskie, J.C. Wilson, Astrophys. J. **644**, 551–559 (2002)

F.W. Very, Astron. Nachrichten **201**(4819-20), 353–400 (1915)

B. Welther, Sky Telesc. **98**(4), 40–44 (1999)

N.J. Woolf, P.S. Smith, W.A. Traub, K.W. Jucks, Astrophys. J. **574**, 430–433 (2002)

# Searching for Signs of Life in the Reflected Light from Exoplanets: A Catalog of Nearby Target Stars

## M.C. Turnbull

Originally published in the journal Space Science Reviews, Volume 135, Nos 1–4.
DOI: 10.1007/s11214-008-9329-0 © Springer Science+Business Media B.V. 2008

**Abstract** Which stars are the best stars to search for habitable planets and signs of life? This is a trick question, because it depends not only on the kind of circumstellar environment we think is likely to be supportive to life as we know it, but it depends also on the technique being used to do the search. For example, the Catalog of Nearby Habitable Stellar Systems was designed for SETI, a search for technological signals. Because this search strategy relies on life forms out-shining their star (at least at certain frequencies), target selection is not complicated by the need to spatially resolve the habitable planets on which these life forms presumably live. On the other hand, because the life forms being sought are technologically advanced, it seems reasonable to assume that their planet had to be continuously habitable for long enough to evolve such biological complexity. Thus the deciding factor for SETI is that of long term habitability. Meanwhile, other missions to directly detect habitable planets (e.g., NASA's TPF and ESA's Darwin) are less worried about long term habitability but must struggle with the competing factors of planet separation from the star and planet brightness relative to the star. This paper outlines a variety of challenges in the search for simple and complex life in the Solar Neighborhood.

**Keywords** Astrobiology · SETI · TPF · Habitable stellar systems

## 1 Introduction to Habstars

The majority of papers in these proceedings discuss in situ methods for detecting life forms and are thus relevant to explorations of the Sun's eight major and many minor planets. But what of the rest of the universe? Currently that is a realm accessible to us only by detection of whatever scant photons these remote life forms send our way. Intelligent life forms may choose to broadcast their presence via narrow band signals, and simple life forms may be remotely detectable by the impact they make on planetary atmospheres. Assuming there will

M.C. Turnbull (✉)
Carnegie Institution of Washington, 5241 BroadBranch Rd NW, Washington, DC, USA
e-mail: turnbull@dtm.ciw.edu

be some way to recognize biological activity, the search for life beyond the Solar System begins with one question: Where should we look?

Based on our miniscule experience with life (i.e., Earth), it seems that life is a phenomenon that thrives on planets where heat from a central star allows for the existence of liquid water on their surfaces in an otherwise frozen universe. One could imagine other "hot spots", like planet interiors, where temperatures could support liquid water, but surface life is more favorable for detecting life forms remotely. Planets that orbit at the right distance from their star to have liquid water on the surface are said to be in the "habitable zone" of their star, and these watery worlds are known as "habitable planets".

Planets, especially Earth-sized planets orbiting within the circumstellar habitable zone, represent at once the most highly sought and difficult to find objects in the nearby universe. Earths are far enough from their star that transits are unlikely, close enough to their star that direct imaging is beyond current technology, and small enough in mass to be lost in the noise of Doppler surveys. So while Doppler surveys make ever swifter progress on finding giant exoplanets and super-Earths (see Marcy et al. 2005; Rivera et al. 2005; Udry et al. 2007), we remain uninformed with regard to terrestrial planets. This situation will hopefully change dramatically in the coming decades, but until then we define a "habitable stellar system" only in terms of the star.

Which, then, are the best stellar systems to search for signs of life? This is a trick question, because it depends not only on the kind of circumstellar environment we think is likely to be supportive to life as we know it, but also on the technique being used to do the search. For example, the Catalog of Nearby Habitable Stellar Systems ("HabCat"; Turnbull and Tarter 2003) was designed for SETI, a search for technological signals. Because this search strategy relies on life forms out-shining their star (at least at certain frequencies), target selection is not complicated by the necessity to spatially resolve the habitable planets on which these life forms presumably live. When making a target list for a mission like NASA's Terrestrial Planet Finder, which is designed to directly detect Earth-like planets and search for spectroscopic evidence of life, one is perhaps less concerned about the habitability of the system over billions of years and more concerned about our ability to directly detect any terrestrial planets in that system.

## 2 Description of a Habstar

Because the life forms being sought by SETI are technologically advanced, it seems reasonable to assume that their planet had to be continuously habitable for long enough to evolve such biological complexity as radio telescopes (in the case of Earth this took 4.5 Gyr). Thus a key factor in target selection for SETI is that of long term habitability, and to create HabCat we defined a habitable stellar system, or "habstar", as a star in which a terrestrial planet could have formed and supported liquid water on its surface throughout the time required to invent radio telescopes (which we took to be $\sim 3$ billion years). As we will describe in this article, implicit in the definition of a habstar are concerns about the star's metallicity, age, mass, variability, trajectory through the Galaxy, and the locations of any stellar or giant planet companions.

To make HabCat we translated the habstar definition into a set of quantitative astrophysical criteria, summarized as follows:

(1) We began with the Hipparcos Catalogue (which, crucially, contains parallax measurements) as our parent list. Parallax data allows a determination of the distance and therefore the intrinsic brightness (luminosity) of each star. Luminosity, in combination with

the star's color (a proxy for temperature), tells us how old the star is, how stable it is, and how long it will survive.

(2) Stars were then eliminated based on age (keeping stars older than 3 Gyr), on evolutionary stage (only stars in the stable hydrogen-burning "main sequence" phase of life were kept), and on main sequence lifetime (must be longer than 3 Gyr).

(3) Stars which vary in their luminosity by more than 3% (the Hipparcos detection limit) were eliminated.

(4) Stars with low metallicity (less than 40% solar) and "thick disk" or "halo" kinematics (see below) were eliminated.

(5) Stars where the orbits of known stellar and planetary companions gravitationally disturb the habitable zone were also ruled out.

Some of these parameters are statistically correlated or anti-correlated with one another (e.g., metallicity, kinematics, and age, as described below). Therefore the stars thus selected, while not all G-type stars, are a uniquely sun-like subset of the Galaxy's population. This HabCat contains 17,129 stars, all of which we have ranked according to their similarity to the Sun in luminosity and color (which translates to a specific mass and age range). Thus our "favorite" stars have main sequence lifetimes of $\sim$ 10 billion years (this shortens for more massive stars), with habitable zones that are $\sim$ 1 AU wide (this shrinks for less massive stars).

## 3 Uncertainties Abound

The translation of habstar requirements into the above specific limits on observable parameters is an uncertain business. As an illustration, consider the metallicity requirement: stars and their planets form out of the same initial material, therefore stars with a higher metal content (where "metal" refers to all elements more massive than helium) would seem more likely to sport terrestrial planets (which are made almost entirely of metals). To get a rough estimate of the minimum metallicity for forming one Earth-mass planet, we can simply say that the Solar System's total terrestrial planet mass ($\sim$ 2 Earth masses) implies a minimum metallicity of about half the solar value, or 1% of the total stellar material ($[\mathrm{Fe/H}] = \log[(\mathrm{Fe/H})_{\mathrm{star}}/(\mathrm{Fe/H})_{\mathrm{sun}}] = -0.3$). Indeed, below this value only a few percent of stars are seen to have giant planet companions, compared to $\sim$ 50% at solar metallicity (Reid 2002). But the likelihood of *terrestrial* planet formation could also:

(1) vary wildly according to the environment in which the system is forming (e.g., radiation from massive stars or dynamical interactions with a companion star may help or hinder formation), or even

(2) decline *above* a certain metallicity (e.g., if giant planets begin to dominate the habitable zone), thereby warranting a *maximum* metallicity cut-off for target selection.

In the making of HabCat, we did not worry about these dubieties and simply included all stars down to about 40% solar ($[\mathrm{Fe/H}] = -0.4$), for the reason that this value corresponds to an easily observed division in stellar kinematics: high metallicity stars tend to have low velocities relative to the Sun (which stays very close to the central plane of the Galaxy), while the population of stars with metallicities below $[\mathrm{Fe/H}] = -0.4$ is dominated by "thick disk" and "halo" stars, components of the Galaxy where stars' orbits are non-circular and make large vertical excursions through the Galactic plane—a level of environmental variability that may affect planetary habitability (Medvedev and Melott 2007).

Indeed, many of the parameters we use to identify habstars are intertwined at some level. Low metallicity usually corresponds to large Galactic velocities, both of which we'd like to rule out. On the other hand, lower metallicity also corresponds to higher age, which is desirable for the reason that life presumably takes some time to get going. The Sun is a member of a specific subset of stars that is high metallicity, old, and kinematically placid.

Likewise, higher stellar mass might be considered desirable because the width of the habitable zone increases strongly with mass. But mass and age are related in that high mass stars do not live as long, so all massive stars are young. Indeed, it is really only the very smallest stars that we can possibly care about in astrobiology: while stars come in masses of 100 that of the Sun, stars with only 50% more mass than the Sun have lifetimes of less than 3 billion years. Fortunately, less massive stars are much more common than higher mass stars, so we still have a few hundred billion to work with in our Galaxy. By ruling out all the young and massive stars, and all the metal poor and kinematically "disturbed" stars, we should be left with a sample of middle-aged metal rich long-lived placidly orbiting suns. "Habstars" are an habitable island in the multidimensional phase space of astrophysics.

## 4 Detecting Habitable Worlds

Another way to search for life among nearby stars is to try to directly image habitable planets and search for atmospheric byproducts of life or signatures of surface organisms. For example, Earth has been broadcasting for at least the last billion years the simultaneous presence of atmospheric oxygen and methane—a combination that persists only because of the biological activity on this planet. The reflection spectrum of our planet's surface has also been altered by photosynthetic organisms, which absorb strongly at optical wavelengths but display a sharp increase in albedo at 7800 angstroms (Clark 1999; Turnbull et al. 2006).

NASA's Terrestrial Planet Finder missions, TPF-C (Levine et al. 2006) and TPF-I (Lawson et al. 2007), and ESA's Darwin mission (Léger et al. 2007), were envisioned as instruments that would detect these signals by directly imaging habitable planets and enabling spectroscopic study of these unresolved "pale blue dots". Because a star's habitable zone is located so near to the star itself, the technology for any mission to image habitable worlds is driven toward extremely high contrast imaging at very small star-planet angular separations. The Earth–Sun separation, seen from 10 parsecs away, is only 0.1 arcseconds. To make matters worse, at optical wavelengths, the Earth shines in reflected sunlight at only one ten billionth the brightness of the Sun. Thus all proposed architectures for habitable-planet imaging missions have therefore involved starlight suppression, in order to reveal the faint light of any orbiting planets (Lawson et al. 2004; Lawson and Dooley 2004). The planet : star contrast ratio is more favorable ($\sim 10^{-6}$) at mid-infrared wavelengths where the planet shines with its own heat, but at these longer wavelengths it is necessary to use larger diameter telescopes or telescope arrays to achieve the necessary angular resolution. This paper focuses on optical wavelength planet finding missions, for which single-dish coronagraphic designs (e.g., TPF-C) are appropriate.

## 5 Target Selection for TPF-C

"Simple" life forms, which may not make an effort to communicate between the stars but can nevertheless have transformative effects on planetary conditions, probably take less time to evolve and inhabit a wider range of environments than their technological counterparts.

Therefore for imaging missions like TPF-C, the habitability requirements can be loosened, at least in terms of timescale (and therefore mass and age). The more important factor is that imaging habitable planets places much stronger demands upon the technology at *our* end of things. Target selection for any mission attempting to image habitable planets is centered on the question of *planet detectability*, which varies strongly from star to star, and with the observation technique.

Aside from the properties of the planet itself (which are not known in advance of the mission), there are two factors that strongly affect planet detectability and which depend only on the target star. One is intrinsic luminosity, and the other is the apparent brightness of the star as seen from Earth (which is proportional to the star's instrinsic luminosity divided by the square of its distance). These two related quantities work against one another in terms of planet detectability. Because a planet in the habitable zone will be the same intrinsic brightness regardless of which star it orbits, a higher contrast ratio must be achieved to detect planets orbiting more luminous stars. For example, an Earth-like planet orbiting an M star (the faintest of main sequence stars) will be a factor of $10^{-8}$ as bright as the star at the outer edge of the habitable zone. This sounds difficult, but consider that for a G star like our sun, planets are only $10^{-10}$ times as bright as the star! Thus the planet:star contrast ratio that must be achieved is inversely proportional to stellar luminosity, and less luminous stars are favored.

On the other hand, the distance of the habitable zone from the star is proportional to the square-root of stellar luminosity. The directly observed quantity, *angular* habitable zone size, goes as this same factor divided by distance. Therefore angular habitable zone size goes as the square root of the apparent brightness as seen from Earth. Because the range of intrinsic stellar luminosities is so great, and because even the nearest stars are still quite far away, this turns out to favor *more* luminous stars[1].

For a planet to be bright enough to detect, and for the habitable zone to be large enough to resolve, we need a happy medium in stellar luminosity. Our detector will be "tuned" to a certain range of luminosities, in that the limiting contrast ratio sets the maximum stellar luminosity for targets, and the minimum resolution sets the minimum apparent brightness for our targets, and it is entirely possible to design a very expensive and technologically advanced instrument for which no stars are suitable targets.

The "inner working angle", or innermost angle at which a $10^{-10}$ contrast ratio can be achieved, is predicted to be $\sim 4\lambda/D$ for favored coronagraph designs, or about 57 mas for an 8-m mirror (at its longest axis) operating at optical wavelengths (Levine et al. 2006). Current laboratory experiments are able to achieve this for laser light (Trauger and Traub 2007), an accomplishment that must be extended to broad spectral windows in order to be useful for planet detection. Figures 1 and 2 show the habitable zone locations for Hipparcos stars within 30 parsecs and explore the number of Earths that could be detectable with this level of capability. We also consider a more optimistic $3\lambda/D$ inner working angle of 42 mas. These specifications determine which stars are viable TPF-C targets, independent of other considerations regarding the habitability of those systems.

Figure 1 indicates, for the 3- and $4\lambda/D$ capabilities, which stars could have detectable Earths at the inner edge of the habitable zone. This paper assumes a circumstellar habitable zone location of 0.7 to 1.5 AU for a star of one solar luminosity, scaled by the square-root of luminosity for other star systems. For Earth-sized planets this is an entirely reasonable range,

---

[1]This selection effect is a consequence of the coronographic technique chosen for TPF-C. Nulling interferometer combining the infrared light from free-flying mirrors (like TPF-I and Darwin), can adapt their baseline to the angular size of the habitable zone around the targeted star.

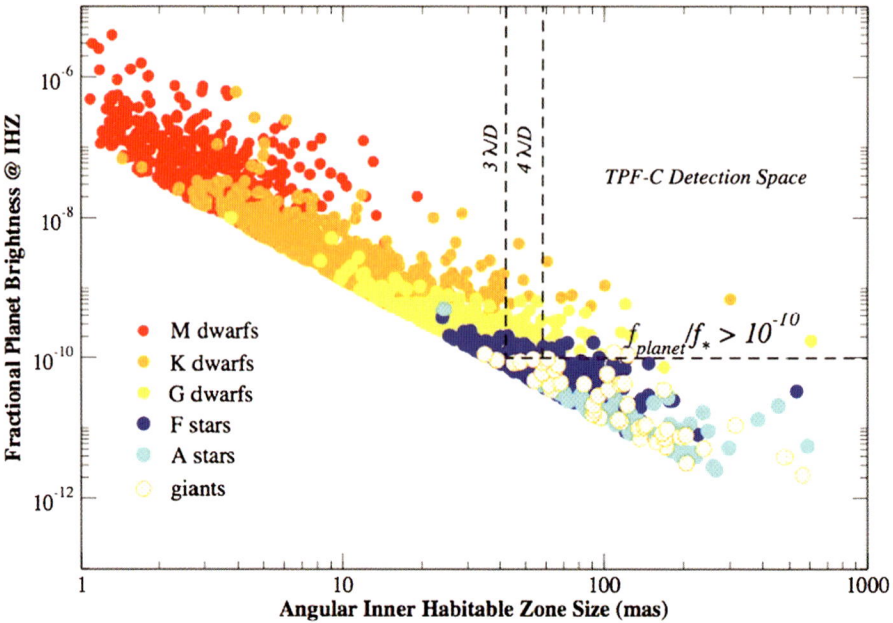

**Fig. 1** Habitable planets imaged in reflected light at the inner edge of the habitable zone

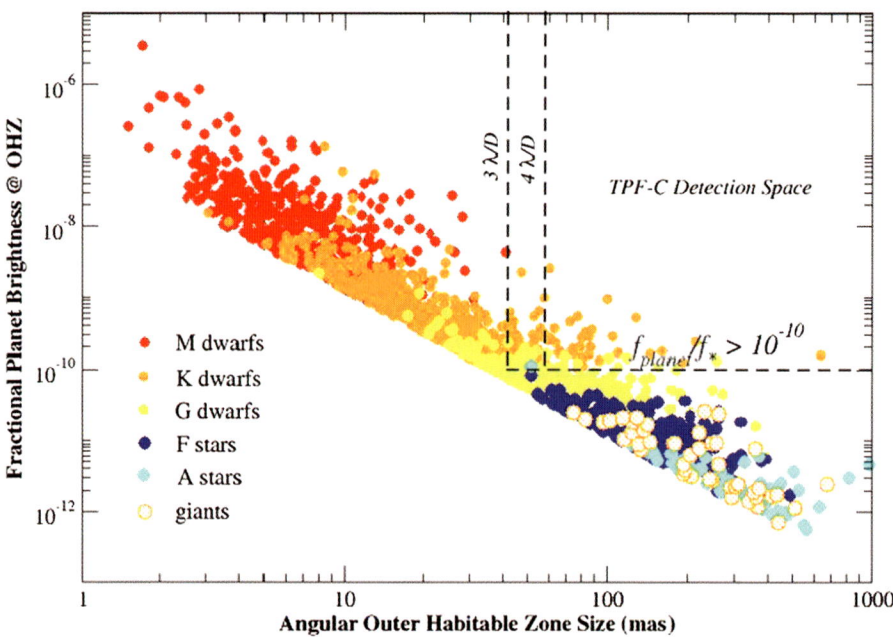

**Fig. 2** Habitable planets imaged in reflected light at the outer edge of the habitable zone

although the exact location of the habitable zone may vary somewhat depending on (a priori unknown) planetary characteristics like albedo, radius, and composition. See Kasting et al. (1996) for a more detailed discussion of habitable zone locations.

The immediate lesson from Fig. 1 should be that the direct detection of habitable planets is indeed a difficult business. Even the magnificent achievement of $10^{-10}$ contrast at 57 milliarcseconds results in only a handful of stars whose habitable zones are imagable. While habitable planets orbiting faint M dwarfs (red dots) are fractionally brighter, they are also closer to the star, making them impossible to detect with TPF-C. And while habitable planets orbiting brighter F and A stars are further out, they are also fractionally fainter, making them impossible to detect with TPF-C. So a $4\lambda/D$, $10^{-10}$ coronagraph is indeed tuned to sunlike stars, at least for the inner edge of the habitable zone.

However, at the outer edge of the habitable zone, planets receive less light and are therefore fainter, as seen in Fig. 2. Most G-type stars are too bright for the detection of Earths at the outer edge of the habitable zone. If we desire to image the *entire* habitable zone of our target stars with TPF-C, then indeed we are left with only ∼10 K stars, and those all within 10 parsecs. Furthermore, working hard to decrease the inner working from $4\lambda/D$ to $3\lambda/D$ angle helps us very little, unless we also deepen the contrast ratio to ∼$10^{-11}$. Recent laboratory results, while acknowledging that much work remains to be done, suggest that this contrast ratio should become possible in the near future (Trauger and Traub 2007).

Ideally, it would be possible to identify habstars for which habitable planets would be detectable and use those as TPF-C targets. If we wish to look for Earth-like planets orbiting stars which seem compatible with the existence of complex life, we find that habstars are not so cooperative. Of the ∼10 K stars mentioned above whose habitable zones are entirely searchable, only one is a habstar (alpha Centauri B), and that one is questionable as a habstar given the proximity of sun-like alpha Cen A. Alpha Cen A will not only shine unwanted light into our detector, but the fact that this double passes within only 11 AU every ∼80 years leads to questions about whether any planets in the habitable zone could have water (Quintana et al. 2002).

The other 9 targets above were ruled out of HabCat. This partly because K stars tend to display more flaring activity than do solar mass stars, a quality viewed negatively by the habstar criteria (but which might not be so bad for microbes), and partly because too much is known about the nearest stars and their little quirks (like faint companions and low-level variability). The more details we find out about the stars in the Solar Neighborhood, the less like the Sun they seem! For the stars that do emerge as being very biologically attractive (delta Pav, 18 Sco), deeper contrast ratios will have to be achieved than what have been proposed for the TPF coronagraph. Given the challenges of ultra-high contrast coronagraphy and space-based nulling interferometry, direct imaging of habitable worlds will have to wait a few more years. Nevertheless, technology is steadily improving, and we are now within sight of that goal. Until then, SETI will continue to search the skies for life.

## 6 Spectral Characterization of Exo-Earths

Despite the technological challenge, the detection of planets in the habitable zones of nearby stars will eventually occur. At that point, what signals might we be able to detect? Planets will not be spatially resolved, and spectroscopy will have very few photons to work with. Figure 3 shows the spectrum we measured for Earth (Turnbull et al. 2006) by looking at the "dark" portion of the crescent moon (see also Arnold 2007). During crescent moon, the Earth shines in reflected sunlight onto the lunar surface, and

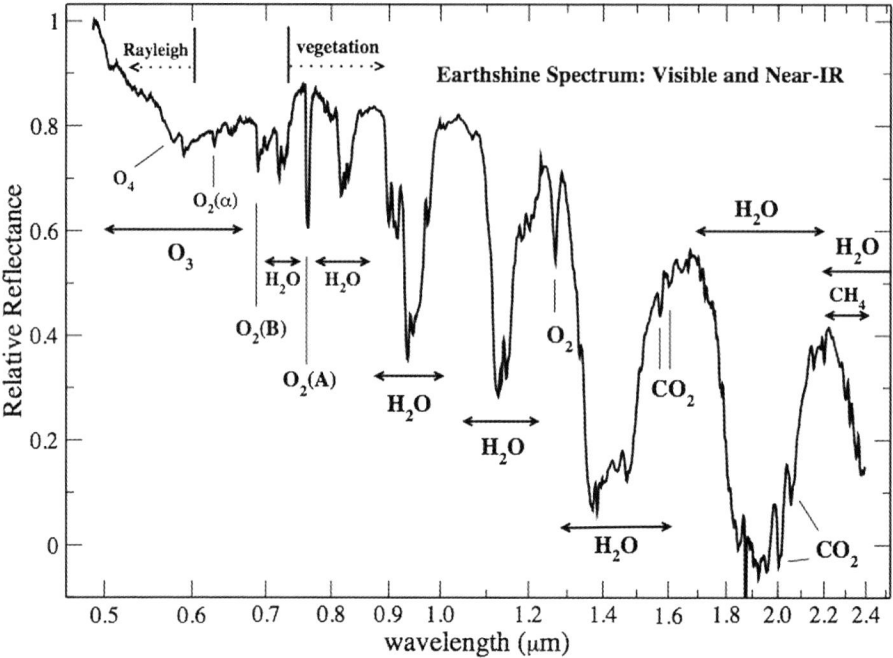

**Fig. 3** The optical and near-IR reflection spectrum of planet Earth as seen in earthshine (from Turnbull et al. 2006)

this provides us with a way to observe the hemispherically integrated spectrum of our world, just as we would observe for an extrasolar planet (see also Woolf et al. 2002; Arnold et al. 2003, and references therein). In the figure, we see many signatures of habitability and life: Rayleigh scattering from the blue sky indicates a significant atmosphere, strong water bands indicate the potential habitability, ozone indicates that the surface is somewhat protected from UV light, and the combination of oxygen and methane (marginally detected here) suggest that the atmosphere is in strong disequilibrium. Indeed no planetary process can sustain an oxygen signal that is as strong as what we have detected here. This signal is entirely due to the presence of plants, which are themselves detected in a rise in reflectivity near 0.75 microns.

While the spectral resolution of TPF-C will necessarily be coarse, and changes due to clouds and planet rotation may wash out surface signatures, this plot suggests that the detection of water, oxygen, ozone and Rayleigh scattering are within reach. Such an accomplishment would usher in a new era in life and planetary science.

## References

L. Arnold, Space Sci. Rev. (2007, this issue). doi:10.1007/s11214-007-9281-4
L. Arnold, S. Gillet, O. Lardiere, P. Riaud, J. Schneider, Int. J. Earth Sci. **8**, 17 (2003)
R.N. Clark, in *Manual of Remote Sensing*, ed. by A. Rencz (Wiley, New York, 1999)
J.F. Kasting, D.C.B. Whittet, W.R. Sheldon, Orig. Life Evol. Biosphere **27**, 413 (1996)
P.R. Lawson, J.A. Dooley, JPL Publication 05-5 (2004)
P.R. Lawson, S.C. Unwin, C. Beichman, JPL Publication 04-014 (2004)
P.R. Lawson, O.P. Lay, K.J. Johnston, C.A. Beichman, TPF-I SWG Report. JPL publication, 2007

A. Léger, T. Herbst, Darwin mission proposal to ESA (2007). arXiv:0707.3385, http://lanl.arxiv.org/abs/0707.
    3385

M. Levine, S. Shaklan, J. Kasting (eds.), JPL Document D-34923, 2006

G. Marcy, R.P. Butler, D. Fischer, S. Vogt, J.T. Wright, C.G. Tinney, H.R. Jones, Prog. Theor. Phys. Suppl.
    **158**, 24 (2005)

M. Medvedev, V. Melott, Astrophys. J. **664**, 879 (2007)

E.V. Quintana, J.J. Lissauer, J.E. Chambers, M.J. Duncan, Astrophys. J. **576**, 982 (2002)

I.N. Reid, in *Bioastronomy 2002: Life Among the Stars*, ed. by R. Norris. ASP Conf. Ser., vol. 213 (ASP, San
    Francisco, 2002)

Rivera et al., Astrophys. J. **634**, 625 (2005)

J.T. Trauger, W.A. Traub, Nature **446**, 771 (2007)

M.C. Turnbull, J.C. Tarter, Astrophys. J. Suppl. **145**, 181 (2003)

M.C. Turnbull, W.A. Traub, K. Jucks, N. Woolf, M. Meyer, N. Gorlova, J. Wilson, M. Skrutskie, Astrophys.
    J. **644**, 551 (2006)

S. Udry, X. Bonfils, X. Defosse, T. Forveille, M. Mayor, C. Perrier, F. Bouchy, C. Lovis, F. Pepe, D. Queloz,
    J.-L. Bertaux, Astron. Astrophys. **469**, L43–L47 (2007)

S. Volonte et al., Darwin: the infrared space interferometer. Tech. Report (ESA), 2000

N.J. Woolf, P.S. Smith, W.A. Traub, K.W. Jucks, Astrophys. J. **574**, 430 (2002)

# Exoplanet Transit Spectroscopy and Photometry

S. Seager

Originally published in the journal Space Science Reviews, Volume 135, Nos 1–4.
DOI: 10.1007/s11214-008-9308-5 © Springer Science+Business Media B.V. 2008

**Abstract** Photometry and spectroscopy of extrasolar planets provides information about their atmospheres and surfaces. From extrasolar planet spectra and photometry we can infer the composition and temperature of the atmospheres as well as the presence of molecular species, including biosignature gases or surface features. So far photometry has been published for three different transiting hot Jupiters (gas giant planets in short-period orbits), opening the era of comparative exoplanetology.

**Keywords** Exoplanets · Atmospheres · Planets

## 1 Introduction

Over 200 extrasolar planets (exoplanets) are now known to orbit main sequence stars. These planets have a nearly continuous range in mass, semi-major axis and eccentricity (see Lovis et al. 2008). With the existence of a wide variety of exoplanets now firmly established it is time to move forward to physically characterizing them. By physically characterize we mean studying the exoplanets' density, temperature, and atmosphere.

There are two paths to studying the physical characteristics of exoplanets: 1) the conventional path of directly imaging an exoplanet and 2) the unexpected approach of observing transiting planets in the combined light of the planet–star system. It is the second path, exoplanet transit photometry and spectroscopy, that is currently yielding all of the data advancing our understanding of exoplanetary properties and atmospheres. Direct imaging for solar-system-aged planets is very difficult because of the brightness contrast between the planet and star which ranges from one million to one billion. In contrast transit studies range from brightness contrasts of one part per hundred to one part per 10,000 and are measurable with existing ground- and space-based telescopes.

S. Seager (✉)
Dept. of Earth, Atmospheric, and Planetary Sciences, Massachusetts Institute of Technology,
77 Massachusetts Ave. 54-1626, Cambridge, MA 01742, USA
e-mail: seager@mit.edu

The probability of a planet to transit its star as seen from Earth is the ratio $R_*/a$ where $R_*$ is the stellar radius and $a$ is the planet's semi-major axis (Sackett 1999). Because the probability to transit grows the closer a planet is to its star, it is "close-in" planets that are most likely to transit and therefore be observed. Indeed each of the fourteen currently known transiting planets are within 0.06 AU of their parent star (where 1 AU is equal to the Earth–Sun distance of $1.5 \times 10^{11}$ m). A planet close to its star also has a very short period to its orbit (i.e., its year is very short) because the circumference of its orbit is small. The period range of the known transiting planets is 1.2 to 4.5 days.[1]

The existence of transiting planets allows many useful measurements. The reason is that the planet can be observed in the combined light of the planet-star system, so that although the planet and star are not separated in space as viewed from a telescope, the brightness of the planet and star can be separated from each other.

The class of exoplanets of most relevance to astrobiology is the habitable planets. Astronomers call a planet habitable if it could have liquid water at its surface—because all life as we know it requires liquid water. No such habitable exoplanets are yet known. The exoplanets accessible for detection and study have up until now been the easiest to detect: either massive planets like Jupiter with no solid surface for life as we know it; or smaller planets so close to the star that they are too hot for liquid surface water. The detailed study of the known exoplanet interiors and atmospheres is a veritable stepping stone to studying the rocky planets more like Earth once they are detected.

In this paper we describe the methods to characterize transiting planets and some recent results.

## 2 Densities of Transiting Exoplanets

The most straightforward physical parameter that can be derived from photometry of a transiting planet is radius. A normal star is at roughly constant brightness, until an exoplanet transits the star. During the exoplanet transit, the starlight drops in the ratio of the planet-to-star area; if the radius of the star is known the exoplanet radius can be determined. The exoplanet radius leads to the exoplanet's density: the radius from transit photometry and the mass from radial velocity measurements. From density the planet's bulk composition can be inferred. At least for solar system planets the dominant materials that different planets are composed of have very different densities (gas giants Jupiter and Saturn are composed mostly of H and He; terrestrial planets Venus and Earth mostly of silicates with some iron; ice giants Uranus and Neptune of water ice, some rock, and a small amount of H and He).

Fourteen short-period transiting giant exoplanets have measured densities (see Fig. 1). Most of these exoplanets have densities consistent with H/He gas giants heated externally by their adjacent parent star. Four transiting exoplanets, however, have unusual densities. Three exoplanets (Bakos et al. 2007; Collier Cameron et al. 2006; Knutson et al. 2006) have very low densities for their mass and age and must have an extra source of internal energy that prevents the exoplanet from cooling and contracting. One exoplanet (HD149026b) has a very high density for its mass, and must be composed of two thirds of heavy elements, comparable to all of the heavy elements in the solar system combined (Sato et al. 2005).

The exoplanet radius uncertainty is quite large for most exoplanets, on the order of ten percent or higher. Some planets have high radius uncertainties due to the difficulty of high-precision photometry on faint stars. Other planets have high radius uncertainties due to the limitation of the radius of the parent star, usually only known to seven to ten percent.

---

[1] http://vo.obspm.fr/exoplanetes/encyclo/catalog-RV.php.

**Fig. 1** Densities of fourteen transiting exoplanets. The exoplanet names are noted. Jupiter and Saturn are shown with a J and S respectively. The *dashed lines* are lines of constant density, with the density labeled in units of g/cm³. For data see papers referenced in http://vo.obspm.fr/exoplanetes/encyclo/catalog-RV.php

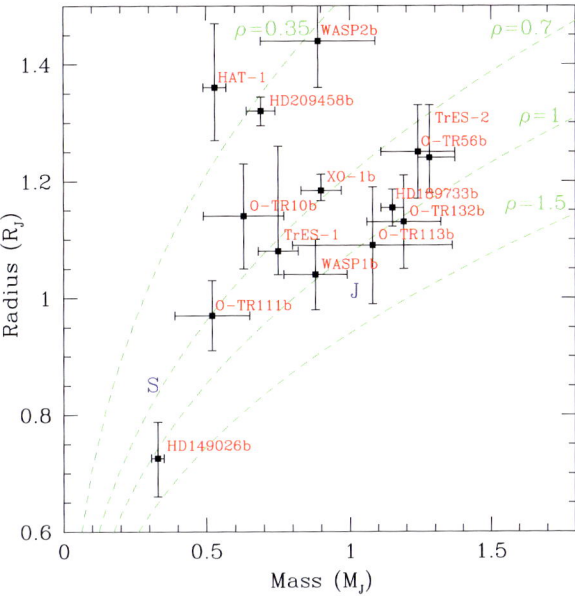

The transit radius signal is approximately the ratio of the planet-to-star area

$$\text{Transit Signal} = \left[ \frac{R_p}{R_*} \right]^2. \qquad (1)$$

For a Jupiter-sized planet transiting a sun-sized star this number is about one percent. The transit signal is as high as three percent in the most favorable case to date of HD189733, a Jupiter-sized planet orbiting a K star.

## 3 Secondary Eclipse Photometry and Spectroscopy

A planet that transits its star also goes behind the star. The secondary eclipse can be exploited to measure reflection or emission from the planet. During secondary eclipse the brightness of the planet is hidden from view and the stellar brightness alone can be measured. Just before and just after secondary eclipse the joint brightness from the planet and star can be measured. These two observations can be subtracted to yield the planet's brightness.

At thermal infrared wavelengths the brightness temperature of an exoplanet can be measured. At the time of this writing three exoplanets have brightness temperatures reported in the literature, both observed with the Spitzer Space Telescope. These are HD209458b: $1130 \pm 150$ K at 24 microns (Deming et al. 2005b), TrES-1b $1160 \pm 50$ K at 8 microns (Charbonneau et al. 2005), and HD189733b $1117 \pm 42$ K at 16 microns (Deming et al. 2006). These temperatures in principle can be used together with atmosphere models to infer a global atmospheric temperature, and atmospheric composition. In practice, more than two observations per planet are needed for any definite conclusions.

The first secondary eclipse spectra of two different exoplanets in the wavelength range $\sim$8 to $\sim$14 microns have been reported (Grillmair et al. 2007; Richardson et al. 2007; Swain et al. 2007).

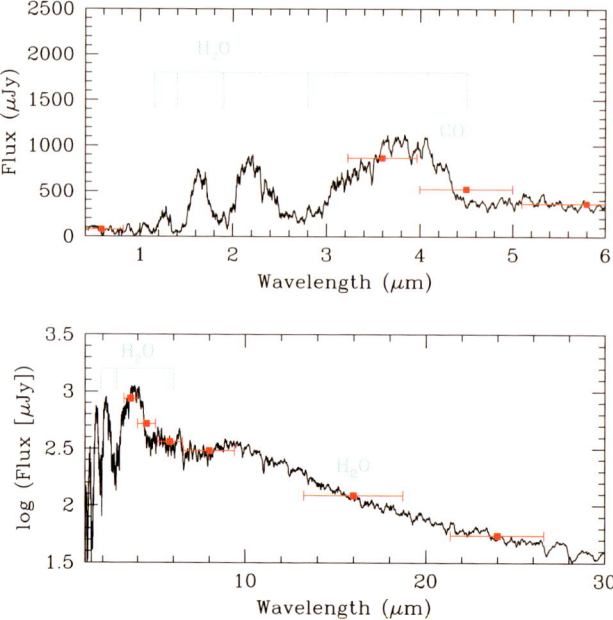

**Fig. 2** Theoretical spectrum of the transiting hot Jupiter HD209458b. This model is for a solar abundance, cloud-free model. Note the different x and y axes on the *top* and *bottom* panels. The data points are also theoretical, and indicate the photometry wavelength range for MOST (the shortest wavelength point) and Spitzer (IRAC, IRS, and MIPS)

The secondary eclipse thermal infrared signal can be estimated as:

$$\text{Thermal Emission Signal} = \frac{T_p}{T_*}\left[\frac{R_p}{R_*}\right]^2. \tag{2}$$

For the hot Jupiters listed above this signal is on the order of a few times $10^{-3}$. (See Fig. 2.) This is a very small signal for thermal infrared measurements, and has so far only been possible from space away from the noisy effects of Earth's atmosphere. We note that the above estimate is technically valid in the Rayleigh-Jeans tail of the blackbody spectrum.

Observations that result in non-detections are still useful to constrain spectral features. HD209458b has one particularly useful upper limit. Richardson et al. (2003) used the ground-based IRTF telescope to constrain the water vapor abundance (Fig. 3). Almost all atmosphere model predictions (e.g., Barman et al. 2001; Sudarsky et al. 2003; Burrows et al. 2005; Fortney et al. 2005) showed very strong water absorption features that are ruled out by these observations. See Seager et al. (2005) for further discussion. Note that ground-based observations are often limited by variations in the terrestrial atmosphere, which make detection of spectral features difficult.

At visible wavelengths the secondary eclipse can also potentially be detected to measure the planet albedo: the amount of incident starlight reflected by the planet. The albedo of the planet reveals whether or not clouds are present and, measured at different wavelengths, the presence of molecules in the atmosphere. While no exoplanet albedos have yet been measured, an upper limit for the albedo for HD209458b has been published: 0.25 (a 3-$\sigma$

**Fig. 3** Theoretical spectrum for HD209458b with atmosphere measurements or upper limits. Shown in the *top panel* is the MOST upper limit (Rowe et al. 2006), and the IRTF constraint on water vapor absorption (Richardson et al. 2003). The *solid lines* are for 1-$\sigma$ upper limits or detections, and the *dashed lines* are for 3-$\sigma$ upper limits or detections

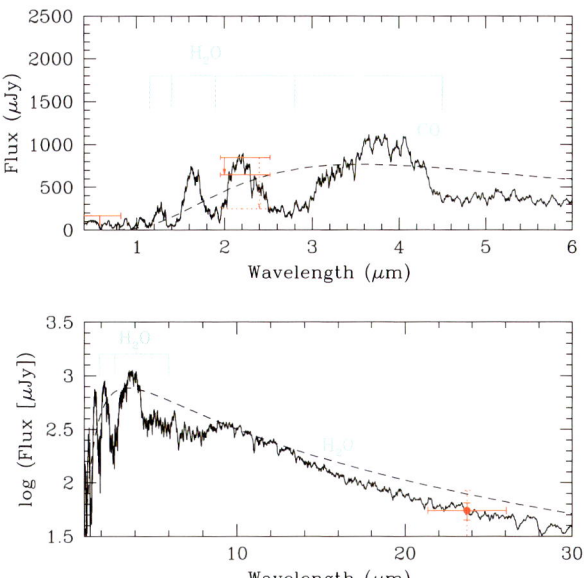

upper limit) (Rowe et al. 2006). Jupiter's albedo (for the same wavelength range as the Rowe et al. (2006) measurement) would be 0.5, so HD209458b is much darker than Jupiter.

The secondary eclipse visible wavelength flux ratio can be estimated as:

$$\text{Reflection Signal} = \frac{F_p}{F_*} = p\left(\frac{R_p}{a}\right)^2. \tag{3}$$

Here $p$ is the planetary geometric albedo and $F_p/F_*$ is the planet-star flux ratio. For hot Jupiters this number is expected to range from a few $\times 10^{-5}$ down to as low as a few $\times 10^{-6}$. If the reflection signal is measured the known exoplanet radius and semi-major axis can be used to infer the albedo.

## 4 Transit Transmission Spectroscopy

When the planet transits its parent star, some of the starlight passes through the planet atmosphere on its route to Earth. This is analogous to a flashlight shining through a fog. Spectral lines from the transiting exoplanet's atmosphere should therefore be superimposed on the stellar spectrum. A spectrum of the star alone can be taken at a time when the planet is not transiting. During the planet transit, the combined light of the star and planet are measured. These two observations can be subtracted to find the planet's spectral lines.

The transit transmission spectra can be estimated as the ratio of the area of the planet annulus to the area of the star.

$$\text{Transmission Spectra Signal} = \frac{\text{Annulus}}{R_*^2}. \tag{4}$$

The planet's spectral lines are miniscule compared to the star's, about 10,000 times smaller. This is because the planet's spectral signal is approximately the ratio of the planet at-

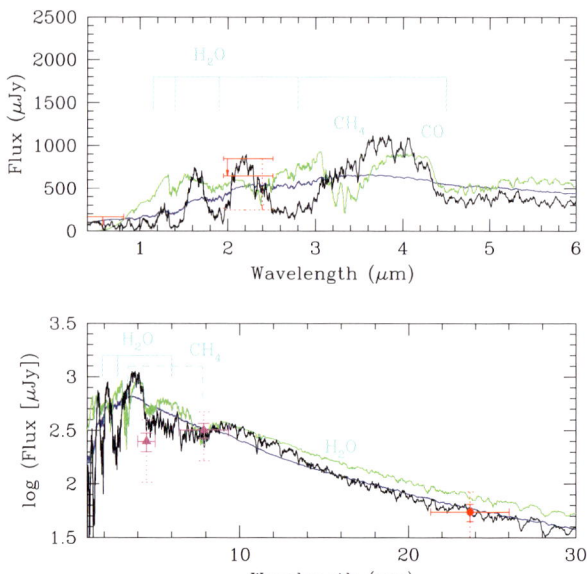

**Fig. 4** Three different theoretical spectra of the hot Jupiter HD209458b. *Black curve*: cloud-free solar abundance model as shown in Figs. 3 and 4. *Blue curve*: cloudy model with solar abundances, where the clouds mute the other absorption features. *Green curve*: a model with the carbon-to-oxygen ratio of C/O = 1.01 (the solar abundance C/O ratio is = 0.5). In this high carbon abundance model, the oxygen is locked up as CO with little O remaining for $H_2O$. Excess C is found in $CH_4$ (methane), even at high temperatures. The *red points* are data points or upper limits as described in Fig. 3. The *magenta points* are Spitzer data points for TrES-1 (Charbonneau et al. 2005), scaled to the flux of HD209458b

mosphere annulus compared to the area of the stellar disk. This is in contrast to secondary eclipse, where the entire disk of the planet can contribute to the spectrum.

To date, one planet, again HD209458b, has been thoroughly observed for transit transmission spectra both from the ground and from space with the Hubble Space Telescope. One spectral feature has been detected in the lower atmosphere: atomic sodium (Charbonneau et al. 2002). Sodium has an overall low abundance in a planetary atmosphere compared to hydrogen, but sodium has a very strong oscillator strength. The sodium detected was approximately a factor of three smaller than expected from models of the atmosphere, suggesting the presence of a high cloud that masks the true sodium abundance. A CO non-detection (i.e., upper limit) with the Keck telescope (Deming et al. 2005a) further reinforces the notion of high clouds in the planet's atmosphere. This is because if the planet had no clouds CO should have been easily detected based on model atmosphere predictions.

Turning to the upper atmosphere of HD209458b, a transit detection at H Lyman-$\alpha$ wavelengths was huge—a 15% drop in stellar intensity during transit, 10 times greater than the transit depth at visible wavelengths (Vidal-Madjar et al. 2003). This implies an extended atmosphere to 3 or 4 Jupiter radii, and suggests that the planet is losing some mass over its lifetime. Detection of hot atomic hydrogen has also been reported (Ballester et al. 2007).

## 5 Atmosphere Phase Curves

Exoplanets change in brightness during their orbital period, and this change in brightness is potentially observable in the combined light of the planet-star system. Phase curves are so interesting because they may be observed for non-transiting exoplanets, thus extending the parameter space of planet atmospheres studies.

At visible wavelengths, the exoplanet will go through illumination phases as seen from Earth due to reflected light. This is similar to the moon's phase changes as seen from Earth. Several ground-based (Leigh et al. 2003) and space-based (Rowe et al. 2006) studies have been unable to detect a visible-wavelength exoplanet phase curve.

At infrared wavelengths, a planet with a strong temperature gradient will have a change in its thermal emission phase curve. This is the case, at 24 microns, for at least one non-transiting hot Jupiter exoplanet, Upsilon Andromedae (Harrington et al. 2006). Other hot Jupiters appear to vary little in temperature (Knutson et al. 2007; Cowan et al. 2006). The hot Jupiters are believed to be tidally-locked, that is presenting the same face to their star at all times. Whether or not they have a hot and cold side or whether atmospheric circulation is effective at redistributing winds is a subject under vigorous study (e.g., Cooper and Showman 2005; Cho et al. 2003).

## 6 Model Atmospheres

Model atmosphere computations are required for an interpretation of exoplanet spectra and photometry data because the planetary fluxes at different wavelengths are shaped by a variety of physical processes.

Model atmosphere codes for exoplanet atmospheres are typically 1D, plane-parallel, local thermodynamic equilibrium models that solve three equations. These equations are: (1) the equation of radiative transfer; (2) the equation for radiative and convective equilibrium; and (3) the equation for hydrostatic equilibrium. The boundary conditions are the stellar radiation at the top of the atmosphere, and the interior entropy at the bottom of the atmosphere. These three equations together with their boundary conditions are solved to derive three unknowns: temperature (as a function of altitude), pressure (as a function of altitude), and the radiation field (as a function of altitude and wavelength). In other words, the model outputs are temperature, pressure, and radiation field (from which the emergent flux can be computed). From the wavelength-integrated flux at the top of the atmosphere, the effective equilibrium temperature can be calculated. The albedo can also be derived from the model output as the reflected flux compared to the incident stellar flux.

A number of input assumptions, including the boundary conditions, are necessary to solve the three equations described above. These include (Seager et al. 2005; Marley et al. 2007):

– Incident stellar irradiation;
– Interior entropy of the planet;
– Atmospheric elemental abundances (metallicities);
– Atmospheric chemistry;
– Cloud model and condensates; and
– Gaseous opacities.

For 1D models we must make an assumption of how the ***incident stellar radiation*** is redistributed across the surface of the planet. We use the parameter $f$ as the proxy for atmospheric circulation: $f = 1$ if the absorbed stellar radiation is redistributed evenly throughout the planet's atmosphere (e.g., due to strong winds rapidly redistributing the heat) and $f = 2$ if only the heated day side reradiates the energy (for the latter $T_{eq}$ is for the day side only). The implementation in the models is achieved by multiplying the incoming stellar radiation[2] by a factor of $f/4$.

For the lower boundary ***interior entropy*** condition we must choose either a total amount of net flux that will pass from the planet interior through the planet atmosphere. Or instead

---

[2]The planet intercepts stellar radiation in a cross sectional area of $\pi R_p^2$ and reradiates the energy into either $2\pi R_p^2$ ($f = 2$) or $4\pi R_p^2$ ($f = 1$).

we may choose a temperature and pressure at the lower boundary. These choices are complicated by the fact that the interior luminosity of the planet is not known, and is age- and mass-dependent.

The *elemental abundances* in exoplanet model atmosphere are often assumed to be "solar," i.e., having the same relative concentrations as the Sun. Some models include atmospheres with higher metallicities. This is appropriate because the solar system giant planets have metallicities from a factor of a few up to factor of 30 higher than solar (Marley et al. 2007). Additionally different planet-hosting stars have relative abundances different from the Sun, in many cases significantly higher. To derive the partitioning of elemental abundances among different molecules, *chemical equilibrium* calculations are used to determine the abundances of the different species as a function of temperature and pressure. For example, given the elemental abundances of carbon and oxygen relative to hydrogen, we can calculate the relative concentrations of methane ($CH_4$) and carbon monoxide (CO). At the high temperatures of hot Jupiters we find that $H_2$ is the dominant species, followed by He, $H_2O$, and CO. Under equilibrium, solar abundance chemistry the planets are too hot for $CH_4$ (e.g., Seager et al. 2000). Non-equilibrium chemistry, including photochemical models (Liang et al. 2003; Liang et al. 2004), could be very important but has not yet been fully explored.

The *opacities* of the expected chemical species in the model atmosphere play a pivotal role in determining the structure of the resulting spectrum. This is because the opacities impede the flow of radiation, resulting in either absorption or emission lines. The strength of the opacity depends on the structure of the atom and molecule, such that, a species of low abundance can create a much stronger spectral absorption feature than a species of higher abundance. This is partly why sodium has been detected in the atmosphere of HD209458b even though, for solar abundance, it is about one million times less abundant than hydrogen. In particular, water, methane, sodium, and potassium all have large spectral signatures and are expected to be present in the atmospheres of hot Jupiters. Opacities are particularly sensitive to choices of metallicity, which species (atomic and molecular) are included, and whether equilibrium or non-equilibrium chemistry is considered. Absorption due to collisions between molecules (called Collision-Induced Absorption, or CIA) also has a measurable effect, and modelers typically have to account for interactions between $H_2$–$H_2$ and $H_2$–He.

*Cloud structure* can play a critical role in controlling the atmospheric spectra because clouds can potentially have high opacity. In hot Jupiters, the clouds are not made of water, for the temperature is too high for water vapor to condense. High temperature condensates such as silicates, iron, aluminum oxide and others may be present as clouds. Clouds are extremely difficult to model and represent one of the greatest uncertainties in exoplanet atmospheric models. The structure, height, and composition of the clouds depends on the transport (horizontal and vertical) of the condensates present in the atmosphere, as well as the local conditions in the atmospheric layer. The type of condensates, the degree of condensation, and the particle size distribution are all free parameters in defining the cloud structure. All exoplanet atmosphere models currently in the literature further assume that the clouds are uniformly distributed over the entire planet. See Ackerman and Marley (2001) and Cooper et al. (2003) for details of exoplanet cloud models.

See Barman et al. (2001), Sudarsky et al. (2003), Burrows et al. (2005), Fortney et al. (2005) for examples of model parameter choices made by other exoplanet atmosphere modelers.

For completeness we here describe the planet's equilibrium temperature. The equilibrium effective temperature $T_{eq}$ is the effective temperature attained by an isothermal planet after it has reached complete equilibrium with the radiation from the parent star. It is a key diagnostic of the planetary atmosphere, and it is defined as

$$T_{eq} = T_* \sqrt{\frac{R_*}{2a}} [f(1 - A_B)]^{1/4}, \tag{5}$$

where $T_*$ is the stellar temperature, $R_*$ is the stellar radius, $a$ is the orbital semi-major axis, and $A_B$ is the Bond albedo. The parameter $f$ is a measure of the atmospheric circulation, as described above.

## 7 Near Term Prospects

The number of transiting exoplanets is growing. Already fourteen exoplanets have measured densities and radii. Ground-based planet transit surveys are expected to uncover a dozen more short-period transiting giant planets in the next few years. The recently launched Corot will find transiting planets ranging from Jupiter-sized down to potentially Earth-sized. These planets will need space-based followup to determine precise planet radii and ground-based measurements to determine planetary mass. NASA's Kepler space telescope is planned for launch in late 2008 and will has as its main goal finding Earth-sized planets in Earth-like orbits around sun-sized stars.

The Spitzer Space Telescope has two more years of cryocooled operation. Already Spitzer has observed more than half a dozen exoplanets with broad-band photometry (see Fig. 2), two planets for spectra, and half a dozen exoplanets for thermal phase curves. In the next few years we expect an emerging picture for hot Jupiter exoplanets. And, we expect this initial understanding to be a stepping stone to the distant future studies of Earth-like planet atmospheres.

## References

A.S. Ackerman, M.S. Marley, Precipitating condensation clouds in substellar atmospheres. Astrophys. J. **556**, 872–884 (2001)

G.Á. Bakos, R.W. Noyes, G. Kovács, D.W. Latham, D.D. Sasselov, G. Torres, D.A. Fischer, R.P. Stefanik, B. Sato, J.A. Johnson, A. Pál, G.W. Marcy, R.P. Butler, G.A. Esquerdo, K.Z. Stanek, J. Lázár, I. Papp, P. Sári, B. Sipőcz, HAT-P-1b: A large-radius, low-density exoplanet transiting one member of a stellar binary. Astrophys. J. **656**, 552–559 (2007)

G.E. Ballester, D.K. Sing, F. Herbert, The signature of hot hydrogen in the atmosphere of the extrasolar planet HD 209458b. Nature **445**, 511–514 (2007)

T.S. Barman, P.H. Hauschildt, F. Allard, Irradiated planets. Astrophys. J. **556**, 885–895 (2001)

A. Burrows, I. Hubeny, D. Sudarsky, Theoretical interpretation of the measurements of the secondary eclipses of TrES-1 and HD209458b. Astrophys. J. **625**, L135–L138 (2005)

D. Charbonneau, T.M. Brown, R.W. Noyes, R.L. Gilliland, Detection of an extrasolar planet atmosphere. Astrophys. J. **568**, 377–384 (2002)

D. Charbonneau, L.E. Allen, S.T. Megeath, G. Torres, R. Alonso, T.M. Brown, R.L. Gilliland, D.W. Latham, G. Mandushev, F.T. O'Donovan, A. Sozzetti, Detection of thermal emission from an extrasolar planet. Astrophys. J. **626**, 523–529 (2005)

J.Y.-K. Cho, K. Menou, B.M.S. Hansen, S. Seager, The changing face of the extrasolar giant planet HD 209458b. Astrophys. J. Lett. **587**, L117–L120 (2003)

A. Collier Cameron, F. Bouchy, G. Hebrard, P. Maxted, D. Pollacco, F. Pont, I. Skillen, B. Smalley, R.A. Street, R.G. West, D.M. Wilson, S. Aigrain, D.J. Christian, W.I. Clarkson, B. Enoch, A. Evans, A. Fitzsimmons, M. Fleenor, M. Gillon, C.A. Haswell, L. Hebb, C. Hellier, S.T. Hodgkin, K. Horne, J. Irwin, S.R. Kane, F.P. Keenan, B. Loeillet, T.A. Lister, M. Mayor, C. Moutou, A.J. Norton, J. Osborne, N. Parley, D. Queloz, R. Ryans, A.H.M.J. Triaud, S. Udry, P.J. Wheatley, WASP-1b and WASP-2b: Two new transiting exoplanets detected with SuperWASP and SOPHIE. ArXiv Astrophysics e-prints (2006)

C.S. Cooper, A.P. Showman, Dynamic meteorology at the photosphere of HD 209458b. Astrophys. J. **629**, L45–L48 (2005)

C.S. Cooper, D. Sudarsky, J.A. Milsom, J.I. Lunine, A. Burrows, Modeling the formation of clouds in brown dwarf atmospheres. Astrophys. J. **586**, 1320–1337 (2003)

N.B. Cowan, E. Agol, D. Charbonneau, Infrared phase variations of hot Jupiters, in *American Astronomical Society Meeting Abstracts*, pp. 163.02–+

D. Deming, T.M. Brown, D. Charbonneau, J. Harrington, L.J. Richardson, A new search for carbon monoxide absorption in the transmission spectrum of the extrasolar planet HD 209458b. Astrophys. J. **622**, 1149–1159 (2005a)

D. Deming, S. Seager, L.J. Richardson, J. Harrington, Infrared radiation from an extrasolar planet. Nature **434**, 740–743 (2005b)

D. Deming, J. Harrington, S. Seager, L.J. Richardson, Strong infrared emission from the extrasolar planet HD 189733b. Astrophys. J. **644**, 560–564 (2006)

J.J. Fortney, M.S. Marley, K. Lodders, D. Saumon, R. Freedman, Comparative planetary atmospheres: Models of TrES-1 and HD209458b. Astrophys. J. **627**, L69–L72 (2005)

C.J. Grillmair, D. Charbonneau, A. Burrows, L. Armus, J. Stauffer, V. Meadows, J. Van Cleve, D. Levine, A Spitzer spectrum of the exoplanet HD 189733b. Astrophys. J. Lett. **658**, L115–L118 (2007)

J. Harrington, B.M. Hansen, S.H. Luszcz, S. Seager, D. Deming, K. Menou, J.Y.-K. Cho, L.J. Richardson, The phase-dependent infrared brightness of the extrasolar planet Andromedae b. Science **314**, 623–626 (2006)

H. Knutson, D. Charbonneau, R.W. Noyes, T.M. Brown, R.L. Gilliland, Using stellar limb-darkening to refine the properties of HD 209458b. ArXiv Astrophysics e-prints (2006)

H.A. Knutson, D. Charbonneau, L.E. Allen, J.J. Fortney, E. Agol, N.B. Cowan, A.P. Showman, C.S. Cooper, S.T. Megeath, A map of the day-night contrast of the extrasolar planet HD 189733b. Nature **447**, 183–186 (2007)

C. Leigh, A.C. Cameron, T. Guillot, Prospects for spectroscopic reflected-light planet searches. Mon. Not. R. Astron. Soc. **346**, 890–896 (2003)

M. Liang, C.D. Parkinson, A.Y.-T. Lee, Y.L. Yung, S. Seager, Source of atomic hydrogen in the atmosphere of HD 209458b. Astrophys. J. Lett. **596**, L247–L250 (2003)

M.-C. Liang, S. Seager, C.D. Parkinson, A.Y.-T. Lee, Y.L. Yung, On the insignificance of photochemical hydrocarbon aerosols in the atmospheres of close-in extrasolar giant planets. Astrophys. J. Lett. **605**, L61–L64 (2004)

Lovis et al., Space Sci. Rev. (2008, this issue)

M.S. Marley, J. Fortney, S. Seager, T. Barman, Atmospheres of extrasolar giant planets, in *Protostars and Planets V*, ed. by B. Reipurth, D. Jewitt, K. Keil, 2007, pp. 733–747

L.J. Richardson, D. Deming, S. Seager, Infrared observations during the secondary eclipse of HD 209458b. II. Strong limits on the infrared spectrum Near 2.2 μm. Astrophys. J. **597**, 581–589 (2003)

L.J. Richardson, D. Deming, K. Horning, S. Seager, J. Harrington, A spectrum of an extrasolar planet. Nature **445**, 892–895 (2007)

J.F. Rowe, J.M. Matthews, S. Seager, R. Kuschnig, D.B. Guenther, A.F.J. Moffat, S.M. Rucinski, D. Sasselov, G.A.H. Walker, W.W. Weiss, An upper limit on the albedo of HD 209458b: Direct imaging photometry with the MOST satellite. Astrophys. J. **646**, 1241–1251 (2006)

P.D. Sackett, Searching for unseen planets via occultation and microlensing, in *NATO ASIC Proc. 532: Planets Outside the Solar System: Theory and Observations*, ed. by J.-M. Mariotti, D. Alloin, 1999, pp. 189–+

B. Sato, D.A. Fischer, G.W. Henry, G. Laughlin, R.P. Butler, G.W. Marcy, S.S. Vogt, P. Bodenheimer, S. Ida, E. Toyota, A. Wolf, J.A. Valenti, L.J. Boyd, J.A. Johnson, J.T. Wright, M. Ammons, S. Robinson, J. Strader, C. McCarthy, K.L. Tah, D. Minniti, The N2K consortium. II. A transiting hot saturn around HD 149026 with a large dense core. Astrophys. J. **633**, 465–473 (2005)

S. Seager, B.A. Whitney, D.D. Sasselov, Photometric light curves and polarization of close-in extrasolar giant planets. Astrophys. J. **540**, 504–520 (2000)

S. Seager, L.J. Richardson, B.M.S. Hansen, K. Menou, J.Y.-K. Cho, D. Deming, On the dayside thermal emission of hot jupiters. Astrophys. J. **632**, 1122–1131 (2005)

D. Sudarsky, A. Burrows, I. Hubeny, Theoretical spectra and atmospheres of extrasolar giant planets. Astrophys. J. **588**, 1121–1148 (2003)

M.R. Swain, J. Bouwman, R. Akeson, S. Lawler, C. Beichman, The mid-infrared spectrum of the transiting exoplanet HD 209458b. ArXiv Astrophysics e-prints (2007)

A. Vidal-Madjar, A. Lecavelier des Etangs, J.-M. Désert, G.E. Ballester, R. Ferlet, G. Hébrard, M. Mayor, An extended upper atmosphere around the extrasolar planet HD209458b. Nature **422**, 143–146 (2003)

# Future Space Missions to Search for Terrestrial Planets

Malcolm Fridlund

Originally published in the journal Space Science Reviews, Volume 135, Nos 1–4.
DOI: 10.1007/s11214-007-9304-1 © Springer Science+Business Media B.V. 2008

**Abstract** Since the first exoplanet was discovered more than 10 years ago, this field has developed rapidly. Currently we know of more than 200 external systems of stars and planets, apart from our own, but what has become the 'holy grail' of exoplanetary research is still eluding us. Here we are, of course, referring to solar systems like our own, containing a number of terrestrial or 'rocky' planets orbiting within the so-called 'habitable zone', and with giant planets such as Jupiter and Saturn—which mostly or totally consist of elements as H or He—at distance much further out from the central star. No such system have to date been discovered around anything resembling our Sun, albeit because of observational biases. Nevertheless, in order to develop the emerging science discipline of Comparative Planetology, we will have to utilize new techniques that will enable us to search for, and then study in detail such systems in an un-biased fashion. This paper describes the emerging techniques and space missions that will allow, finally, the investigation of planets capable of hosting life as we know it.

**Keywords** Exoplanets · Terrestrial planets · Space missions · Nulling interferometry

## 1 Introduction

The idea that worlds like our own exist external to our solar system has its roots in philosophical speculations made already during antiquity. Even that these worlds would be inhabited with life forms that would be as it was put "*some like us and some otherwise*" (Epicurus, 300 B.C.) was theorized. Since the dawn of recorded history, imaginative people have returned to the issue for a multitude of reasons, but, during several millennia, they have been left with just that—philosophical speculation. Our generation is the first that can see the beginning of actual scientific research being brought to bear on the idea. During the middle of the last century, some prescient researchers pointed out that we were now becoming able to detect

M. Fridlund (✉)
Research and Scientific Support Department, ESTEC, P.O. Box 299, Noordwijk 2200AG,
The Netherlands
e-mail: Malcolm.Fridlund@esa.int

planets orbiting other stars. Particularly, Struve (1952) hypothesized that very large planets (Jupiter sized or larger) would, if orbiting very close to solar type stars be detectable by the deflection they would cause in the radial velocity signature in the stellar spectrum. It took, however, 37 years before this prediction became observable fact when first Latham et al. (1989) and then Mayor and Queloz (1995) and Butler and Marcy (1996) discovered such bodies. The continued work of these and other researchers have since then lead quickly to the understanding that bona fide exo-planets are to be considered as discovered, and thus, the first steppingstone on the path to finding true 'exo-life' has been obtained.

Since then, the field of exo-planets have developed with remarkable speed. Currently we know of more than 260 planets in more than 200 exo-systems. What still eludes us is what has come to be regarded as the 'holy grail' of Life Detection outside the Solar System, namely 'true' Terrestrial (or 'rocky') planetary bodies like our own Earth orbiting within their stars so-called 'Habitable Zone'. HZ—The Habitable Zone is defined as the region around a specific star where it is *in principle* possible to find the conditions that make life possible on our own Earth. Usually, it is taken to be the volume in space where water would be found in liquid form without postulating special conditions on the planetary surface, like e.g. a very strong greenhouse effect. A related definition is the Continuously Habitable Zone, where one take into account stellar evolution and the consequently changing of the luminosity of the central source.

Although we have thus found plenty of planetary systems around solar type stars in the vicinity of our Sun, and also taking into account that discoveries are continuing at an accelerated rate, what is currently lacking are any indication of *systems* like our own, e.g. giant planets in circular orbits found *outside* the HZ. The discovery of such systems would at least allow for the possibility of terrestrial bodies being in the HZ although current technologies do not allow for their detection. This situation is of course at least partly caused by an observational bias. The ground based methods used so far—detection of the abovementioned radial velocity deflection caused in the stellar spectrum by a sub-stellar mass orbiting the primary, or the occultation of part of the stellar light when the planetary body passes between us and the star—has so far not had the sensitivity to either detect Earth-size objects, nor has the time been long enough to pick up planets similar to our Jupiter or Saturn (periods of 12 and 29 years). Progress is being made continuously, however, and objects with (minimum) masses of order 8–10 Earth masses are now being picked up with some regularity. Note, that the masses of most planets detected so far are minimum masses since the methods used mostly measures only one component of the velocity of the star (see below). Recently, even more exciting results have been obtained through the technique of gravitational lensing (e.g., Beaulieu et al. 2006). In a first detection of a possibly 'rocky' planet, a 5.5 Earth mass object ('absolute mass') has been detected orbiting some astronomical units away from a M-dwarf star that is itself located several kiloparsec away from the Earth. Note that this planet is orbiting well outside the HZ of its primary. Albeit, there are several sources of uncertainties in this observation, it unambiguously show the power of the method, and the ability to detect planets down to Earth mass makes it the prime candidate for first detection of worlds like our own. Nevertheless, this method also contain large uncertainties and will have to be used with carefulness (see below).

So where do we go from here. The sky is literally the limit. The reasons for modern man to spend significant amounts of effort and money are multiple and important. Paramount, it is of as great an importance today, as it has been all through history, for us to find out man's place in the universe. Are we—as life forms—alone and unique in the Universe or is it inevitable that life arise when the conditions are the right ones? In order to find that out we need to touch on several fundamental aspects of our understanding of nature. We need to find out if other worlds like our own exist; we must determine if conditions are similar to what has been the case on the Earth historically and presently. We have to trace the evolution of exo-planetary systems and compare this with what we know about the Solar System, and we need actually to look for such signatures of life—so-called biomarkers—that can be traced over interstellar distances. This is a truly remarkable task.

While the ground based techniques that has hitherto been used continue to be developed, it has, however, recently become abundantly clear that to progress towards the direct detection and study of exoplanets the size of our own world will require instruments deployed in space.

The scientific rationale given for exo-planetary space missions has been put into the broader context of the European Space Agency's (ESA) new science plan for the period 2015–2025. This plan—designated Cosmic Vision—divide the major scientific questions to be addressed by European space science during the next few decades into 4 themes, the first of which is "What are the conditions for Planet formation and the emergence of life?" Addressing the discovery and census of Terrestrial planets around nearby stars, as well as a first determination of their physical parameters—including their habitability is the challenging objective of Cosmic Vision Theme 1.

## 2 Formulation of the Problem and High Level Scientific Requirements

As far as we can tell, from our limited statistics of one, and our (still) very poor understanding of what life really is and how it forms, we need a planetary surface in order to expect the processes that have apparently taken place on our planet. It has been hypothesized, by e.g., Carl Sagan, that life could originate develop and continue to exist in the atmosphere of gas giant planets like our Jupiter. This is described in Sagan's fascinating TV series "Cosmos" from the beginning of the 1980's and its accompanying book (Sagan 1980), as well as in the paper of Sagan and Salpeter (1976). As far as we understand today, however, none of the life forms we are aware of could have done so—which is why the Galileo spacecraft was crashed into Jupiter's atmosphere in order to make it impossible for it to later to contaminate the Jupiter moons—especially Europa—with any hitchhiking terrestrial life form. In order to make any progress in this area we need to properly understand how planets form and evolve, and particularly how a Solar System like our own come to be, and why so many of the systems found so far external to our own look completely differently. This will mean that we will have to understand the physical processes that are involved much better than today. In turn, this will compel us to fully understand, at least empirically, the birth, evolution and death of stars in more detail, as well as a much more thorough understanding of the fundamental physical processes involved.

In Fridlund (2000), the high level scientific requirements were described in the context of space missions, as being the answer to the following questions:

– Are we alone in the Universe?
– How unique is the Earth as a planet?
– How unique is life in the Universe?

A feasibility study carried out by ESA between 1997 and 2000, in order to investigate if it was currently technically possible to address these questions led—after a number of detailed technical investigations—to what is currently being represented by the mission concept of a 'nulling interferometer' (Bracewell 1978; see also below), which is designated Darwin and that is being developed to establish the following:

1. Search a large number of nearby stars for Terrestrial planets to find their frequency and location.
2. Detect planets within the so-called 'Habitable Zone'. In the Darwin studies, the HZ was considered only in terms of black body temperature. No provision was made to take into account atmospheric pressure, etc.
3. Determine the planets orbital characteristics (period, eccentricity, inclination etc.).
4. Observe the spectrum of the planet. Detection of the presence of an atmosphere, effective temperature and diameter of planet (through the albedo).
5. Determine the composition of the atmosphere, viz. the presence of water, ozone/oxygen in an Earth type planet, mainly inert gases in a Mars/Venus type planet and Hydrogen/Methane atmospheres in Jupiter type planets or 'primordial' Earth-like planets.

These high level scientific requirements are translated into specific observational requirements that can be converted into mission requirements. The process is detailed in Fridlund et al. (2006), but can be summarized briefly as:

1. Minimum number of single, solar type (F-K main sequence) stars to be surveyed for Terrestrial exoplanets in the HZ during primary mission is to be 165 to 500. M-dwarfs, the most common stars in the Galaxy are also considered.
   a. 165 under the added condition that significant amounts of dust (more than 10 times the level found in our solar system) is present in every object.
   b. 500 (this is essentially a complete sample of single F, G and K stars found out to 25 pc. Added to this is a number of M dwarfs) under conditions of similar levels of dust as found in the solar system.
2. Completeness of survey (probability that one has not missed a planet in the HZ for a specific star) to be better than 90%.

The reasons for these mission specific requirements are, of course, so that a negative answer (i.e. the ***non-detection*** of any Earth-like world) would be meaningful in a scientific sense.

3. Spectral signatures to be observed for each planet in any detected exo-planetary system— e.g. the presence of the so-called bio-markers $CO_2$, $H_2O$, $CH_4$, and $O_3$.

These are very demanding requirements and as a consequence the Darwin mission concept (see below) is arguably the most ambitious scientific space experiment ever contemplated. It is therefore apparent that a stepwise approach with simpler systems answering (fundamental) parts of the issues (such as e.g., the frequency of Earth size planets, etc.) could (and should) be implemented in preparation for the ultimate 'flagship' mission(s). This would not only save resources and funds, but it would also alleviate the risk when finally implementing the more complex systems.

## 3 Comparative Planetology

Comparative Planetology is currently rapidly acquiring the status of a new science. Originating from within the disciplines of Astronomy & Astrophysics, it is now becoming truly cross-disciplinary incorporating elements of biology (astrobiology), geophysics, chemistry, meteorology, etc., and thus also attracts interest from researchers in all these disciplines. Many of these individual scientists are of course attracted by the possibility that, through comparison, we will make advances in areas of research, relating to our own planet and existence that have been blocked or is very difficult to carry out here on Earth. An important example would be the origin of life itself on our planet. We do not today know, neither when, nor how and where this occurred on our planet. The evolving state of the planets crust has obliterated direct evidence, and consequently we do not know if life arose spontaneously and easily (as would be indicated by an early occurrence), or if it is under only very special circumstances (as may be implied from the extremely long, subsequent, development time of complex life forms). Observing a large number of exo-planetary systems of a young age and finding (or not finding) signs of e.g., chemical disequilibrium in their atmospheres would allow us to draw some conclusions in this area.

When we use the term Comparative Planetology, we have to realize that this label have hitherto only been used for comparisons between objects within our own Solar System, e.g., Venus and the Earth. It is, however, now being used in the much broader sense of a comparison between different Solar Systems (and their components of planets comets and dust). These studies of separate systems, having different histories and evolution, (original) physical conditions, different formation processes, etc. and then making full use of empirical science to more completely understand our own system. When we compare planets within the Solar System with each other, one makes instead a cross section through time at the current status of objects which have had the same origin and have experienced the same evolution, and where the major differences between bodies are probably caused only or mostly by their different location within the system. Expanding our knowledge to other stars and their accompanying planets eliminate these limitations. This is already evidenced by the amount of information acquired over the last 10 years since the first bona fide exoplanet was discovered (Mayor and Queloz 1995). Always remembering that we here are still limited by ground based observational bias.

We have so far for a long time (essentially since the days of Copernicus) made a fundamental assumption that we soon will be able to prove. We have assumed that the Earth is an average planet around an average star—and with an average history and evolution. Here we note that we do not really know this. After searching for a true solar analogue for more than 50 years, with continuously more refined methods, and failing to find an exact copy, we must realise that our Sun is only average in the same way as e.g., people can be considered average. When we take the 'finger prints' of stars, although we find very similar stars, we cannot find a single identical solar analogue. Thus we must draw the conclusion that the Sun is 'special' like people are 'special'. What we still do not know is how important these small differences in chemistry and evolution is when investigating e.g., how common small rocky planets like our own is, or if these differences would impact on life's ability to arise on a planet with a slightly different chemical make-up or on such a planet's ability continue to host life over long periods of time.

## 4 Methods and the Need to Go to Space

The prime methods that have been used so far (from the ground) are:

1. The radial velocity method
2. Occultation's of a star by a planetary body
3. Gravitational lensing
4. Astrometry.

All of these methods have delivered tremendous results as can most clearly be seen from an inspection of *The Extrasolar Planets Encyclopaedia* web pages (http://exoplanets.eu) maintained by Jean Schneider at Meudon Observatory in Paris. Nevertheless each method has severe limitations imposed by the location on the surface of a planet.

The first method, the radial velocity method (e.g., Udry et al. 2007), measures the deflection of the spectral lines in a stellar spectrum, caused by the gravitational tug imposed upon its surface by an orbiting body. It is of course a function of both the mass of the orbiting body and its distance from the stellar surface, and therefore the method is naturally biased towards (as pointed out by Struve in the above quoted paper from 1952) massive planets orbiting very close to the star. *It is interesting to notice that Struve also pointed out that there were no a priori reasons NOT to expect such planets, just because they are non-existent in our own Solar System.* As we want to detect planets like our own, we realize that an Earth-mass body, orbiting 1 astronomical unit away from a G2V (our Sun) star, will cause a deflection (amplitude) in the radial velocity curve of 0.1 m/s over a period of one year. This is significantly smaller than the amplitude change caused by the 5-minute period acoustical oscillations in the solar atmosphere (so-called p-modes), the modes of which have life times of order half a year. On top of these signals (RV + p-modes) we also have the disturbances introduced by the solar activity, and which have amplitudes that are similar or larger. Recent results from the CoRoT mission indicate that these 'noise' sources are indeed serious when addressing the RV signature of a planet the size of the Earth. If we search for an Earth within the zone where we expect to find life, around smaller, less luminous, solar type stars (K-dwarf stars), the situation is alleviated as what regards the first two of these problems (larger planetary RV amplitude and the p-modes are likely to have lifetimes different from the orbital period since the latter is shorter and the p-modes have longer life-times) but not in the third area (activity) where the situation may actually be somewhat worse. Taken together with the technical difficulties, it is clear that while not impossible, it will be hard to detect an Earth analogue from the ground with the RV method. It will require dedicated large telescopes, equipped with spectrographs with significantly better performance than what is presently state-of-the-art (Udry et al. 2007). The observing runs will be very long (years). Further, since we only measure one component (along the line of sight) of the velocity, the mass of the planet that we determine will be a ***minimum*** mass. Some other method (see below) can exceptionally be used to determine the other component and therefore leads to an exact mass (only depending on the estimate of the stellar mass).

The second method in importance (so far), the occultation method (Rauer and Eriksson 2007) has already delivered results in a number of cases. This is also a method that shows strong promise of being important in space based applications—see the CoRoT and Kepler mission descriptions below. In this method we detect the drop in luminosity of the star as the planet passes between us, and the star, and draw conclusions about both star and planet from the shape of the light curve. What is actually measured is the ratio of stellar to planetary radius, and an assumption about the one has to be made in order to determine the other. The problem here is of course that it is also biased towards large planets orbiting very close to a (small) star—a situation not found in our own Solar system. A Jupiter-size planet passing between us and a solar type star will cause a drop in luminosity of about 1–3%, while an Earth size body will cause the light to diminish by a factor of about $10^{-4}$. Further, the occultation lasts for some hours, which means that the shorter the period (the first detected occultation by an exo-planet repeat every 3.5 days) the easier it is to detect it. Then we have to take into account the random orientation of exoplanetary orbital planes.

It can be shown (Charbonneau et al. 2007) that the probability that an occultation occurs is:

$$F_{\mathrm{tr}} = 0.0045 \, (1 \text{ AU}/a)((R_{\mathrm{star}} + R_{\mathrm{planet}})/1 \, R_{\mathrm{Sun}})((1 + e\cos(\pi/2 - \omega))/(1 - e^2)),$$

where $e$ is the eccentricity and $\omega$ is the longitude of the periastron.

This result in about 1–10% chance of an occultation to happen for planets close to the star (0.3 AU to 0.05 AU for a solar analogue and with a planet significantly smaller than Jupiter), while the probability is significantly smaller for planets orbiting in the Habitable Zone of a solar type star. This means that the method need to be applied either to very large samples of stars, simultaneously, or that one has very strong reasons to suspect that the planetary orbital plane is to cross the line of sight. This was the case for the first detected occultation—that of HD 209458b (Charbonneau et al. 2000). Using wide-field telescopes, either from the ground (where an additional limitation is the Earth's rotation which requires networks of telescopes) or from space, one can observe large numbers of stars at the same time. It is then a powerful method mostly limited by the disturbances in photometric precision induced by our atmosphere.

Detection of exoplanets through gravitational lensing (Beaulieu et al. 2006) is a very promising method. This method is no-doubt going to be significantly developed during the next decade. It is a potentially very powerful method for determining statistics of exoplanets. The drawback is that for this method to work, one needs both a lensed object as well as a lensing system lined up along the line of sight. In order to have any significant chance of detecting planets we require a large number of candidates, which makes both the foreground and background objects faint since they will then by necessity also be very distant. We then have difficulty assigning types and luminosities to both the lensed object and the lenser. The planetary mass will depend on both of these parameters. In the last case quoted above (Beaulieu et al. 2006), we assume that the blended image of the lensed object close to the galactic centre and the lenser at about half this distance are of F and M type respectively. These assumptions lead to an estimate of the planetary mass of 5.5 $M_{Earth}$. We will not know if the mass is correct for many years since this depends on the stellar type of the two objects and a better knowledge will not come until the nearer star have moved out of the line of sight to the lensed object.

The fourth method, the astrometric method—where one measures the proper motion changes of a star as it travels across the sky, and interpret deviations from its predicted motion, is so far less successful. A number of ground-based multi-epoch observations carried out during the last century—mainly with long focus refractors—reported the discovery of large planets around some nearby stars. One can mention 70 Ophiuchi, where Reuyl and Holmberg (1943) found a 10 Jupiter mass body orbiting the star. None of these objects have been confirmed today.

Finally, astrometric observations, utilizing mainly the fine guidance sensors on the Hubble Space telescope have been used to determine the deflection in the plane of the sky in a few cases. In the case of the planet Gliese 876b, the deflection is about 25 milli-arcseconds, which taken together with a well determine parallax leads to a planetary mass of $1.89 \pm 0.34$ Jupiter masses, the largest uncertainty being the assumption about the stellar mass (Benedict et al. 2002). Dedicated space missions, with a capability of determining proper motions of a few micro-arcseconds, will make possible systematic surveys for planets of sizes down to maybe 10 $M_{Earth}$ within the foreseeable future (see below).

Finally, we say something about the capability of Extremely Large Telescopes (ELT's) to directly detect exo-planets. The problem here is to have a wave front arriving on the imaging detector of such a quality, that one can detect contrast differences of between $10^5$ and $10^{10}$. This a distance from the optical centroid of the star will be less than 0.5 arcseconds for Jupiter and 0.1 arcseconds for the Earth if viewed at a distance of 10 pc. The requirement that one need to be able to detect the planet in the residual of the airy disk at these angular distances is daunting enough without adding the noise sources provided by the atmosphere and either the segmentation of very large telescope mirrors or in the case of monolithic mirrors irregularities in the mirror surface. Referring the reader to the paper by Chelli (2005), we only here conclude, that a telescope with 100 m diameter, and with a correction of the wave front arriving on the detector to a precision 2–3 orders of magnitude better than what is currently achievable is required to detect the Earth, orbiting in the HZ around an early G-type star. The limiting distance for a search would be 10 pc (Chelli 2005) to 18 pc (Gilmozzi, private communication). Within 18 pc we find a total of 39 single (multiple stars would be a problem because of the extra scattered light) G-type stars (Kaltenegger et al. 2007). Further, most of these are found towards the later classes of the G types. Thus the majority of stars within the 18 pc limit will have their HZ at much smaller angular distances from the central object, and thus be significantly harder to detect. Add to this the problem that when searching for a planet similar to our own hosting life as we know it, the spectral signatures to search for are also in our atmosphere—but enormously much stronger.

While thus not excluding the possibility of the detection of planets similar to our Earth from the ground in the future, we may safely conclude that this is a very difficult task, fully in class with the most complicated of space missions. Further, it appears that the scientific case is going to be limited (small number of available targets of which we have no indication how common terrestrial planets are). This being said it is on the other hand also quite clear that the large (30 m to 40 m) class ELT's currently being planned for operation beyond the year 2020 will make significant contributions to the studies of gas giants—particularly those of the kind found in our own solar system. These objects will be bright and well separated from their primaries, and spectroscopy will not be a problem since the features to study are very different from those found in our atmosphere. Radial velocity observations and other indirect detections of small rocky planets will also be feasible.

## 5 Space Missions

Thus, if we want to truly address the scientific requirements outlined above, we will need to go to space. This has become more and more apparent over the last 10 years through a large number of very ambitious studies and projects developed in Europe, The United States and Canada. As a consequence, a number of projects/missions of increasing level of complexity and scope have been and are being developed. One, the Canadian mission MOST is in orbit since several years. The CoRoT spacecraft was successfully launched less than a year ago and has been taking excellent data since February 2007. Kepler and GAIA are under construction or approaching launch, and the 'flagships'—Europe's Darwin and NASA's TPF are being developed both scientifically and technically. We now briefly describe these missions one bye one.

## 6 MOST—The First Step

MOST (Microvariability & Oscillations of Stars) is a suitcase sized mini satellite launched into orbit by the Canadian Space Agency (CSA) on a Russian Rockot (ex-ICBM) into a low polar orbit around the Earth on June 30, 2003. It deploys a 15 cm telescope feeding two 1024 by 1024 pixel CCDs. It measures very high precision variability in white light. A number of target stars have already been searched for exoplanetary signals, with negative results so far. Although it is a small telescope and the maximum dwell time on any target star is only 58 days, nevertheless, it has been possible to exclude e.g., 2–5 Earth radii planets in several orbits around the known occulting star

HD 209458, as well as providing upper limits to the albedo of the known planet HD 209458b (http://www.astro.ubc.ca/MOST/milestones.html#results and references therein). MOST only observe its targets one by one, and thus do not have the multiplexing advantage of observing large numbers (up to hundreds of thousands) of stars like CoRoT and Kepler.

## 7 CoRoT—The First True Exo-Planetary Mission

MOST was designed as an asteroseismological mission, that could also search for exoplanetary occultation's under some circumstances. CoRoT (Convection, Rotation & planetary Transits), on the other hand, is designed from the beginning with such transits as one of the two primary objectives, and with most of the observing time allocated to this objective. A mini satellite from the PROTEUS series of small spacecraft initiated by the French space agency, CNES, it is a true international mission with contributions from the European Space Agency (ESA), Brazil, Belgium, Spain Austria and Germany. It consist of an afocal telescope with a 27 cm main mirror and a field of view of 2.8 by 2.8 square degrees. The focal plane contains 4 CCD's consisting of 2048 by 2048 pixels each. Two each of the CCD's are allocated to the two main objectives. In the exo-planetary program, CoRoT is pointed towards the same field of view for 150 consecutive days with up to 12000 targets brighter than magnitude 15.5 being observed. A grism in front of the exoplanetary detector produces a short spectrum of each target and the data of the (pre-selected) occultation targets are individually read-out and transmitted to the ground as a unique 150 d long essentially uninterrupted light curve (the duty cycle in orbit is >97%). Having a short spectrum of each target allow us to discriminate between (flares of) activity and true planetary transits. In orbit verification has demonstrated the telescope to be essentially photon noise limited for the magnitude interval of the exoplanetary channel. The size of the planet that can be discerned depend on how close to its primary star it is. In principle, if the star is in an orbit with a period shorter than a few weeks, planets the size of the Earth can be detected. For longer periods, of more than a few tens of days, radii between 1.5 and 2 times our own planet are observable.

CoRoT was launched from Russia in December of 2006. In-orbit verification was terminated in February and the spacecraft has been taking data since then. The first exo-planet—a 'hot Jupiter' named CoRoT Exo 1b—was discovered and confirmed within a few weeks and was presented in a press release. The orbit is a polar circular orbit of 896 km altitude. The nominal mission is 2 ½ years with possibilities of extending the mission to more than double that. A major program of follow-up with spectroscopy carried out bye the HARPS instrument on the 3.6 m reflector of the European Southern Observatory (ESO) at La Silla in Chile, the Observatoire de Haute Provence in France and the Tautenburg 2 m telescope near Jena in Germany, as well as several photometric telescopes worldwide, has been initiated. The photometric observations are required in order to exclude the possibility of the variability being caused by confusing objects within the CoRoT point-spread-function. With the radial velocities determined through these means, and taken together with the inclination of the planetary orbital plane from the actual measured light curve we will be able to determine the true mass (and not only the minimum mass as we get when we only have one part of the data like the radial velocity curve). Taken together with the diameter (which we get again from the shape of the light curve together with the diameter of the star—the latter parameter either measured with interferometers or through modelling), we can get the planets average density. Further follow-up observations with large telescopes provide information about the atmosphere of the planet through spectra or colours.

A large program of supporting observations is also required in order to learn as much as possible about each host star (and the total number of stars during an extended mission could reach 100 000)—with either spectroscopic observations or Johnson and Strömgren photometry or both.

## 8 Kepler—The Exploration Continues

CoRoT will be followed in 2009 or later by the National Aeronautics and Space Administrations (NASA's) Kepler mission. This is essentially a larger version of the former mission with a 1 m class Schmidt telescope. This mission will be deployed in a so-called drift-away orbit, travelling along the Earth's orbit while slowly drifting away (after ½ year in orbit the spacecraft is several tenths of astronomical units away from the Earth). This provides very benevolent circumstances for the mission, and allow the essentially uninterrupted pointing towards a dedicated star field for a period of no less than 4 years. It now becomes possible to detect Earth-sized planets orbiting within their primaries Habitable Zones (1+ year orbital period for an early G-type star, about ½ year for a late K-type object). Sometime around 2011–2012, these objects will be accessible for a detailed follow-up from the ground along the same guidelines as mentioned above for CoRoT. We must remember, however, that we are dealing with objects many hundred's of parsecs away, and the direct detection of these planets will be highly unlikely—at least with presently envisaged techniques. Nevertheless, we will learn much from these missions, and one of the most important pieces of information will be the value of $\eta_{Earth}$, which designates the percentage of Solar Type stars that have planets like our own (Beichman 2000). A first estimate of this number will have been determined by CoRoT, and will be very important when designing missions with the most ambitious goal of them all—that of directly detect an Earth-like planet at interstellar distances, and analyzing the light from it spectroscopically in order to determine its characteristics.

## 9 The Role of Herschel in the Search for Other Earth's

When ESA's Far-InfraRed (FIR) space mission Herschel launches in 2008, it will be for years to come, the world's largest space telescope (the main mirror has a diameter of 3.5 m). As such, it will play an important role in the fulfilment of the objectives outlined above.

Herschel has a unique capability in detecting cold dust representative of the temperatures found in the Solar Systems Kuiper-Edgeworth belt. For really nearby targets, within 15–20 pc, it will be possible for the first time, to detect systems with amounts of dust similar to that which is found in the outer parts of our system. Distributed between about 10 AU and 100 AU, this dust component is probably being provided by collisions between cometary (or asteroidal) objects within this zone. Without a continuous re-supply through repeated collisions, and/or larger, planetary bodies that lock the dust into resonances where it can not be ejected through e.g., radiative pressure or falling into the Sun through the Poynting-Roberson effect, this dust component would disappear in a relatively short time. The detection of a dust component similar to our own would thus indicate the presence of a KE belt and/or large outer planets, and would be a clear indicator of which systems, among the nearby stars, should be searched first when the capability of direct planetary detection becomes available (see below).

## 10  GAIA—The First Global Survey Instrument and the Cosmic Census

GAIA is an ambitious mission, under development by ESA, and with the goal of charting a three-dimensional map of a large part of our Galaxy. Building on the highly successful HIPPARCOS mission, GAIA will provide never before realized positional and radial velocity measurements of more than one billion stars in the Milky Way galaxy. Also stars in galaxies located in the Local Group, i.e. the cluster of galaxies that the Milky Way and the Andromeda galaxy (among others) are members of. Launched in 2011 or 2012, and operating for 5 years, each of its three telescopes will observe the billion stars about 100 times measuring its position with down to about 20 microarcseconds angular accuracy. At the same time a spectrometer, operating in a narrow band in the visual wavelength range will obtain about 40 spectra of each object, and thus providing the velocity along the line of sight (see above).

Placed in an orbit around the Sun–Earth L2 libration point, it will be undisturbed by the presence of either the Earth or the Moon, and can observe all the sky many times.

GAIA will play an important role for exoplanetary research in two ways. First, it will provide a database of relatively large (about 15 Earth mass or larger) planets around all detectable stars out to a distance of several hundred parsecs. Based on the knowledge of the exo-planets found since 1995, we can expect literally thousands of new objects. This will allow statistical investigations several order of magnitude more accurately than at present.

Secondly, GAIA will detect all unknown nearby (within 25 pc) stars of all spectral types. It is clear from the work on the target lists for missions like Darwin (see below) that we have a very poor knowledge about late K-type and M-dwarf star in the intended sample. These stars are intrinsically weak, and therefore does not exist in the parallax databases (e.g. HIPPARCOS). Since GAIA will operate down to 23 magnitude and weaker in the R-band, the parallaxes of several hundred microarcseconds will be detectable already 1 ½ year into the mission.

## 11  SIM—Planet Quest

SIM (Space Interferometer Mission) is NASA's version of GAIA. Using interferometric methods (GAIA has three separate telescopes spinning across the sky, measuring the distances between individual stars globally), the relative distance between objects are measured extremely accurately (maybe a factor of 10 better than GAIA or close to 1 microarcsecond at best). Both parallaxes and proper motions can be measured in this way. The price you pay is in that you only measure a fraction of the number of objects that GAIA can do plus a limited sensitivity, as well as significantly increased complexity and cost. On the other hand, it is in principle possible to see the deflection in the motion across the sky caused by Earth-sized planets, for a handful of the very closest stars.

Further, SIM will be the first interferometric system in space, and as such will be a precursor to Darwin and TPF (see below)—the missions dedicated to direct observation of planets like our own. The technology of optical interferometry is in its early phases here on Earth, and it would be extremely valuable to test at least elements of this technology in space before one launches gigantic systems like Darwin with all it entails.

SIM will also study a number of important galactic and extragalactic targets, where the highly accurate measurements will provide new insight into the physical mechanisms. The added name "Planet Quest", however, demonstrate the order of priority of the scientific targets.

## 12 Darwin and the Terrestrial Planet Finder(s)—Other Worlds with Life as We Know It

The direct detection of Earth-like exo-planets orbiting nearby stars, and the characterization of such planets—particularly as what concerns their evolution, their atmospheres and their ability to host life as we know it, and to understand how planets form and life emerges— is the ultimate goal of Theme 1 in the ESA's Cosmic Vision 2015–2025 science plan. The same objectives can be found in the long-term plans of NASA as evidenced by the program suggested by their (previous) decadal committee. Thus both of the world's major space agencies have as one of their most important goals the placing of life in the Solar System into the context of the rest of the Universe.

The space missions, capable of meeting the challenges posed by these goals, are designated respectively Darwin (ESA—Legér et al. 1996a, 1996b; Fridlund 2000; Cockell et al. 2007) and the Terrestrial Planet Finder or TPF (NASA—Beichman 2000).

As mentioned above, the direct detection of a planet, with the size of the Earth, orbiting around its parent star within the HZ constitutes a challenging problem, since the signal detected from the planet is between about $10^{10-11}$ (visual wavelength range) and $10^{6-7}$ (mid-IR spectral range) times fainter than the signal received from the nearby star. Selecting the appropriate spectral region in which to attempt detection is governed by this contrast problem, *and* the selection of a region in which the characterization of the planet and its habitability is optimum. This problem has recently been addressed by a number of researchers e.g., Selsis (2002), Traub (2003) and Kaltenegger and Selsis (2007). The European mission Darwin has selected the spectral region between 6 μm and 20 μm, a region that contain (among others) the $CO_2$, $H_2O$, $CH_4$, and the $O_3$ spectral features found in the terrestrial atmosphere. The presence or absence of these spectral features would indicate similarities or differences with respect to the atmospheres of known telluric planets such as Venus, Earth and Mars.

ESA has selected a so-called 'Nulling Interferometer' operating by the principle of destructive interferometry, and in the mid-IR wavelength range for detailed study and possible implementation. It utilizes free flying telescopes (i.e. each telescope on a separate satellite and thus no connected structures). This is what is commonly referred to as the Darwin mission (Fridlund 2000). It is thus implementing the new technology of nulling (or destructive on-axis) interferometry (Bracewell 1978; Bracewell and McPhie 1979). The basic concept here is to sample the incoming wave front from the star and its planet(s) with several ($\geq 2$) telescopes that individually do not resolve the system. By applying suitable phase shifts between different telescopes in this interferometer array, destructive interference is achieved on the optical axis of the system in the combined beam. At the same time, constructive interference is realized a short distance away from the optical axis. Through the appropriate choice of configurations and distances, one can for the specific case, place areas of constructive interference on regions representative of the HZ and so achieve the required contrast at this location. The first practical demonstration of 'nulling' (from the ground) was undertaken in February 1998 (Hinz et al. 1998). Using the Multiple Mirror telescope on Mount Hopkins, Arizona, these authors were able to cancel out the image of a star: $\alpha$-Orionis, while detecting faint circumstellar dust. The ability of the interferometer to suppress the entire Airy pattern was demonstrated and in this case the nulled image had a peak in intensity of 4.0% and a total integrated flux of 6.0% of the constructive image. Both in Europe and the US, a light suppression in excess of $10^5$ have been achieved in the laboratory.

Planet finding missions like the Darwin mission or the Terrestrial Planet Finder could be implemented in a wide variety of different nulling interferometer architectures (e.g., Karlsson et al. 2004; Coulter 2003) and configurations, constrained by the number of tele-

**Fig. 1** An artist's conception of how the Darwin mission would look when orbiting in the L2 libration point in the Sun–Earth system, 1.5 million km out from the Earth. In this configuration, four telescopes, each of a diameter of 4 m are positioned in a "stretched X-array" around a central beam combiner (second from right). This system could search at least 500 of the closest (within 25 pc) stars for an Earth-size planet orbiting within the Habitable Zone

scopes and the necessary background and starlight suppression. Figure 1 shows a so-called stretched X-array, which consists of 4 telescopes in an X-shape and with a beam combiner in the centre. By 'stretching' the X in one direction we create several sub-interferometers, the output of each which can be combined or subtracted from each other, thus creating a modulation in the transmission pattern on the sky, which effectively removes the background (which is significantly stronger than the planetary signal). Recent work suggests that requirements on the shape of the null can be relaxed when one takes into account unavoidable noise contributions introduced by instrumental errors (Kaltenegger and Karlsson 2004; Dubovitsky and Lay 2004). Thus configurations with fewer telescopes (3 or 4 as compared with the original 5 or 6) are now investigated at system level as candidates for the Darwin mission reducing complexity and cost of the mission. Further, it may not be necessary to have the beam-combiner in the same plane as the telescopes. The combination process can be significantly simplified if all the beams from the telescope are injected, with the appropriate phase delays, simultaneously into a beam-combiner about 1 km away from the telescope array. In this scenario, the telescopes become also much easier to construct and are essentially flux-collectors (Absil et al. 2007).

Both ESA and NASA have carried out ambitious technology development programs. In these programs, every technological challenge has been addressed—at least to some level of development. Here can be mentioned: Achromatic phase shifters (to control the required path lengths in the interferometer to the very high precision required over the complete spectral band), Optical fibres operating within the required band (to perform the required spatial filtering relaxing the requirements on accuracy of the optical surfaces with 2–3 orders of magnitude) and integrated optics ('optics on a chip'), i.e. mixing and splitting of signals on an integrated chip instead of with bulk optics, and with no moving parts—something with a potential to reduce the mass and complexity significantly.

NASA has also been investigating the possibility of implementing TPF as a corono-graphic telescope operating in the visual wavelength range. The investigations have demon-strated that a telescope with an elliptic primary mirror of 3.5 m by 8 m could screen be-tween 50 and 150 nearby stars for Earth-sized planets in the HZ. It is somewhat biased towards brighter stars (F-type and Early G-type). Given that an Earth sized planet is discov-ered around one of these stars, this mission—which has been designated TPF-C to separate it from the interferometer which is TPF-I—could detect bio-markers if the object is one of the 30 most suitable targets.

Darwin and TPF-I have been developed in parallel over the last ten years. Already in 1998, an agreement was made between ESA and NASA that information should be ex-changed with the ultimate goal of a joining of the two missions. Being very complex and thus probably also expensive, it makes definite sense to join the two missions since very little (except for reasons of redundancy) could be gained by having two identical systems observing the same, relatively few targets.

## 13 Further Future—Mission Accomplished?

The success of space missions like those described above, especially Darwin/TPF will no doubt impact on a number of scientific disciplines, and maybe change them beyond recog-nition. The successful results will allow us to address questions within these subjects about not only how common our system is, but also indicate its history and in what direction our planet and its geo/bio-sphere will evolve. Among the 1000 closest stars, we have a nice randomized distribution of ages spanning the range from 10 million to 10 billion years.

Particularly, the successful detection of 'true' Earth analogues will of course influence our understanding of how life itself arose and evolved on our planet. Empirical evidence, long ago destroyed by our planets dynamic surface, can be observed on other worlds—ultimately in a time-lapse movie fashion.

Eventually, more detailed observations will be required. It is clear that instruments more complex and ambitious than what have been discussed in this chapter will not be seriously considered before we know if a 'cousin' or more exist in the vicinity of the Sun. The fasci-nating aspect of our times is that this 'first step' is being contemplated right now. Success in this endeavour will lead to the design of larger systems currently only being in their first experimental phase (Le Coroller et al. 2004), but involving maybe literally dozens of 10 m class space telescopes flying in formation with inter-satellite distances of several hundred km, and with the focal plane maybe 5000 km away from the telescope units. Such an in-strument could take 'true' pictures of the surface of the Earth at distances of 10–20 pc, with resolutions of 10–50 elements and using white light. It would then become possible to draw proper maps of continents, oceans and . . . .

We live truly in exciting times.

## References

O. Absil, D. Defrére, T. Herbst, C.V.M. Fridlund, in Worlds like our own – The report of ESA's Terrestrial Exoplanet Science Advisory Team (ESA SP, Noordwijk, The Netherlands, 2007) (in preparation)

J.-Ph. Beaulieu, D. Bennet et al., Nature **439**, 437 (2006)

C. Beichman, Darwin and astronomy – The infrared space interferometer, ESA SP-451, 2000, p. 239

G.F. Benedict et al., Astrophys. J. **581**, L115 (2002)

R.N. Bracewell, Nature **274**, 780 (1978)

R.N. Bracewell, R.H. McPhie, Icarus **38**, 136 (1979)

P. Butler, G. Marcy, Astrophys. J. Lett. **464**, L153 (1996)

D. Charbonneau, T.M. Brown, D.W. Latham, M. Mayor, Astrophys. J. **529**, L45 (2000)

D. Charbonneau, T.M. Brown, A. Burrows, G. Laughlin, in *Protostars and Planets V*, ed. by B. Reipurth, D. Jewitt, K. Keil (University of Arizona Press, Tucson, 2007), p. 701

A. Chelli, Astron. Astrophys. **441**, 1205 (2005)

C.S. Cockell, T. Herbst, A. Legér, O. Absil et al., Darwin – An experimental astronomy mission to search for extrasolar planets. Exp. Astron. (2007, submitted)

D.R. Coulter, in *Proceedings of the Conference on "Towards Other Earths: DARWIN/TPF and the Search for Extrasolar Terrestrial Planets*, 22–25 April 2003, Heidelberg, Germany. Eds. M. Fridlund, T. Henning, complied by H. Lacoste. ESA SP-539 (ESA Publications Division, Noordwijk, The Netherlands, 2003), p. 47

S. Dubovitsky, O.P. Lay, in *Proceedings of SPIE*, vol. 5491 (2004), p. 284

M. Fridlund, ESA-SCI (2000), 12

M. Fridlund et al. (2006, in preparation)

P.M. Hinz, J.R.P. Angel, W.F. Hoffman, McCarth et al., Nature **395**, 251 (1998)

L. Kaltenegger, A. Karlsson, in *Proceedings of SPIE*, vol. 5491 (2004), p. 275

L. Kaltenegger, C. Eiroa, M. Fridlund (2007, in preparation)

L. Kaltenegger, F. Selsis, in *Extrasolar Planets*, ed. by R. Dvorak (Wiley-VCH, Weinheim, 2007), p. 79

A.L. Karlsson, O. Wallner, J.M. Perdigues Armengol, O. Absil, in *Proceedings of SPIE*, vol. 5491 (2004), p. 831

D.W. Latham, R.P. Stefanik, M. Mazeh, T. Mayor, G. Burki, Nature **339**, 38 (1989)

H. Le Coroller et al., Astron. Astrophys. **426**, 721 (2004)

A. Legér, J.M. Mariotti, B. Mennesson, M. Ollivier et al., Icarus **123**, 249 (1996a)

A. Legér, J.M. Mariotti, B. Mennesson, M. Ollivier et al., Astrophys. Space Sci. **241**, 135 (1996b)

M. Mayor, D. Queloz, Nature **378**, 355 (1995)

H. Rauer, A. Eriksson, in *Extrasolar Planets*, ed. by R. Dvorak (Wiley-VCH, Weinheim, 2007), p. 207

D. Reuyl, E. Holmberg, Astrophys. J. **97**, 41 (1943)

C. Sagan, E.E. Salpeter, Astrophys. J. Suppl. **32**, 737 (1976)

C. Sagan, in *COSMOS* (Random House Inc., 1980), ISBN 0-394-50294-9

F. Selsis, in *Proceedings of the First European Workshop on Exo-Astrobiology*, 16–19 September 2002, Graz, Austria, ed. by H. Lacoste, ESA SP-518 (Noordwijk, The Netherlands, 2002), p. 365

O. Struve, The Observatory **72**, 199 (1952)

W. Traub, in *Proceedings of the Conference on "Towards Other Earths: DARWIN/TPF and the Search for Extrasolar Terrestrial Planets*, 22–25 April 2003, Heidelberg, Germany. Eds. M. Fridlund, T. Henning, complied by H. Lacoste. ESA SP-539 (ESA Publications Division, Noordwijk, The Netherlands, 2003), p. 231

S. Udry, D. Fischer, D. Queloz, in *Protostars and Planets V*, ed. by B. Reipurth, D. Jewitt, K. Keil (University of Arizona Press, Tucson, 2007), p. 915

# Summary

# "Strategies of Life Detection": Summary and Outlook

Oliver Botta · Jeffrey L. Bada · Javier Gomez-Elvira ·
Emmanuelle Javaux · Franck Selsis · Roger Summons

Originally published in the journal Space Science Reviews, Volume 135, Nos 1–4.
DOI: 10.1007/s11214-008-9357-9 © Springer Science+Business Media B.V. 2008

**Keywords** Biosignatures · Spectroscopy · In-situ measurements · Organic chemistry ·
Meteorites · Solar system · Exploration

O. Botta (✉)
International Space Science Institute, Hallerstrasse 6, 3012 Bern, Switzerland
e-mail: botta@issibern.ch
url: www.issibern.ch

J.L. Bada
Scripps Institution of Oceanography, University of California at San Diego, 9500 Gilman Dr., La Jolla,
CA 92093-0212, USA
e-mail: jbada@ucsd.edu
url: http://exobio.ucsd.edu/bada.htm

J. Gomez-Elvira
Robotics & Planetary Exploration Laboratory, Centro de Astrobiología (INTA/CSIC), Instituto
Nacional de Técnica Aeroespacial, Crtra. Ajalvir km.4 - 28850 Torrejón de Ardoz, Madrid 28850, Spain
e-mail: gomezej@inta.es

E. Javaux
Department of Geology, University of Liege, 17 Allée du 6 Août, Bât. B18, Sart-Tilman Liège 4000,
Belgium
e-mail: ej.javaux@ulg.ac.be

F. Selsis
Laboratoire d'Astrophysique de Bordeaux, CNRS, Université Bordeaux 1, BP89, 33270, Floirac, France
e-mail: franck.selsis@obs.u-bordeaux1.fr

R. Summons
Department of Earth, Atmospheric and Planetary Sciences, Massachusetts Institute of Technology,
77 Massachusetts Ave., Cambridge, MA 02139, USA
e-mail: rsummons@mit.edu
url: eaps.mit.edu/geobiology

## 1 The Search for Extraterrestrial Life

The question of whether we are alone in the Universe is as old as when humans have started looking at the stars. Since the old Greeks, this question has been a frequent companion to science and philosophy. The heliocentric system of Copernicus, the use of telescopes and spectroscopes for the observations of galaxies, stars and planets, the discovery of the principle of evolution by natural selection by Darwin, and finally the technologies of the space age that enabled robotic spacecraft to visit other planets in the solar system has reversed our anthropocentric view and moved Earth, and with it humanity, into a less and less relevant cosmic context. However, despite this new point of view, we still have not detected any traces or signals that would indicate that life exists or did exist anywhere beyond the Earth. The Copernican revolution has occurred in astronomy (abundances of planets, stars and galaxies), chemistry (abundances of heavy elements) and in physics (abundance of baryonic matter), but it has *not* occurred in respect to biological phenomena since the development of the Darwinian theory of evolution.

The search for life in the Solar System requires a mixed approach with regard to mission scenarios and detection methods, primarily because of the large distances and travel times for planetary probes to the outer solar system. *In-situ* "life-detection missions" to the inner, terrestrial planets can in principle consist of mobile vehicles (rovers, airplanes, balloons), which can be relatively simple and be launched at reasonable costs. Sample return missions can also be conceived, however, their enabling technologies are significantly more complex (ascent vehicles, autonomous orbital rendezvous, etc.). Launch opportunities are relatively frequent and communications bandwidths are relatively high. Of course, such missions have to and will be complemented by ground- and space-based astronomical observation campaigns. On the other hand missions to the outer planets will probably focus on spectroscopic observations either from the ground or from space. *In-situ* exploration is possible, but these missions will be relatively complex (and therefore expensive, see for example the Cassini-Huygens mission), launch opportunities are less frequent and communications are more difficult (i.e. more autonomy will be required). Possible future missions to the outer solar system with astrobiological relevance, for which the question of biosignatures could be relevant, are a Europa orbiter and/or lander, a long-term orbiter/lander/rover mission to Titan or a mission to Enceladus.

With this background, both spectroscopic and *in-situ* biosignatures are highly relevant for the search for life in the solar system and should be explored by theoretical (for example with regard to energy efficiency of living matter) as well as experimental investigations.

### 1.1 The Search for Organic Compounds on Mars

Mars has always been a prime candidate to search for extraterrestrial life. There is a very strong public interest in, and enthusiasm for, space exploration and, technically, it also one of the easiest targets to reach with space missions, with launch windows every 26 months and launch masses such that a significantly sized spacecraft can be placed on the surface. The success of the current generation of Mars Exploration Rovers and orbiting satellites gives significant credibility for the potential of future missions. Scientifically, and also with regard to possible human exploration, the question whether or not life exists or ever existed on Mars should be answered before any human being will set foot on Mars.

In the 1970's each of the two Viking landers carried a biology experiment package (VBI) and a gas chromatograph–mass spectrometer (GC-MS) to the Martian surface to search for signs of life based on metabolism as well as to detect presence of organic molecules

(Schuerger and Clark 2008). Although the results of individual experiments (in particular the Labeled Release Experiment) did not fulfill all criteria for an abiotic explanation, taken together the results provided by the VBI/GC-MS instrument packages provided evidence that appears to exclude the presence of life at least at the Viking landing site within 30 cm to the surface. After these confounding results the exploration of Mars was significantly reduced until, in 1996, the claim of the detection of traces of ancient life in the Martian meteorite ALH84001 (McKay et al. 1996) re-ignited the interest in the search for life on the Red Planet. Simultaneously, there has been a paradigm shift from metabolic measurements as the most desirable for an initial Mars life detection mission (Klein 1974), towards molecular, isotopic and morphological biosignatures. In addition, there was a realization that the geological and geochemical context in which these biosignatures would be detected needs to be clearly and unambiguously understood. A fleet of orbiters from both NASA and ESA is currently providing surface imaging at unprecedented detail as well as mineralogical and geochemical information. The two Mars Exploration Rovers (MERs) *Spirit* and *Opportunity* have provided clear geochemical, sedimentological and imaging evidence for a warmer, wetter history of Mars (Squyres et al. 2004), supporting the view that life could have originated there at least once (provided it was similar to terrestrial life as we know it). Although these rovers did not carry any instruments that were designed to search for life, the Mars Science Laboratory (MSL) and the ExoMars instrument packages will be equipped to detect the organic compounds that should be there from the accumulation of meteoritic matter over the lifetime of the planet (Bland and Smith 2000).

The investigation of organic matter in meteorites, in particular of carbonaceous chondrites, provides an established means to determine the identities and isotopic compositions of free and polymeric organic compounds that is the result of abiotic processes on the asteroidal parent body (Sephton and Botta 2005, 2008). The molecular distributions and C-, H- and N-isotopic compositions of these compounds unambiguously identify them as being extraterrestrial products of non-biological chemical synthetic processes (Cronin and Chang 1993; Botta and Bada 2002; Sephton 2002, for reviews). On the other hand, aromatic compounds identified in Murchison and other carbonaceous chondrites, such as PAHs and nucleobases, show less obvious structural signatures of abiotic origin, but by inference are assumed to be extraterrestrial as well. This observed composition of abiotic organic compounds, both soluble and insoluble, provides a "background" against which a biogenic signature would have to be identified (see papers by McKay 2008; Summons et al. 2008).

## 1.2 Exploration of the Outer Solar System

With the exception of asteroids and comets, from which fragments reach the Earth naturally in the form of meteorites and micrometeorites, and the *in-situ* GC-MS data from the Huygens probes, all existing information about organic compounds in the outer Solar System (beyond Mars) has been obtained through spectroscopic identifications. These spectroscopic signatures of gaseous components in the atmospheres can be found over the whole range of the electromagnetic spectrum.

As a reference, the reflected solar spectrum of the Earth exhibits signatures of $CO_2$, $H_2O$, CO and traces of $CH_4$, $O_3$, $N_2O$. On Venus and Mars, the atmospheres are dominated by $CO_2$ with a few percent $N_2$. Minor atmospheric species on Venus have been detected on the night side between 1 and 4 μm and include CO, $H_2O$, HDO, $SO_2$, OCS, HCl, HF (Bézard et al. 1990). On Mars, minor atmospheric constituents are $O_3$ (detected at around 10 μm), $H_2O_2$ (at 8 μm) as well as methane, which were tentatively detected both from the ground (Krasnopolsky et al. 2004; Mumma et al. 2004) and from orbit (PFS on Mars Express, Formisano et al. (2004). Although the three observations are inconsistent in both measured total methane abundance as well as spatial distribution on Mars, and despite the fact that there are geochemical pathways that could produce $CH_4$ in the Martian subsurface, the discussions in the literature have shown that methane is considered a prime candidate for a spectroscopic biosignature, at least for Earth-like planets. However, it is crucial that another carbon-bearing species, such as CO or $CO_2$, constitutes a major component of the atmosphere. Methane and ammonia are also components of the atmospheres of the giant planets, which are dominated by primordial $H_2$ and He, but they are obviously not considered as biosignatures on these planets. The thermal spectrum of the giant planets (between 7 and 12 μm) is a mixture of emission (stratospheric) and absorption (tropospheric) lines (Encrenaz 2008). All emission lines are due to hydrocarbons, including $C_2H_2$, $C_2H_6$ and others.

Another object in solar system that might be interesting from the prebiotic chemistry point of view is Saturn's moon Titan if a liquid water body was present in the subsurface (Raulin 2008). Hydrocarbons have been detected in its atmosphere of in the 1980's with the flybys of the two Voyager spacecraft and later from the ground and the ISO satellite. These measurements have formed the basis for a suite of complex photochemical models that lead to the synthesis of these organics abiotically on Titan. These theoretical models have been supported and extended by simulation experiments in the laboratory, with more than 150 different organic molecules identified in such mixtures (Coll et al. 1999). The composition of the stratosphere shows a distribution that can be explained using such chemical model networks and no compound stands out in such as way that it could be interpreted to be a monomeric building block of or a product for life.

Solid organic mixtures produced in laboratory simulation experiments, called "tholins", are the "artificial" analogues of Titan's aerosols. The composition of tholins cannot be determined spectroscopically and requires *in-situ* analysis. The Cassini-Huygens mission carried a GC-MS system with ACP instrument that had the capability to determine the organic components in these aerosols, which were collected during the descent of the probe through Titan's atmosphere. The results show that these aerosol particles are made of refractory organic nucleus that is covered with condensed volatile compounds (Raulin 2008).

Although the surface temperatures on Titan are too low to provide sufficient thermal energy, a hypothetical subsurface ocean could be as warm as 260 K and at depths of 200 km a pressure of 5 kbar would not be incompatible with life. If life would exist in such a subsurface location it may be possible that there exists a exchange with the atmosphere through diffusion of volatile species such as methane through the mantle. Assuming that such life would, like on Earth, produce an enrichment in $^{12}C$ during metabolism, this methane would be decrease the $^{12}C/^{13}C$ ratio of the atmospheric methane. Huygens GC-MS measured a value of 82 for methane in Titan (compared to the Earth value of 89) indicating an abiotic origin of the methane (Niemann et al. 2005). However, it is crucial to compare this value against an "internal" standard on Titan, which is not available.

## 2 Biosignatures

### 2.1 Molecular and Isotopic Biosignatures

Stable isotope ratios of carbon ($^{13}C/^{12}C$ or, in a different notation, $\delta^{13}C$ values) have been at the center of the search for the very earliest ($>3.5$ Ga) traces of life on Earth (Mojzsis et al. 1996; Van Zuilen et al. 2002; Van Zuilen 2008) just as they have been used to determine the extraterrestrial origin of organic compounds found in carbonaceous meteorites (Botta and Bada 2002). In a certain sense these are indirect indicators of metabolism and therefore a fundamental kind of biosignature. A strong kinetic isotope effect as, for example, seen in the carbon assimilation associated with oxygenic photosynthesis, is responsible for the negative $\delta^{13}C$ values of molecules derived from cyanobacteria, algae and vascular plants. More recently, multiple sulfur-isotope systematics have been shown to be a powerful tool to investigate the bacterial processing within the sulfide-sulfur- sulfate system (Ono 2008). Oxygen, hydrogen and nitrogen are other elements that undergo significant biogeochemical isotopic fractionation. Because of their investigation of light element isotopic patterns, is one of the most readily.

At a more complex level, molecular biosignatures are used in terrestrial biogeochemistry for the identification of traces of life in the epochs of the Earth. For example, distinctive compounds isolable from living organisms and their related fossil derivatives, which are derived from the original molecules through geochemical processes, are called biomarkers and can even be assigned to certain kinds of organisms (Summons et al. 2008). These established biomarkers may be used as guides for the search for molecular biosignatures on other planets. The molecular attributes of these biosignatures can be broadly classified into two types, namely stereochemical and structural. Stereochemical attributes include enantiomeric excesses in chiral molecules and diastereoisomeric preferences in molecules with multiple chiral centers. Structural attributes include the preference for a limited number of constitutional (structural) isomers among all those possible, the repetition of constitutional sub-units systematic isotopic ordering at molecular and intramolecular levels and finally the uneven distribution patterns (or clusters) of structurally related compounds (Summons et al. 2008).

These phenomena reflect the overwhelming tendency for terran biology to exploit structural exactness and stereochemical purity in practically all the molecules it makes. As far as is known, all biology exhibits and exploits the phenomenon known as homochirality (Barron 2008). Small molecules such as amino acids and sugars exist almost exclusively in one of the possible chiral forms. In other words, these small molecules exhibit 'handedness'. Moreover, biopolymers containing these small molecules, proteins and nucleic acids for example, also exhibit a corresponding handedness in their spiral secondary structures. Thus homochirality is a signature of life and has long been recognized as a quality that should be (Barron 2008) detectable wherever biology is or has been active. Unfortunately, chirality is not a chemically stable quality and tends not to be preserved over geological timescales. Key chemical bonds in the optically pure forms of biomolecules can progressively break and reform ultimately generating 50:50 mixtures of the two forms in the process known as racemization. In fact, the process is sufficiently predictable in amino acids that it forms the basis of a dating method provided that the mode of preservation entails known constant environmental conditions. Thus, while the homochirality of past life might become lost over vast stretches of time, it remains a potential gold standard for recognizing extant biology.

In addition to the stereochemical (spatial) phenomena that result from precision with which enzymes construct molecules, generic biosignatures could also be founded on the observation that terran biology makes its structural, operational and informational macromolecules from a limited subset of universal precursor building blocks (i.e. the Lego®-principle put forward by McKay 2004; Summons et al. 2008). All proteins are constructed from just twenty amino acids, all DNA is built around just four nucleobases and all lipids are constructed from acetate or isopentenyl-pyrophosphate precursors. This fact leads to the repeating patterns long recognized by organic chemists as the hallmark of biological molecules.

On the other hand, there is a 'randomness' to the structures, and an absence of stereochemical preference, in the structures of organic molecules that have been identified in carbonaceous meteorites and as the products of abiological chemosynthetic processes. It is difficult to imagine any form of life that would not have evolved to employ some form of replication in its building large and complex molecules. Accordingly, any patterns or recurring motifs in the chemistry of small to moderate-sized organic compounds might constitute a powerful and generic approach to biosignature recognition.

One of the most persistent problems in identifying and interpreting biosignatures in Earthly materials relates to our inability to always discriminate the origins of different components within complex mixtures. Virtually any sample of organic matter we encounter will comprise molecules derived from a multitude of discrete sources. For example, these could encompass complex assemblages of contemporaneous organisms or mixtures of ancient and modern materials. The latter situation has confounded many attempts to study ancient life on Earth due to the pervasiveness and tenacity of modern biology and the importance of recycling at every level of terrestrial biogeochemistry. One example would be the role of modern bacteria in degrading the organic matter that was formed by organisms living tens to hundreds of millions of years previously (e.g., Petsch et al. 2005). Approaches to addressing this problem are discussed in detail be Eigenbrode (2008).

As a basis for approaching the problem of identifying sedimentary records of extra-terrestrial life, Eigenbrode looks at the molecular signals of life encoded into rocks formed in the early part of Earth history and asks how might we best ascertain whether or not a particular kind of molecule is either native to the material being analyzed or a contaminant. Precise measurements of numerous variables related to the organic matter and the host rock itself opens the way to statistical treatments. In turn, methods such as Principle Components Analysis can expose relationships that would most likely only arise if the two have a shared history. In this way Eigenbrode provides powerful evidence of the syngeneity of certain kinds of fossil hydrocarbons by relating their distribution patterns to immutable aspects of the host rock lithology and geochemistry.

Enormous advances in the way informational molecules are analyzed for their nucleotide sequences potentially opens the door to identifying the most compelling type of biosignature. The simple fact of detecting extra-terrestrial DNA would be a profound first step. The ability to clone and sequence, on the other hand, exposes the possibility to investigate forms of life with a shared evolutionary history as discussed by Gogarten et al. (2008). Just as the genomic revolution allows us to be confident in our appreciation of the relatedness of all known life on Earth, it may also lead to answers about common origins, or otherwise, of other forms of life that might ultimately be encountered in the exploration of the solar system and beyond.

## 2.2 Morphologic Biosignatures

Geobiological investigations in recent and Precambrian environments, as well as laboratory experiments, are essential to determine the criteria for biogenicity of macroscopic and microscopic morphological signatures of life. Three main aspects need to be considered:

- The preservational environments: under what conditions cells with varying biochemical properties can be preserved?
- The taphonomy: how do processes of degradation and preservation retain, alter or erase original biological properties?
- The criteria for biogenicity: how can we tell biological from non-biological?

Studies in recent environments (such as the Spanish river Rio Tinto, the Antarctic Dry Valleys, or the Yellowstone hot springs) displaying a variety of physico-chemical conditions analogue to some of the conditions that prevailed on early Earth or on extraterrestrial environments, show that organic matter is rarely preserved, and that traces of life may include macroscopic biosedimentary structures (coniform structures, coated streamers or filament bundles, pinnacles and other irregular mat surface morphologies) and microscopic cellular casts and moulds by various minerals. In most cases the biogenicity of such traces in ancient rocks stays ambiguous without the presence of organic matter, and the presence of organic matter itself is not sufficient to prove biogenicity but needs to be considered with other criteria including a good understanding of the environmental conditions. Laboratory studies including artificial cell degradation, bioalteration of minerals, and mineralization experiments might reveal biotic patterns and preservable properties, although these studies may not reflect the full complexity of natural environmental conditions. Documentation of abiotic patterns and morphologies is also essential, but is difficult outside the lab in our biological planet. Palaeontological investigations of a range of marine (intertidal to basinal) environments from the Archean through the Proterozoic preserve biosignatures in various lithologies (see papers by Krumbein 2008; Westall 2008; Altermann 2008). The preservable biological properties include particulate organic matter, biomarkers (fossils molecules), isotopic fractionation, macroscopic biosedimentary structures, biominerals, and microfossils, however all these features can be ambiguous in some environmental conditions that need to be investigated further. Microscopic and microchemical analyses of fossils isolated from shales or embedded in various rocks are needed to demonstrate a range of preservable biological properties, as well as to determine the taphonomic processes specific to the preservational environment and to the original biology.

The biogenicity criteria summarized by Westall (2008) and illustrated by examples from the Archean rock record include: (i) the morphological aspects of individual microorganisms (size, shape, cell division, cell death, cell envelope texture, flexibility); (ii) colonial and biofilm characteristics (association with a number of other organisms of the same species, association with different species in the vicinity (consortium), association with microbially produced polymer (extracellular polymeric substances, EPS)); (iii) evidence of interaction of individual microorganisms and their biofilms with their microenvironment (biolaminations; and microbial mineral precipitation); and (iv) biogeochemical characteristics including (a) biomarkers molecules, (b) bio-elements, such as N, S, and P (c) associated heavy metals and Rare Earth Elements (REEs), (d) C, S, N, and Fe isotopes fractionation. Other important observations such as the pigment and biopolymer composition of individual walls or sheaths, the abundance (population) and distribution in the rock, the orientation recording behaviour and motility of benthic organisms (for example mats with vertical filaments erected towards light) and very importantly, the taphonomy (producing features such as collapsed folded hollow vesicles, sinuous segmented filamentous shape, pigmented surfaces of colonies) need to be considered as well (Hofmann 2004; Knoll and Golubic 1992).

Krumbein (2008), in an original approach, imagines an alien exploration team trying to assess the habitability of a rocky planet at different scales of observations. This set-up allows him to detail the diversity of microbially induced sedimentary structures (MISS) or microbially altered rock surfaces (MARS) or other microbial growth structures in carbonates and siliciclatic rocks. Krumbein proposes to consider global biogenic structures and processes at the planetary scale and defines the new term of "geophysiology", i.e. "the integrated study of the phenomena and processes of living planets".

Altermann (2008) reviews Archean stromatolite occurrences, their morphological classification and their modes of growth in the context of sedimentary facies. The varying stromatolite morphologies, ranging from large, kilometre scale to minute, sub-mm scale shapes are considered as an expression of biology acting under differing physical (hydrodynamic) conditions.

However, the early Archean record of stromatolites and of microscopic biosignatures is still strongly debated, illustrating the difficulty (or impossibility) to define unambiguous traces of life in very old rocks, moreover on another planet where knowledge of the geological background is limited. Therefore, the search for traces of life in ancient and extraterrestrial materials requires using a multitude of complementary studies in order to reach an acceptable level of reliability in any interpretation of biogenicity, as Westall (2008) underlines.

The terrestrial recent and fossil record demonstrates clearly that biosignatures will have characteristics strongly depending on the preservational environments and the original biology. Study of the preservational environments and processes help to make predictions of the types of biosignatures to search for in potential past or present extraterrestrial habitats, important when deciding landing sites for exobiological missions *in situ* and for returned samples.

## 3 *In-situ* Instrumentation

For *in-situ* life detection space missions, the instrument best suited to distinguish between abiotic and biogenic organic molecules is a chromatograph coupled to a mass spectrometer. In fact, any chromatographic method is suitable for this task (for example high-performance liquid chromatography (HPLC) or capillary electrophoresis (CE)). The choice for *in-situ* instruments is primarily directed by their potential for miniaturization. As mentioned earlier, each of the Viking landers actually carried such an instrument, the pyrolysis-GC-MS. The system was able to analyze three samples (using three ovens) on a single chromatographic column coupled to a magnetic sector MS with a mass range of $m/z$ 11.5 to 215. The package was the size of a shoebox and weighed about 20 kg. In total, four samples (two by each Viking lander) were analyzed and none of them were found to contain any organic compounds at levels of a few parts-per-billion (ppb). This unexpected result was considered "a real wipe out" for the possibility for life on Mars (Klein 1977) and difficult to understand given the fact that the Martian surface is exposed to the same influx of chondritic carbonaceous matter as Earth (Bland and Smith 2000). Although the instrument packages functioned perfectly as designed the result can be understood in the context or Mars surface chemistry which would lead to progressive oxidation of exposed organics (Benner et al. 2000).

For the forthcoming MSL and ExoMars missions, two instruments are under development that will continue the search for organic carbon on Mars: Sample Analysis at Mars (SAM) and UREY, respectively. SAM is a GC-MS system with improved performance as compared to Viking especially in respect to its capacity to conduct more than 70 sample analyses using any of six chromatographic columns with varying compound separation capabilities and to employ chemical derivatization on a subset of soils. The SAM instrument also includes a Tunable Diode Laser (TDL) system to measure carbon isotopic compositions of atmospheric gases and volatiles released from soil samples (Mahaffy 2008). UREY combines a subcritical water extractor with a sublimation system, chemical derivatization, and a μ-CE separation module (Bada et al. 2008). Both of these instruments should have the sensitivity to detect at least the meteoritic component in the Martian surface, provided that the rover can access material that was not subjected to oxidation in the near-surface environment.

Compound-specific isotope ratios for high-mass organic molecules ($> 100$ amu) are probably beyond the realm of *in-situ* instruments on robotic spacecraft since the measurement requires high resolution mass spectrometry, which is very difficult to miniaturize. For such measurements, which are highly valuable for the determination of different reservoirs in the carbon cycle, a sample return mission is probably the best strategy.

## 4 Remote Sensing of (Exo)Planets

The last decade of extrasolar planet observation has unveiled an unexpected diversity of giant planets: Jupiter- and Neptune-sized planets at extremely small orbital distance, planets with high eccentricity, compact multiple systems, planets with an extremely low or extremely high density. The discovery of these objects that have no equivalent in the Solar system has produced a revolution in planetary sciences.

The study of exoplanets has recently reached two fascinating new stages. First, we are now able to probe the physical and chemical nature of the atmosphere of some of the detected exoplanets (Seager 2008). Detecting atmospheric species, measuring the temperature, constraining the structure and dynamics of the atmosphere was so far restricted to Solar System planets (Encrenaz 2008) but now became feasible for the exoplanets transiting around nearby stars. The second remarkable achievement is that we now stand ready to explore the population and the properties of planets with masses below 10 $M_{Earth}$. Radial velocity observations, transit searches and microlensing surveys do already provide the capabilities to detect planets of a few Earth masses, at *habitable* distance from their star. Within two decades, we should have the technology to search for *exoEarths* around nearby stars, and to perform a spectral characterization of these new worlds (Fridlund 2008). Beside the fantastic progress this achievement will represent for the study of terrestrial planets and their atmospheres, it will also allow us to search for spectral signatures of biological activity.

Thanks to photosynthesis, the Earth's biosphere converts solar energy into chemical energy and keeps the upper layers of the planet (the surface, the ocean, the atmosphere) in a chemical steady state that could not be sustained by abiotic processes alone. As a consequence, and for at least the last 2.5 billion years, our planet has been exhibiting spectral features that are directly or indirectly inherited from life. Some of these features can be detected from astronomical distances, even with a low spectral resolution, as soon as the photons from the Earth can be distinguished from those from the Sun (Arnold 2008). Searching for similar biosignatures in the spectrum of nearby extrasolar planets will be achievable by the next generation of telescope. In this perspective, it is important to study the numerous stars in the vicinity of the Sun in order to select the prime targets around which Earth-like companion could be searched for (Turnbull 2008).

## References

Altermann, Space Sci. Rev. (2008, this issue)
Arnold, Space Sci. Rev. (2008, this issue)
Bada, et al. Space Sci. Rev. (2008, this issue)
Barron, et al. Space Sci. Rev. (2008, this issue)
S.A. Benner et al., The missing organic molecules on Mars. Proc. Nat. Acad. Sci USA **97**, 2425–2430 (2000)

B. Bézard, C. de Bergh, D. Crisp, J.-P. Maillard, The deep atmosphere of Venus revealed by high-resolution nightside spectra. Nature **345**, 508–511 (1990)

P.A. Bland, T.B. Smith, Meteorite accumulation on Mars. Icarus **144**, 21–26 (2000)

O. Botta, J.L. Bada, Extraterrestrial organic compounds in meteorites. Surv. Geophys. **23**, 411–467 (2002)

P. Coll et al., Experimental laboratory simulation of Titan's atmosphere: Aerosols and gas phase. Planet. Space Sci. **47**, 1331–1340 (1999)

J.R. Cronin, S. Chang, Organic matter in meteorites: Molecular and isotopic analyses of the Murchison meteorite, in *Chemistry of Life's Origins*, ed. by J.M. Greenberg, V. Pirronello (Kluwer, Dordrecht, 1993), pp. 209–258

Encrenaz, Space Sci. Rev. (2008, this issue)

Eigenbrode, Space Sci. Rev. (2008, this issue)

V. Formisano, S.K. Atreya, T. Encrenaz, N. Ignatiev, M. Giuranna, Detection of methane in the atmosphere of Mars. Science **306**, 1758–1761 (2004)

Fridlund, Space Sci. Rev. (2008, this issue)

Gogarten et al., Space Sci. Rev. (2008, this issue)

H.J. Hofmann, Archean microfossils and abiomorphs. Astrobiology **4**, 135–136 (2004)

H.P. Klein, Automated life-detection experiments for the Viking mission to Mars. Origins of Life **5**, 431–441 (1974)

H.P. Klein, The Viking Biological Investigation: General aspects. J. Geophys. Res. **82**, 4677–4680 (1977)

A.H. Knoll, S. Golubic, Proterozoic and living cyanobacteria, in *Early Organic Evolution: Implications for Mineral and Energy Resources*, ed. by Schidlowski et al. (Springer, Berlin, 1992), pp. 450–462

V.A. Krasnopolsky, J.-P. Maillard, T. Owen, Detection of methane in the Martian atmosphere: evidence for life? Icarus **172**, 537–547 (2004)

Krumbein, Space Sci. Rev. (2008, this issue)

Mahaffy, Space Sci. Rev. (2008, this issue)

C.P. McKay, What is life – and how do we search for it on other worlds? Public Libr. Sci. Biol. **2**, 1260–1263 (2004)

McKay, Space Sci. Rev. (2008, this issue)

D.S. McKay et al., Search for past life on Mars: Possible relic biogenic activity in Martian meteorite ALH84001. Science **273**, 924–930 (1996)

S.J. Mojzsis et al., Evidence for life on Earth before 3,800 million years ago. Nature **384**, 55–59 (1996)

M.J. Mumma et al., Detection and mapping of methane and water vapor on Mars. Bull. Am. Astron. Soc. **36**, 1127 (2004)

H.B. Niemann et al., The abundances of constituents of Titan's atmosphere from the GCMS instrument on the Huygens probe. Nature **438**, 779–784 (2005)

Ono, Space Sci. Rev. (2008, this issue)

S.T. Petsch, K.J. Edwards, T.I. Eglinton, Microbial transformations of organic matter in black shales and implications for global biochemical cycles. Paleogeogr. Paleoclimate Paleoecology **219**, 157–170 (2005)

Raulin, Space Sci. Rev. (2008, this issue)

Seager, Space Sci. Rev. (2008, this issue)

M.A. Sephton, Organic compounds in meteorites. Nat. Prod. Rev. **19**, 292–311 (2002)

M.A. Sephton, O. Botta, Recognising life in the solar system: Guidance from meteoritic organic matter. Int. J. Astrobiol. **4**(3–4), 269–276 (2005)

Sephton, Botta, Space Sci. Rev. (2008, this issue)

Schuerger, Clark, Space Sci. Rev. (2008, this issue)

S.W. Squyres et al., In situ evidence for an ancient aqueous environment at Meridiani Planum, Mars. Science **306**, 1709–1714 (2004)

Summons et al., Space Sci. Rev. (2008, this issue)

Turnbull, Space Sci. Rev. (2008, this issue)

M.A. Van Zuilen, Space Sci. Rev. (2008, this issue)

M.A. Van Zuilen, A. Lepland, G. Arrhenius, Reassessing the evidence for the earliest traces of life. Nature **418**, 627–631 (2002)

Westall, Space Sci. Rev. (2008, this issue)

# Space Science Series of ISSI

1. R. von Steiger, R. Lallement and M.A. Lee (eds.): *The Heliosphere in the Local Interstellar Medium*. 1996                                                              ISBN 0-7923-4320-4
2. B. Hultqvist and M. Øieroset (eds.): *Transport Across the Boundaries of the Magnetosphere*. 1997                                                              ISBN 0-7923-4788-9
3. L.A. Fisk, J.R. Jokipii, G.M. Simnett, R. von Steiger and K.-P. Wenzel (eds.): *Cosmic Rays in the Heliosphere*. 1998                                                              ISBN 0-7923-5069-3
4. N. Prantzos, M. Tosi and R. von Steiger (eds.): *Primordial Nuclei and Their Galactic Evolution*. 1998                                                              ISBN 0-7923-5114-2
5. C. Fröhlich, M.C.E. Huber, S.K. Solanki and R. von Steiger (eds.): *Solar Composition and its Evolution – From Core to Corona*. 1998                                                              ISBN 0-7923-5496-6
6. B. Hultqvist, M. Øieroset, Goetz Paschmann and R. Treumann (eds.): *Magnetospheric Plasma Sources and Losses*. 1999                                                              ISBN 0-7923-5846-5
7. A. Balogh, J.T. Gosling, J.R. Jokipii, R. Kallenbach and H. Kunow (eds.): *Co-rotating Interaction Regions*. 1999                                                              ISBN 0-7923-6080-X
8. K. Altwegg, P. Ehrenfreund, J. Geiss and W. Huebner (eds.): *Composition and Origin of Cometary Materials*. 1999                                                              ISBN 0-7923-6154-7
9. W. Benz, R. Kallenbach and G.W. Lugmair (eds.): *From Dust to Terrestrial Planets*. 2000                                                              ISBN 0-7923-6467-8
10. J.W. Bieber, E. Eroshenko, P. Evenson, E.O. Flückiger and R. Kallenbach (eds.): *Cosmic Rays and Earth*. 2000                                                              ISBN 0-7923-6712-X
11. E. Friis-Christensen, C. Fröhlich, J.D. Haigh, M. Schüssler and R. von Steiger (eds.): *Solar Variability and Climate*. 2000                                                              ISBN 0-7923-6741-3
12. R. Kallenbach, J. Geiss and W.K. Hartmann (eds.): *Chronology and Evolution of Mars*. 2001                                                              ISBN 0-7923-7051-1
13. R. Diehl, E. Parizot, R. Kallenbach and R. von Steiger (eds.): *The Astrophysics of Galactic Cosmic Rays*. 2001                                                              ISBN 0-7923-7051-1
14. Ph. Jetzer, K. Pretzl and R. von Steiger (eds.): *Matter in the Universe*. 2001
                                                              ISBN 1-4020-0666-7
15. G. Paschmann, S. Haaland and R. Treumann (eds.): *Auroral Plasma Physics*. 2002
                                                              ISBN 1-4020-0963-1
16. R. Kallenbach, T. Encrenaz, J. Geiss, K. Mauersberger, T.C. Owen and F. Robert (eds.): *Solar System History from Isotopic Signatures of Volatile Elements*. 2003
                                                              ISBN 1-4020-1177-6
17. G. Beutler, M.R. Drinkwater, R. Rummel and R. Von Steiger (eds.): *Earth Gravity Field from Space – from Sensors to Earth Sciences*. 2003                                                              ISBN 1-4020-1408-2
18. D. Winterhalter, M. Acuña and A. Zakharov (eds.): *"Mars" Magnetism and its Interaction with the Solar Wind*. 2004                                                              ISBN 1-4020-2048-1
19. T. Encrenaz, R. Kallenbach, T.C. Owen and C. Sotin: *The Outer Planets and their Moons*                                                              ISBN 1-4020-3362-1
20. G. Paschmann, S.J. Schwartz, C.P. Escoubet and S. Haaland (eds.): *Outer Magnetospheric Boundaries: Cluster Results*                                                              ISBN 1-4020-3488-1
21. H. Kunow, N.U. Crooker, J.A. Linker, R. Schwenn and R. von Steiger (eds.): *Coronal Mass Ejections*                                                              ISBN 978-0-387-45086-5

Springer – Dordrecht / Boston / London